Regenerative Medicine

Corey W Hunter • Timothy T. Davis
Michael J. DePalma

Editors

Regenerative Medicine

A Complete Guide for Musculoskeletal and Spine Disorders

Editors
Corey W Hunter
Ainsworth Institute of Pain Management
Mount Sinai Hospital Ainsworth Institute of Pain Management
New York, NY
USA

Timothy T. Davis
Source Healthcare and Source Surgery Center
Santa Monica, CA
USA

Michael J. DePalma
Virginia iSpine Physicians
Richmond, VA
USA

ISBN 978-3-030-75516-4   ISBN 978-3-030-75517-1   (eBook)
https://doi.org/10.1007/978-3-030-75517-1

This Springer imprint is published by the registered company Springer Nature Switzerland AG
The registered company address is: Gewerbestrasse 11, 6330 Cham, Switzerland

I would like to thank all of my mentors (Tim Deer, Tim Davis, Robert Levy, Doug Beall, Mike DePalma, Sudhir Diwan, Rick Paicius, Stan Golovac, Salim Hayek, and Keith Boettiger) for showing me the way, my professional colleagues (Amit Gulati, Dawood Sayed, Kasra Amirdelfan, Jason Pope, Steven Falowski, Rob Heros, Houman Danesh, Ed Rubin, John Formoso, and so many others) for making sure I stayed on the path, my personal friends for keeping me grounded along the way, and most of all my family (my amazing wife Courtney, Alexandra, Liam, Harley, and my big brother Bill) for giving me the strength to make sure I got there. I am who I am because of all of you, and I dedicate this book to all of you. – Corey W Hunter, MD

I dedicate this textbook to my cherished colleagues who graciously respond to provide insight, direction, and humor. May our field advance and innovate partly by works such as this textbook to share viewpoints, knowledge, and stimulus for future research. – Timothy T. Davis, MD

Dedicated to my daughter Elina and son Taj. – Michael J. DePalma, MD

Preface

Regenerate (*re·gen·er·ate*; ri-ˈje-nə-rət): to generate or produce anew; to replace (a body part) by a new growth of tissue; to restore to original strength or properties.

The field of medicine has come a long way since the days of opium and willow tree bark. While eras of time are earmarked by diseases and illnesses that plagued the living of that period, they are equally remembered for the scientific breakthroughs they gave way to that allowed for our species to survive and persevere. Each discovery rendered treatments that allowed mankind to treat the untreatable. Fortunately, the minds of luminaries never grew content, ever searching for the next "cure" or an advancement that would allow us to continue to push the boundaries of what was possible. But for each problem that was solved, another obstacle was revealed; the obstacle of our time has been finding the means to make the body "fix itself."

The human body is a resilient machine that can repair itself after withstanding impressive amounts of damage, yet there are certain instances where it cannot, relegating it to an impasse that it will never cross. For as far as we have progressed, we still have no way to make the brain regenerate lost tissue, repair a damaged spinal cord in quadriplegic, regrow lost cartilage in an arthritic joint, or even repair a degenerative disc in the spine. For these and many other conditions where damage appears to be irreversible, we have turned to the idea of regenerative medicine in an attempt to "trick" the body into regenerating itself and repairing the injury.

Regenerative medicine has gone by many names over the years: biologics, stem cell therapy, embryonic stem cells, platelet rich plasma, prolotherapy, amniotic…whatever the name or label, many view the field like to be the "holy grail" or even the final frontier of medicine due to the seemingly endless possibilities for its utility. Whether it be repairing the nigrostriatal pathway of the brain in a person with Parkinson's, regrowing lost islet cells in the pancreas of a person with diabetes, repairing retinal damage in an eye, or simply treating pain in a damaged tendon, regenerative medicine may very well hold the key to helping mankind move beyond the obstacles of our era.

As the field has grown, so has its use in everyday practice – particularly in the specialties of pain and orthopedics. Despite the groundswell in its popularity, a lack of adherence to evidence-based medicine and essential standards has developed, for which it is utilized in basic practice. Contrary to virtually every other therapy and/or medical treatment currently being utilized, regenerative medicine is not part of an educational curriculum within any field or specialty within medicine; rather it is relegated to "word of mouth" or weekend courses whereby one medical professional will merely share their personal experiences for a price to others that wish to bring these methods into their own practices. Often, there is little regard for evidence or best practices within these "educational" offerings, simply ways to maximize profits. Consequently, the belief that regenerative therapies are pixie dust or some form of magical treatment that can cure all ailments has developed which has created a schism between reality and marketing fiction.

Regenerative medicine is impressive and revolutionary by its right, without any need for embellishment or exaggeration – if it is ever to reach its full potential, it will need to stand on its own merit with real data and factual evidence as the foundation. The purpose of this book was to bring together the world's experts in regenerative medicine and consolidate that evidence into a first-of-its kind resource that will give future practitioners an evidence-based resource on how to best implement this therapy in the real world.

It is our hope that this book will be a beacon of light for the field that highlights not only the importance of data but the continual need for even more, as well as becoming call-to-action that will drive others to build upon what we present here by publishing similar works predicated on real data. As time moves on, and the future becomes the past, we endeavor to make this offering a "living document." that will be continually updated as more data enters the fold and future therapies come into play.

We are extremely grateful to the many authors who made this offering possible, especially the tireless efforts of our section editors (Douglas P. Beall, Aaron Calodney, and George C. Chang Chien) for putting this publication on their backs and helping to carry it to the finish line.

To our readers: please use this book with one and one goal only in mind – do no harm. As a wise man once said, "Be a good doctor, and everything else will follow." (Timothy Ray Deer, MD)

New York, NY, USA   Corey W Hunter
Santa Monica, CA, USA   Timothy T. Davis
Richmond, VA, USA   Michael J. DePalma

Contents

Contributors

Editors

Corey W Hunter, MD Ainsworth Institute of Pain Management, New York, NY, USA

Physical Medicine & Rehabilitation, Icahn School of Medicine at Mount Sinai Hospital, New York, NY, USA

Timothy T. Davis, MD Source Healthcare and Source Surgery Center, Santa Monica, CA, USA

Michael J. DePalma, MD Virginia iSpine Physicians, PC, Interventional Spine Care Fellowship Program, Richmond, VA, USA

Section Editors

Douglas P. Beall, MD Interventional Surgical Services of Oklahoma, Department of Interventional Radiology, Edmond, OK, USA

Aaron Calodney, MD, FASA, FIPP, ABIPP Precision Spine Care, Texas Spine and Joint Hospital, Tyler, TX, USA

George C. Chang Chien, DO GCC Institute, Department of Musculoskeletal Medicine and Medical Aesthetics, Newport Beach, CA, USA

Contributors

Joel A. Aronowitz, MD University Stem Cell Centers, Cedars-Sinai Medical Center, Department of Plastic Surgery, Los Angeles, CA, USA

Steve M. Aydin, DO Zucker School of Medicine at Hofstra University/Northwell Health, Department of Medicine and Rehabilitation, Manhasset, NY, USA

Alexander Bautista, MD University of Louisville, Department of Anesthesiology and Perioperative Medicine, Louisville, KY, USA

Lora L. Brown, MD, DABIPP TruWell, PLLC, St Petersburg, FL, USA

Nathan Cai, BS University Stem Cell Center, Los Angeles, CA, USA

Aaron Calodney, MD, FASA, FIPP, ABIPP Precision Spine Care, Texas Spine and Joint Hospital, Tyler, TX, USA

Kenneth D. Candido, MD Department of Anesthesiology, Department of Anesthesia and Pain Medicine, Advocate Illinois Masonic Medical Center, University of Illinois College of Medicine-Chicago, Chicago, IL, USA

Cameron Cartier, DO, MBA Jason Attaman, PLLC, Department of Pain Medicine, Bellevue, WA, USA

Christopher J. Centeno, MD Centeno-Schultz Clinic, Broomfield, CO, USA

George C. Chang Chien, DO GCC Institute, Department of Musculoskeletal Medicine and Medical Aesthetics, Newport Beach, CA, USA

Kenneth B. Chapman, MD Pain Medicine, Northwell Health Systems, Department of Pain Medicine, New York, NY, USA

Joseph Clayton, BS University Stem Cell Center, Los Angeles, CA, USA

Arianna Cook, MD, MPH University of North Carolina, Department of Anesthesiology, Chapel Hill, NC, USA

Houman Danesh, MD The Mount Sinai Hospital, Department of Anesthesiology, New York, NY, USA

Ian D. Dworkin, MD, FAAPMR Board Certified in Physical Medicine and Rehabilitation and Pain Medicine, Department of Interventional Pain Physiatry, Newport Care Medical Group, Newport Beach, CA, USA

Timothy Ganey, PhD Vivex Biomedical, Inc., Miami, FL, USA

Alexander Ghatan, DO UCLA/VA of Greater Los Angeles, Department of Physical Medicine and Rehabilitation, Los Angeles, CA, USA

Amitabh Gulati, MD Memorial Sloan Kettering Cancer Center, Department of Anesthesiology and Critical Care, New York, NY, USA

Matthew Hyzy, DO Centeno-Schultz Clinic, Broomfield, CO, USA

Elise M. Itano, MD Boulder Medical Center, Interventional Sports and Spine, Department of Physical Medicine and Rehabilitation, Louisville, CO, USA

Mairin A. Jerome, MD Centeno-Schultz Clinic, Broomfield, CO, USA

Jason Kajbaf, DO David Geffen School of Medicine—UCLA, Department of Physical Medicine and Rehabilitation, Los Angeles, CA, USA

Naveen S. Khokhar Virginia iSpine Physicians, PC, Interventional Spine Care Fellowship Program, Richmond, VA, USA

Roy R. Liu, MD Icahn School of Medicine at Mount Sinai, Department of Anesthesiology, Perioperative and Paine Medicine, New York, NY, USA

Matthew Lucas, DO Peak Orthopedics and Spine–A Division of Orthopedic Centers of Colorado, Englewood, CO, USA

Gregory Lutz, MD Hospital for Special Surgery, Regenerative Sports Care Institute, New York, NY, USA

Tennison Malcolm, BS, MD Brigham and Women's Hospital, Department of Anesthesiology, Perioperative and Pain Medicine, Boston, MA, USA

Tory L. McJunkin, MD Arizona Pain Specialists, Scottsdale, AZ, USA

Roya S. Moheimani, MD UCLA/GLA VA, Department of Physical Medicine and Rehabilitation, Los Angeles, CA, USA

Maxim Moradian, MD Interventional Physiatrist, iSCORE (Interventional Spine Care and Orthopedic Regenerative Experts, PC), Arcadia, CA, USA

Matthew B. Murphy, PhD Murphy Technology Consulting, Austin, TX, USA

Daniel Oheb, BS Tower Outpatient Surgery Center, Los Angeles, CA, USA

Raj Panchal, DO NYU Langone Medical Center, Rusk Rehabilitation, Department of Physical Medicine and Rehabilitation, New York, NY, USA

Kevin Joseph Pauza, MD Baylor Scott and White, Texas Spine and Joint Hospital, Tyler, TX, USA

Regenerative Sportscare Institute, Senior Physician, New York, NY, USA

Asli Pekcan, BS University Stem Center, Los Angeles, CA, USA

Theodore T. Sand, PhD Sand Consulting, Poway, CA, USA

Edward L. Swing, PhD Phoenix Children's, Department of Medical Education, Phoenix, AZ, USA

H. Thomas Temple, MD Department of Orthopaedic Surgery, HCA Healthcare Inc., Miami, FL, USA

Anthony Tran, MD New York-Presbyterian/Columbia and Cornell, Department of Rehabilitation Medicine, New York, NY, USA

Andrew T. Vest, BS, DO (Expected Class of 2020) University of North Texas Health Science Center, Dallas, TX, USA

Clairese M. Webb, MD University of Oklahoma Health Sciences Center, Department of Anesthesiology and Pain Medicine, Edmond, OK, USA

Christopher J. Williams, MD Emory University, Department of Rehabilitation Medicine, Atlanta, GA, USA

Bridget Winterhalter, PA-C University Stem Cell Centers, Cedars-Sinai Medical Center, Department of Plastic Surgery, Los Angeles, CA, USA

Allan Zhang, DO University of Connecticut, Department of Radiology, Farmington, CT, USA

Part I

Basics of Regenerative Medicine

Introduction to Regenerative Medicine

1

Timothy Ganey and H. Thomas Temple

Regenerative medicine has been one of the frontiers for understanding human biology for centuries. Long before it fell under the category of medical research, or was assigned the regenerative moniker, or even considered medical practice, humans have sought to understand the basic context of how the body emerges with such complexity and near error-less synchrony to produce the dividends of interdependent function. When that elegant system fails and disease or degeneration breaks into perfection, however, the challenge then becomes to isolate the weakness and either replace or regenerate the affected tissue. As physicians and biologists seek to "unbrick" the wall, to isolate the piece or pieces from the whole, the evolution of knowledge has shifted the balance of understanding to seek the indivisible rather than to reintegrate the fragments as a functional system.

Medical practice also comes under the aegis of governing approvals and oversight that ensures that both safety and efficacy are attained. With both health and commercial practices collaborating and competing to accommodate patient care without compromising the economics of reimbursement, technologies have been developed that are broadly reductionist and guided by the trajectory of regulatory approval. Seeking "mechanism of action," strategies aligned to identify niche assets of a biological process that are economically viable and scientifically accurate, companies have strived for therapeutic advantages for patient care and have evolved along with the principles of action–reaction understanding.

It has been clarified that with respect to cell-based matrices, the market and regulatory bodies have accepted that living cells can be included in allograft for use in repairing bone. With a source that is allogeneic, i.e., from a donor that will be used for homologous use, several products have come to market that have living cells and are marketed under FDA guidelines that regulate them as Human Cells, Tissues, and Cellular and Tissue-Based Products (HCT/Ps). To meet the threshold that defines that categorization, there are several criteria that must be met [1]. Without broad discussion, four tenets have defined the cornerstone of dialogue with the FDA and the discussions for product development for commercial purposes:

- The HCT/P is minimally manipulated;
- The HCT/P is intended for homologous use only, as reflected by the labeling, advertising, or other indications of the manufacturer's objective intent;
- The manufacture of the HCT/P does not involve the combination of the cells or tissues with another article, except for water, crystalloids, or a sterilizing, preserving, or storage agent, provided that the addition of water, crystalloids, or the sterilizing, preserving, or storage agent does not raise new clinical safety concerns with respect to HCT/P; and
- Either:
  - The HCT/P does not have a systemic effect and is not dependent upon the metabolic activity of living cells for its primary function; or
  - The HCT/P has a systemic effect or is dependent upon the metabolic activity of living cells for its primary function, and:
    - Is for autologous use;
    - Is for allogeneic use in a first-degree or second-degree blood relative; or
    - Is for reproductive use.

If the definition of the cells, in particular, does not remain within the margins imposed by this standard, then the proposed product is regulated as a "Biologic" and requires different stringency for FDA approval prior to marketing in the United States. For companies trying to commercialize regenerative medicaments for therapeutic use, such additional

T. Ganey (✉)
Vivex Biomedical, Inc., Miami, FL, USA

BonePharm, LLC, Tampa, FL, USA
e-mail: tim@bonepharm.com

H. T. Temple
Department of Orthopaedic Surgery, HCA Healthcare Inc., Miami, FL, USA

© Springer Nature Switzerland AG 2023
C. W Hunter et al. (eds.), *Regenerative Medicine*, https://doi.org/10.1007/978-3-030-75517-1_1

regulatory approval often creates an economic barrier that is insurmountable.

With regard to viable cell allografts, cell-based matrices, and other living tissues, the FDA to date has permitted the use of cells having metabolic function in the transplant. In the field of regenerative medicine, stem cells have been shown to be able to self-renew but also give rise to daughter cells committed to lineage-specific differentiation. To achieve this remarkable task, they undergo an intrinsically asymmetric cell division whereby they segregate cell fate determinants into only one of the two daughter cells. Stem cells have been typed with a specific "**c**luster of **d**ifferentiation," "**c**luster of **d**esignation," or "**c**lassification **d**eterminant" that is usually abbreviated as **CD** for identification. Once identified, extensive in vitro investigations have been carried out to determine specific cell functions under precise conditions in the scientific method of changing single variables and measuring outcomes. Elaborate cellular mechanisms that orchestrate the processes required for asymmetric cell divisions are shared between stem cells and other asymmetrically dividing cells. These cells demonstrate that asymmetry/polarity is guided by varying degrees of intrinsic and extrinsic cues and intracellular machineries that divide the desired orientation into a balance of asymmetry/polarity.

Regenerative medicine has utilized cells with the CD designations from a variety of sources, including bone marrow, adipose tissue, peripheral blood, placental tissues, etc. to produce cell lines that have purposeful uses and specific cell phenotypes. When isolated and expanded cells (called A in this example) are steeped in science (called B), cells with the same phenotype present in cellular matrices (labeled as C) are then presumed supported by the same science. In this case of transitive equivalence, A = B, and A = C, so therefore B = C. The argument, empiric in nature, may be essentially accurate but likely understates the science and underestimates the integration of cell, cell factors, cell polarity, asymmetry, immune integration, and a myriad of yet undiscovered complications needing further explication.

A colleague offered me insight into this context a few years ago stemming from his appreciation of music. Although our discussion focused on spectrums of incident light and electromagnetic frequency on adaptation, its analogous value to music was the space between the notes. His awareness was fresh and reverent to an understanding widely ascribed to the French composer Claude Debussy, a prominent musician in the style commonly referred to as "Impressionist Music." Neither likely intended to link a harmonic guiding a psychophysical phenomenon, remarkable in part because the perception of periodicities, namely pulse and meter, arise from stimuli that are not periodic [2]. One possible function of such a transformation is to enable synchronization between individuals through perception of a common abstract temporal structure (e.g., during music performance). Understanding

that the underlying brain processes are a fundamental aspect of human perception enables communication between neural areas such as auditory and motor cortices. Should we think that the organization and integration of tissue interfaces are any less driven by a similar message? I think that is one of the keys that accelerates the interest and defines the concept of regeneration—essentially aligning that the periodicity of the asymmetry of the tissues and cells is a more orchestrated construct than one relying on the simplicity of "**CD**" designation and tissue composition.

Curiosity drives the human mind to find out more and to look for additional factors, but each evolving inner analog offers less information that contributes to a better understanding of the whole. Using bone as an example of a tissue that for the most part retains restorative potential throughout life, it remains opportune that regenerative medicine engages the subsets of understandings that have been found in reducing its parts as we make attempts to further the regenerative techniques we have gleaned from this reduction (Fig. 1.1).

The concept is straightforward; for every point on a line, there is a space between, and within that space exists something unmeasured, something assumed to be average or represented by the adjacent known entities, but still vastly unknown. From a classic perspective of molecular metrics first demonstrated by Kees Boeke in 1957 [3], the lay public was offered that insight in the seminal work of *Powers of Ten* by Philip and Phyllis Morrison [4]. Both depictions collapse a logarithmic trek from the cosmic outer limits to the ocean of the universe within a carbon atom, with humans serving as a mere intercept along the journey, a placeholder, or milestone to a personalized awareness. Coupling the musician's awareness of the silence between the notes that brand the music, the challenge to biologists is to understand the space between the defined but arbitrary scales of life and investigate the depths of the dark space to differentiate determinants of illness from measures of health. A better sense of that space should help facilitate understanding and translate an unknown into a meaningful therapeutic intervention.

A Holy Grail of modern stem-cell research is the recreation of a functioning organ. The vital importance of achieving this goal is all too clear. In the United States alone, nearly 9% of patients with liver failure die waiting for a new organ. An example of a much broader need is the organ transplant services, where from December 1988 through February 7, 2019, more than 758,000 transplants have successively been performed [5]. With the demand for transplantable organs far exceeding supply, the need for regeneration therapies has never been greater. This translates into a significant opportunity to repair, restore, and regenerate organs before the need for replacement imposes a life or death mandate.

Among the earliest attestation to regenerative medicine emerges from the Greek literature in the myth of Prometheus. Each day, an eagle would feast on his liver, and each night

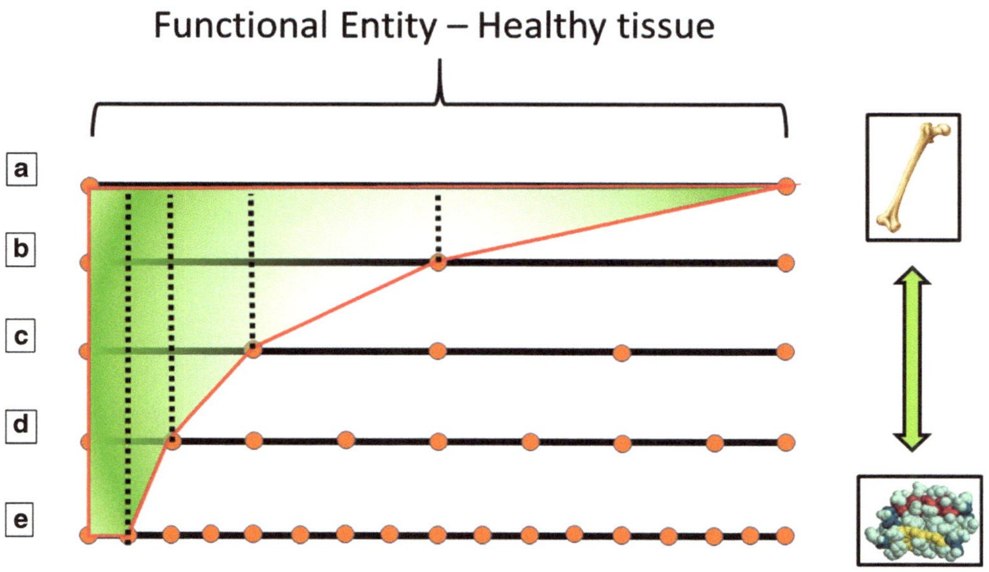

Functional Entity – Healthy tissue

Fig. 1.1 Functional entity—healthy tissue. While it is possible to know ever more distinct areas of a system, it is more challenging to fully integrate individual aspects of their actions into a predictable scheme. The science of regenerative medicine has been paved in individual bricks that appear to offer both dimension and direction. This cartoon depicts the ever-increasing complexity that defines a sector, but at the same time independent of the connection might not fully characterize the science attending the conclusion. (**a**) Bone is a living tissue that provides skeletal support. "Bone" is the whole. (**b**) Skeletal support is dynamic and interdependent on mechanical stimulation for modeling. Interdependent and analog spaces are "bone" and "load." (**c**) Mechanical modeling of bone depends on adequate blood supply, endocrine interaction, and nutrition. "Bone" has now four derivatives: load, blood supply, endocrine, and nutrition. (**d**) Each Linnean reduction comes with a subset of its own reductions, and if a fraction of blood supply is further divided, the logic of asymptotic understanding is assured. In the instance of blood supply, the additions of endothelial lining, sympathetic tone, growth factor activity, endogenous regulation, and repair are just the start. (**e**) Furthering those strands of knowledge, say fibroblast growth factor as an example, is it possible to extrapolate FGF in vascular homeostasis as a meaningful prediction of the whole organism?

his liver would regrow in time for the eagle's return. When hearing this tale, it is tempting to consider that the ancient Greeks had witnessed the amazing capacity of the liver to restore itself and noted the cruel and incremental penance as a substantiation of the immortality of the gods as, in fact, it was Zeus who had deemed this his punishment. This possibility fascinates those engaged in regeneration research, and for some, it is the seminal reference to a cultural understanding of regenerative powers by the Greeks [6–8]. Authors assume that the Greeks knew about the liver's regenerative powers [9] or adopted an agnostic attitude through uncited logic in exceptional journals [10, 11]. An extensive discussion of the regenerative awareness of ancient civilizations suggests that early human anatomists trailed the myths by more than 1500 years and that the more likely scenario of culinary prowess, a belief in organ vitality, and the subsequent blurred lines of myth and time perhaps have led to more confusion than convincing evidence [12].

The literature is replete with notations of what constitutes attempts by the body to make the system whole. Since the time that it was observed and long before it was documented that limb regeneration occurs in amphibians, inquisitive individuals sought a remedy for loss and a solution to the need for restoration. There is little argument that regenerative medicine harbors the potential to restore tissues and organs

and reconstitute their function, yet the tenets of agreement rapidly diverge with broad tentacles that tack an immense number of strategies. Limitations of technology did not blur early insight but reduced many of the scientific merits to musing. In what is a limitless framework of observation, experimentation and communication, key elements can be drawn together to formulate a basic understanding of the potention for regeneration and how it can be utilized in legitimate medicine. It is also possible to append the analog between the cardinal points to better perceive, if not correct, the trajectory of pathology.

Modern therapeutic remedies are guided in the framework of regulation and under the auspices of what is safe and efficacious and what the main risks are. Perhaps a more rigorous evaluation would engage an overview of how regeneration differs from generation. Ernst Haeckel coined the phrase that each acolyte in the sciences is exposed to—*Ontogeny recapitulates Phylogeny*, which is akin to "The Biogenetic Law" that assigns a context where evolution added new stages to produce new life forms. Thus, embryonic development became a record of evolutionary history. The single cell corresponded to amoeba-like ancestors, developing eventually into a sea squirt, a fish, and so on.

By the turn of the century, discoveries were made that defied Haeckel's so-called law. Initially cast as exceptions, the rise of genetics and the modern synthesis has since explained the rate and direction of embryonic development. Individual genes can mutate and cause different changes to the way embryos grow, either adding or taking away new stages at any point along their path or altering the speed of development. This science of epigenetics is the foundation of regenerative medicine, and although somewhat guided through the tiers by a Lamarckian notion that evolution has direction, the challenges of regenerative integration compared with generative development are vastly different (Fig. 1.2).

In the framework intersecting science and experience as this book touts, there are foundations that assure certainty and others that remain to be conquered. It is not richly imaginative to appreciate the fact that all life on Earth shares a common

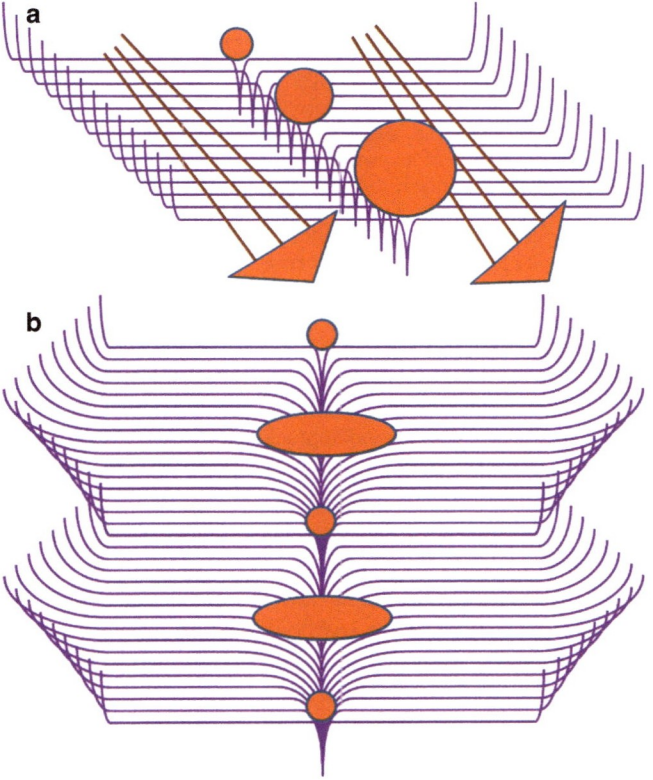

Fig. 1.2 The concept of ontogeny recapitulating phylogeny is rooted to the attribution of Ernst Haeckel, who suggested that an individual organism's biological development parallels and summarizes its species' evolutionary development. (**a**) Represented are the stages of development an organism proceeds through, with the orange sphere denoting the path of development. In this illustration, the course is singular and successive and directional as growth. (**b**) During a regenerative event, development, integration, and achieving appropriate size decorate an existing grid rather than establish a new one. In this example, the epigenetic influences of the existing scaffold, cell activity, and organism age serve as architects of the new potential, and the distortion or the variation between the generative and regenerative dimensions is illustrated as orange spheres that are at once both inconstant and responsive to the morphologenic field by which they are imposed. (**a**) Linear development isometric; (**b**) epigenetic and shaping influences

ancestor, a cell that arose from bacterial progenitors nearly 4 billion years ago. Whether it was from a freak accident, divine intervention, or the perseverance of a change that remained while other experiments failed remains to be determined. It is an interesting exercise to wonder how many attempts were made to unify the efficiency of a colony as a resonating single cell that could divide, diversify, and then reassemble the colony with singular and plural cell versatility expressing physical characteristics and traits that were diversified within the organism. Those cells emanating from a common ancestor have become a fundamental aspect of the science of biology and the core foundation of regenerative medicine. Distinguishing the cells as a core feature, it is important to determine the cues that shift the diversity and sort the reaction to stimulus and symptomatic change. Are the subtle signals standardized to the single cell, or does a synchrony dictate the cross-talk and exchange that can be part of the translation? Better tools, more extensive thought, shrinking dimensions of the space between the dots of knowledge bring us into a nexus that allows cells to be nearly infinitely sorted as a taxonomy in a style that Linnaeus would envy [13].

An argument could be made that the genesis of DNA discovery contributed to an evolving increase in gaps that are parallel in scope and number to the points learned. Erwin Schrodinger made two key points in his 1944 book *What is life?* [14]. Relevant to the topic of regeneration, he noted that life somehow resists the universal tendency to decay, a process that is otherwise known as entropy and stipulated in the second law of thermodynamics and second, that the secret to life's evasion of entropy lied in the genes. Years before Crick and Watson inferred the sequence of bases carried the genetic information, Schrodinger proposed that the lack of nonrepeating bases could act as a "code-script"—the first use of the term in the biologic literature [12, 15], which has become the basis of modern biology. The realms of code, 3 billion letters in our case, reads like a novel of enchanted, coherent stories and vast swaths of repetition that result in a 2% coding for proteins, a larger portion for regulatory functions, and the remainder still assigned to the cliché of needing a better understanding. Understanding the structure of the code has created the ultimate conundrum for regenerative therapeutics as genomes do not predict the future but recall the past. They reflect the exigencies of history and the containment of the environment.

What does regeneration look like in the context of tissue where information previously in equilibrium finds itself not only disrupted but unconstrained? Are those tissues able to recapitulate the origin, pass through, and return via a stable state that is differentiated to function, facilitated to anatomy, and fostered with sufficient receptors that will balance and check re-integration? As a starting point, and short of the replacement of entire extremities and organs, what features have guided the science, established a hierarchy of ethical domains, and regulated the industry? In this chapter, the goal is to establish basic tenets of regenerative approaches, in particular, a potential that cells maintain for self-replacement, lineage multiplicity, and informed exchange to guide the function of complex tissues. In that regard, all tissues require a metabolic supply, and nearly all function from a vascular supply. Recognizing the regulation of the vascular system in and of itself is incomplete science, but given the appearance of angioblasts and a cardiac beat at 21 days from conception, its role must be carefully calibrated in the morphogenesis of tissues. From the simplest of consideration of pressure and shear forces in vessels, or as advocated by the mathematician and computer scientist Alan Turing, in 1952, it was the molecular diffusion of nutrients [16]. While none of these hypotheses were absolutely wrong, continuing work has demonstrated that factors released affect activation, perfusion, dimension, and flow dynamics that are paracrine, cytokine, and hormonal. It is safe to note that vasculogenesis is the formation of early vessels laid down by programming that is genetically deep and that satisfies a quorum of conditions to ensure competent and controlled inherent expression. At the periphery is another consideration that demonstrated mindfulness can change the response and that placebo invigoration is in itself a medicine. This consideration that functional metabolic responses follow that course is interesting but perhaps a bit peripheral. A concept that likely warrants a brief mention is the current understanding that the placebo response and gene variants in catechol-O-methyltransferase (*COMT*) gene may act as risk factors for psychopathology [17].

This is interesting for several reasons, two of which are noted here. For some time, it has been known that the brain and matter are inextricably linked by brain peptides, emotions, and physical expression of symptoms. The mind–body connection has been known in academia long before it has developed into a mainstream awareness. This is critical to understanding regenerative medicine as the psychophysiological manifestations might offer avenues of insight into repair, variation of response, and restrictions to healing that are inherent to various disease processes [18, 19]. In some instances, individuals with multiple personality disorders display symptoms that vary with each personality such as allergy to cats, diabetes, and so on. This suggests that what we know of matter and mind are surely integrated in a regenerative frontier. What are the primal signals for degeneration that guide the scope of recognizing a need for regeneration, and if it is simply insufficiency, why does the body not respond and compensate?

Regenerative medicine is a field that involves replacing, engineering, or regenerating human cells, tissues, or organs to establish, restore, or enhance normal function. It is an area with great promise that includes cell therapies, therapeutic tissue-engineering products, human cells, and scaffolds upon which cells can grow [20]. Recently, there has been much interest specifically in the potential of adult stem cells to address a wide variety of conditions. A process of renewal, restoration, and growth regeneration allows genomes, cells, organisms, and ecosystems to attain resiliency to natural fluctuations or events that cause disturbance or damage. Every species is capable of regeneration, from bacteria to humans. Regeneration can either be complete, where the new tissue is the same as the lost tissue, or incomplete, where in the process of repair, the lost tissue is replaced by fibrotic tissue or scar formation.

At its most elementary level, regeneration is mediated by the molecular processes of gene regulation, adequate proliferation, and balanced structuring of tissues with an accompanying metabolic support. Regeneration in biology, however, mainly refers to the morphogenic processes that characterize the phenotypic plasticity of traits allowing multicellular organisms to repair and maintain the integrity of their physiological and morphological states. Everyone is familiar with the concept of debridement, the process of removing unhealthy tissue from the body. The affected tissue may be necrotic (dead), infected, damaged, and contaminated, or there may be a foreign body in the tissue that requires removal. In the context of regenerative medicine and tissue regeneration, how does the body recognize the boundaries of healthy tissue and preserve and annotate the morphogenetic field for integrated replacement. How does the body understand sufficiency and not overreach and replace an entire area during what might be intended to be a focal repair?

Regeneration somehow balances the extant or existing tissue, recognizes the errant or injured tissue, and in some primal manner determines what is repairable and what is expendable and then aligns a paradigm of repair mechanisms to make the tissue whole. Deposition, modeling, cues of repair, charge, density, permeability, porosity, and morphology are all factors that are considered.

The goal is one of normalizing self, recognizing limits of volume, cellularity, cell density, cell inhibition, adequate metabolic demands, and sympathetic restoration. How are the setpoints for the repair integrated and satisfied on the whole? What is the equilibrium that measures and weights the variants that differentiate absence and abundance? At what point does the relevance of metabolism merge into the controlled moderation genetic hierarchies, immune and injury response, tissue repair, and remodeling until function has been restored to a pattern that is more physiological than pathological.

It is common to think of tissue replaced in vast definable anatomies. Apart from large tissue regeneration or repair, the ongoing focal replacements should also be understood as regeneration. Humans and animals lose tissues and organs due to congenital defects, trauma, and disease at all times. The human body has a low regenerative potential as opposed to the urodele amphibians commonly referred to as salamanders. Traditionally, transplantation of intact tissues and organs has been the treatment method to replace damaged and diseased parts of the body. Today variations of that technique expand the capabilities of traditional tissue banking and transplantation options. One asset of the human tissue allograft has been the inclusion of cells as a viable allograft. Through refinements in technology, techniques for sorting and collecting cells that afford renewed vigor to a host needing tissue have been developed. Variations in methods, cell sources, and carrier scaffold imbue not only viability and vitality but bioavailability as well. That availability has the paracrine factors that enable cell–cell communication at a cellular and subcellular level. This understanding is not a new one as paracrine and cytokine communication has been known for quite some time. Chapter 9 discusses exosomes as the depth of exosome understanding deserves a greater depth of discussion. With the clear evidence that it offers the basis of epigenesis, exosomes perform the transfer of genetic cargo from cell to cell. Stem cell populations can secrete various bioactive compounds, including exosomes, extracellular vesicles, and an entire secretome that effects change and, in some cases, restores equilibrium. Successful isolation of these complexes (which contain a variety of active signaling agents) and their subsequent administration might be an alternative strategy to stimulate the functions of host cell populations in damaged tissue sites. Properly delivered active signaling molecules could subsequently facilitate ECM deposition, the tissue remodeling process, and tissue regeneration.

Strategies for repair including resorting the tissue to its previous ability to take stress and strain along with the neural integration and regaining function are challenging but connected components of any therapeutic strategy. Many studies on the mechanisms of regeneration have led to the identification of cytokines, growth factors, and signal transducers that are produced by cell types within the organ being replaced or transported to the tissue repair site by vascular, lymph, and interstitial transport. These cytokines and growth factors are thought to cause cell expansion and proliferation, resulting in functional recovery. The details of such mechanisms, however, have not been sufficiently elucidated, and the practical applicability of regeneration based on the action of cytokines and growth factors is still unclear. Various cells and organs are involved in the regeneration process, which proceeds as a result of the coordination of many factors. Exposure to cytokines alone might be a trigger, but it would be naïve to assume that exchange is dormant and waiting for physiologic inspiration to activate.

Adult stem cells are important for the normal maintenance and repair of wounded tissues through their ability to differentiate, remodel the extracellular matrix (ECM), modulate the immune response, and secrete growth factors

and cytokines that stimulate cell migration and neovascularization [21, 22]. Mesenchymal stem cells (MSCs) originate in many tissues, but bone marrow and adipose tissue-derived stem cells (ADSCs) are the most available for harvest. Mesenchymal stem cells are known to differentiate down several cell lineage pathways to form cartilage, fat, muscle, and connective tissue, but they are also actively involved in the regulation of wound healing [23]. Mesenchymal stem cells have also been shown to regulate the immune response and inflammation in wounds through the secretion of immunomodulatory cytokines. These cells also cause enhanced proliferation, migration, and secretion of biologically active molecules by a process known as paracrine signaling [24]. Studies suggest that the paracrine activity of MSCs significantly enhances responsiveness and migration of macrophages, epithelial cells, and endothelial cells [25]. As a result, autologous and allogeneic stem cell therapies have been considered a form of treatment to stimulate healing of wounds, both chronic and acute. In consideration of what a wound is, every surgery is technically a wound requiring a healing event. It is not surprising that cellular, peripheral vascular, and other adjunctive assets have been employed in the mechanisms of aiding healing and hastening the process of regeneration. Coming back to the conjunction of the viable allografts/cellular bone matrices, the activity of the cells is one of the measurable metabolic parameters and one of the catalysts that produces byproducts of cell disintegration that sends the sum of its parts as a signal. In regenerative medicine, there is little evidence to associate autologous cell activity with grafting placement, and a similar abbreviated understanding needs further elucidation.

In addition to humoral factors, the autonomic nervous system is also involved in the regeneration process as noted in human liver repair [26]. Studies examining the direct feedback relationship between the liver and brain are transmitted via the afferent sympathetic nervous system to the ventromedial region of the hypothalamus and then to the lateral region of the hypothalamus. They then pass through the dorsal nucleus of the vagus nerve in the medulla oblongata, after which they return to the liver [27, 28]. It appears that the autonomic nervous system first activates the afferent sympathetic nerves in the damaged liver, which transduces the signal to the center of the autonomic nervous system in the brain and then to the efferent vagus nerve. This results in the activation of cell proliferation in various organs inside the abdominal cavity, such as the liver, gastrointestinal tract organs, and pancreas. Despite this known process, no study has focused on the effect of this system on liver regeneration. Therefore, the system as an effector of liver regeneration plays as much of a role as the local tissue. Given the large focus on cell therapy intended for applications in mesenchymal-derived tissue, is it likely that vascular, endocrine, immune, or innervation also plays an important role and is no less involved in constitutive repair.

As the body and tissue bend to breathe the autonomic nervous system as an integrated response is attuned to both cellular and biochemical functions, how does knowing the margins of whole reintegrate the control, posture, and strength in accord with the demands of a challenging tissue like musculoskeletal? What are the signals that interpret flexibility and stiffness as sufficient, or what connotations of electrophysiology affect a membrane resonance that confers a loss of polarity and presses for proliferation and migration over matrix attachment, cell grounding, and matrix elaboration? How do the cells that might have been transplanted attenuate their placement, attach to the matrix, and reintegrate appropriate function and location? What guides the polarity of cells during asymmetric division and resonates a functional status that comprises rather than compromises that biologic expression? These questions challenge scientists, biologists, and tissue engineers to better appreciate the complexity and authorize process identity that nurtures the repair not as an alternative but as an expectation.

Fate of Transplanted Cells

The human body consists of billions of cells that exist together as an intricately organized and mutually supportive community. This cell community is a dynamic system that is maintained by a well-regulated balance between cell proliferation and death. Medical science surrounding regenerative

intervention speaks to stem cell delivery and identifying the cell lineages, but little discussion goes beyond the perception of viability and appropriate markers that suggest pluripotential before placement. Is it conceivable that by placing a bolus of cells into a defined tissue it will provide immediate integration and sustain the ligands and cell markers that have been validated in the process or is it more likely that the cascade of response directly results from the cytokine exchange and evolving phenotype in both the cells delivered as well as in the cells in the host tissue? It seems unlikely as well that cells remaining viable after placement would assemble as might be analogous to a 4-segment "Tetris" puzzle (Fig. 1.3).

Do cells migrate after they are placed and what evidence that has been defined in vitro can be applicable to in vivo validation? It is clear that cell migration plays a central role in a wide variety of biological phenomena such as in embryogenesis where cellular migrations are a recurring theme in important morphogenic processes ranging from gastrulation to development of the nervous system. Migration remains prominent in the adult organism as well as is seen in both normal physiology and with pathologic processes. In the inflammatory response, leukocytes immigrate into areas of insult, where they mediate phagocytic and immune functions [29]. Migration of fibroblasts and vascular endothelial cells is essential for wound healing and, to the extent that even the finest surgical interventions result in wounds, many considerations of the wound field are applicable. Finally, cell migration is crucial to technological applications such as tissue engineering, playing an essential role in colonization of biomaterials scaffolding. While under the aegis of regenera-

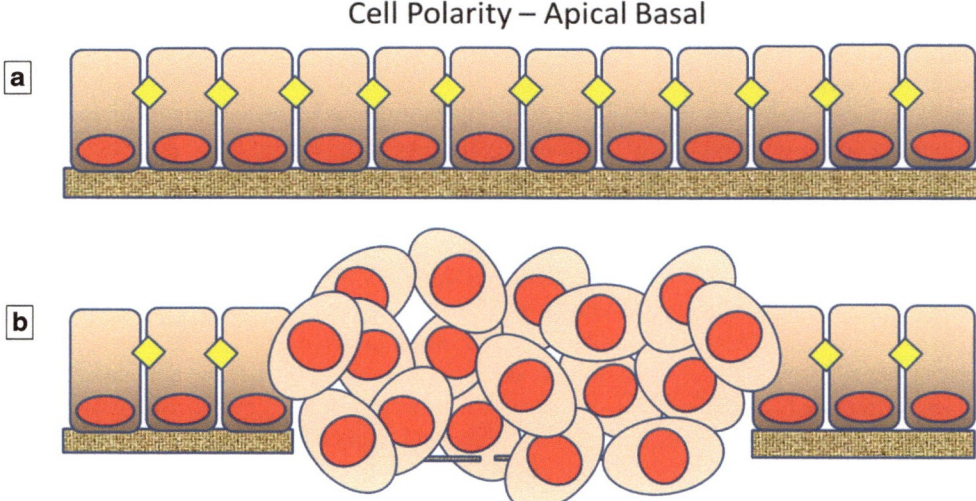

Cell Polarity – Apical Basal

Fig. 1.3 Organizing. (**a**) Normal simple epithelium is comprised of a monolayer of individual cells that display distinct apical-basal polarity. Cells are tightly packed and connected to each other by the apical junctional complexes (yellow), which separate apical (light pink) and basolateral attachment to the basement membrane domain. (**b**) With deposition, delivery, or wound healing, individual cells devoid of polarity must achieve a dimension defining basement membrane repair to guide their attachment and integrate as an aligned tissue within the extant tissue

tion, the restorative requirement of volume replacement in some injuries will require a scaffold that protects the space of the injury. However, when this balance is skewed by injury, repair, dysregulation of activity, or even cell accumulation, predictable outcomes are less assured. Individuals citing theoretical risks turn to the possibility of tumor development or the potential death of the entire cell community due to inflammation or uncontrolled biologic processes.

As with many other cellular processes, the molecular components involved in cell migration are being identified at a rapid rate, including the determination of how they participate in migration. The manner in which these components work together, like most other cell functions, as a dynamic integrated system to give rise to migration is only beginning to be studied. Understanding cell migration as an integrated process requires an appreciation of chemical and physical properties of multicomponent structures and assemblies, including their thermodynamic, kinetic, and mechanical characteristics, because migration is a process that is physically coordinated both spatially and temporally. Only when it is understood as an integrated system will its alteration via genetic, pharmacologic, or materials-based interventions acquire a truly rational basis where placement, scaffold, density, identity, and intention are balanced by the initial delivery—much the same as noted by the analogy of music and perception. A sense of assembly hears an orchestra where science is still recognizing the instruments.

From experiments performed more than two decades ago, a compilation of existing data emerged that maximal cell migration speed tends to correlate inversely with contractile force [30]. Further refinement of those observations contrasted contractual-migration interfaces that would imply that the optimal cell–substratum adhesive strengths yielding maximal migration speeds for fibroblasts, neutrophils, and keratocytes, respectively, would be in descending order with approximately tenfold interval decreases [29]. This dependence of cell locomotion speed on overall cell–substratum adhesive strength and the degree of spatial asymmetry suggests one means by which the various molecules regulating adhesion complexes can effectively control migration. The mechanical strength of protein–protein bonds is logarithmically related to their biochemical affinities, so alteration of the affinities of linkages within adhesion complexes by covalent modifications can "tune" overall adhesiveness as well as a spatial adhesiveness differential [31]. Regarding regenerative cells, the variations in migration, the association with ligand, and the course of spatial appropriation all appear to be time dependent as well as signal reliant.

More current discovery has identified a cell migration-dependent mechanism for releasing cellular contents, wherein a cell will leave retraction fibers behind it and vesicles at the intersections of retraction fibers. Coined "migrasomes" by the investigators, these migration nodes contain numerous smaller vesicles, with diameters of about 50–100 nm [32]. During migrasome biogenesis, an initial phase of rapid growth and extension is followed by a relatively stable period. Most compelling to the observation is the subsequent integration and involvement following the retraction. Migrasomes are released into the medium or directly taken up by surrounding cells. The migrasome formation is an integrated conglomeration of specific integrin-coupled microvesicle exosomes that depend on integrin pairing with tetraspanin identities. The idea of trailing edge vesicle release as a primer or as a footprint for other cells to connect is not a new one [33]. It is unique that vesicle release of cellular contents renders location-specific footprinting that other cells can home to. This authenticates the possibility for spatial and biochemical information from outgoing cells, or from host cells to direct regenerative effort, or for cells delivered with extensive potency to map in time and space a scaffold that is at once both connected and polarized by attachment that spatial and biochemical information from outgoing cells can be acquired by incoming cells. Given that many important physiological functions, such as the formation of neuronal networks and innate and adaptive immune responses require localized communication between cells to achieve not only polarity but morphogenic integrity, this importance of this process becomes obvious.

Summarizing a few points in this discussion, cell polarity is a driver of activity and dissociation of charge can activate a differentiation of cell phenotype. In the context of multicellular organisms, and in particular to the essential goal of regenerative medicine, it is critical to keep in mind that cells communicate with each other utilizing chemical messengers and that the pharmacology of release is dependent on membrane polarity, which in turn affects transcription, which in turn generates exosome release that constitutes a connection if not a scaffold. For many of these messenger molecules, the membrane is an insurmountable barrier. New techniques have been developed to examine binding to surface proteins that can measure ligand binding with high temporal resolution and on a single cellular level [34]. With insights into the change of membrane, the ability to reciprocate voltage changes or imbue a capacitive coupled inference to the membranes might be possible. This insight has been similarly guiding strategies to not only define but to detect system change in many tissues by many attempts [35–37]. Possibly, it is accommodated by inherent growth factors accompanying cell and concentrated plasma products, but this connection and intention to regenerate is nourished by factors not yet completely understood. Strategies for repair are to recognize, respond, resolve differences, regenerate material, and/or supplement a scaffold that enhances generation and to provide recognition that allows integration.

Although it is likely that the assurance of composition trumps the eventual continuity of geometry, observations of simple wound healing on skin are obvious to any who have seen restitution of tissue following a cut or scratch where pigmentation and surface erase any obvious trace of the wound. It is clear that it takes more time to guide the remodeling than it does the replacement. Biodynamics and basic physiology including the pharmacology of receptors, half-life of molecular forms, and the interaction and migrations of cells have guided the course of regeneration from the foundations of embryology. Other chapters within this book broadly discuss current regenerative therapies such as platelet-rich plasma that are not unexpectedly replete with platelet-derived growth factor (PDGF). For some time, it has been known that PDGF has a positive effect on the stimulation of mesenchymal stem cells [38]. Within a short time following that report, further studies of PDGF elucidated its receptor and biochemistry and defined the temporal context of its binding half-life [39]. Not surprisingly, the receptor half-life is less than that of the growth factor, as one might imagine the imposition of locked stimulation that was in any way errant in the magnitude of response. Using just this single growth factor as an example and considering the discrete families of growth factors and the large number of variants, the ability to responsibly expect cause and effect never strays from predictive variability. Strategies emerging from the example of PDGF have sought to block action, selectively repress signal, or in other ways to sequence activity in the time–space coordination of tissue development [40, 41]. Determining the fate or defining the relative contributions of therapy presents different challenges.

A revolutionary insight at the time was the tracking of cells using quail–chick chimeras [42]. Nicole Le Douarin, in many ways, pioneered the science defining cell fate, origin in the context of migration, and demonstrated with unusual clarity cell fate, cell migration, and anatomical emergence of distinct tissues. Her prescient observation was that the nucleolus was particularly large and conspicuous in quail mesenchymal cells. Although her work was initially given to evaluating hepatocyte cells and liver development, one characteristic she observed in the differentiation of hepatocytes was the enlargement of the nucleolus. What became intriguing to her and informative to the many who have looked at cell fate during development was that the large nucleolus was evident not only in the hepatocytes but also in the mesenchymal cells of the chimeric liver lobes that developed in culture. These observations were aided by histochemical techniques such as the Feulgen–Rossenbeck's procedure that stains DNA and a method for staining the RNA components of the nucleolus. Thus, as far as the structure of its nucleolus was concerned, the quail species appeared as an exception. This particularity made quail cells easily recognizable from chick cells at the single-cell level and at any developmental stage. Using this method of identification, determinants from early differentiation to mature tissues could be tracked—an example of the chimeric tissue where the quail nucleoli of the neural crest region are stained and the subsequent migration led to the reconstitution of contributions of cell source to mature tissue (Fig. 1.4). The quail–chick marker system enhanced significantly the value of the avian embryo as a model for embryological research in developmental biology, combining the advantages of the availability of the embryo with observation and manipulations during the entire period of development with molecular methods [43–45].

Findings from the experiments provided some of the first evidence of pluripotency and of multiple potential contributions to various tissues including skeletal, nerve, and endocrine tissue. A demonstration of the considerable contribution of the neural crest to the vertebrate head—to the facial and visceral arch skeletal and connective structures, the skull, and the cardiovascular system were furthered beyond the likely considerations that the peripheral nerve system was sourced as well. These notions were new, and the notion of plasticity of the neural crest cells fated to build up the ganglia and nerves of the peripheral nervous system (PNS) largely depended upon environmental cues arising from the tissues in which they differentiate at the end of their migration [46]. A striking feature of the neural crest observation work was the fact that it gave rise to a large number of different cell types and that the feedback during development was regionally and spatially cued. Although the neural crest is regionalized into several distinct areas yielding different PNS structures in normal development, spatial disturbances of this preexisting order did not result in major abnormalities in PNS ontogeny, meaning that one neural crest area can be substituted for another to provide the embryo with sensory, sympathetic, parasympathetic, and enteric ganglia. In the context of regenerative medicine, coordinated repair, cell plasticity, and a sense of morphogenetic patterning seem to process an inexact but equilibrated dimension where tissue emerges not only where the anatomy would dictate but in a form that is appropriate for functional resonance with the surrounding tissue. This underscores the uncertainty of predictability in the morphogenesis of form but may not fully resonate with regenerative capabilities.

What is the remedy for random? Does order dictate and define the direction and dimension of the repairs? Accepting the context of biological "clay," does the science of regenerative medicine have sufficient scope of understanding to predict not only the physical but the deeper and less predictable odds of the analog between the digits—what is the primal biologic utterance? Does the overarching structure impose a simplicity that is amplified in alignment rather than forcing an alignment based on an offset of constraints that are likely triggered by a symphony of cells, growth factors, cytokines, charges, cell surface kinetics, and even the polarity of the individual cells within the emerging tissues? Based on common pharmacology understanding, the location of ligands, or

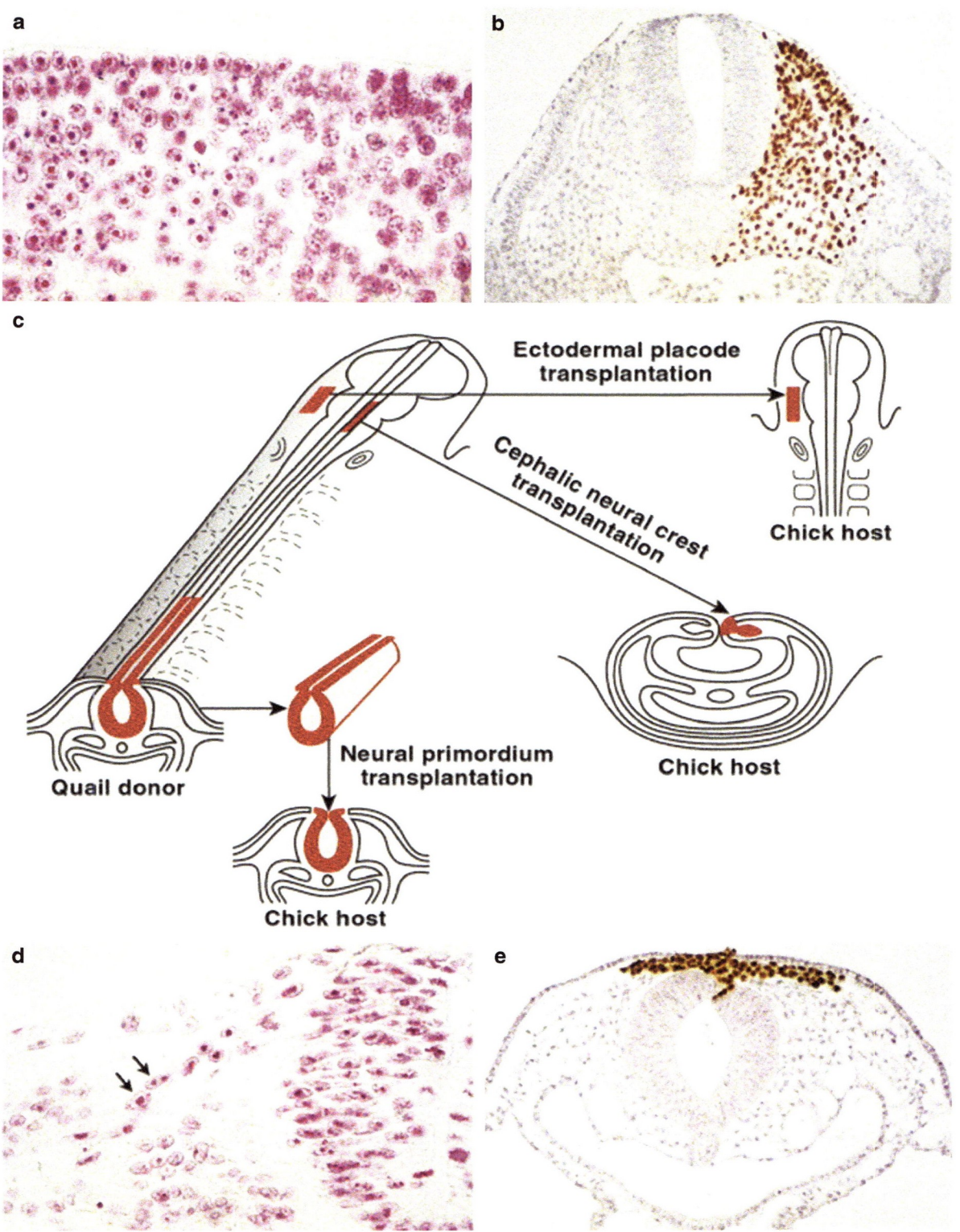

Fig. 1.4 The quail–chick marker system (**a, b**): two means for recognizing quail from chick cells. (**a**) Feulgen staining of DNA shows a large mass of heterochromatin in the center of the nucleus, which is associated with the nucleolus in quail cells (left). In chick cells, the heterochromatin is evenly distributed (right). (**b**) Staining of quail cells (half a somite on the right) grafted into a chick embryo with a monoclonal antibody raised against a quail nuclear antigen (produced by Carlson and Carlson, University of Michigan). (**c**) Different types of grafts from chick to quail (or vice versa) embryos at the same developmental stages. The graft may involve the placodal ectoderm, the neural fold at the head level, or the neural tube including the neural folds prior to the onset of NCC emigration. (**d, e**) NCC migrating from a neural tube quail graft at the trunk level (**d**) or from a neural fold (right) graft at the cephalic level. Note that a unilateral NC fold graft expands on both sides during migration. (Le Douarin [48]. Used with Permission from Elsevier)

the ability for ligands to promote attachment are not surface agnostic and are highly ordered and directed to polarity. In terms of polarity, regenerative medicine still has to define the context of where the lead in active regeneration is precipitated and reactive, what is prescribed and responsive to inner cues, or what is defined in terms of static spaces that offer analog life as part of the procession.

Precipitation Viewed as a Nonsolute Quality

The alteration of proteins and changing the biologic charges and dynamics are some of the most important changes that can alter the course of the living organism. The interactions between linear polysaccharides and proteins can be regulated by additional modifications such as sulfation or branch formation of polysaccharide chains. Although the atomic details have been reported for certain heparin-binding proteins including antithrombin, FGF2, and annexin, 3D structural information on the recognition mode of sulfated polysaccharides remains limited. The effect that stabilizing the environment of a living cell has on the activity of the tissue in terms of differentiation and growth that results from regenerative integration has to do with what drives the fit and guides the morphology. Whether a cell forms a tumor or repairs itself appears to be generative at every step, but somehow the restitution to sameness is lost, and insufficiency becomes the faux of excess.

Mechanical vs. Medical

One of the inherent challenges of regenerative medicine is considering a closed system where all the factors are identified and defined in the hopes of achieving biologic closure. One might argue that accepting that science as defining the dots of understanding does not define the analog space between the concepts and that the mystery forces guiding living organisms to maintain cell membrane voltage, define material space, and evolve against all entropic predictions is a sufficient impetus to guide regenerative efforts even if they are yet insufficiently understood or explained.

The guiding light of medical treatment has always enlisted the applied principles of tissue engineering for years, transplanting and shifting matrices within patients to promote regenerative potential. The advent of new technology offers even greater promise and brings unbridled enthusiasm that full regenerative potential of tissue and whole organ systems can be achieved in the near future.

The implicit goals of regenerative medicine are to achieve restitution of space, mechanical solidarity, and functional continuity. Often, the biological signals do not provide sufficient stimulus to attain a full repair. The therapeutic goal is to omit compliance features such as strain tolerance, reduced stiffness, and attenuated strength and instead promote primary tissue formation within the physical approximation of a wound or diseased tissue. Future bio-engineering strategies will combine several favorable properties of identified biologic processes in an effort to support tissue differentiation without shielding capacity for integrated modeling. Ideally, tissue compatibility that minimizes patient morbidity is optimal for healing. In scope, regenerative implantation will offer structurally enhancing solutions that are inductively optimum for predictable tissue formation. What is clear from the literature and multiple congresses that are held on an annual basis is that the prescriptions for the ultimate answer to regenerative medicine are as numerous as the opinions presenting them. Unlike the six blind men of Indostan who were unable to recognize an elephant except for the terms under which they had been exposed, the words of the Russian playwright and physician, Anton Checkhov, should be considered. Noting in the Cherry Orchard, "If many remedies are prescribed for an illness, you may be sure that the illness has no cure" [47]. Based on the broad use of regenerative adjuncts of cells, matrices of both autologous and allogeneic, platelet concentrates, and placental tissue products, the remedies themselves have become the numerator over a lesser number of denominating potentials. Being able to not only accept the promise but the performance of these therapeutic adjuncts is critical. Forced to seek commercial opportunities that provide meaningful clinical applications, much of the patient risk is deferred to testimonials or personal recommendations. When meaningful data have been synthetically collected rather than abstracted without regard to the number of participants, regimen of therapy, functional improvement, then regenerative medicine will be able to shed the cloak of the alternative and tailor itself to an impressive fit in various clinical applications. In the acceptable practice of physician oversight and practice with a data-driven understanding of performance, expectations should meet the intentions of regeneration as a science and tool for clinical care.

References

1. https://www.fda.gov/downloads/biologicsbloodvaccines/guidance-complianceregulatoryinformation/guidances/cellularandgenetherapy/ucm585403.pdf
2. Large EW, Snyder JS. Pulse and meter as neural resonance. Ann N Y Acad Sci. 2009;1169:46–57.
3. Boeke K. The universe in forty jumps. New York: John Day Company, Inc.; 1957.
4. Morrison P, Morrison P, Office of Charles and Ray Eames. Powers of ten: about the relative size of things in the universe. Stuttgart: Scientific American Books, Holtzbrinck; 1982.
5. Organ Procurement and Transplantation Network; Transplants by Organ Type January 1, 1988—December 31, 2018, Based on OPTN data as of February 7, 2019, Rockville.

6. Koniaris LG, McKillop IH, Schwartz SI, Zimmers TA. Liver regeneration. J Am Coll Surg. 2003;197:634–59. [PMID: 14522336].

7. Michalopoulos GK, DeFrances MC. Liver regeneration. Science. 1997;276:60–6. [PMID: 9082986].

8. Spyridonidis A, Zeiser R, Follo M, Metaxas Y, Finke J. Stem cell plasticity: the debate begins to clarify. Stem Cell Rev. 2005;1:37–43. [PMID: 17132873].

9. van Gene D. Prometheus, Pandora, and the myths of cloning. Hum Life Rev. 2006;32:15–27.

10. Schneider MD. Regenerative medicine: Prometheus unbound. Nature. 2004;432:451–3. [PMID: 15565135].

11. Rosenthal N. Prometheus's vulture and the stem-cell promise. N Engl J Med. 2003;349:267–74. [PMID: 12867611].

12. Power C, Rasko J. Whither Prometheus' liver? Greek myth and the science of regeneration. Ann Intern Med. 2008;149:421–6.

13. Linnaeus, Carl (1755) [1751]. Philosophia botanica: in qua explicantur fundamenta botanica cum definitionibus partium, exemplis terminorum, observationibus rariorum, adiectis figuris aeneis. originally published simultaneously by R. Kiesewetter (Stockholm) and Z. Chatelain (Amsterdam). Vienna: Joannis Thomae Trattner (1755) [1751].

14. Schrodinger E. What is life? Cambridge: Cambridge University Press; 1944.

15. Watson JD, Crick FHC. Genetical implications of the structure of deoxyribonucleic acid. Nature. 1953;171:964–7.

16. Fleury V. Branching morphogenesis in a reaction-diffusion model. Phys Rev E. 2000;61:4156–60, PMID 11088210.

17. Servaas MN, Geerligs L, Bastiaansen JA, et al. Associations between genetic risk, functional brain network organization and neuroticism. Brain Imaging Behav. 2016;11(6):1581–91.

18. Marmar CR, Weiss DS, Schlenger WE, Fairbank JA, Jordan BK, Kulka RA, Hough RL. Peritraumatic dissociation and post-traumatic stress in male Vietnam theater veterans. Am J Psychiatr. 1994;151:902–7.

19. Putnam FW. The psychophysiologic investigation of multiple personality disorder. Psychiatr Clin N Am. 1984;7:31–9.

20. Mason C, Dunnill P. A brief definition of regenerative medicine. Regen Med. 2008;3:1–5.

21. Maxson S, Lopez EA, Yoo D, Danilkovitch-Miagkova A, Leroux MA. Concise review: role of mesenchymal stem cells in wound repair. Stem Cells Transl Med. 2012;1:142–9.

22. Chen L, Tredget EE, Wu PY, Wu Y. Paracrine factors of mesenchymal stem cells recruit macrophages and endothelial lineage cells and enhance wound healing. PLoS One. 2008;3:e1886.

23. Kim WS, Park BS, Sung JH, Yang JM, Park SB, Kwak SJ, Park JS. Wound healing effect of adipose-derived stem cells: a critical role of secretory factors on human dermal fibroblasts. J Dermatol Sci. 2007;48:15–24.

24. Aggarwal S, Pittenger MF. Human mesenchymal stem cells modulate allogeneic immune cell responses. Blood. 2005;105:1815–22.

25. Hocking AM, Gibran NS. Mesenchymal stem cells: paracrine signaling and differentiation during cutaneous wound repair. Exp Cell Res. 2010;316:2213–9.

26. Kamimura K, Inoue R, Nagoya T, et al. Autonomic nervous system network and liver regeneration. World J Gastroenterol. 2018;24(15):1616–21.

27. Kiba T, Tanaka K, Numata K, Hoshino M, Inoue S. Facilitation of liver regeneration after partial hepatectomy by ventromedial hypothalamic lesions in rats. Pflugers Arch. 1994;428:26–9.

28. Hendricks MT, Thuluvath PJ, Triger DR. Natural history of autonomic neuropathy in chronic liver disease. Lancet. 1992;339:1462–4.

29. Luster AD, Alon R, von Andrian U. Immune cell migration in inflammation: present and future therapeutic targets. Nat Immunol. 2005;6:1182–90.

30. Oliver T, Lee J, Jacobson K. Forces exerted by locomoting cells. Semin Cell Biol. 1994;5:139–47.

31. Kuo S, Lauffenburger DA. Relationship between receptor/ligand binding affinity and adhesion strength. Biophys J. 1993;65:2191–200.

32. Ma L, Li Y, Peng J, Wu D, Zhao W, et al. Discovery of the migrasome, an organelle mediating release of cytoplasmic contents during cell migration. Cell Res. 2015;25:24–38.

33. Kriebel PW, Barr VA, Rericha EC, Zhang GF, Parent CA. Collective cell migration requires vesicular trafficking for chemoattractant delivery at the trailing edge. J Cell Biol. 2008;183:949–61.

34. Burtscher V, Hotka M, Sandtner W. Detection of ligand-binding to membrane proteins by capacitance measurements. Bio Protoc. 2019;9(1):e3138.

35. Basset C, Pawluk R, Pilla A. Augmentation of bone repair by inductively coupled electromagnetic fields. Science. 1974;184:575–7.

36. Brighton C, Wang W, Seldes R, et al. Signal transduction in electrically stimulated bone cells. J Bone Joint Surg. 2001;83-A:1514–23.

37. Fitzsimmons RJ, Gordon S, Kronberg J, Ganey TM, Pilla AA. A pulsing electric field (PEF) increases human chondrocyte proliferation through a transduction pathway involving nitric oxide signaling. J Orthop Res. 2008;6:854–9.

38. Ek B, Westermark B, Wasteson A, Heldin CH. Stimulation of tyrosine-specific phosphorylation by platelet-derived growth factor. Nature. 1982;295(5848):419–20.

39. Keating MT, Williams LT. Processing of the platelet-derived growth factor receptor. biosynthetic and degradation studies using anti-receptor antibodies. J Biol Chem. 1987;262:7932–7.

40. Shulman T, Sauer FG, Jackman RM, Chang CN, Landolfi NF. An antibody reactive with domain 4 of the platelet-derived growth factor beta receptor allows BB binding while inhibiting proliferation by impairing receptor dimerization. J Biol Chem. July 1997;272(28):17400–4.

41. Elangovan S, D'Mello SR, Hong L, Ross RD, Allamargot C, Dawson DV, Stanford CM, Johnson GK, Sumner DR, Salem AK. The enhancement of bone regeneration by gene activated matrix encoding for platelet derived growth factor. Biomaterials. 2014;35(2):737–47.

42. Le Douarin N. Particularités du noyau interphasique chez la Caille japonaise (Coturnix japonica). Utilisation de ces particularités comme "marquage biologique" dans des recherches sur les interactions tissulaires et les migrations cellulaires au cours de l'ontogenèse. Bull Biol Fr Belg. 1969;103:435–52.

43. Funahashi J, Okafuji T, Ohuchi H, Noji S, Tanaka H, Nakamura H. Role of Pax-5 in the regulation of a mid-hindbrain organizer's activity. Develop Growth Differ. 1999;41:59–72.

44. Nakamura H, Watanabe Y, Funahashi J. Misexpression of genes in brain vesicles by in ovo electroporation. Develop Growth Differ. 2000;42:199–201.

45. Creuzet S, Martinez S, Le Douarin NM. The cephalic neural crest exerts a critical effect on forebrain and midbrain development. Proc Natl Acad Sci U S A. 2006;103:14033–8.

46. Le Douarin NM, Teillet M-A. Experimental analysis of the migration and differentiation of neuroblasts of the autonomic nervous system and of neurectodermal mesenchymal derivatives, using a biological cell marking technique. Dev Biol. 1974;41:162–84.

47. Chekhov A. Collection of the association "Knowledge" for 1903. Book Two. SPb, 1904, pp. 29–105.

48. Le Douarin NM. The avian embryo as a model to study the development of the neural crest: a long and still ongoing story. Mech Dev. 2004;121(9):1089–102.

Prolotherapy

Jason Kajbaf

Introduction

Regenerative medicine, the idea of creating new healthy tissue to replace old damaged tissue, such as replenishing the articular cartilage of an osteoarthritic joint, or repairing damaged tendons, has become an increasingly researched field [1]. Currently, a significant amount of resources and time are being put into platelet-rich plasma (PRP) and stem cell injections for various musculoskeletal disorders (which will be discussed in a later chapter); however, prior to PRP and stem cells, regenerative medicine began to make its claim as a legitimate treatment option via prolotherapy.

Prolotherapy is essentially the first form of modern regenerative medicine, having been used for the past 100 years, and later further described by Dr. George Hackett [2]. The overarching theory is to introduce a noxious stimulus, most commonly hypertonic dextrose, to induce cell death, which would in turn result in the release of local inflammatory chemicals as well as growth hormones [3]. Subsequently, this local inflammatory response has been thought to stimulate the production of new cells and help heal chronically damaged nearby tissues [3]. Even though prolotherapy has been around for many years, and thought to have been effective based on anecdotal evidence, there has been an increase in the numbers of recent randomized clinical trials that have further supported the use of this prolotherapy, which will be discussed throughout this chapter.

Prolotherapy is most commonly used for chronic musculoskeletal conditions, typically with tendinoses, ligament sprains, or articular cartilage damage, as in knee osteoarthritis [4]. This is all of particular importance because chronic musculoskeletal disorders are currently the leading cause of chronic pain in the United States [5]. It is estimated that roughly one in two people over the age of 18 is affected by a musculoskeletal condition, and this number is even greater in

the elderly, affecting about three of every four people over the age of 65 [6]. And with the United States' aging population, the number of people affected is expected to grow. Thus, it should not come as a surprise that the economic impact of chronic musculoskeletal disorders is massive, with recent estimates being roughly $874 billion dollars annually [6]. However, according to NIH estimates of funding for research for various medical conditions, chronic pain (with musculoskeletal disorders being the major cause of chronic pain) did not crack the top 50 in 2017 [7]. Ultimately, the significant economic impact, effect on individuals' quality of life, and overall prevalence of musculoskeletal disorders have led to greater interest in viable treatment options, a la prolotherapy.

Prolotherapy vs. Platelet-Rich Plasma (PRP)

Platelet-rich plasma (PRP), which will be further discussed in a later chapter, is in brief, precisely what the name suggests, the injection of platelet-rich plasma that is obtained via the centrifugation of a person's own blood, and the platelet rich layer is removed and injected [61]. This procedure is thought to work via the release of platelet-derived growth factor (PDGF) among other growth factors and cytokines, which are stored and released by platelets, and stimulate the production of local stem cells [61]. Thus, it has been studied that PRP may help with the treatment of the various chronic conditions (i.e., knee OA) that prolotherapy has long been used to treat. Recently, a study by Rahimzadeh et al. directly compared PRP to hypertonic dextrose prolotherapy injections for the treatment of chronic knee pain due to osteoarthritis. They found that though both injections provided pain relief and functional improvement, the PRP group had statistically significantly more pain relief and functional improvement than the prolotherapy group [61]. Table 2.1 reviews this and another study that compared prolotherapy to PRP injec-

J. Kajbaf (✉)
David Geffen School of Medicine—UCLA, Department of
Physical Medicine and Rehabilitation, Los Angeles, CA, USA

© Springer Nature Switzerland AG 2023
C. W Hunter et al. (eds.), *Regenerative Medicine*, https://doi.org/10.1007/978-3-030-75517-1_2

Table 2.1 Summary of prolotherapy vs. platelet-rich plasma (PRP)

Study	Study design	Population	Intervention	Primary outcome measure	Results	Conclusions
Rahimzadeh et al. [61]	Prospective randomized double-blinded clinical trial	$N = 42$; patients with knee pain due to OA	Group 1: Injection of 7 cc PRP Group 2: Injection of 7 cc hypertonic dextrose prolotherapy	WOMAC scores at baseline, 1-, 2- and 6-month follow-ups	Improved WOMAC scores from baseline at 1, 2, and 6 months ($P < 0.001$) Greater improvement in PRP group vs prolotherapy group at 2 and 6 months ($P < 0.05$)	Both provide significant pain relief, but PRP may provide longer, more significant pain relief and functional improvement
Kim et al. [62]	Prospective randomized single-blinded clinical trial	$N = 21$; Patients with chronic plantar fasciitis	Group 1: 2 injections of 2 cc PRP at 2-wk intervals Group 2: Injection of 2 cc hypertonic dextrose prolotherapy at 2-wk intervals	Foot functional index *(measures pain, function, and disability)* at baseline, 2 weeks, 2- and 6-month follow-ups	Both groups had significant reduction in all subscales in the FFI Greater improvement in function and disability in PRP group vs prolotherapy group at 2 weeks and 2 months ($P < 0.05$). No difference in pain at all intervals	Dextrose prolotherapy and PRP both result in clinically significant improved pain, function, and stiffness

tions. Overall, it appears that although both procedures are effective treatment options, PRP may provide significantly more relief.

Indications

Chronic pain continues to compose a large percentage of annual doctor's visits, a majority of the chronic pain being due to musculoskeletal disorders [11]. A majority of chronic musculoskeletal pain conditions, such as various tendinoses or chronic sprains, have been attributed to repetitive microtrauma or various single traumatic events that damage the soft tissue to such an extent that in turn leads to poorly healed ligaments and tendons. Thus, in an attempt to improve or perhaps even reverse these chronic degenerative changes that ensue, prolotherapy has been increasingly researched on various joints or soft tissues in order to provide a viable, safe, and cost-effective treatment option.

Lower Extremities

Knee Osteoarthritis

The human knee is put through constant, daily repetitive stress, and in a healthy joint, with healthy tendons, ligaments, and articular cartilage, this repetitive pounding is withstood. However, as knees age, these same soft tissues begin to deteriorate due to the same daily stresses and trauma and may eventually lead to, by way of osteoarthritis or tendon/ligament damage, chronic knee pain and/or instability. Symptomatic knee osteoarthritis is among the most common cause of knee pain and is seen in about 13% of women and 10% of men [20].

Current therapy for knee osteoarthritis is traditionally a combination of topical or oral analgesics, steroid and hyaluronic acid injections, and physical therapy, prior to proceeding with surgery [21]. However, within the field of regenerative medicine come treatments including prolotherapy, platelet-rich plasma, and stem cell injections, which have begun to change the treatment algorithm of knee osteoarthritis.

There has been a recent surge in research looking at prolotherapy and knee OA, and although there has been general heterogeneity in protocols of these studies, they continue to demonstrate significant improvement in symptoms. A recent single-arm study by Rabago et al. in 2012 demonstrated improvement of pain scores and Western Ontario and McMaster Universities Osteoarthritis Index (WOMAC) scores 4 weeks after the initial injection, as well as significant improvement at 52-week follow-up [22]. These results were then later reproduced by a randomized controlled trial, again by Rabago et al. in 2013, comparing prolotherapy with hypertonic dextrose to conservative management with a home exercise program. This study used the same injection protocol as the earlier single-arm trial and had very similar results, with significantly improved WOMAC scores that persisted through 52 weeks [23]. A similar study using hypertonic dextrose yielded similar improvements in pain, as well as improvements in knee buckling and anterior cruciate ligament (ACL) laxity, thus further suggesting that there may be a possible role for prolotherapy in cases where ACL disruption may be present [24]. These results are further summarized in Table 2.2. Subsequently, a meta-analysis in 2016 by Sit et al. reviewed four RCTs and concluded that prolotherapy provides significant pain relief and improved WOMAC scores when compared to controls of home exercise programs and saline injections [25].

Table 2.2 Summary of prolotherapy for treatment of knee osteoarthritis

Study	Study design	Population	Primary outcome measure	Results	Conclusions	Adverse events
Rabago et al. [22]	Prospective single-arm trial with hypertonic dextrose	N = 36; moderate-severe OA patients	WOMAC scores	Improved WOMAC scores (P < 0.001)	Dextrose prolotherapy may result in improved pain, function, and stiffness	None
Rabago et al. [23]	Prospective randomized double-blinded placebo-controlled trial Normal saline injection vs. hypertonic dextrose injections vs. exercise alone	N = 90; patients with at least 3 months of moderate to severe knee pain due to OA	WOMAC scores	Greater WOMAC score improvement of dextrose prolotherapy compared to controls (P < 0.05)	Dextrose prolotherapy results in clinically significant improved pain, function, and stiffness	None
Reeves et al. [24]	Prospective randomized double-blinded placebo-controlled trial 10% dextrose +0.075% lidocaine in bacteriostatic water vs. 0.075% lidocaine in bacteriostatic water	N = 68; patients with at least 6 months of knee pain due to knee OA	Visual Analog Scale (VAS) for pain and frequency of knee buckling, and KT1000-measured anterior displacement difference (ADD) for objective analysis of ACL laxity	Improved pain and buckling frequency in dextrose injection group (P < 0.015); improved ACL laxity (P = 0.021)	Prolotherapy injection with 10% dextrose resulted in clinically and statistically significant improvements in knee osteoarthritis, and ACL laxity	None

Patellar Tendinopathy

Patellar tendonitis, inflammation of the patellar tendon typically due to repetitive jumping or knee bending activities (i.e., basketball, volleyball, and high jump), has been known to affect up to 20% of jumping athletes [26]. However, there is very little research studying the use of prolotherapy for patellar tendinopathy. There is a non-randomized single-arm pilot study that used hypertonic dextrose injections for patellar tendonitis refractory to conservative management, which found that there was both a clinical improvement in patient pain scores and improvement of the structural integrity of the patellar tendon that was evident by increased neovascularity under ultrasound.

Anterior Cruciate Ligament (ACL) Laxity

ACL injuries are commonly quite debilitating injuries seen most predominantly in athletes, resulting in significant time away from sports, due to treatment typically with surgery and subsequent rehab [17]. Currently, acute ACL injuries are managed surgically; however, in cases of chronic knee pain associated with ACL laxity, prolotherapy has been suggested to serve as a potential treatment option [24]. This study is further detailed in Table 2.2. Ultimately, there is a paucity of clinically significant research studying the use of prolotherapy for ACL injuries.

Osgood-Schlatter Disease

Osgood-Schlatter disease has long been considered a disease of young male athletes, most commonly between the ages of 9 and 14, and is usually benign and self-resolves once the growth plate ossifies [27, 28]. Although the pain is self-limiting and resolves with conservative management in most cases, in rarer situations, chronic pain may persist, and prolotherapy may have a role in its treatment. One study compared prolotherapy to conservative management and lignocaine injections and found that prolotherapy provided statistically significant improvements in both short- and long-term pain relief when compared to non-prolotherapy groups. Furthermore, they found that the prolotherapy group also had a higher asymptomatic return to sports rate than the other treatment groups, thereby demonstrating its positive effects on overall quality of life and level of function [29].

Achilles Tendinopathy

Achilles tendonitis is found commonly among the general population, though it is more prevalent among athletes. Up to 24% of athletes have a lifetime incidence of Achilles tendinopathy [30]. Risk factors include stop-and-go sports (i.e., basketball, soccer), history of Achilles tendinopathy, a sudden increase in exercise, male gender, obesity, and poor running mechanics, among others [30]. There have been many

Table 2.3 Summary of prolotherapy for treatment of Achilles tendinopathy

Study	Study design	Population	Primary outcome measure	Results	Conclusions	Adverse events
Maxwell, et al. [32]	Prospective case series Ultrasound-guided Achilles tendon prolotherapy injections (25% dextrose +1% lignocaine) every 6 weeks	$N = 36$; mean age 54.0	VAS pain scores: VAS 1 at rest VAS 2 during normal activity VAS 3 during sport activity Tendon thickness measured by ultrasound	Improvements in VAS 1, 2, & 3 ($P < 0.001$) Decreased Achilles tendon diameter ($P < 0.007$)	Dextrose prolotherapy results in statistically improved pain and reduces tendon inflammation as seen by reduced tendon diameter	None
Ryan et al. [33]	Prospective case series Ultrasound-guided Achilles tendon prolotherapy injections (25% dextrose +1% lignocaine) every 6 weeks	$N = 99$; mean age 54.0	VAS pain scores: VAS 1 at rest VAS 2 during normal activity VAS 3 during sport activity Tendon thickness measured by ultrasound	Improvements in VAS 1, 2, & 3 ($P < 0.001$) No change in Achilles tendon diameter	Dextrose prolotherapy results in statistically significant reduction of pain from Achilles tendinopathy	None

studies as of late looking at the use of prolotherapy for Achilles tendinopathy; however, a recent systematic review identified several studies that were of moderate to good quality, which have been further detailed in Table 2.3 [31]. One of these studies by Maxwell et al. found statistically significant reduction in pain scores at rest, with normal daily activity, and even after strenuous activity or sports after ultrasound-guided prolotherapy injections [32]. Additionally, they demonstrated statistically significant reductions in Achilles tendon diameter following prolotherapy injections, further signifying an improvement of the tendinitis. This was then later corroborated by a study by Ryan et al. who followed the same injection protocol and had similar statistically significant reductions in pain scores during rest, normal activity, and sports [33].

Plantar Fasciitis

There are approximately one million people affected annually by foot pain caused by plantar fasciitis, and more than half of them will seek medical treatment. Current practice is to initially treat this conservatively with shoe inserts or heel pads, and various stretches and exercises; however, in cases of refractory plantar fasciitis, injections have been shown to

be helpful [34]. The initial injection commonly practiced is steroid injections, though like other chronic ligamentous or tendinous disorders mentioned above, prolotherapy may have a role in plantar fasciitis treatment. Two separate studies, deemed to have moderate to good quality of evidence by a recent systematic review, have both demonstrated significant reduction of symptoms and are reviewed in more detail in Table 2.4.

Upper Extremities

Rotator Cuff Tendinopathy

Shoulder pain is extremely common among the general population, even more so in the physically active population, and among the most common reasons for shoulder pain is rotator cuff tendinopathy [37]. Rotator cuff tendinopathy is commonly treated conservatively with physical therapy and oral analgesics, and when refractory or if rotator cuff impingement syndrome is suspected, then a subacromial steroid injection may be beneficial. However, a recent systematic review compared steroid injections with NSAIDs vs NSAIDs alone, and there was no difference in various outcomes

Table 2.4 Summary of prolotherapy for treatment of plantar fasciitis

Study	Study design	Population	Primary outcome measure	Results	Conclusions	Adverse events
Ryan et al. [35]	Prospective case series Ultrasound-guided plantar fascia prolotherapy injections (25% dextrose +1% lignocaine) every 6 weeks	$N = 20$; mean age 51.2	VAS pain scores: VAS 1 at rest VAS 2 during normal activity VAS 3 during sport activity	Improvements in VAS 1, 2, & 3 ($P < 0.001$) at 28-week follow-up	Dextrose prolotherapy results in statistically improved long-term pain relief	None
Kim et al. [36]	Prospective single-blinded randomized clinical trial Ultrasound-guided plantar fascia platelet-rich-plasma (PRP) injections vs. prolotherapy (15% dextrose +1.25% lignocaine) injections	$N = 21$; mean age 37.0	Foot functional index (*measures pain, function, and disability*) at baseline, 2 weeks, 2- and 6-month follow-ups	Both groups had significant reduction in all subscales in the FFI Greater improvement in function and disability in PRP group vs. prolotherapy group at 2 weeks and 2 months ($P < 0.05$) Significant reduction of pain, with no difference between groups	Dextrose prolotherapy and PRP both result in clinically significant improved pain, function, and stiffness	None

Table 2.5 Summary of prolotherapy for treatment of rotator cuff tendinopathy

Study	Study design	Population	Primary outcome measure	Results	Conclusions	Adverse events
Seven et al. [39]	Randomized clinical trial Ultrasound-guided 25% dextrose +1% lidocaine (prolotherapy) injection vs. physical therapy alone	$N = 101$; 35 females, 42 males	Visual analog scale (VAS) for pain Shoulder pain and disability index (SPADI) Western Ontario Rotatory Cuff (WORC) Index Patient satisfaction Shoulder range of motion	Improvements in VAS, SPADI, WORC and shoulder range of motion in both groups Statistically significant greater improvements in VAS, SPADI, WORC, and shoulder abduction, flexion, and internal rotation Greater patient satisfaction in prolotherapy group (92.9% vs. 56.8%)	Dextrose prolotherapy provides significant improvements in pain and level of function for chronic rotator cuff tendinopathy	None

including, pain level, abduction range of motion, and overall level of function [38]. Therefore, other options, such as prolotherapy, should be considered for the treatment of chronic rotator cuff tendinopathy when common first-line measures have failed. As detailed in Table 2.5, prolotherapy has been proven to be as efficacious, and even better in some aspects, than physical therapy. Ultimately, although the data is currently limited, prolotherapy does appear to provide beneficial results for chronic rotator cuff tendinopathy.

Glenohumeral and Acromioclavicular Joint Osteoarthritis

Unlike osteoarthritis of the knee, there is currently no significant literature looking at the efficacy of prolotherapy for either glenohumeral or acromioclavicular joint osteoarthritis.

Lateral Epicondylosis

Chronic lateral epicondylosis, commonly referred to as "tennis elbow," is estimated to be the cause of about 2.4 per 1000 annual doctor's visits [40]. The treatment of chronic lateral epicondylosis is currently on the conservative side, with evidence present for long-term benefits with physical therapy. Steroid injections, on the other hand, although have been shown to provide immediate short-term relief, do not provide significant long-term relief, but rather have been shown to have poorer outcomes than either physical therapy or with no treatment [41]. Thus, there have been multiple studies looking at the use of prolotherapy for cases of refractory chronic lateral epicondylosis, and two in particular have been reviewed in Table 2.6. Essentially both studies, including various other studies not included here, have demonstrated statistically significant functional and pain improvement

Table 2.6 Summary of prolotherapy for treatment of lateral epicondylosis

Study	Study design	Population	Primary outcome measure	Results	Conclusions	Adverse events
Rabago, et al. [42]	Three-arm randomized clinical trial Ultrasound-guided dextrose prolotherapy vs. dextrose-morrhuate prolotherapy vs. watchful waiting (wait-and-see)	$N = 36$ elbows (26 adults); mean age 48.2	Patient-Related Tennis Elbow Evaluation (PRTEE) composite score	Improved PRTEE score in both prolotherapy groups compared to control group ($P < 0.05$)	Prolotherapy results in safe, significant improvement in elbow pain and level of function	None
Scarpone et al. [43]	Double-blinded randomized clinical trial 50% dextrose +5% sodium morrhuate +4% lidocaine +0.5% sensorcaine + normal saline (prolotherapy) injection vs. normal saline injection	$N = 24$	Resting elbow pain	Improved resting elbow pain at 8-, 16-, and 52-week follow-up in prolotherapy group when compared to baseline and control ($P < 0.001$)	Dextrose prolotherapy results in statistically significant reduction of pain from lateral epicondylosis	None

with prolotherapy injections. Overall, there appears to be a beneficial role for prolotherapy injections in chronic lateral epicondylosis.

First Carpometacarpal Joint Osteoarthritis/Hand Osteoarthritis

Hand osteoarthritis (OA) is an extremely common disease, and its prevalence is known to increase with age, as with OA of other joints in the human body. It has been estimated that about 13% of men and 26% of women over the age of 70 have symptomatic hand OA, which becomes a primary concern given the aging population [44]. Furthermore, hand OA can be quite debilitating since symptoms of pain with finger flexion and reduced range of motion would limit the ability to carry out activities of daily living, thus significantly affecting one's quality of life. Current treatment of hand OA is typically with a combination of physical or occupational therapy, adaptive equipment, braces and splints, and NSAIDs [45]. Once a patient has failed conservative management, a corticosteroid injection may be trialed, particularly with first carpometacarpal (CMC) joint OA. However, a recent systematic review in 2016 compared intra-articular steroid injections with placebo for first CMC joint OA and found no differences in pain or other secondary outcomes (i.e., grip strength, pinch strength, or joint stiffness) at 26-week follow-up [45]. Thus, given the apparent lack of success with corticosteroid injections, it becomes even more key to find alternative interventions.

There are currently two double-blinded randomized clinical trials that have studied the efficacy of prolotherapy for hand OA. One compared prolotherapy with control injections among various joints in the fingers, and another study compared prolotherapy with steroid injections for the treatment of first carpometacarpal (CMC) joint osteoarthritis. Table 2.7 further reviews these studies in additional detail

and demonstrates that prolotherapy may provide significant long-term pain relief for osteoarthritis among interphalangeal and CMC joints.

Axial

Lumbar Ligament Sprains and Facet Joint Capsule Laxity

As discussed several times already in this chapter, musculoskeletal disorders are extremely common complaints among the general population; however, the most common musculoskeletal complaint among the general population is low back pain [48]. However, given the multitude of potential causes of low back pain, it takes an astute clinician to make the appropriate diagnosis of the underlying pathology, whether it is ligamentous, musculotendinous, or bony. This becomes of particular interest when considering injections such as prolotherapy, as this would guide the location of injections (i.e., interspinous ligament versus iliolumbar ligament versus facet joint capsule). However, there are currently no significant studies looking at prolotherapy for various lumbar ligament sprains. A study by Klein et al. compared prolotherapy with local anesthetic injections; however, the results were not statistically significant. Moreover, both treatment and control groups received a gluteal steroid injection, which would in theory nullify the inflammatory cascade expected in a prolotherapy injection [49]. Similarly, though a study by Ongley, et al. did provide statistically significant results, they gave concomitant prolotherapy injections and gluteal steroid injections to the treatment group only, again likely nullifying the effects of the prolotherapy injections [50]. There is, however, a more recent study that is further detailed in Table 2.8, which injected facet joint capsules with prolotherapy that pro-

Table 2.7 Summary of prolotherapy for treatment of hand osteoarthritis

Study	Study design	Population	Primary outcome measure	Results	Conclusions	Adverse events
Reeves et al. [46]	Double-blinded placebo-controlled randomized clinical trial 10% dextrose +0.075% xylocaine + bacteriostatic water (prolotherapy) injection vs. 0.075% xylocaine + bacteriostatic water (control) injection Injected symptomatic PIP, DIP, and first carpometacarpal (CMC) joints	$N = 27$ patients (total of 150 osteoarthritic hand joints)	VAS for pain at rest, with joint movement and grip, and flexion range of motion	Statistically significant improvement in pain during joint movement ($P = 0.027$) Improved range of motion in prolotherapy group ($P < 0.003$)	Prolotherapy results in safe, significant improvement in joint pain with movement and range of motion in osteoarthritic finger joints	Minimal—not elaborated in study
Jahangiri et al. [47]	Double-blinded randomized clinical trial 20% dextrose +2% lidocaine (prolotherapy) injections vs. 40 mg methylprednisolone acetate (0.5 ml) + 2% lidocaine (0.5 ml) injection Injected first carpometacarpal (CMC) joints	$N = 55$; mean age 63.6	VAS for pain, intensity of tenderness (standardized by Fischer's pressure algometer; applied 40 N/cm^2 for all participants)	Greater pain reduction at 1 month with steroid group, comparable at 2 months between groups, and greater pain reduction after 6 months in prolotherapy group ($P = 0.02$)	Dextrose prolotherapy results in statistically significant reduction of pain in the long term for patients with pain due to first CMC joint OA	None

Table 2.8 Summary of prolotherapy for treatment of low back pain due to facet capsule laxity

Study	Study design	Population	Primary outcome measure	Results	Conclusions
Hooper et al. [51]	Case series 20% dextrose injections into the facet capsules of the cervical, thoracic, lumbar spine (weekly injections up to 3 weeks and 1 month later if needed)	$N = 147$, with more than 6 months of low back pain	Neck disability index (NDI), Patient-Specific Functional Scale (PSFS), Roland Morris Disability Questionnaire (RMDQ)	Improvements in all disability scales at 1-year follow-up ($P < 0.0001$)	Prolotherapy results in statistically improved pain and level of function

vided long-term pain relief [50]. Unfortunately, there is currently very limited quality data for prolotherapy as a treatment option for back pain due to the various causes listed above.

Sacroiliac (SI) Joint Dysfunction

Low back pain, as discussed above, is an extremely common complaint among the general population; however, the particular cause of low back pain requires an astute clinician to use the proper physical examination techniques to pinpoint. Pain due to SI joint dysfunction, for instance, is a very common cause of low back pain, and its prevalence has been estimated to be about 22.5% in patients with low back pain [52]. Conservative measures to treat acute to subacute pain due to SI joint dysfunction should include the use of topical and/or oral analgesics in combination with physical therapy in order to restore normal SI joint mechanics [53]. However, if con-

servative treatment fails, then current practice is to consider intra-articular steroid injections or radiofrequency ablation [53]. On the other hand, though, a recent study suggests the possible role of prolotherapy for chronic SI joint dysfunction. Kim, et al. performed a randomized clinical trial comparing the efficacy of intra-articular prolotherapy injections to corticosteroid injections. They found that both groups provided significant pain relief in the short term; however, the prolotherapy group provided better long-term pain relief [54]. This may be due to the fact that sacroiliac pain is typically not solely due to the joint itself but also commonly due to sprains of the surrounding ligamentous structures (i.e., posterior sacroiliac ligament), for which prolotherapy has been commonly used, as described throughout this chapter. Thus, future research should be done to study the effects of prolotherapy for sacroiliac pain due to chronic sacroiliac ligaments sprains.

Cervical Ligament Sprains and Facet Joint Capsule Laxity

The annual incidence of neck pain in the general population is roughly between 30 and 50%, and current treatment modalities of chronic neck pain are similar to the treatment of chronic low back pain. This includes oral and topical analgesics, physical therapy, and at times more interventional treatments with medial branch blocks, radiofrequency ablations, or facet steroid injections, among other procedures [55]. A large and common cause of chronic neck pain has been attributed to laxity of the cervical spine capsular ligaments, the supporting ligaments of the facet joints. This laxity may be caused by single traumatic events or by repetitive microtrauma [55]. Eventually, capsular ligament laxity may lead to cervical spine instability, which in turn commonly leads to overlying myofascial pain and muscle spasms [55]. Thus, an agent that corrects ligamentous laxity, such as prolotherapy, may pose to be a successful form of treatment for chronic neck pain due to cervical instability.

There are currently no randomized clinical trials that studied the efficacy of prolotherapy for neck pain; however, the study by Hooper et al. discussed in section 2.8a had a subset of patients with neck pain attributed to cervical instability related to facet capsule laxity. As discussed earlier, and further reviewed in Table 2.8, dextrose injections provided significant long-term pain relief for these patients [51].

Head & Face

Temporomandibular Joint (TMJ) Disorder

Jaw pain related to TMJ dysfunction can be quite a debilitating disorder, one that can decrease one's quality of life and level of function significantly. The prevalence of TMJ disorder in the United States is estimated be roughly 5%, affecting women more so than men [56]. Current algorithms for the treatment of TMJ disorder involve starting with patient education to avoid exacerbating behaviors (i.e., nail biting), physical therapy, and the use of mouth splints [57]. If these conservative measures fail, then the next step in management may involve various injections, such as botulinum toxin injections, intra-articular corticosteroid injections, or acu-

puncture, though large-scale randomized trials are lacking to definitely claim efficacy of these treatments [58].

An alternative to the treatments mentioned above includes the use of prolotherapy injections. A recent single-arm study found statistically significant improvements in pain, maximum mouth opening (MMO), clicking, and frequency of locking in patients with diagnosed TMJ disorder (Table 2.9) [59]. Though the data is limited, prolotherapy may be considered as a treatment option in cases of refractory TMJ disorder.

Microanatomy and Biochemistry

The proposed mechanism of action of prolotherapy is to introduce a local irritant in order to cause cell death, thereby promoting the release of various growth factors. The most commonly used prolotherapy injectate, hypertonic dextrose (ranging from 12.5 to 25% dextrose), has been shown to cause the death of nearby cells by changing the osmotic gradient, which then attracts macrophages and other granulocytes to promote healing. Furthermore, this localized hypertonic environment also affects platelets, which are of particular interest given the number growth factors they contain, and thereby released upon their destruction [4].

However, when discussing prolotherapy, there are a multitude of different injectates used, including phenol-glucose-glycerin (P2G) and sodium morrhuate [8]. P2G has been shown to work via alkylation of proteins on the surface of cells, thereby damaging the cells directly or via an antigenic affect, attracting local granulocytes [9]. Sodium morrhuate, on the other hand, is composed of the precursor to various natural chemotactants, such as prostaglandins and leukotrienes, and as such is thought to work as a chemotactic agent, and promote the attraction of local inflammatory cells [9]. Additionally, sodium morrhuate has also been known to act as a sclerosing agent. Sclerotherapy, the process of causing endothelial damage, is commonly performed for varicose veins or telangectasias. In the world of chronic musculoskeletal disorders, on the other hand, scleropathy has been considered to be effective due to its damaging effects on the neovessels that accompany the nerve fibers, essentially destroying the nerve fibers transmitting pain from a chronic tendinopathy [10].

Table 2.9 Summary of prolotherapy for treatment of temporomandibular Joint disorder

Study	Study design	Population	Primary outcome measure	Results	Conclusions	Adverse events
Refai H. [59]	Single-arm prospective study 10% dextrose, 2% mepivacaine 4 injections in each TMJ, spaced 6 weeks apart	$N = 61$	Pain score MMO in cm between the incisal edges of the upper and lower incisors Frequency of clicking sound Frequency of locking	Improvements in all outcome measures just prior third injection ($P < 0.01$ for all) Improvements in pain and clicking 3 months after last injection ($P < 0.001$)	Dextrose prolotherapy provides significant and sustained reduction of pain and associated symptoms with TMJ disorder	None

Various in vitro studies have further demonstrated that there is a significant release of growth factors when chondrocytes and fibroblasts (cells seen in cartilage, ligaments, and tendons) are exposed to prolotherapy solutions; more specifically, it has been demonstrated to stimulate the production or release of platelet-derived growth factor (PDGF), TGF-beta, epidermal growth factor, basic fibroblast growth factor, connective tissue growth factor, and insulin-like growth factor. These growth factors in turn are thought to strengthen cartilage, ligaments, and tendons via the production of type 1 and 3 collagen [11].

Moreover, in addition to the biochemical process initiated after a specific injectate is introduced, the introduction of a needle into a tendon or ligament, has also been shown to improve to pain [12]. This procedure is known as needle tenotomy, dry needling, or tendon fenestration. The proposed mechanism is to cause an acute injury to a chronic one via the introduction of a needle into tendons and/or ligaments, which subsequently causes the formation of a local inflammatory process that increases the body's inherent ability to heal itself [12]. Tenotomy has been demonstrated to be beneficial in various locations, including the common extensor tendon of the elbow, Achilles tendon, as well as tendons around the hip, such as the gluteus medius insertion tendon. Jacobson et al. performed tenotomy under ultrasound on patients with gluteus medius, gluteus minimus, proximal hamstring, and/or tensor fascia lata tendinosis, and found significant improvement in their symptoms [13]. This improvement was attributed to the tendon trauma caused by the needle, which changes the chronic injury to an acute one, thereby promoting an increase in local growth factors to promote healing [13]. Also, prolotherapy injections also commonly incorporate the needle tenotomy technique during the injection process in order to further add to the local inflammatory process. Ultimately, it has been proposed that ligament and tendon laxity, as well as degenerative changes seen in osteoarthritis, may improve with prolotherapy injections, via the various mechanisms described above.

Hypertonic Dextrose and Neurolysis

In addition to the mechanisms above, however, there is another method by which hypertonic dextrose, in particular, has been proposed to function for pain relief. It has been demonstrated that the introduction of a hyperosmolar solution (>1000 mOsm/l) may induce the separation of myelin lamellae in myelinated nerve fibers, as well as cause the destruction of unmyelinated nerve fibers, thus functioning as a neurolytic agent [14]. For instance, a common prolotherapy mixture of 50% dextrose with 1% lidocaine has an osmolarity of 1388 mOsm/l, which would thus function as a hyperosmolar neurolytic block [14]. This would theoreti-

cally provide immediate and sustained pain relief, which is what Miller et al. discovered when they performed intradiscal prolotherapy using hypertonic dextrose for the treatment of discogenic low back pain. They concluded that the immediate and sustained pain relief after the injections was due to the neurolytic effect of the hypertonic dextrose, rather than its regenerative effects, because the patients had significant pain relief in a timeframe that was too rapid for tissue regeneration to have occurred [15].

Animal Studies

Various animal studies have looked at the regenerative effects of prolotherapy injections among different structures, including Achilles tendons, knee medial collateral ligaments, and femoral articular cartilage.

Rat Achilles Tendon and Prolotherapy

Prolotherapy has been studied on healthy rat Achilles tendons in order to help demonstrate its safety and regenerative effects discussed earlier in this chapter. One study compared hypertonic dextrose injections with saline and found no difference in tensile strength of the Achilles tendon at 0, 5, or 10 days, as well as no significant changes in histological appearance that would suggest tendon degeneration, as is the concern with steroid injections [16]. A separate study further demonstrated an increase in fibroblast counts 4 weeks after prolotherapy injections compared with no injections [17]. Lastly, a study by Kim et al. subsequently demonstrated an increase in gross tendon diameter, as well as fibroblast counts [18].

Rat Medial Collateral Ligament (MCL) and Prolotherapy

MCL injuries and sprains are the second most common knee ligamentous injuries, with ACL injuries being the most common, or at least most commonly reported [19]. However, there is little data looking at the efficacy of prolotherapy for MCL injuries in human patients. An animal study by Jenson et al. in 2008 compared the response of rat MCLs to hypertonic dextrose injections to saline injections, and they found that the test group had increased overall MCL cross-sectional area; however, there was no change in ligament laxity [8]. Otherwise, there is no significant current literature or research that looks at the efficacy of prolotherapy on human models.

Rabbit Femoral Cartilage and Prolotherapy

An animal study by Kim et al. performed 2 mm punch lesions in adult rabbit femoral cartilage and subsequently injected them with hypertonic dextrose prolotherapy, platelet-poor plasma, or no injection as the control. They found that both the prolotherapy and platelet-poor plasma groups had significant chondrocyte filling after 6-week evaluation when compared to the control group, thus further demonstrating the possible regenerative role of prolotherapy injections [18].

Overall, the studies discussed above all appear to demonstrate significant regenerative effects with prolotherapy on various soft tissue structures; however, additional studies should be performed to further corroborate these limited findings.

Basic Concerns and Contraindications

Prolotherapy injections are commonly thought of as safe procedures, with relatively few contraindications. The few absolute contraindications include local abscess, cellulitis, or a septic joint. Furthermore, if a patient is on anticoagulation, the physician should yield caution with injections [60].

Common adverse effects include pain with the injection itself and various levels of soreness in the couple days following the injection; however, given that the injected solution is meant to stimulate an inflammatory response, this is to be expected [60]. If pain is severe and intolerable, then pain control should be sought with acetaminophen rather than anti-inflammatory medications (i.e., NSAIDs) or even ice, which is also known to have anti-inflammatory effects, as either of these could potentially diminish the efficacy of the procedure. Other less common adverse events include possible pneumothorax (depending on injection location), nerve damage, or significant bleeding (predominantly seen in patients on anticoagulation or with a history of bleeding disorders), this according to a survey conducted by Dagenais et al. [60]

Preoperative Considerations

- Informed consent and proper explanation of all potential complications.
- Anti-coagulation—if prolotherapy is being used for a neuraxial procedure, standard anti-coagulation precautions should be followed. Hemarthrosis is a consideration for intra-articular procedures and should be discussed with the patient beforehand.
- Physical examination of the area for infection, skin ulceration or necrosis, and extent of disease.
- Free of systemic infections for a minimum of 1 week (i.e., common cold, sore throat, cough, etc.). Any antibiotics should be completed prior to the procedure.
- If using radiographic guidance, evaluate the patient for any contrast allergies prior to the procedure.

Radiographic Guidance

If the intended targets for prolotherapy are in the spine or sacroiliac joint, the authors recommend using live fluoroscopy. Soft tissue and intra-articular targets can easily be performed "blind"; however, image guidance is recommended due to the increased precision and ability to directly visualize the target region. Prolotherapy is highly inflammatory and should therefore be avoided in healthy tissue if at all possible. Moreover, injecting hypertonic dextrose into a blood vessel or peripheral nerve can have dangerous consequences.

- *Fluoroscopy*—when injecting into/along/adjacent to the spine, it is the standard of care to utilize fluoroscopic imaging to ensure proper needle placement. In the case of an intra-articular injection into a medium to large joint (i.e. elbow, knee, shoulder, or hip, one can perform an arthrogram under fluoroscopy to confirm needle placement within the joint capsule prior to injecting, thus optimizing the likelihood of delivering the maximum amount to the desired location.
- *Ultrasound*—this allows the operator the ability to evaluate underlying soft tissue of a painful area in an attempt to locate any pathology that could ultimately be the pain generator, thus ascertaining what the target for prolotherapy will be. Simultaneously, when injecting into tendons, ligaments, and other soft tissue with ultrasound, one can directly visualize, in real time, the target area, position of the needle tip, and spread of the injectate, thus making ultrasound a must for these types of prolotherapy injections.

Equipment

- 50% Dextrose
- 1% Lidocaine
- 10 cc syringe (for targeting multiple areas in one sitting)
- 5 cc syringe (for large joints)
- 1 cc syringe (for TMJ, first CMC or interphalangeal joints, and facet joints)
- 18-gauge blunt tip needle
- 25-gauge 1.5-inch needle (all targets except spinal and SIJ)
- 22-gauge 3.5-inch spinal needle (axial procedures, including SIJ)

Technique

Relatively, prolotherapy injections are not unlike a traditional corticosteroid injection. The primary difference is the injectate, which may vary among providers, but traditionally, a solution containing half 1% lidocaine and half 50% dextrose [50:50 mixture of 50% dextrose and 1% lidocaine (unbuffered) has been used and has been shown], as detailed earlier in this chapter, to provide significant pain relief for various pathologies noted above. For larger joints, such as shoulders, knees, and hips, a solution containing 3–4 cc of dextrose and lidocaine may be used; however, for smaller joints, such as the TMJ, first CMC or interphalangeal joints, and facet joints, 0.5 cc suffices. The following will briefly review suggested technique for various prolotherapy injections.

Knee prolotherapy injections may be performed without imaging by using anatomical guidance as historically performed, with the patient seated, knee flexed 90 degrees, and via a medial or lateral infrapatellar approach, a 25-gauge 1.5 inch needle is inserted at approximately a 30-45degree angle, towards the opposite femoral condyle. However, it has been suggested that the use of image guidance may make intra-articular knee injections more accurate. Berkoff et al. found that the use of ultrasound guidance for knee injections resulted in significantly improved accuracy than anatomical guidance (95.8% versus 77.8%, $P < 0.001$) [70]. An ultrasound-guided intra-articular knee injection may be performed in multiple methods; however, a lateral suprapatellar approach has been commonly used to access the suprapatellar bursa, which joins to the knee joint synovial space.

Shoulder prolotherapy injections, again like knee and most other musculoskeletal injections, may be done with anatomic or image guidance. However, when performing a prolotherapy injection for regenerative purposes on a partial supraspinatus tear, for example, then ultrasound guidance would make for a much more accurate injection. By placing the patient in a modified Crass position, the supraspinatus can be visualized by ultrasound. Extending and externally rotating the glenohumeral joint and placing the hand on the ipsilateral iliac crest with the elbow pointing posteriorly can achieve modified Crass position. Once in this position, the ultrasound probe can be placed on the anterosuperior aspect of the shoulder, parallel to an imagined line from the opposite shoulder to the ipsilateral ASIS. Then, using an in-plane approach, a needle can be directed toward the supraspinatus muscle.

Post-procedure Considerations

After a prolotherapy injection, patients must be informed of what to expect. Due to the proinflammatory mechanism of prolotherapy, increased soreness of the injection site may be expected for several days [71]. During that time, it is important to avoid anti-inflammatory medications (for at least 2 weeks) as the creation of targeted inflammation was the specific goal here, as taking such medications would be counterproductive and potentially negate the treatment itself.

Furthermore, light activity for the first 3–4 days is recommended, and normal activity and exercise may be resumed after 4–5 days. Additionally, unlike corticosteroid injections, prolotherapy injections may be done monthly for up to 3–5 months in order to achieve maximal pain relief. Thus, patients ought to be informed about the length of treatment this injection series may require in order to help set expectations.

Potential Complications and Pitfalls

There is an inherent risk with any medical procedure, regardless of how minimally invasive it claims to be—even with the use of lidocaine mixed with sugar as the injectate.

- Irritation at the injection site is common due to the intended creation of inflammation.
- Localized soreness or discomfort from the injection itself.
- Increased pain or inflammation in the area where the prolotherapy was deposited due may last up to 10 days post-procedure. In many cases, this can be due to the body intentionally creating some inflammation in and around the area where the injectate was delivered. This inflammation is the body's response to the presence of the dextrose into an injured area and attempting to offer additional assistance by luring growth factors and inflammatory markers to deal with this newly found inflammation.
- Infection—this can occur due to the nature of a simple injection if aseptic technique is not strictly adhered to or from contamination of the blood during the transfer process.

Clinical Pearls

Annually, there is an exponential rise in the number of studies looking at prolotherapy for the treatment of common musculoskeletal conditions, and a significant majority of these studies have demonstrated positive results with regard to both pain and functional improvements. The mechanism by which prolotherapy functions is primarily due to its creation of local inflammation via osmotic, chemotactic, or alkylating effects (depending on the specific injectate used); however, there also appears to be a neurolytic effect of hypertonic dextrose that may provide immediate pain relief. As reviewed throughout this chapter, prolotherapy may serve as a safe and viable nonsurgical treatment option for patients with the chronic musculoskeletal conditions discussed above.

- 50:50 mix of 50% dextrose and 1% lidocaine (buffered).
- 0.5 cc for small joints, 3–4 cc for intermediate to large joints.

- Dextrose induces inflammation, unlike corticosteroids which relieve it.
- Image guidance is recommended to limit the spread of injectate into healthy areas.

Suggested Reading

1. Nih Fact Sheets – Regenerative Medicine. https://report.nih.gov/nihfactsheets/ViewFactSheet.aspx?csid=62.
2. Hackett GS, Hemwall GA, Montgomery GA. Ligament and tendon relaxation treated by prolotherapy. 5. Gustav A. Hemwall: Oak Park; 1993.
3. Rabago D, Slattengren A, Zgierska A. Prolotherapy in primary care practice. Prim Care. 2010;37:65–80.
4. Hauser RA, Lackner JB, Steilen-Matias D, Harris DK. A systematic review of dextrose prolotherapy for chronic musculoskeletal pain. Clinical medicine insights. Arthritis Musculoskelet Disord. 2016: 9.
5. US Department of Health and Human Services, Centers for Disease Control and Prevention, National Center for Health Statistics Health. With Chartbook on trends in the health of Americans. Hyattsville: National Center for Health Statistics; 2006. DHHS Publication No 2006 1232
6. United States Bone and Joint Decade. The burden of musculoskeletal diseases in the United States. 1st ed. Rosemont: American Academy of Orthopaedic Surgeons; 2008. p. 42.
7. Estimates of Funding for Various Research, Condition, and Disease Categories (RCDC). 2017. Retrieved 23 Apr 2018, from https://report.nih.gov/categorical_spending.aspx.
8. Jenson K, Rabago D, Best T, Patterson J, Vanderby R Jr. Response of knee ligaments to prolotherapy in a rat model. Am J Sports Med. 2008;36(7):1347–57.
9. Banks AR. The rationale for prolotherapy. J Orthop Med. 1991;13.
10. Morath O, Kubosch EJ, Taeymars J, Zwingmann J, Konstantinidis L, Südkamp NP, Hirschmüller A. The effect of sclerotherapy and prolotherapy on chronic painful Achilles tendinopathy-a systematic review including meta-analysis. Scandinavian J Med Sci Sports. 2017;28(1):4–15. https://doi.org/10.1111/sms.12898.
11. Jensen KT, Rabago DP, Best TM, Patterson JJ, Vanderby R. Response of knee ligaments to prolotherapy in a rat injury model. Am J Sports Med. 2008;36(7):1347–57.
12. Housner JA, Jacobson JA, Misko R. Sonographically guided percutaneous needle tenotomy for the treatment of chronic tendinosis. J Ultrasound Med. 2009;28(9):1187–92. https://doi.org/10.7863/jum.2009.28.9.1187.
13. Jacobson JA, Rubin J, Yablon CM, Kim SM, Kalume-Brigido M, Parameswaran A. Ultrasound-guided fenestration of tendons about the hip and pelvis. J Ultrasound Med. 2015;34(11):2029–35. https://doi.org/10.7863/ultra.15.01009.
14. Torres F. An assessment on the ultrasound guided dextrose prolotherapy for persistent coccygeal pain: a case series and review of literature. Altern Integr Med. 2014;03(02) https://doi.org/10.4172/2327-5162.1000155.
15. Miller MR, Mathews RS, Reeves KD. Treatment of painful advanced internal lumbar disc derangement with intradiscal injection of hypertonic dextrose. Pain Phys. 2006;9(2):115–21.
16. Martins CA, Bertuzzi RT, Tisot RA, et al. Dextrose prolotherapy and corticosteroid injection into rat Achilles tendon. Knee Surg Sports Traumatol Arthrosc. 2012;20:1895–900.
17. Ahn KH, Kim HS, Lee WK, et al. The effect of the prolotherapy on the injured Achilles tendon in a rat model. J Korean Acad Rehabil Med. 2002;26:332–6.
18. Kim HJ, Jeong TS, Kim WS, et al. Comparison of histological changes in accordance with the level of dextrose-concentration in experimental prolotherapy model. J Korean Acad Rehabil Med. 2003;27:935–40.
19. Bollen S. Epidemiology of knee injuries: diagnosis and triage. Br J Sports Med. 2000;34(3).
20. Heidari B. Knee osteoarthritis prevalence, risk factors, pathogenesis and features: part I. Caspain J Int Med. 2011;2(2).
21. Michael J, Schlüter-Brust K, Eysel P. The epidemiology, etiology, diagnosis, and treatment of osteoarthritis of the knee. Dtsch Arztebl Int. 2010;107:152–62.
22. Rabago D, Zgierska A, Fortney L, et al. Hypertonic dextrose injections (prolotherapy) for knee osteoarthritis: results of a single-armed uncontrolled study with 1-year follow-up. J Altern Complement Med. 2012;18(4):408–14.
23. Rabago D, Patterson JJ, Mundt M, Kijowski R, Grettie J, Segal NA, Zgierska A. Dextrose prolotherapy for knee osteoarthritis: a randomized controlled trial. Ann Fam Med. 2013;11(3):229–37.
24. Reeves KD, Hassanein K. Randomized prospective double-blind placebo controlled study of dextrose prolotherapy for knee osteoarthritis with or without ACL laxity. Altern Ther Health Med. 2000;6(2):68–74. 77-80
25. Sit RW, Chung VC, Reeves KD, Rabago D, Chan KK, Chan DC, et al. Hypertonic dextrose injections (prolotherapy) in the treatment of symptomatic knee osteoarthritis: a systematic review and meta-analysis. Sci Rep. 2016;6(1).
26. Jumper's Knee. 2017. Retrieved 07 Apr 2018, from https://emedicine.medscape.com/article/89569-overview#a8.
27. Kujala UM, Kvist M, Heinonen O. Osgood-Schlatters disease in adolescent athletes. Am J Sports Med. 1985;13(4):236–41.
28. Wall EJ. Osgood-Schlatter disease: practical treatment for a self-limiting condition. Phys Sportsmed. 2010;38(4):29–34.
29. Topol GA, Podesta LA, Reeves KD, Raya MF, Fullerton BD, Yeh H. Hyperosmolar dextrose injection for recalcitrant Osgood-Schlatter disease. Pediatrics. 2011;128(5):e1121–8.
30. Maughan K. n.d. Achilles tendinopathy and tendon rupture. Retrieved 07 Apr 2018, from https://www.uptodate.com/contents/achilles-tendinopathy-and-tendon-rupture.
31. Sanderson LM, Bryant A. Effectiveness and safety of prolotherapy injections for management of lower limb tendinopathy and fasciopathy: a systematic review. J Foot Ankle Res. 2015;8(1).
32. Maxwell NK, Ryan M, Taunton J, Gillies JH, Wong A. Sonographically guided intratendinous injection of hyperosmolar dextrose of the Achilles tendon: a pilot study. Am J Roentgenol. 2007;189(4):W215–20.
33. Ryan M, Wong A, Taunton J. Favourable outcomes after sonographically guided inratendinous injection of hyperosmolar dextrose for chronic insertional and midportion Achilles tendinosis. Am J Roentgenol. 2010;194:1047–53.
34. Goff J, Crawford R. Diagnosis and treatment of plantar fasciitis. Am Fam Physician. 2011;84(6):676–82.
35. Ryan M, Wong A, Gillies J, Wong J, Taunton J. Sonographically guided intratendinous injections of hyperosmolar dextrose/lidocaine: a pilot study for the treatment of chronic plantar fasciitis. Br J Sports Med. 2009;43:303–6.
36. Kim E, Lee J. Autologous platelet rich plasma versus dextrose prolotherapy for the treatment of chronic recalcitrant plantar fasciitis. Phys Med Rehabil Int. 2014;6:152–8.
37. Van der Windt D, Koes B, De Jong B, Bouter L. Shoulder disorders in general practice: incidence, patient characteristics, and management. Ann Rhem Disease. 1995;54(12):959–64.
38. Buchbinder R, Green S, Youd JM. Corticosteroid injections for shoulder pain. Cochrane Database Syst Rev. 2003;(1):Cd004016.
39. Seven MM, Ersen O, Akpancar S, Ozkan H, Turkkan S, Yıldız Y, et al. Effectiveness of prolotherapy in the treatment of chronic ro-

tator cuff lesions. Orthop Traumatol Surg Res. 2017;103(3):427–33. https://doi.org/10.1016/j.otsr.2017.01.003.

40. Lebiedziński R, Synder M, Buchcic P, Polguj M, Grzegorzewski A, Sibiński M. A randomized study of autologous conditioned plasma and steroid injections in the treatment of lateral epicondylitis. Int Orthop. 2015;39(11):2199–203.

41. Bisset L, Beller E, Jull G, et al. Mobilization with movement and exercise, corticosteroid injection or wait and see. BMJ. 2006;333:939. [PubMed: 17012266]

42. Rabago D, Lee KS, Ryan M, Chourasia AO, Sesto ME, Zgierska A, et al. Hypertonic dextrose and morrhuate sodium injections (prolotherapy) for lateral epicondylosis (tennis elbow). Am J Phys Med Rehabil. 2013;92(7):587–96. https://doi.org/10.1097/phm.0b013e31827d695f.

43. Scarpone M, Rabago DP, Zgierska A, Arbogast G, Snell E. The efficacy of prolotherapy for lateral epicondylosis: a pilot study. Clin J Sport Med. 2008;18(3):248–54. https://doi.org/10.1097/jsm.0b013e318170fc87.

44. Zhang Y, Niu J, Kelly-Hayes M, Chaisson C, Aliabadi P, Felson D. Prevalence of symptomatic hand osteoarthritis and its impact on functional status among the elderly: the Framingham study. Am J Epidemiol. 2002;156(11):1021–7. https://doi.org/10.1093/aje/kwf141.

45. Kroon F, Rubio R, Schoones J, Kloppenburg M. Intra-articular therapies in the treatment of hand osteoarthritis: a systematic literature review. Osteoarthr Cartil. 2016;24 https://doi.org/10.1016/j.joca.2016.01.952.

46. Reeves KD, Hassanein K. Randomized, prospective, placebo-controlled double-blind study of dextrose prolotherapy for osteoarthritic thumb and finger (DIP, PIP, and trapeziometacarpal) joints: evidence of clinical efficacy. J Altern Complement Med. 2000;6(4):311–20. https://doi.org/10.1089/10755530050120673.

47. Jahangiri A, Moghaddam FR, Najafi S. Hypertonic dextrose versus corticosteroid local injection for the treatment of osteoarthritis in the first carpometacarpal joint: a double-blind randomized clinical trial. J Orthop Sci. 2014;19(5):737–43. https://doi.org/10.1007/s00776-014-0587-2.

48. Feldman, M. D. (2004). Prolotherapy for the treatment of chronic low back pain. California Technology Assessment Forum.

49. Klein RG, Eek BC, DeLong WB, Mooney V. A randomized double-blind trial of dextrose-glycerine phenol injections for chronic, low back pain. J Spinal Disord. 1993;6:23–33.

50. Ongley MJ, Klein RG, Dorman TA, Eek BC, Hubert LJ. A new approach to the treatment of chronic low back pain. Lancet. 1987;2(8551):143–6.

51. Hooper RA, Yelland M, Fonstad P, Southern D. Prospective case series of litigants and non-litigants with chronic spinal pain treated with dextrose prolotherapy. Int Musculoskelet Med. 2011;33(1):15–20.

52. Bernard TN Jr, Kirkaldy-Willis WH. Recognizing specific characteristics of nonspecific low back pain. Clin Orthop. 1987;217:266–80.

53. Frost S, Wheeler M, Fortin J, Vilensky J. The sacroiliac joint: anatomy, physiology and clinical significance. Pain Physician. 2006;9:61–8.

54. Kim WM, Lee HG, Jeong CW, Kim CM, Yoon MH. A randomized controlled trial of intra-articular prolotherapy versus steroid injection for sacroiliac joint pain. J Altern Complement Med. 2010;16(12):1285–90. https://doi.org/10.1089/acm.2010.0031.

55. Danielle S, C., M. J., Ross, H., Barbara, W., & Sarah, S. Chronic neck pain: making the connection between capsular ligament laxity and cervical instability. Open Orthop J. 2014;8(1):326–45. https://doi.org/10.2174/1874325001408010326.

56. Slade GD. Epidemiology of temporomandibular joint disorders and related painful conditions. Mol Pain. 2014;10(Suppl 1):O16.

57. Ebrahim S, Montoya L, Busse JW, Carrasco-Labra A, Guyatt GH. The effectiveness of splint therapy in patients with temporomandibular disorders. J Ame Dental Assoc. 2012;143(8):847–57. https://doi.org/10.14219/jada.archive.2012.0289.

58. Scrivani SJ, Mehta NR. 2016. Temporomandibular disorders in adults. Retrieved 22 Apr 2018, from https://www.uptodate.com/contents/temporomandibular-disorders-in-adults.

59. Refai H. Long-term therapeutic effects of dextrose prolotherapy in patients with hypermobility of the temporomandibular joint: a single-arm study with 1-4 years' follow up. Br J Oral Maxillofac Surg. 2017;55(5):465–70. https://doi.org/10.1016/j.bjoms.2016.12.002.

60. Dagenais S, Ogunseitan O, Haldeman S, et al. Side effects and adverse events related to intraligamentous injection of sclerosing solutions (prolotherapy) for back and neck pain: a survey of practitioners. Arch Phys Med Rehabil. 2006;87:909–13.

61. Rahimzadeh P, Imani F, Faiz SH, Entezary SR, Zamanabadi MN, Alebouyeh MR. The effects of injecting intra-articular platelet-rich plasma or prolotherapy on pain score and function in knee osteoarthritis. Clin Interv Aging. 2018;13:73–9. https://doi.org/10.2147/cia.s147757.

62. Kim E, Lee JH. Autologous platelet-rich plasma versus dextrose prolotherapy for the treatment of chronic recalcitrant plantar fasciitis. Pm&r. 2014;6(2):152–8. https://doi.org/10.1016/j.pmrj.2013.07.003.

63. Ryan M, Wong A, Rabago D, Lee K, Taunton J. Ultrasound-guided injections of hyperosmolar dextrose for overuse patellar tendinopathy: a pilot study. Br J Sports Med. 2011;45(12):972–7.

64. Labella CR, Hennrikus W, Hewett TE. Anterior cruciate ligament injuries: diagnosis, treatment, and prevention. Pediatrics. 2014;133(5).

65. Dorman TA. Treatment for spinal pain arising in ligaments–using prolotherapy: a retrospective survey. J Orthop Med. 1991;13:13–9.

66. Dorman T. Whiplash injuries: treatment with prolotherapy and a new hypothesis. J Orthop Med. 1999;21:13–21.

67. Iseminger T, Palaikis R. A retrospective study of prolotherapy. San Luis Obispo: Cal State Polytechnic, Biological Sciences Department; honors paper (undergraduate); 1995:15.

68. Lopez RG, Jung H. Achilles tendinosis: treatment options. Clin Orthop Surg. 2015;7(1):1. https://doi.org/10.4055/cios.2015.7.1.1.

69. Center for Disease Control and Prevention. National ambulatory medical care survey. 2010. Available at http://www.cdc.gov/nchs/ahcd.htm.

70. Block J, Berkoff M. Clinical utility of ultrasound guidance for intra-articular knee injections: a review. Clin Interv Aging. 2012:89–95.

71. Rabago D, van Leuven L, Benes L, Fortney L, Slattengren A, Grettie J, Mundt M. Qualitative assessment of patients receiving prolotherapy for knee osteoarthritis in a multimethod study. J Altern Complement Med. 2016;22:983–9.

Steve M. Aydin

Platelet-Rich Plasma

History

The idea of platelet-rich plasma (PRP) was initially introduced in the 1970s, but it was not until the 1990s that PRP began gaining popularity and its applications were being used for sports injuries and other painful conditions [1]. Over the last decade, its applications have been used in multiple fields and there has been a prominent increase in the number of publications focusing on PRP [2].

With the availability of mobile tabletop centrifuges and office-based ultrasound equipment, the application of platelet-rich plasma in the treatment of musculoskeletal issues became more widespread and has allowed for the introduction and expansion of regenerative medicine [3, 4]. In the past, imaging guidance of therapies related to the soft and connective tissues was difficult, but ultrasound imaging has allowed for this to happen. This has resulted in less invasive and additional interventional treatment options for patients [5–8]. With many patients looking for nonsurgical options and the desire to avoid the chronic administration of steroids either within a joint or in a peritendinous location, the application of platelet-rich plasma products has offered another option [5].

This evolution has facilitated the utilization of orthobiologic products that have come to the forefront of multiple medical specialties. Its applications have generated a significant amount of clinical and academic interest and PRP and other orthobiologic products are now employed on the global medical level [9, 10]. Those who are practicing and utilizing regenerative medicine and incorporating it into their practices are often quite optimistic about its use. As the medical treatment paradigm advances, regenerative medicine will continue to propagate and physicians will need to have an awareness of this so that these treatments may be considered in their algorithm of treatment [11].

Preparation and Obtaining PRP

Producing platelet-rich plasma is the process of obtaining blood from a patient and concentrating the platelets to greater than four times the normal level. Often, this can be done with the use of multiple collecting tubes and a centrifuge, but several commercial companies have created systems and kits that can automate the collection, concentration, and separation process [12–14] (Fig. 3.1).

In most cases, the process will involve a sterile venous puncture with a collection of about 40 mL of whole blood in anticoagulated tubes. If the tube does not have an anticoagulant present, the addition of an anticoagulant such as EDTA or acid citrate dextrose (ACD) needs to be done [5, 14]. This will act as an anticoagulant to prevent the platelets from clotting and clumping and preventing it from becoming an unusable sample. Once the sample has been obtained, it is then processed with a centrifuge. This can have some variations in regard to the centrifuge speed and the length of spin time. In general, a first spin or centrifuge process is performed with the initially drawn whole blood. This will often be done at $600g$ for 7 minutes and is considered the first specimen. After the completion of the first spin, the volume below the platelet border is aspirated. This is the upper layer of plasma along with the buffy coat [15–19] (Fig. 3.1).

This first separation will include platelet-poor plasma, the desired portion of PRP, and the buffy coat along with some red blood cells. Then a second centrifuge spin is conducted. This will often be at $2000g$ for 5 minutes. This process will allow for the separation of the desired PRP product from the platelet-poor plasma (PPP) [5, 20].

Often the initial spin is considered a "soft spin" allowing for the blood products to be separated out. The second specimen is then considered a "hard spin" allowing for the

S. M. Aydin (✉)
Zucker School of Medicine at Hofstra University/Northwell Health, Department of Medicine and Rehabilitation, Manhasset, NY, USA

Fig. 3.1 This is the process of blood draw, preparing with an anticoagulant agent, and the centrifuge process. The centrifuge is often done with a 600G spine, followed by a 2000G spin to getting the layers out and a final platelet-rich plasma product

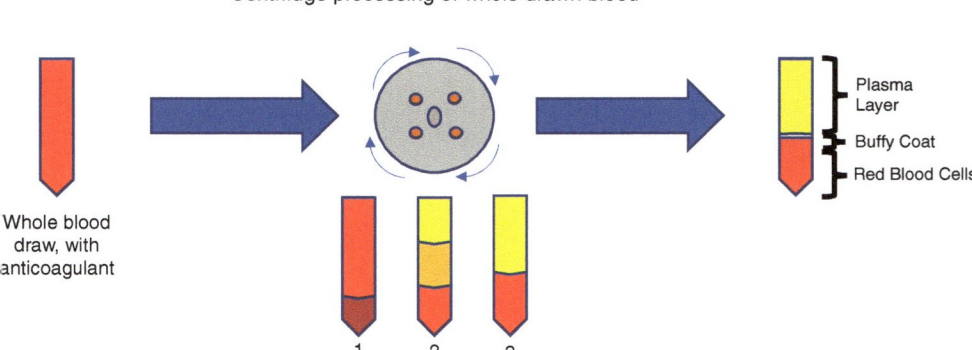

separation of the remaining red blood cells and PPP from the desired platelet-rich plasma [19, 21–23].

After the completion of the second spin, the PPP is often discarded and the remaining product is considered to be the desired platelet-rich plasma. This will require the removal of the upper fluid layer containing the PPP. The volume that is most often leftover is between 2 and 5 mL from the initial draw of 40 mL. The volume that is recovered from the second spin can certainly be variable from individual to individual, technique to technique, and/or system to system [24–27].

When considering blood and its products, a basic understanding of blood components is important. Blood is primarily made up of red blood cells, white blood cells, platelets, and plasma. The volume of plasma is usually about 55% and primarily consists of water. The cellular components will make up about 45% of the blood volume [2, 28–30]. The cellular components are often distributed throughout the buffy coat and among the red blood cells. The buffy coat is the portion of the blood after centrifuging process that remains between the plasma and the red blood cells. This will contain the white blood cells and platelets. The plasma will contain blood proteins, such as fibrinogen, albumin, globulin, electrolytes, hormones, and water [30–33].

Another method to obtain the buffy coat involves storing the whole blood at 20–24°C prior to the centrifuge process. After the room temperature storage, the whole blood is placed in a high-speed centrifuge and the plasma and buffy coat are separated out and subsequently removed. This is then centrifuged at a lower speed to separate the white blood cells from the red blood cells [32, 33].

PRP Types and Classifications

With the different preparation processes for PRP, there can be variability in the type of PRP that is obtained. This may be intentional in many cases and based on the type of injection therapy or treatment the practitioner would like to render. However, in certain cases, this may be purely based on the individual's blood components and concentration along with the type of processing or machine automation [20, 34–36]. The overarching purpose of preparing PRP is to isolate out the platelets and remove the red blood cells. Given this, many practitioners focus on having a very yellow-colored final product indicating very little red blood cell content, although in certain cases a red-tinged or red product may be obtained or desired. There are five main types of platelet-rich plasma (PRP). This is based on the cell content and the presence of other blood products [36, 37].

The first type is red PRP. Platelet-rich plasma that appears red has a low concentration of platelets. It will purposely have a certain content of red blood cells that are thought to have an inflammatory reaction following an injection. It is not clear as to the clinical role of red PRP, but ongoing studies are being conducted to determine its utility in specific conditions [2, 20, 37–39].

When the prepared PRP has more of a yellowish color, the concentration of platelets in the solution can be either high or low and can be called either high-concentration or low-concentration PRP. Low-concentration or low-density PRP is yellowish or amber in color with lower levels of platelets. This solution is typically poor and devoid of white or red blood cells. It is thought that this preparation is less reactive in regard to producing an inflammatory process and results in less of a chemotactic response following injection compared with that of red PRP. This is a newer approach and sometimes is considered to be "pure" PRP or P-PRP. It will also have a low density of fibrin in this preparation [2, 20, 30, 31].

PRP that is prepared with leukocytes present is the most common type of PRP and is often prepared in the automated systems and kits that are available. This will also have a low density of fibrin and is known as leukocyte-rich PRP or L-PRP [36, 40].

A fourth type is a pure platelet-rich fibrin or leukocyte-poor/platelet-rich fibrin preparation that is thought to have no leukocytes but has a high density of fibrin products in it. This is called pure platelet-rich fibrin or P-PRF [20, 40].

Finally, there is a leukocyte-rich and platelet-rich fibrin PRP. This is thought to represent the second generation of PRP products with a high density of fibrin and leukocytes mixed in with the platelets and its other factors. This is known as L-PRF [2, 19, 30].

Proposed Mechanisms of Action

The purpose of injecting or providing platelet-rich plasma is to increase the concentration of growth factors and platelets at that site compared to what is available in whole blood [32, 35]. The injected PRP is thought to have a healing and/or regenerative effect on the environment it is injected into. There remain many unknowns regarding the appropriate concentrations of blood products in the PRP preparations that would produce the maximum desired effect on the different tissues [19].

The PRP preparation also contains many growth factors as well as platelets. Platelets have a combination of dense and alpha granules that, when activated, will release these factors into their medium or surrounding environment [11]. The growth factors that are considered to be important are transforming growth factor-beta (TGF-β), platelet-derived growth factor (PDGF), vascular endothelial growth factor (VEGF), epidermal growth factor (EGF), fibroblast growth factor II (FGF2), and insulin growth factor (IGF). These factors are considered to have stimulating properties for enhancing both tissue and bone healing. Others are thought to stimulate osteoblast proliferation and differentiation, as well as to stimulate epidermal cell proliferation. It is also believed that these factors are involved in stimulating collagen synthesis, as well as angiogenesis and revascularization of tissue [22, 31, 32, 41] (Table 3.1).

Further understanding of the properties of platelet-rich plasma is facilitated by adequate comprehension of the normal healing process. The repair response in musculoskeletal injuries will often start with a blood clot and degranulation of the platelets [8]. This process will then release the platelet growth factors present in the alpha and dense granules after they burst and will, in turn, result in the chemotactic effect, which promotes the migration of inflammatory cells and proliferation of pro-generator cells [2, 11, 31] (Fig. 3.2).

When we consider this and look at structures in the musculoskeletal system, one can see that vascular supply plays a very important role in this process. The muscle tissues are very well vascularized and often demonstrate a remodeling and rapid healing process similar to that of other organ systems with an optimal vascular supply such as the integumentary system [14, 16]. When we consider other connective tissues, however, vascular supply is often not as prevalent as in muscle or skin tissue. Tissues with a vascular supply that is less replete will often heal slower and are

Table 3.1 Common factors present in platelet-rich plasma and the proposed mechanism of action

Name		Function
Platelet-derived growth factor	PDGF	Enhances collagen synthesis, proliferation of bone cells, fibroblast chemotaxis, and proliferative activity; macrophage activation
Transforming growth factor β	TGF-β	Enhances synthesis of type I collagen; promotes angiogenesis; stimulates chemotaxis of immune cells; inhibits osteoclast formation and bone resorption
Vascular endothelial growth factor	VEGF	Stimulates angiogenesis, migration, and mitosis of endothelial cells; increases permeability of the vessels; stimulates chemotaxis of macrophages and neutrophils
Epidermal growth factor	EGF	Stimulates cellular proliferation and differentiation of epithelial cells; promotes cytokine secretion by mesenchymal and epithelial cells
Insulin-like growth factor	IGF	Promotes cell growth, differentiation, and recruitment in bone, blood vessel, skin, and other tissues; stimulates collagen synthesis together with PDGF
Fibroblast growth factor	FGF	Promotes proliferation of mesenchymal cells, chondrocytes, and osteoblasts; stimulates the growth and differentiation of chondrocytes and osteoblasts

more prone to chronic inflammatory processes and slower healing [9, 25]. With the application of platelet-rich plasma, the growth factors present can increase the cell activity and products that are locally available in those tissue environments that are otherwise limited or devoid of an ample vascular supply [29].

Applications in Musculoskeletal Medicine

Platelet-rich plasma is most commonly used for musculoskeletal applications. Although the number of overall treatments for all organ systems is vast, its utilization in the musculoskeletal accounts for the greatest number of therapies by far. The applications of PRP in the musculoskeletal tissues can be segmented into regions or tissue types, including injections into muscles, tendons, ligaments, and joints [12, 19, 40].

When considering skeletal muscle injuries such as sprains, strains, and muscle tears, there are certain barriers to recovery that exist. Some limitations can be from scar formation or fibrosis development within the muscle as well as repeated stress and use that prolongs the period of recovery [40, 41]. Platelet-rich plasma has been demonstrated to have a role in modulating the inflammation and healing process in muscle. There have been studies that have demonstrated the application of PRP compared with saline for muscle tears in animal models showing an increase in satellite cell activation

Fig. 3.2 Demonstration of the healing cascade and timeline for tissue remodeling

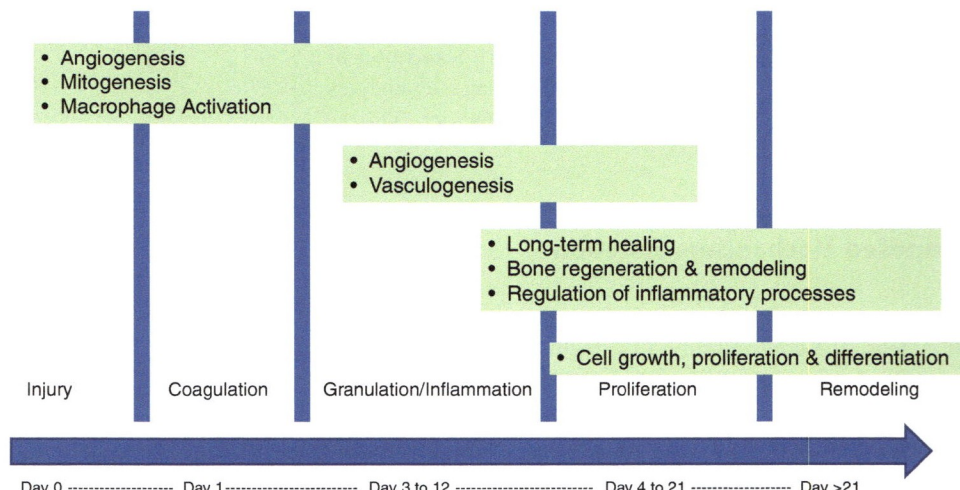

- Angiogenesis
- Mitogenesis
- Macrophage Activation

- Angiogenesis
- Vasculogenesis

- Long-term healing
- Bone regeneration & remodeling
- Regulation of inflammatory processes

- Cell growth, proliferation & differentiation

| Injury | Coagulation | Granulation/Inflammation | Proliferation | Remodeling |

Day 0 ------------------- Day 1----------------------- Day 3 to 12 ------------------------- Day 4 to 21 ------------------ Day >21

postinjury for those animals injected with PRP [36, 39]. Several studies have demonstrated the different factors that may be responsible for this. This certainly can have an impact on certain types of injuries particularly to the muscle that demonstrate delayed healing or are refractory to other certain types of therapies [30, 32, 36, 39].

Tendons and ligaments are often subject to high stress and shear forces and can be injured with areas of rupture, partial tear, or complete tear. These injuries are often seen in a background of chronic inflammation. Several laboratory studies have shown beneficial effects of PRP on the tendon healing process [21–23, 28, 31–33]. The healing mechanism involves stimulation of cell proliferation and total collagen production in those injected with PRP or PPP compared with those that were not injected. Much of the literature regarding the tendons and ligaments has been primarily focused on lateral epicondylitis. These data have been applied to other tendon injuries with the hope that analogous outcomes would occur given the findings and the research that supports lateral epicondylitis. Several studies have demonstrated histologic changes and extracellular matrix responses following PRP injections [32]. These have been demonstrated at the 7–10-day postinjection interval as well as at histological evaluations at the 6–12 week postinjection timeframe. These histologic evaluations showed robust cellular responses to the injections and overall increased cellular activity [27, 28, 30, 31, 34].

There are several studies ongoing for different applications of platelet-rich plasma. Many of these studies are focused on cartilage and joint injury and/or degeneration [31, 33, 34, 36]. There has been good literature support for PRP injections to treat intra-articular injuries as well as for tendon, ligament, and muscle use. In the intra-articular applications, the exact mechanism of action and the appropriate concentration of blood products are not well understood nor are they well known [36]. Further research including ran-

domized controlled trials are needed to determine the effects and impact of regenerative techniques on musculoskeletal injuries. What is known is that the use of platelet-rich plasma in different environments has demonstrated some clinical promise and positively affects the injured tissue in a way that is conducive to healing [2, 20, 33].

Platelet-Poor Plasma with Alpha-2-Macroglobulin

Introduction

As previously described in the preparation of platelet-rich plasma, there are many factors that are present within whole blood. Each component can have a specific regenerative medicine role, but it is not always clear as to what the right concentration needs to be to accomplish a specific purpose or which component is most helpful to accomplish a certain treatment goal [42, 43].

Platelet-poor plasma (PPP) is the preparation that is devoid of platelets and is defined as plasma with a concentration of platelets of less than 10×10^3 per mL. Platelet-poor plasma was often used as a control in comparison to platelet-rich plasma, but during its utilization, the studies have demonstrated the PPP solution to have elevated levels of fibrinogen. This high fibrinogen content has shown to have the ability to form and activate fibrin-rich clots and to assist with wound healing [43, 44].

As more techniques developed different filters to be used in isolation processes, specific proteins and factors were successfully isolated and extracted from various PPP and PRP products. Alpha-2-macroglobulin is a plasma protein found in the blood, concentrated in the plasma, and is mainly produced by the liver as well as synthesized by macrophages, fibroblasts, and adrenocortical cells [42, 44].

Properties

Platelet-poor plasma is often discarded when the preparation of PRP is obtained, but it has shown to have a role in certain cases of wound healing. It also has a role in the promotion of hemostasis and blood clotting [43].

Alpha-2-macroglobulin (A2M) is a large plasma protein present in the blood. It requires a special filtration process to obtain and concentrate A2M. Alpha-2-macroglobulin is primarily present in the plasma portion of PRP and PPP. To obtain the A2M product, a special filtration process must be employed to obtain a higher concentration of this product and isolate it from its PRP counterpart [44].

It acts as an anti-protease and is able to inactivate many different kinds of proteinases. It primarily functions to inhibit plasmin and kallikrein that are responsible for degrading many blood plasma proteins including fibrin clots. It also functions to inhibit thrombin that has a crucial role in both the coagulation cascade and in the activation of platelets [42]. Its application to regenerative medicine thus far has been specifically focused on the matrix of cells responsible for the breakdown of cartilage within joints. The use of A2M has been shown to inhibit the action of matrix metalloproteinases (MMPs), which have been shown to break down tissues within the joint capsule [43]. These MMPs have been demonstrated to have a role in the process of developing osteoarthritis in joints by breaking down the articular cartilage. The application of A2M can interfere with this and may have some beneficial properties in preventing the progression of osteoarthritis [44].

Indications

The applications of PPP have been primarily in wound management and cosmetic surgery patients. Additional uses are being studied with applications in patients with degenerative arthritis, degenerative disc disease, and numerous other conditions.

The application of A2M is currently primarily intra-articular, and its role is thought to be specifically to prevent the progression of osteoarthritis. In many cases, a fibronectin aggregation (FAC) G3 complex test is conducted prior to A2M injection to determine if a patient is eligible for the therapy. This is done by first doing a joint aspiration and then sending the fluid off for FAC testing [44]. If this determination is positive, then early intervention with A2M can be conducted for the purpose of trying to mitigate the arthritic process and preserving the integrity of the articular cartilage. Its application as of broad-spectrum protease inhibitor has demonstrated good results and efficacy as a treatment for osteoarthritis. At present, there are limited data and a few manuscripts that have examined the use of A2M in humans. Given its promising early results, additional studies are important to demonstrate its full spectrum of use and applications [42–44].

Orthokine, Regenokine, Interleukin Receptor Antagonist Protein

Introduction

With many of the autologous products including preparation of platelet-rich plasma, other filtering and isolation processes can be done to obtain concentrations of other factors that may be more helpful in specific injuries and environments [45]. Interleukin receptor antagonist protein (IRAP) is one product that can be obtained from whole blood and the PRP preparation process. It is also commercially known as Orthokine and/or Regenokine [46].

The process of using this product was initially introduced in the early 2000s and first made its presence in a high level playing athletes who sought out this treatment for chronic injuries that were hindering their play. Initially developed and offered in Germany, IRAP therapy is now utilized worldwide, but many athletes and patients still travel to Europe to be treated [47].

History

The IRAP production is an isolation process that focuses on isolating anti-inflammatory factors. Dr. Peter Wehling pioneered the process of isolating this factor in the 1980s. In 2003, the process was approved for use in Germany [45].

It is less invasive than surgical treatments and has anti-inflammatory properties making it a less painful treatment option for those receiving it. The popularity of IRAP was facilitated by its use in high-level athletes who would initially fly from the United States of America to Germany for treatment and then return home to play [44, 46].

Properties and Relation to PRP

Orthokine is a product that is obtained from PRP. As mentioned previously, PRP is a blood product obtained from whole blood drawn from the patient to be treated. In the process of obtaining Orthokine, 60 cc of the patient's blood is obtained and then incubated at a body temperature for 24 hours. Following this, it is centrifuged at $2100g$ for 10 minutes. Then, through a filtration process via syringes, the product is obtained and utilized [44].

Table 3.2 Comparison of several commercial systems

Company	Blood volume (mL)	PRP volume (mL)	WBC concentration factor	PRP concentration factor	Centrifuge time (min)
Peak	27	3	5×	6.8×	1
Arthremex	11	4	0.5×	2.0×	5
Biomet	27	3	5×	4.7×	15
Harvest	27	3	3.5×	4.9×	15
Arteriocyte	27	3	2×	6.0×	17

The systems noted here have been compared with one another and have published data on their results and concentrations of PRP. Several other companies are present in the market; however, not all have been compared with a standard or with other systems available on the market
PRP platelet-rich plasma, *WBC* white blood cells

The Orthokine product and procedure are closely related to the preparation of PRP but with a few more steps and modifications. The final product obtained is the Interleukin 1 receptor antagonist protein (IRAP). The function of IRAP is to bind to the IL-1 receptor and to block signaling, which promotes an immune and inflammatory response. The IRAP binds the IL-1 receptor and prevents inflammatory cascades and potentially harmful inflammatory responses in the environment being used (i.e., within a joint) [44, 45].

Role and Applications

In its purest form, IRAP is thought to be only anti-inflammatory. Its uses have shown good clinical outcomes in patients who have undergone the treatment. There are strong 2-year data demonstrating beneficial outcomes for the treatment of osteoarthritis of the knees, and most of the applications have been focused on intra-articular use and the treatment of degenerative arthritis [47].

That being said, it is not clear if its use is disease modifying, chondroprotective, or chrondroregenerative. Further data are needed to demonstrate the specific circumstances this product would best be used in [46, 47].

Commercially Available Products

Introduction

Several systems exist in the commercial market space for platelet-rich plasma. Certain factors should be present when considering a system, and these may include cost, whether the system is open versus closed, and the ability to manipulate the product for a higher or lower number of platelets that can be obtained. One should evaluate each system and make their own determination based on the specific clinical practice needs [2, 20, 31, 48].

Available Products and Vendors

Several vendors and companies have systems that can produce platelet-rich plasma. When considering a system, the platelet concentration is certainly the most important thing. However, other factors such as whether the system is closed or open are important to consider [2, 3, 31, 49]. An open system is one in which the blood is drawn and transferred to the machine for processing and then removed from the machine for patient use. A closed system is one where the blood is drawn from the patient into the system where it is not touched and does not leave the container until it is utilized for the patient. Additionally, certain systems can be adjusted to alter the concentration of white cells, blood cells, platelets, and other factors based on different filtration processes [50] (Table 3.2).

References

1. Hackett GS, Hemwall GA, Montgomery GA. Ligament and tendon relaxation treated by prolotherapy. 5th ed. Institute in Basic Life Principles: Oak Brook; 1991.
2. Mishra A, Pavelko T. Treatment of chronic elbow tendinosis with buffered platelet-rich plasma. Am J Sports Med. 2006;34(11):1774–8.
3. Alderman D. Prolotherapy: platelet rich plasma in prolotherapy. Pract Pain Manag. 2009;9(1):68–9.
4. Hauser RA, Phillips HJ, Maddela H. Prolotherapy: platelet rich plasma prolotherapy as first-line treatment for meniscal pathology. Pract Pain Manag. 2010;10(6):53–640. Publishing Co., Inc; 2005.
5. Marx R, Kevy SV, Jacobson MS. Platelet rich plasma (PRP): a primer. Pract Pain Manag. 2008;8(2):46–7.
6. Marx R. Platelet-rich plasma: evidence to support its use. J Oral Maxillofac Surg. 2004;62(4):489–96.
7. Creaney L, Hamilton B. Growth factor delivery methods in the management of sports injuries: the state of play. Br J Sports Med. 2008;42(5):314–20.
8. Foster TE, Puskas BL, Mandelbaum BR, Gerhardt MB, Rodeo SA. Platelet-rich plasma: from basic science to clinical applications. Am J Sports Med. 2009;37(11):2259–72.
9. El-Sharkawy H, Kantarci A, Deady J, et al. Platelet-rich plasma: growth factors and pro- and anti-inflammatory properties. J Periodontol. 2007;78(4):661–9.

10. Zuk PA. The adipose-derived stem cell: looking back and looking ahead. Mol Biol Cell. 2010;21(11):1783–7.

11. Caplan A, Fink D, Goto T, et al. Mesenchymal stem cells and tissue repair. In: Jackson DW, editor. The anterior cruciate ligament: current and future concepts. New York: Raven Press; 1993. p. 405–17.

12. Caplan A. Mesenchymal stem cells. J Orthop Res. 1991; 9(5):641–50.

13. Fraser JK, Wulur I, Alfonso Z, Hedrick MH. Fat tissue: an unappreciated source of stem cells for biotechnology. Trends Biotechnol. 2006;24(4):150–4.

14. Centeno CJ, Busse D, Kisiday J, Keohan C, Freeman M, Karli D. Increased knee cartilage volume in degenerative joint disease using percutaneously implanted, autologous mesenchymal stem cells. Pain Physician. 2008;11(3):343–53.

15. Horwitz EM, Gordon PL, Koo WK, et al. Isolated allogeneic bone marrow-derived mesenchymal cells engraft and stimulate growth in children with ontogenesis imperfect: implications for cell therapy of bone. Proc Natl Acad Sci U S A. 2002;99(13):8932–7.

16. Alexander RW. Use of platelet-rich plasma (PRP) in autologous fat grafting. In: Shiffman M, editor. Autologous fat grating. Berlin: Springer; 2010. p. 140–67.

17. Alexander RW. Fat transfer with platelet-rich plasma for breast augmentation. In: Breast augmentation: principles and practice. 1st ed. Berlin: Springer; 2009: Chapter 56.

18. Epstein-Sher S, Jaffe DH, Lahad A. Are they complying? Physicians' knowledge, attitudes & readiness-to-change regarding low Back pain treatment guideline adherence. Spine (Phila Pa 1976). 2017;42(4):247–52.

19. Centeno CJ, Bashir J. Safety and regulatory issues, regarding stem cell therapies: one clinic's perspective. PMR. 2015;7(4 Suppl):S4–7.

20. United States Food and Drug Administration. Code of Federal Regulations. Title 21, Part 1271, 2013.

21. Murrell W, Anz AW, Badsha H, Bennett WF, Boykin RE, Caplan AI. Regenerative treatments to enhance orthopedic surgical outcomes. PMR. 2015;7(4 Suppl):S41–52.

22. Monfet M, Harrison J, Boachie-Adjei K, Lutz G. Intradiscal platelet-rich-plasma (PRP) injections for discogenic low back pain: an update. Int Orthop. 2016;40(6):1321–8.

23. Tuakli-Wosornu YA, et al. Lumbar intradiscal platelet rich plasma (PRP) injections: a prospective, double-blind. Randomized Controlled Study PMR. 2016;8(1):1–10.

24. Zhang W, Ouyang H, Dass CR, Xu J. Current research on pharmacologic and regenerative therapies for osteoarthritis. Bone Res. 2016;4:15040.

25. Mautner K, et al. Treatment of tendonopathies with platelet rich plasma. Phys Med Rehabil Clin N Am. 2014;25(4):865–80.

26. Hartman S, et al. Analysis of performance-based functional test in comparison with the visual analog scale for post-operative outcome assessment following lumbar spondylodesis. Eur Spine J. 2016;25(5):1620–6.

27. Varatharajan S, et al. Are non-invasive interventions effective for the management of headaches associated with neck pain? An update of the Bone and Joint Decade Task Force on Neck Pain and Its Associated Disorders by the Ontario Protocol for Traffic Injury Management (OPTIMa) Collaboration. Eur Spine J. 2016;25(7):1971–99.

28. Nguyen RT, Borg-Stein J, McInnis K. Applications of platelet rich plasma in musculoskeletal and sports medicine: an evidence based approach. PMR. 2011;3(3):226–50.

29. Anitua E. Plasma rich growth factors: preliminary results of use in the preparation of future implants. Int J Oral Maxillofac Implant. 1999;14(4):529–53.

30. Mishra A, Harmon K, Woodall J, Vieira A. Sports medicine applications of platelet rich plasma. Curr Pharm Biotechnol. 2012;13(7):1185–95.

31. Mishra A, et al. Buffered platelet-rich plasma enhances mesenchymal stem cell proliferation and chondrogenic differentiation. Tissue Eng Part C Methods. 2009;15(3):431–5.

32. Nazempour A, Van Wie BJ. Chondrocytes, mesenchymal stem cells, and their combination in articular cartilage regenerative medicine. Ann Biomed Eng. 2016;44(5):1325–54.

33. Nguyen C, Lefèvre-Colau MM, Poiraudeau S, Rannou F. Evidence and recommendations for use of intra-articular injections for knee osteoarthritis. Ann Phys Rehabil Med. 2016;59(3):184–9.

34. Castillo TN, Pouliot MA, Kim HJ, Dragoo JL. Comparison of growth factors and platelet concentration from commercial platelet-rich plasma separation systems. Am J Sports Med. 2011;39(2):266–71.

35. DePalma MJ, Gasper JJ. Cellular supplementation technologies for painful spine disorders. PMR. 2015;7(4 Suppl):S19–25.

36. Ko GD, Mindra S, Lawson GE, Whitmore S, Arseneau L. Case series of ultrasound-guided platelet-rich plasma injections for sacroiliac joint dysfunction. J Back Musculoskelet Rehabil. 2017;30(2):363–70.

37. Peerbooms JC, Sluimer J, Bruijn DJ, Gosens T. Positive effect of an autologous platelet concentration in lateral epicondylitis in a double-blind randomized controlled trial: platelet rich plasma versus corticosteroid injection with a 1-year follow up. Am J Sports Med. 2010;303(2):144–9.

38. Iqbal J, Pepkowitz SH, Klapper E. Platelet rich plasma for the replenishment of bone. Curr Osteoporos Rep. 2011;9(4):258–63.

39. James IB, Coleman SR, Rubin JP. Fat, stem cells, and platelet-rich plasma. Clin Plast Surg. 2016;43(3):473–88.

40. Tsai CH, Hsu HC, Chen YJ, Lin MJ, Chen HT. Using the growth factors-enriched platelet glue in spinal fusion and its efficiency. J Spinal Disord Tech. 2009;22(4):246–50.

41. Weiner BK, Walker M. Efficacy of autologous growth factors in lumbar intertransverse fusions. Spine (Phila Pa 1976). 2003;28(17):1968–70.

42. Andersen GR, Koch TJ, Dolmer K, Sottrup-Jensen L, Nyborg J. Low resolution X-ray structure of human methylamine-treated alpha 2-macroglobulin. J Biol Chem. 1995;270(42):25133–41.

43. Zhang Y, Wei X, Browning S, Scuderi G, Hanna LS, Wei L. Targeted designed variants of alpha-2-macroglobulin (A2M) attenuate cartilage degeneration in a rat model of osteoarthritis induced by anterior cruciate ligament transection. Arthritis Res Ther. 2017;19(1):175. https://doi.org/10.1186/s13075-017-1363-4.

44. Tchetverikov I, Lohmander LS, Verzijl N, Huizinga TW, TeKoppele JM, Hanemaaijer R, DeGroot J. MMP protein and activity levels in synovial fluid from patients with joint injury, inflammatory arthritis, and osteoarthritis. Ann Rheum Dis. 2005;64(5):694–8.

45. Baltzer AW, Moser C, Jansen SA, Krauspe R. Autologous conditioned serum (Orthokine) is an effective treatment for knee osteoarthritis. Osteoarthr Cartil. 2009;17(2):152–60. https://doi.org/10.1016/j.joca.2008.06.014. Epub 2008 Jul 31

46. Fox BA, Stephens MM. Treatment of knee osteoarthritis with Orthokine-derived autologous conditioned serum. Expert Rev Clin Immunol. 2010;6(3):335–45.

47. Alvarez-Camino JC, Vázquez-Delgado E, Gay-Escoda C. Use of autologous conditioned serum (Orthokine) for the treatment of the degenerative osteoarthritis of the temporomandibular joint. Review of the literature. Med Oral Patol Oral Cir Bucal. 2013;18(3):e433–8. Review.

48. Wehling P, Evans C, Wehling J, Maixner W. Effectiveness of intra-articular therapies in osteoarthritis: a literature review. Ther Adv Musculoskelet Dis. 2017;9(8):183–96.

49. Rim YA, Nam Y, Ju JH. Application of cord blood and cord blood-derived induced pluripotent stem cells for cartilage regeneration. Cell Transplant. 2018;25:963689718794864.

50. Fitzpatrick J, et al. Analysis of platelet rich plasma extraction. Orthop J Sports Med. 2017;5(1):2325967116675272.

Bone Marrow-Derived Stem Cells and Their Application in Pain Medicine

4

Christopher J. Centeno, Matthew Hyzy,
Christopher J. Williams, Matthew Lucas, Mairin A. Jerome,
and Cameron Cartier

Introduction

In recent years, there has been growing interest in the use of bone marrow in the treatment of various musculoskeletal disorders based on its possible regenerative capabilities. The most common type of therapy uses bone marrow concentrate (BMC) obtained by isolating the buffy coat found within centrifuged bone marrow aspirate [1]. Bone marrow is a good source of mesenchymal stem cells (MSCs), which play a vital role in repair process for damaged musculoskeletal tissues [2]. MSCs have been shown to play a role in tissue healing through their ability to mobilize to the site of damaged tissue and differentiate into other mesenchymal precursors, as well as signal neighboring cells to assist in repair. Early clinical data show the clinical use of MSCs in the treatment of knee, hip, and shoulder osteoarthritis as well as intervertebral disc disease [3–12].

Incorporating BMC into clinical treatment options has the potential to create a shift in the treatment of musculoskeletal injuries, from traditional orthopedic surgery focused on removing or modifying tissue to precise, image-guided injections to facilitate healing of injured or damaged soft tissue and bone. The potential advantages of using a regenerative approach to treat musculoskeletal conditions include decreased procedural risk when compared with surgical alternatives, lessened post-procedural morbidity, and decreased healthcare cost. This approach has many implications for pain management clinicians as their interventional skill sets allow for the precise administration of BMC preparations into a specific structure of need.

Microanatomy and Biochemistry

The following three properties help describe stem cells:

- Undifferentiated
- Capable of cell differentiation
- Capable of cell division through mitosis.

Bone marrow was first discovered to be a source of mesenchymal stem cells in the 1960s [13]. Since then, there have been many advances in our understanding of the MSC's role in tissue repair. In addition, several other bone marrow cell types have been studied, all of which may have significant clinical implications in the future.

- *Mesenchymal stem cells (MSCs):* adult stem cells which are multipotent, capable of dividing into progeny that give rise to all skeletal tissue types including cartilage, bone, tendon, ligament, and connective tissue [14].
  - MSCs are derived from other mesodermal tissues and are also known as marrow stromal cells and later assayed and renamed "colony-forming fibroblasts" in the 1970s [14].
  - Numbered by colony-forming units (CFUs)
  - MSCs are a heterogeneous population of similar cells rather than one distinct cell type.

C. J. Centeno (✉) · M. Hyzy · M. A. Jerome
Centeno-Schultz Clinic, Broomfield, CO, USA
e-mail: centenooffice@centenoschultz.com

C. J. Williams
Emory University, Department of Rehabilitation Medicine,
Atlanta, GA, USA

M. Lucas
Peak Orthopedics and Spine–A Division of Orthopedic Centers of
Colorado, Englewood, CO, USA

C. Cartier
Jason Attaman, PLLC, Department of Pain Medicine,
Bellevue, WA, USA

© Springer Nature Switzerland AG 2023
C. W Hunter et al. (eds.), *Regenerative Medicine*, https://doi.org/10.1007/978-3-030-75517-1_4

- Several international groups have provided criteria for identifying MSCs in the research and clinical setting. A mesenchymal stem cell must demonstrate [15]
  - Adherence to plastic
  - Cell surface markers specific to MSCs
  - Multi-lineage mesodermal tissue differentiation.
- There are several unique properties of MSCs which provide a physiologic basis for their clinical application in regenerative medicine for orthopedic applications.
  - MSCs respond to local environmental stimuli, signaling them to differentiate into their various terminal cell types (for example, culturing these cells with ascorbic acid, inorganic phosphate, or dexamethasone could differentiate cells to osteoblasts, while exposure to TGF-beta caused cells to differentiate into chondrocytes) [16].
  - MSCs also participate in paracrine signaling prompting neighboring cells to participate in tissue repair [2, 17].
  - They have also shown to be capable of mobilizing through the peripheral circulation to distant sites of injury in a mouse model [18]
- *Hematopoietic stem cells (HSCs)*: primarily give rise to nucleated cells of the blood and may be secondarily involved in muscle repair [19].
  - Satellite cells recruit HSCs to the local area from the bone marrow reservoir when muscle repair is incomplete.
- *Endothelial progenitor cells (EPCs):* recruited from bone marrow to facilitate vascular homeostasis and neovasculogenesis [20].
  - Musculoskeletal tissue that has suffered chronic injury and is unable to completely heal may have poor blood supply. EPCs may aid in re-establishing vascularity through secreting vascular endothelial growth factor (VEGF).
- *Pericytes*: located near blood vessels and recruited from bone marrow to promote neovasculogenesis and tissue repair [21].
  - Research suggests pericytes may differentiate into MSCs when injury is detected [22].
- *Osteochondral reticular cells (ORCs)*: recently discovered and concentrated in the metaphysis of long bones. Hence, these are not found in BMA, but may be found in other bone marrow procedures that involve bone grafts.
  - ORCs differentiate into osteoblasts, chondrocytes, and reticular marrow stromal cells [23].
- *Multilineage Differentiating Stress Enduring (MUSE) Cells:* capable of differentiating between all three embryonic layers (endoderm, mesoderm, and ectoderm).
  - Activated by physical stress, MUSE cells act as a progenitor reserve cell source, in part because they survive longer and harsher environments than many other cell types. They are also involved in regenerative homeostasis and tissue repair.

BMC vs. Adipose

Controversy exists as to which tissue type provides the best source of mesenchymal stem cells. Several studies suggest that adipose tissue contains a higher stem cell count when compared to bone marrow [24, 25]. However, this is largely a misconception due to difference in interpretation of cell content between the two tissues.

- Adipose tissue has a higher percentage of MSCs as compared to nucleated cells.
  - Adipose tissue: 1–5% of nucleated cells are MSCs.
  - Bone marrow: 0.01–0.5% of nucleated cells are MSCs.
- However, bone marrow has approximately 100 times more total nucleated cells (TNCs) than adipose tissue [26] per volume.
- Also, adipose tissue contains significantly fewer HSCs, which give rise to nucleated blood cells and play a role in muscle repair as mentioned above.
  - Generally, bone marrow contains the same or more total stem cells per unit volume compared to adipose tissue.

Indications

As of August 2019, the total number of patients treated with bone marrow stem cells for orthopedic conditions that have been published in the U.S. Library of Medicine therapy was 11,467. The number was obtained by summing the *n* of all clinical studies that used either bone marrow concentrate or culture expanded MSCs.

The following indications represent the majority of clinical outcome data available using BMC:

- Osteonecrosis
  - Hernigou et al. published the largest study to date (*n* = 342) using core decompression + autologous BMC in treatment of osteonecrosis of the hip [27]
    - ARCO grade 1–2: showed approximately an 80% long-term likelihood of not requiring hip arthroplasty.
    - ARCO grade 3–4: there was declining success.
- Knee Osteoarthritis:
  - Vagness et al. reported approximately one in four patients demonstrated an increase in meniscus size [28].
  - Vega found significant improvement in cartilage signal on follow-up T2 MRI sequences [29]

- Centeno et al. published a large case series demonstrating improved pain/functional outcomes regarding knee OA. Also, it was found that the addition of a fat graft does not improve outcomes over injecting BMC alone [30].
- Hernigou has published two works focused on intraosseous injection of BMC [31, 32]. In one randomized trial, he injected knee osteoarthritis patients on one side with intra-articular BMC injection versus the other side with intraosseous BMC injection (IO). The IO injection had fewer patients convert to knee arthroplasty at 15 years. In a second trial, he compared the efficacy of IO BMC injection to knee arthroplasty at 15 years and found good results for the majority of those treated with BMC. Those patients with more bone marrow edema fared more poorly on long-term follow-up (Table 4.1).
- Hip Osteoarthritis (Table 4.2):

- Based on the author's experience and unpublished registry data, severe disease yields lower response rates, on average.
- Centeno et al. reported on a case series of 196 patients treated with BMC injection. Poorer outcomes were found for patients over 55. It is suspected that these patients likely had more severe underlining disease [42].
- Emadedin et al. performed a small case series of five patients treated with culture expanded bone marrow MSCs and reported functional improvement [40].
- Rivera has published a prospective comparison of surgical BMC use to treat hip femoroacetabular impingement to a retrospective cohort of surgically treated patients. The author found more efficacy for the BMC-treated group [41].

- Shoulder Rotator Cuff (Table 4.3):

Table 4.1 Summary of selected published research using bone marrow concentrate or culture expanded bone marrow mesenchymal stem cells for knee osteoarthritis

Author	Study type	Intervention	Patient n	Stem cells used	Functional improvement	Notes
Vangsness [28]	Randomized controlled trial	Partial meniscectomy with MSC injection vs. placebo	55	Allogeneic cultured bone marrow MSCs	Yes	1 in 4 patients with increased meniscus volume
Centeno [33]	Randomized controlled trial	Image-guided injection vs. physical therapy	48	Autologous bone marrow concentrate	Yes	Cross over with physical therapy at 3 months
Centeno [30]	Prospective case series	Image-guided injection	840	Autologous bone marrow concentrate	Yes	2/3 of patients were knee arthroplasty+ candidates
Kim [34]	Prospective case series	Injection	49	Autologous cultured bone marrow MSCs	Yes	Full-thickness chondral lesions <6 cm responded best
Vega [29]	Randomized controlled trial	Injection of MSCs vs. hyaluronic acid	30	Allogeneic cultured bone marrow MSCs	Yes	Improved cartilage signal on MRI T2 mapping
Teo [35]	Prospective case series	Surgical implantation of MSCs vs. first generation autologous chondrocyte implantation	36	Autologous cultured bone marrow MSCs	Yes	MSCs equivalent to autologous chondrocyte implantation
Mautner [36]	Case series	Injection of BMC vs. microfragmented adipose tissue (Mfat)	41	Autologous bone marrow concentrate	Yes	No difference between BMC and Mfat
Gobbi [37]	Case series	Surgical implantation of bone marrow concentrate plus hyaluronic acid	23	Autologous bone marrow concentrate	Yes	Results not dependent on chondral lesion size or location
Kim [38]	Case series	High tibial osteotomy with surgical implantation of MSCs alone vs. MSCs+ allogeneic cartilage implant	80	Allogeneic cultured bone marrow MSCs	Yes	MSCs+cartilage better than MSCs alone
Hernigou [31]	Randomized controlled trial	One knee injected intra-articular and the other injected intraosseous	60	Bone marrow concentrate	Yes	Intraosseous alone superior for preventing the need for knee arthroplasty at 15 years
Hernigou [32]	Randomized controlled trial	One knee had knee arthoplasty and the other intraosseous injection	140	Bone marrow concentrate	Yes	Intraosseous alone helped most patients avoid the need for TKA on the non-operated side

MSCs mesenchymal stem cells

Table 4.2 Summary of published research using bone marrow concentrate or cultured expanded bone marrow mesenchymal stem cells for hip osteoarthritis

Author	Study type	Intervention	Patient n	Stem cells used	Functional improvement	Notes
Centeno [39]	Prospective case series	Image-guided injection	196	Bone marrow concentrate	Yes	Majority of patients were hip arthroplasty candidates
Emadedin [40]	Prospective case series	Unknown	5	Culture expanded	Yes	Severity of hip osteoarthritis unknown
Rivera [41]	Prospective case series compared to retrospective cohort	Arthroscopy with or without bone marrow concentrate	80	Bone marrow concentrate	Yes	Adding bone marrow concentrate helped surgical outcomes

Table 4.3 Summary of published research using bone marrow concentrate for shoulder osteoarthritis and rotator cuff tear

Author	Study type	Intervention	Patient n	Stem cells used	Functional improvement	Notes
Centeno [3]	Prospective case series	Image-guided injection	105	Bone marrow concentrate	Yes	Patients failed conservative management
Centeno [43]	Randomized controlled trial	Image-guided injection	25	Bone marrow concentrate	Yes	Bone marrow concentrate better than physical therapy
Hernigou [8]	Prospective case controlled	Arthroscopic rotator cuff repair with bone marrow concentrate vs. surgical repair only	45	Bone marrow concentrate	Yes	100% healing of tendon on MRI vs. 67% in control group at 6 months and intact tendon in 87% vs. 44% at 10 years

- In a comparison trial of surgical repair with and without injected BMC, patients injected with BMC experienced a re-tear rate 50% less than the surgery only group [8].
- Centeno et al. demonstrated significant reductions in pain and increases in validated functional metrics through a case series of 102 patients who had both shoulder OA and a rotator cuff tear [3] In a follow-on RCT, the same author demonstrated good results when using bone marrow concentrate to treat rotator cuff tears in a cross-over RCT with physical therapy alone as a comparator (see Fig. 4.1) [43].
• Lumbar Intervertebral Disc—Degenerative Disc Disease (DDD) (Table 4.4):
- Pettine et al. published showed that higher MSC (CFUs) doses corresponded to the best outcomes at 1- and 2-year results [10, 11].
- Orozco et al. treated ten patients with chronic low back pain and disc degeneration with culture-expanded MSCs from BMC and found statistically significant improvements in pain and function which were sustained at 1 year [9].
- In another study, nine patients were injected with autologous BM-MSCs that were co-cultured with nucleus pulposus cells, into Pfirrmann grade III degenerated discs adjacent to spinal fusion levels. It showed that there was no progression of disc degeneration in adjacent segments to spinal fusion over a 3 -year follow-up time period [9, 44, 46].
- Finally, Noriega injected degenerative discs with allogeneic MSCs and found that a responder cohort of about 40% reported significant decreases in pain and improve-

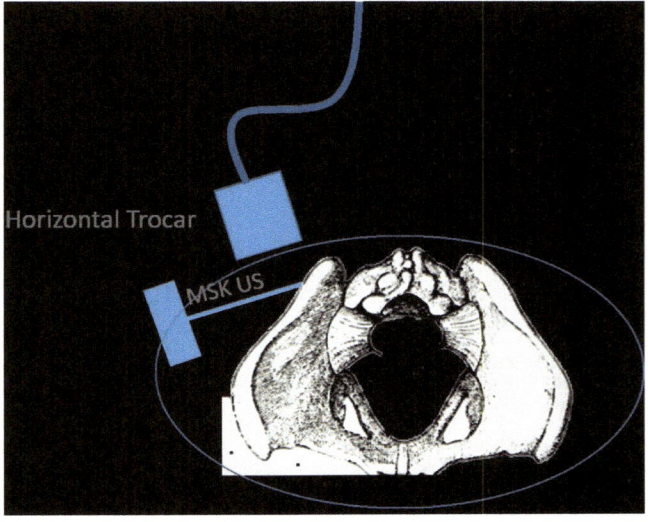

Fig. 4.1 Illustration of ideal ultrasound linear probe and trocar placement for identification of the posterior superior iliac spine during a bone marrow aspiration procedure

ments in function [49]. This concept of a "responder cohort" for DDD patients treated with MSCs is also consistent with non-peer-reviewed data presented via press release by Mesoblast, a company pursuing FDA approval for allogeneic bone marrow MSCs.
• Ankle Disorders:
- Emadedin et al. treated ankle osteoarthritis with cultured MSCs and reported a significant reduction in pain as well as subchondral edema on MRI 6 months post-procedure. In addition, there was improved function [40].

Table 4.4 A summary of BMC and culture expanded mesenchymal stem cells and other cell type-treated degenerative disc disease patients

Author	Study type	Intervention	Patient n	Stem cells used	Functional improvement	Notes
Mochida [44]	Prospective case series	Surgical implant	9	Autologous nucleus pulposus cells	No-minimal MRI changes	Safety study
Pettine [45]	Prospective case series	Image-guided intradiscal injection	26	Bone marrow concentrate	Yes	Possible slight changes in MRI, but within error of degenerative disc disease grading scale
Pang [46]	Prospective case series	Surgical implantation	2	Allogeneic cord blood MSCs	Yes	No imaging
Orozco [9]	Prospective case series	Image-guided intradiscal injection	10	Autologous culture expanded bone marrow-MSCs	Yes	No improvement in disc height, some decrease in T2 signal
Centeno [47]	Prospective Case series	Image-guided intradiscal injection	33	Autologous culture expanded bone marrow-MSCs	Yes	Improvements in disc bulge size
Elabd, Centeno [48]	Prospective Case series	Image-guided intradiscal injection	5	Autologous culture expanded bone marrow-MSCs	Yes	Improvements in disc bulge size
Noriega [49]	Randomized controlled trial	Image-guided intradiscal injection vs. sham injection into paravertebral muscles	24	Allogeneic culture expanded bone marrow MSCs	Yes	Only a group of "responders" at 40% of the treatment cohort had positive results

MSCs mesenchymal stem cells

- Hernigou et al. published a large study comparing 86 diabetic ankle fracture non-union patients treated with BM-MSCs vs. 86 treated traditionally with iliac bone graft. Patients receiving traditional treatment with iliac crest bone graft had a 62% healing rate, whereas those treated with BM-MSCs had a success rate of 82% and fewer complications [50].
- Epicondylitis:
  - Singh et al. performed a case series of 30 patients treated with a single injection of BMC for lateral epicondylitis. The report showed a significant reduction in symptoms at short and medium follow-up intervals [51].

Safety Profile and Contraindications

Two large studies have demonstrated the safety of orthopedic conditions treated with BMC.

- In 2013, Hernigou et al. published findings on 1873 patients that had been monitored for an average of 12.5 years and found incidence of neoplasm in the area of BMC injection [52].
- In 2016, Centeno et al. published findings for 2372 patients who had been treated at multiple clinic sites with either BMC or culture expanded MSCs and followed for up to 9 years regarding all adverse events. They reported a 1.5% incidence of serious adverse events and a lower incidence of neoplasm over the course of follow-up than that occurs in the general population [53].

Contraindications include

- Anemia
- Coagulopathy
- Active or history of neoplasm (Relative contraindications)
  - Cancer patients treated with BMC injections for orthopedic conditions did not show any increase in new neoplasm rates [52].

There is a theoretical risk that injection of MSCs into or near tumor cells or malignancy could act to promote tumor growth and cell proliferation though this remains controversial [54].

Preoperative Considerations

- Patient needs to be aware of the following:
  - Potential complications of BMA
    - Procedure site pain, infection, bleeding/hematoma, post-aspiration anemia, potential injury to surrounding structures, and embolic event in at-risk patients (cluneal nerves).
  - Risks associated with intended target procedure (example: inadvertent dural puncture in disc procedure).
  - Alternative treatments.
- Provider needs to be aware of the following:
  - Pertinent medical issues or active infections that may increase procedural risk.
  - Hematocrit levels should be assessed to estimate max BMA volume that can be harvested. For example, taking 60 mL of bone marrow aspirate in a small anemic female may be ill advised.

- Anti-coagulation status or bleeding disorders that could complicate normal clotting after penetration of the periosteum.
- If patient has a history of heparin-induced thrombocytopenia (HIT), then ACD (acid-citrate-dextrose) should be used to avoid blood clots during the aspiration.
- Provider needs to perform the following:
  - Physical exam of the harvest area to assess for infection, skin ulcerations, or signs of injury.

Radiographic Guidance

Proper use of BMC in the treatment of musculoskeletal conditions requires image guidance both for precise administration of the injectate to the area of pathology, but also to perform a safe bone marrow aspiration that optimizes the amount and quality of MSCs obtained. Either fluoroscopy or ultrasound may be utilized, and both have their benefits and limitations. Attempting a BMA without imaging guidance is below the standard of care. Imaging guidance helps prevent significant complications from inappropriate trocar placement. The specific area of cannulation needs to be visualized to monitor cannula placement and to avoid areas of thin bone marrow cavity (Fig. 4.2).

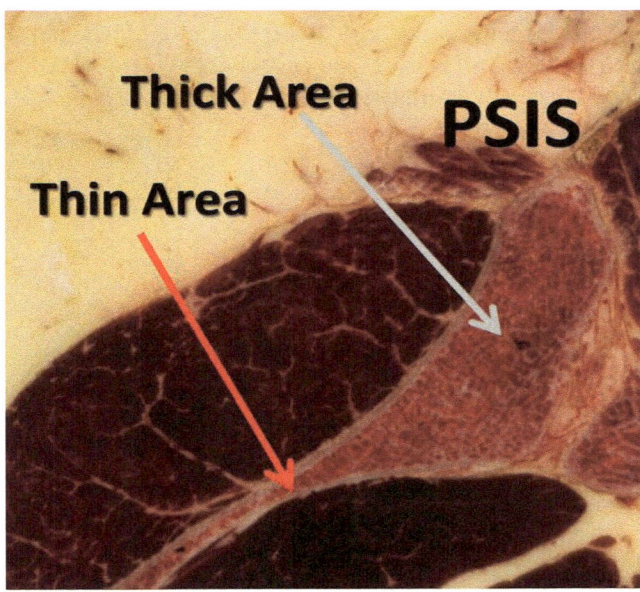

Fig. 4.2 A slice through the bony pelvis from the digital human project showing two marrow draw angles. The first through the "thin area" or the area identified as more radio-lucent. This is a thin area of the pelvis where the likelihood of passing through the marrow space is very high. The "thick area" noted here is the more radio-opaque area shown on the prior slide. This area has a large marrow space with less risk of passing through the marrow rich area and much higher likelihood of drawing whole marrow

Ultrasound

- PRO: Visualizing superficial and soft tissues as well as neurovascular structures.
- CON: Structures deep to bone are not able to be visualized.
- Example: Recommended to inject the rotator cuff of the shoulder, but not recommended when injecting the ACL of the knee due to the ligaments being inside the bony trochlear groove.

Fluoroscopy

- PRO: Visualizing bone and other deeper structures with the use of contrast.
- CON: Unable to image superficial soft tissues. Radiation exposure and cost.
- Example: Recommended for injecting stem cells into an osteonecrosis lesion of the hip, but would be less appropriate to inject a rotator cuff tear.

Key points to maximize MSC yield from BMC:

- The posterior superior iliac spine (PSIS) contains significantly more nucleated cells than other bone aspiration sites [55].
- Focus on drawing small volumes (5 mL per site) rather than drawing a large volume (over 20 cc) from a single bone site reduces [55].
- Multiple aspiration sites may yield more MSCs that reside in subcortical areas as well as pericytes that are located close to blood vessels.
- Ropivacaine 0.25% or less is *highly recommended* when providing local anesthesia. Any amount of bupivacaine or lidocaine can be toxic to MSCs [56, 57].

Equipment

- 30 g or 27 g needle
- 25 g 3.5-inch spinal needle
- Sterile 11-gauge disposable trocar (one for each side of access)
- 10–15 cc of 1% Lidocaine or 0.25% Ropivacaine
- 5000 IU Vial of heparin
- 20,000 IU and 10,000 IU vials of heparin
- Preservative free normal saline
- 5 cc syringe
- 30 cc syringes.[1]

---

[1] Hernigou et al. suggested using multiple 5–10 cc syringes may increase MSC yield [58].

Technique

Harvesting Risk

Using the following bone marrow aspiration procedure guidelines, BMA is a safe and reliable procedure. A large U.K. registry reported an incidence of serious adverse event rate of 15 in a total of 20,323 procedures [59].

The steps for a BMA are as follows:

- The patient is positioned prone on a procedure table.
- After sterile prep, the skin is anesthetized with 10–15 mL local anesthetic. Ropivacaine 0.25% is highly recommended. If 1% Lidocaine is used, make certain that it does not contact the BMA.
  - Imaging guidance is critical during the injection of anesthetic. The skin, surrounding soft tissues, and periosteum need to be adequately anesthetized. If not, the patient may experience significant discomfort.
- After anesthetizing the skin and deep tissues, focus on drawing up the remaining medications to allow sufficient time for the local anesthetic to take effect.
- Draw 1 cc of 5000 IU/cc heparin into 5 cc syringe, and dilute it with an additional 4 cc of preservative free normal saline to make a 1000 IU/cc concentration (or follow the instructions of the point of care automated centrifuge).
- Draw 30,000 units into each 30 cc syringes intended for use, with a remaining concentration of 10,000 IU/cc.
- See Figs. 4.3, 4.4, and 4.5 to help guide angle of entry depending on imaging modality used. A shallow angle is used when using ultrasound (Fig. 4.3), and a steeper angle is used when using fluoroscopy (Fig. 4.4). Using these angles when approaching the PSIS (Fig. 4.5) optimizes draw sites where the bone marrow is best accessed in the safest fashion.
- Pass the trocar through anesthetized skin and soft tissues until contact is made at the bone cortex. Forward pressure is used while the device is turned clockwise/counterclockwise at the trocar handle, using the angled tip to bore a hole in the bone. Advancing another 5–10 mm will help seat the trocar in the cortex. The trocar may have incremental measurements to help gauge depth.
- Ensure the trocar is properly seated in the bone by wiggling the trocar handle gently. If it feels loose, further advancement will be needed, no more than 1 cm at a time and reassessing with another wiggle test. If the trocar resists any movement, no further advancement is needed.
- Remove the stylet from the trocar, and ensure the trocar is still well seated. Re-inserting the stylet and further advancing 1 cm at a time are not uncommon until adequate depth is achieved.

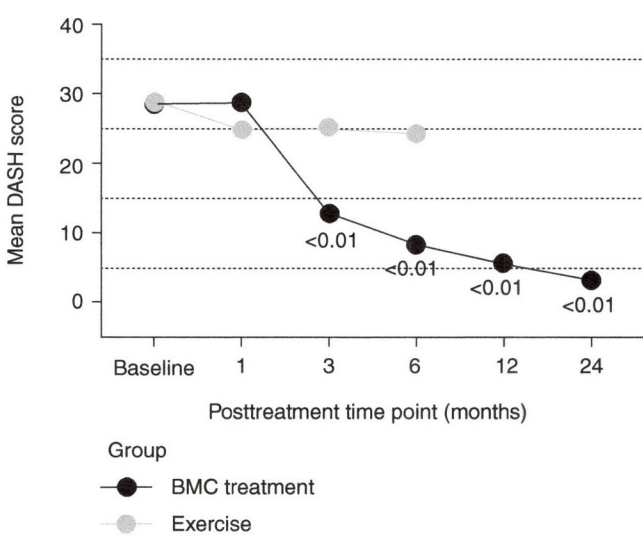

Fig. 4.3 Published randomized controlled trial results of rotator cuff tears treated with bone marrow concentrate injection versus exercise therapy

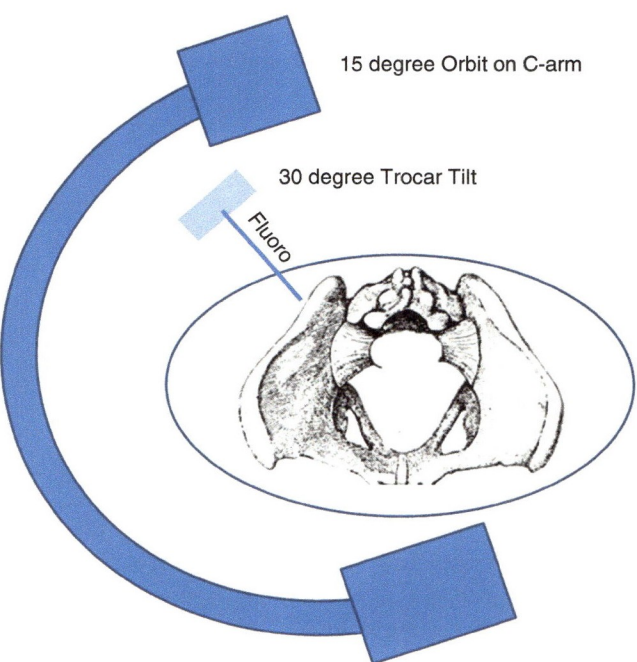

Fig. 4.4 Illustration of the ideal fluoroscopic C-arm and trocar placement for identification of the posterior superior iliac spine during a bone marrow aspiration procedure

- After the stylet is removed, attach the 5 cc syringe with 1000 IU/cc heparin and inject approximately 500–750 units to help prevent clotting. This step is important to prevent MSC trapping within a potential clot. This is performed for each bone site entered.

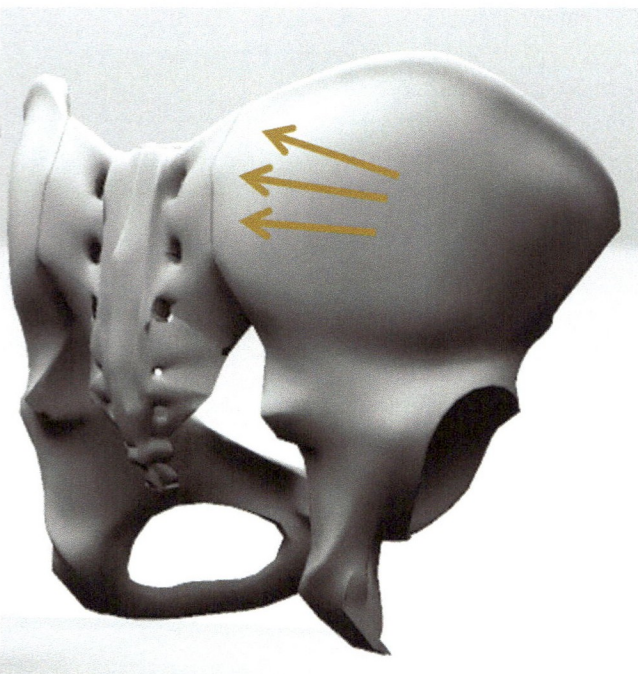

Fig. 4.5 Depiction of the posterior superior iliac spine (PSIS) located on the posterior pelvis. This is the ideal area for 3–4 draw sites from each PSIS

Table 4.5 A comparison between commercial bedside centrifuges and manual processing used to process bone marrow aspirate

	Commercial bedside centrifuge (510 K Approved)	Manual processing
Pros	Easy to learn Easy to use Lower start-up costs	Precise volumes, concentrations, cell counts Variable injectates are possible
Cons	Little versatility, variability Imprecise treatment protocols	Requires biologic safety cabinet Higher start-up costs More training required

- Attach the 30 cc draw syringe to the trocar. Pull back on the plunger according to patient tolerance. As BMA enters the syringe, gently agitate the syringe to help mix the heparin with the BMA to help mix the heparin and prevent clotting.
- Restrict the draw to 5–15 cc per site. Pull back and redirect the trocar without removing the trocar from the skin and reengage another bone cortex site. Any redirection needs to be performed under ultrasound or fluoroscopic guidance.
- Patient weight, number, and size of areas to be treated all help to determine the total BMA volume.
  - Females <47 kg, total volume should not exceed 50 cc.
  - Females >47 kg pounds but <54 kg, total volume should not exceed 60 cc.
  - Males or females >54 kg but <68 kg, total volume should not exceed 90 cc.

  - Male >68 kg, total volume should not exceed 120 cc.

Processing

The goal of BMA concentration is to isolate the buffy coat: the small, gray, middle section in a centrifuged BMA sample. Most providers injecting BMC utilize a commercial bedside centrifuge to concentrate the buffy coat rather than manual processing and lab technicians. There is limited third-party research available comparing these concentration devices. Table 4.5 helps describe the positive and negative aspects of each technique that are known.

510 K Approved Bedside Centrifuge Systems:

- Accelerate: Autologous Platelet Concentrating System
- Accelerated Biologics: BC 60 and BC 120 Pure
- Arthrex Angel
- BioCUE by Biomet
- Celling ART BMC
- CellPoint-ISTO Biologics
- Emcyte 544E
- Emcyte PureBMC
- GenesisCS Component Concentrating System
- Harvest Technologies SmartPrep 2
- ISTO CellPoint.

Dosing Bone Marrow Concentrate

Dosing of BMC can be quantified as follows:

- Colony-forming unit (CFU) assay: BMC is cultured in monolayer and incubated until colonies of MSCs form that adhere to plastic. The total number provides a rough metric of MSC content [60].
  - CFUs are primarily useful in the research setting rather than clinical, as the time needed for cell culture testing is not conducive to clinical practice setting.
- Flow cytometry: BMC cells are stained with fluorescent antibodies to MSC specific cell surface markers and processed through a flow cytometer. The International Society for Cellular Therapy issued a position statement, defining minimal criteria to identify an MSC. MSCs must express CD105, CD73, and CD90, but not CD34, CD45, CD14, CD11b, CD79alpha, CD19, or HLA-DR [61].
  - The cost and expertise required to run and analyze the results also makes this impractical in most clinic settings.
- Total Nucleated Cell (TNC) Count: the number of nucleated cells in BMC can be used as an indirect measurement, or proxy, of MSC content given the MSC/TNC ratio discussed above (0.01–0.5% of nucleated cells are MSCs).

– TNC is most convenient for clinical use. A manual hemocytometer or a commercial automated counting system is required (Peters and Watts 2016).

Research shows that better clinical outcomes is associated with higher CFU or TNC counts [11].

Post-operative Considerations

There are several medications known to impair MSC function and viability, and ultimately alter cell culture results. It is recommended that the following medications should be held for 2–3 serum half-lives before and at least 2–4 weeks after a BMC procedure to optimize clinical outcomes:

- Non-steroidal Anti-Inflammatories (NSAIDs) [62]
- Corticosteroids [63]
- ACE Inhibitors [64]
- Statins [65].

Potential Complications and Pitfalls

- Several local anesthetics, including Marcaine, Bupivacaine, and Lidocaine, are toxic to MSCs at low concentrations, and therefore, administering these in conjunction with BMC will significantly reduce cell viability. Ropivacaine at low concentrations of 0.125–0.25% is safe to use with MSCs [56, 57].
- It is very important to anesthetize not only the skin and subcutaneous soft tissue but also the periosteum. Incomplete anesthesia of the periosteum can lead to intense pain and even neuralgia.
- The clinician MUST provide adequate time for the local anesthetic to take effect (typically 3–5 minutes) prior to starting the procedure.
- BMAs using single site draws/collections with high volume aspiration (60 cc or more) will dramatically reduce cellular yield (please see Sect. 4.8).
- Preventing clots in the bone marrow aspirate sample is important to optimize cellular yield. Thus, it is imperative to pre-heparinizing the syringes used for sample collection as well as using heparin at the draw sites (the authors suggest using heparin—more effective anti-coagulant than ACD (anticoagulant citrate dextrose)).
  - Heparin must be used in the BMA draw syringe (see above) and should be gently shaken/mixed with the first BMA sample as soon as aspirated (it will not efficiently mix through diffusion).
  - During draw, immediately inject small amount of heparin (500–750 units) immediately after cannulating the cortex AND after each advancement of the trocar prior to aspirating.

Clinical Pearls

- It is important to remember that adipose tissue does not necessarily yield higher counts of stem cells.
- Forming a standardized routine is essential to proper BMA and patient comfort/safety.
  - Start with injecting local anesthetic to soft tissue and periosteum, step away and heparinize syringes, prepare trocar, set up image guidance, mark skin boundaries, etc. prior to starting the procedure.
- Remember to identify key anatomic landmarks when performing with fluoroscopy, prior to the procedure, to define target area.
- When using ultrasound guidance for imaging, the authors suggest using a sterile surgical marker on the skin to define safe borders for aspiration as well as to mark the previously anesthetized areas and prior draw sites.
- Use of a multi-site draw technique with several smaller aspiration volumes at each site will allow for higher cell yields.

References

1. Sampson S, Botto-van Bemden A, Aufiero D. Autologous bone marrow concentrate: review and application of a novel intra-articular orthobiologic for cartilage disease. Phys Sportsmed. 2013;41(3):7–18.
2. Tedesco FS, et al. Repairing skeletal muscle: regenerative potential of skeletal muscle stem cells. J Clin Invest. 2010;120(1):11–9.
3. Centeno CJ, et al. A prospective multi-site registry study of a specific protocol of autologous bone marrow concentrate for the treatment of shoulder rotator cuff tears and osteoarthritis. J Pain Res. 2015;8:269–76.
4. Daltro GC, et al. Efficacy of autologous stem cell-based therapy for osteonecrosis of the femoral head in sickle cell disease: a five-year follow-up study. Stem Cell Res Ther. 2015;6:110.
5. Gobbi A, et al. One-step surgery with multipotent stem cells and Hyaluronan-based scaffold for the treatment of full-thickness chondral defects of the knee in patients older than 45 years. Knee Surg Sports Traumatol Arthrosc. 2017;25(8):2494–501.
6. Gobbi A, Whyte GP. One-stage cartilage repair using a hyaluronic acid-based scaffold with activated bone marrow-derived mesenchymal stem cells compared with microfracture: five-year follow-up. Am J Sports Med. 2016;44(11):2846–54.
7. Havlas V, et al. Use of cultured human autologous bone marrow stem cells in repair of a rotator cuff tear: preliminary results of a safety study. Acta Chir Orthop Traumatol Cechoslov. 2015;82(3):229–34.
8. Hernigou P, et al. Biologic augmentation of rotator cuff repair with mesenchymal stem cells during arthroscopy improves healing and prevents further tears: a case-controlled study. Int Orthop. 2014;38(9):1811–8.
9. Orozco L, et al. Intervertebral disc repair by autologous mesenchymal bone marrow cells: a pilot study. Transplantation. 2011;92(7):822–8.
10. Pettine K, et al. Treatment of discogenic back pain with autologous bone marrow concentrate injection with minimum two year follow-up. Int Orthop. 2016;40(1):135–40.
11. Pettine KA, et al. Percutaneous injection of autologous bone marrow concentrate cells significantly reduces lumbar discogenic pain through 12 months. Stem Cells. 2015;33(1):146–56.

12. Zhao D, et al. Autologous bone marrow mesenchymal stem cells associated with tantalum rod implantation and vascularized iliac grafting for the treatment of end-stage osteonecrosis of the femoral head. Biomed Res Int. 2015;2015:240506.

13. Friedenstein AJ, Chailakhjan RK, Lalykina KS. The development of fibroblast colonies in monolayer cultures of guinea-pig bone marrow and spleen cells. Cell Tissue Kinet. 1970;3(4):393–403.

14. Caplan AI. Mesenchymal stem cells. J Orthop Res. 1991;9(5):641–50.

15. Horwitz EM, et al. Clarification of the nomenclature for MSC: the International Society for Cellular Therapy position statement. Cytotherapy. 2005;7(5):393–5.

16. Koga H, et al. Local adherent technique for transplanting mesenchymal stem cells as a potential treatment of cartilage defect. Arthritis Res Ther. 2008;10(4):R84.

17. Quintero AJ, et al. Stem cells for the treatment of skeletal muscle injury. Clin Sports Med. 2009;28(1):1–11.

18. Ferrari G, et al. Muscle regeneration by bone marrow-derived myogenic progenitors. Science. 1998;279(5356):1528–30.

19. Stromberg A, et al. Bone marrow derived cells in adult skeletal muscle tissue in humans. Skelet Muscle. 2013;3(1):12.

20. Szmitko PE, et al. Biomarkers of vascular disease linking inflammation to endothelial activation: part II. Circulation. 2003;108(17):2041–8.

21. Bergmann CE, et al. Arteriogenesis depends on circulating monocytes and macrophage accumulation and is severely depressed in op/op mice. J Leukoc Biol. 2006;80(1):59–65.

22. Crisan M, et al. A perivascular origin for mesenchymal stem cells in multiple human organs. Cell Stem Cell. 2008;3(3):301–13.

23. Worthley DL, et al. Gremlin 1 identifies a skeletal stem cell with bone, cartilage, and reticular stromal potential. Cell. 2015;160(1–2):269–84.

24. Mafi R, et al. Sources of adult mesenchymal stem cells applicable for musculoskeletal applications - a systematic review of the literature. Open Orthop J. 2011;5 Suppl 2:242–8.

25. Zuk PA, et al. Multilineage cells from human adipose tissue: implications for cell-based therapies. Tissue Eng. 2001;7(2):211–28.

26. Sakaguchi Y, et al. Comparison of human stem cells derived from various mesenchymal tissues: superiority of synovium as a cell source. Arthritis Rheum. 2005;52(8):2521–9.

27. Hernigou P, et al. Cell therapy of hip osteonecrosis with autologous bone marrow grafting. Indian J Orthop. 2009;43(1):40–5.

28. Vangsness CT Jr, et al. Adult human mesenchymal stem cells delivered via intra-articular injection to the knee following partial medial meniscectomy: a randomized, double-blind, controlled study. J Bone Joint Surg Am. 2014;96(2):90–8.

29. Vega A, et al. Treatment of knee osteoarthritis with allogeneic bone marrow mesenchymal stem cells: a randomized controlled trial. Transplantation. 2015;99(8):1681–90.

30. Centeno C, et al. Efficacy of autologous bone marrow concentrate for knee osteoarthritis with and without adipose graft. Biomed Res Int. 2014;2014:370621.

31. Hernigou P, et al. Subchondral bone or intra-articular injection of bone marrow concentrate mesenchymal stem cells in bilateral knee osteoarthritis: what better postpone knee arthroplasty at fifteen years? A randomized study. Int Orthop. 2021;45(2):391–9.

32. Hernigou P, et al. Human bone marrow mesenchymal stem cell injection in subchondral lesions of knee osteoarthritis: a prospective randomized study versus contralateral arthroplasty at a mean fifteen year follow-up. Int Orthop. 2021;45(2):365–73.

33. Centeno C, et al. A specific protocol of autologous bone marrow concentrate and platelet products versus exercise therapy for symptomatic knee osteoarthritis: a randomized controlled trial with 2 year follow-up. J Transl Med. 2018;16(1):355.

34. Kim SH, et al. Intra-articular injection of mesenchymal stem cells for clinical outcomes and cartilage repair in osteoarthritis of the knee: a meta-analysis of randomized controlled trials. Arch Orthop Trauma Surg. 2019;139(7):971–80.

35. Teo AQA, et al. Equivalent 10-year outcomes after implantation of autologous bone marrow-derived mesenchymal stem cells versus autologous chondrocyte implantation for chondral defects of the knee. Am J Sports Med. 2019;47(12):2881–7.

36. Mautner K, et al. Functional outcomes following microfragmented adipose tissue versus bone marrow aspirate concentrate injections for symptomatic knee osteoarthritis. Stem Cells Transl Med. 2019;8(11):1149–56.

37. Gobbi A, Whyte GP. Long-term clinical outcomes of one-stage cartilage repair in the knee with hyaluronic acid-based scaffold embedded with mesenchymal stem cells sourced from bone marrow aspirate concentrate. Am J Sports Med. 2019;47(7):1621–8.

38. Kim YS, et al. Implantation of mesenchymal stem cells in combination with allogenic cartilage improves cartilage regeneration and clinical outcomes in patients with concomitant high tibial osteotomy. Knee Surg Sports Traumatol Arthrosc. 2020;28(2):544–54.

39. Centeno C, Pitts J, Al-Sayegh H, Freeman M. Efficacy and safety of bone marrow concentrate for osteoarthritis of the knee; treatment registry results for 196 patients. J Stem Cell Res Ther. 2014;4(10):242.

40. Emadedin M, et al. Long-term follow-up of intra-articular injection of autologous mesenchymal stem cells in patients with knee, ankle, or hip osteoarthritis. Arch Iran Med. 2015;18(6):336–44.

41. Rivera E, et al. Outcomes at 2-years follow-up after hip arthroscopy combining bone marrow concentrate. J Investig Surg. 2020;33(7):655–63.

42. Centeno CJ, Jitts P, Al-Sayegh H, Freeman MD. Efficacy and safety of bone marrow concentrate for osteoarthritis of the hip; treatment registry results for 196 patients. J Stem Cell Res Ther. 2014;4:242.

43. Centeno C, et al. A randomized controlled trial of the treatment of rotator cuff tears with bone marrow concentrate and platelet products compared to exercise therapy: a midterm analysis. Stem Cells Int. 2020;2020:5962354.

44. Mochida J, et al. Intervertebral disc repair with activated nucleus pulposus cell transplantation: a three-year, prospective clinical study of its safety. Eur Cell Mater. 2015;29:202–12; discussion 212.

45. Pettine KA, et al. Autologous bone marrow concentrate intradiscal injection for the treatment of degenerative disc disease with three-year follow-up. Int Orthop. 2017;41(10):2097–103.

46. Pang X, Yang H, Peng B. Human umbilical cord mesenchymal stem cell transplantation for the treatment of chronic discogenic low back pain. Pain Physician. 2014;17(4):E525–30.

47. Centeno C, et al. Treatment of lumbar degenerative disc disease-associated radicular pain with culture-expanded autologous mesenchymal stem cells: a pilot study on safety and efficacy. J Transl Med. 2017;15(1):197.

48. Elabd C, et al. Intra-discal injection of autologous, hypoxic cultured bone marrow-derived mesenchymal stem cells in five patients with chronic lower back pain: a long-term safety and feasibility study. J Transl Med. 2016;14:253.

49. Noriega DC, et al. Intervertebral disc repair by allogeneic mesenchymal bone marrow cells: a randomized controlled trial. Transplantation. 2017;101(8):1945–51.

50. Hernigou P, et al. Percutaneous injection of bone marrow mesenchymal stem cells for ankle non-unions decreases complications in patients with diabetes. Int Orthop. 2015;39(8):1639–43.

51. Singh A, Gangwar DS, Singh S. Bone marrow injection: a novel treatment for tennis elbow. J Nat Sci Biol Med. 2014;5(2):389–91.

52. Hernigou P, et al. Cancer risk is not increased in patients treated for orthopaedic diseases with autologous bone marrow cell concentrate. J Bone Joint Surg Am. 2013;95(24):2215–21.

53. Centeno CJ, et al. A multi-center analysis of adverse events among two thousand, three hundred and seventy two adult patients undergoing adult autologous stem cell therapy for orthopaedic conditions. Int Orthop. 2016;40(8):1755–65.

54. Lee HY, Hong IS. Double-edged sword of mesenchymal stem cells: Cancer-promoting versus therapeutic potential. Cancer Sci. 2017;108(10):1939–46.

55. Marx RE, Tursun R. A qualitative and quantitative analysis of autologous human multipotent adult stem cells derived from three anatomic areas by marrow aspiration: tibia, anterior ilium, and posterior ilium. Int J Oral Maxillofac Implants. 2013;28(5):e290–4.

56. Breu A, et al. Cytotoxicity of local anesthetics on human mesenchymal stem cells in vitro. Arthroscopy. 2013;29(10):1676–84.

57. Dregalla RC, et al. Amide-type local anesthetics and human mesenchymal stem cells: clinical implications for stem cell therapy. Stem Cells Transl Med. 2014;3(3):365–74.

58. Bain BJ. Morbidity associated with bone marrow aspiration and trephine biopsy - a review of UK data for 2004. Haematologica. 2006;91(9):1293–4.

59. Hernigou P, et al. Benefits of small volume and small syringe for bone marrow aspirations of mesenchymal stem cells. Int Orthop. 2013;37(11):2279–87.

60. Franken NA, et al. Clonogenic assay of cells in vitro. Nat Protoc. 2006;1(5):2315–9.

61. Dominici M, et al. Minimal criteria for defining multipotent mesenchymal stromal cells. The International Society for Cellular Therapy position statement. Cytotherapy. 2006;8(4):315–7.

62. Chang JK, et al. Effects of anti-inflammatory drugs on proliferation, cytotoxicity and osteogenesis in bone marrow mesenchymal stem cells. Biochem Pharmacol. 2007;74(9):1371–82.

63. Wyles CC, et al. Differential cytotoxicity of corticosteroids on human mesenchymal stem cells. Clin Orthop Relat Res. 2015;473(3):1155–64.

64. Durik M, Seva Pessoa B, Roks AJ. The renin-angiotensin system, bone marrow and progenitor cells. Clin Sci (Lond). 2012;123(4):205–23.

65. Izadpanah R, et al. The impact of statins on biological characteristics of stem cells provides a novel explanation for their pleiotropic beneficial and adverse clinical effects. Am J Physiol Cell Physiol. 2015;309(8):C522–31.

Adipose-Derived Stromal Stem Cells

Lora L. Brown

Introduction

Autologous adipose-derived stem cells are an important source of therapeutic cells for patients suffering from traumatic, degenerative, or inflammatory disease processes. Clinical data have identified adipose tissue as an alternative source of mesenchymal stem cells (MSCs). Stromal vascular tissue derived from adipose tissue contains a subset of tissue that is different from that found in blood cells. Adipose stromal tissue contains a subset of multipotent progenitor cells with adipogenic, chondrogenic, osteogenic, and myogenic differentiation potential [1].

Adipose tissue is abundant, easily accessible, and easily obtainable via lipoaspiration with little patient discomfort. Adipocytes make up the bulk of adipose tissue. A heterogeneous cell population called the stromal vascular fraction (SVF) surrounds the mature adipocytes. The SVF includes adipose stromal stem/progenitor cells (ASCs), pericytes, mature and immature vascular endothelial cells, fibroblasts, and hematopoietic-lineage cells [2] (Table 5.1).

A large body of in vitro research shows that adipose-derived stem cells are located within the perivascular niche within the stromal vascular fraction. The stromal vascular fraction (SVF) parallels the mononuclear cell fraction obtained from bone marrow–derived stem cells [3] (Table 5.2). Both tissue sources possess regenerative cellular potential, but 1 mL of adipose tissue contains 300 to 500 times more MSCs than 1 mL of bone marrow aspirate [4]. The cell populations present in the SVF include hematopoietic-lineage cells (stem and progenitor cells, granulocytes, monocytes, lymphocytes), endothelial cells, pericytes, and stromal cells. Collectively, these cell populations possess many advantageous characteristics, including immunomodulatory, anti-inflammatory, antiapoptotic, angiogenic, and mitogenic properties. They also resist scar cascade initiation. These cells accomplish regenerative functions via complex secretion and signaling of growth factors and cytokines. These paracrine effects, as well as direct cell-to-cell interactions, exert great effects on local tissue repair by activating endogenous progenitor cells previously dormant in the affected tissue [1, 5–8]. Consequently, there is a decrease in inflammation and pain, as well as regeneration of tissue in the damaged areas.

It should be noted that stem cell paracrine potential is thought to vary based upon cell tissue origin. Cell surface markers of mesenchymal stem cells have demonstrated

Table 5.1 Commonly used markers to characterize cell populations in SVF

Cell type	Phenotype	Proportion of nonheme (CD45−) nucleated cells
Stromal/ preadipocytes	CD31−, CD34+, CD146−/+, CD90+	67.6 ± 29.7%
Endothelial progenitor	CD31+, CD34+, CD146+, CD90+	5.2 ± 6.1%
Endothelial mature	CD31+, CD34−, CD146−, CD90−	Variable with harvest technique
Pericytes	CD31−, CD34−, CD146+, CD90+	0.8 ± 0.7%

Table 5.2 Comparison of bone marrow–derived and adipose-derived stem cells

Bone marrow aspirate concentration (BMAC)	Adipose-derived stem cells (SVF)
Easy to obtain	Moderate difficulty to obtain
Bone marrow aspiration	Tumescent liposuction
Centrifuge and remainder of materials come in commercially available kits	Flow hood, incubator, tissue culture hood, plus equipment that is typically purchased a la carte
Takes less than an hour to harvest cells, process, and inject to target region	Can take an hour just to harvest cells
Lower nucleated cell concentrations	Higher nucleated cell concentrations
Progenitor and stem cell concentrations unpredictable and typically lower	Progenitor and stem cell concentrations predictable and much higher

L. L. Brown (✉)
TruWell, PLLC, St Petersburg, FL, USA

region-specific variation. Understanding the metabolic activity mechanisms within different stem cell tissue (SVF) niches is a current area of research interest that may provide a clearer understanding of the cellular maintenance of mesenchymal stem cells as well as their regenerative and pro-angiogenic potential.

Although limited, human studies involving MSCs for the treatment of osteoarthritis are promising. Mesenchymal stem cells derived from bone marrow aspirate and percutaneously injected into subjects with MRI-proven degenerative joint disease of the knee showed statistically significant cartilage and meniscus growth on MRI, as well as increased range of motion and decreased modified Visual Analog Scale (VAS) pain scores at 21 weeks after the injection [9]. Emadedin et al. treated six female subjects with osteoarthritis of the knee who were candidates for knee replacements with bone marrow–derived MSCs and found improvements in pain, functional status, and walking distance 6 months post-injection [10]. MRI images at baseline and 6 months postinjection demonstrated an increase in cartilage thickness, extension of repair tissue over the subchondral bone, and a considerable decrease in the size of edematous subchondral patches. In a similar study, autologous MSCs derived from adipose tissue were administered to 18 patients with osteoarthritis of the knee. The results showed that intra-articular injection of 1.0×10^8 adipose-derived MSCs into the osteoarthritic knee improved function and pain of the knee joint without causing adverse events, and it reduced cartilage defects by regeneration of hyaline-like articular cartilage [11].

Another area of interest in regenerative medicine is the treatment of degenerative disc disease. Researchers have demonstrated that intervertebral discs contain an endogenous stem cell population of skeletal progenitor cells displaying osteogenic, adipogenic, and chondrogenic characteristics, which are the same characteristics shared by MSCs derived from both bone marrow and adipose tissue. Mesenchymal stem cell implantation has been shown to stimulate nucleus pulposus cell proliferation and MSC chondrogenic differentiation, as well as increasing production of cytokines, particularly transforming growth factor-beta [12, 13].

Animal studies for the treatment of disc degeneration have demonstrated that MSCs injected into the nucleus pulposus not only survive but proliferate in canine, porcine, and rabbit models. The results of these studies also showed that the transplanted stem cells influenced the production of extracellular matrix proteins, including aggrecan, proteoglycans, and type I and type II collagen. Most importantly, these injections resulted in the preservation of both water content and height in the damaged disc [14–17].

Human studies utilizing stem cells for the treatment of degenerative disc disease are promising. Orozco et al. conducted a pilot study utilizing autologous culture-expanded bone marrow mesenchymal cells for intervertebral disc repair [18]. Ten subjects were followed for 1 year to evaluate back pain, disability, and quality of life. Magnetic resonance (MR) imaging measurements of disc height and fluid content were also performed. Results confirmed feasibility and safety. Patients exhibited rapid improvement of pain and disability at 85% of maximum in 3 months. MRI scans showed that although disc height was not recovered, water content was significantly elevated at 12 months.

Pettine et al. investigated the use of autologous bone marrow concentrate for the treatment of discogenic pain [19]. Twenty-six subjects received percutaneous injections in one or two intervertebral discs and were evaluated using MR imaging, the Oswestry Disability Index (ODI), and VAS. Results showed a substantial reduction in pain of 69.5% on the ODI and 70.6% on the VAS. Eight of 20 patients improved by one modified Pfirrmann grade at 1 year. Furthermore, recent basic research and preclinical studies have revealed that the use of adipose-derived MSCs in regenerative medicine is not limited to mesodermal tissue but extends to ectodermal and endodermal tissues and organs as well [20].

Although there are little data to support the wide array of disease processes treated with stem cell therapy, the evidence is growing exponentially. Physicians around the world utilize adipose-derived MSCs to treat some of the most troubling maladies. Today these therapies are limited to "last resort treatments" for those who can afford them, but some day, regenerative therapies will likely be at the forefront of advanced medical therapies.

Indications

In the field of musculoskeletal medicine, adipose stem cell therapy has been used in the treatment of muscle, tendon, and ligament injuries as well as joint arthritis. Painful degenerative disc disease, facet arthritis, and sacroiliac joint pain are also reasonable applications for this therapy.

Although there are no clear treatment protocols defined for the use of adipose stem cell therapy, the current standard of care preserves this treatment for those patients who have failed conventional treatment options or who are not candidates for conventional treatment options.

Musculoskeletal Conditions Treated with Adipose-Derived MSCS

- Joint osteoarthritis and rheumatoid arthritis
- Tendon, ligament, or meniscal incomplete tears
- Shoulder or hip labral tears

- Rotator cuff disease
- Degenerative disc disease
- Facet and sacroiliac joint disease

An evolving body of evidence suggests adipose-derived stem cells are also therapeutic for systemic autoimmune and inflammatory diseases. Although these diseases may fall outside the scope of this book, it important to understand the breadth of potential therapeutic applications of this treatment.

Chronic Conditions Treated with Adipose-Derived MSCS

- Osteoarthritis
- Rheumatoid arthritis
- COPD
- Heart failure
- Multiple sclerosis
- Alzheimer's disease
- Parkinson's disease.
- ALS
- Ulcerative colitis
- Poorly healing wounds
- Spinal cord injury
- Post-stroke
- Diabetic neuropathy
- Erectile dysfunction

Microanatomy and Biochemistry

The mesenchymal stem cells (MSCs) in adult adipose tissue are powerful progenitor cells that have the amazing capacity to differentiate into specific cell types that generate mesenchymal tissue including bone, cartilage, tendon and ligament, muscle, fat, dermis, and other connective tissues. These cell types include osteoblasts, chondrocytes, myoblasts, and fibroblasts, the very lineages that evolve to many of the musculoskeletal tissues targeted in regenerative medicine [1, 6–8, 20].

The characterization of adipose-derived MSCs has been described in the literature [21, 22]. The stromal vascular fraction (SVF) is composed of the following:

- Hematopoietic stem cells, 2%
- Pre/endothelial cells, 7%
- Pericytes/smooth muscle cells, 2%
- Fibroblasts, 47%
- Other (macrophages, various blood cells), 33%
- Adipose-derived stem cells, 2–5%

Adipose-derived MSCs have trophic, immunomodulatory, and antimicrobial functions. Included in the trophic functions are angiogenic, mitogenic, antiapoptotic, and antiscarring properties [1, 7, 8, 20–22].

Some of the cytokines found in the adipose-derived SVF include high levels of expression of several growth factors:

- Hepatocyte growth factor (HGF): Plays a major role in embryonic organ development; in adult, organ regeneration and wound healing.
- Vascular endothelial growth factor (VEGF): Stimulates growth of new blood vessels.
- Placental growth factor (PGF): Involves in angiogenesis and vasculogenesis.
- Transforming growth factor-beta (TGFβ): Controls proliferation, cellular differentiation, and other functions.

Also found are moderate levels of expression of other factors:

- Fibroblast growth factor (FGF-2): Involves in wound healing and angiogenesis.
- Angiopoietin (Ang-1 and Ang-2): Promotes angiogenesis and formation of blood vessels.

Mesenchymal stem cells demonstrate the ability to release bioactive molecules that are immunoregulatory. They respond to environmental signals that are tissue specific. In response to these signals, MSCs secrete a wide array of paracrine factors that create a regenerative milieu that possesses trophic regenerative properties. Consequently, it is felt that the beneficial impact of adipose-derived MSCs on various tissues and organs may be due to soluble factors produced by the cells, rather than to their tissue differentiation capabilities. Moreover, it has also been shown that the soluble factors secreted by adipose-derived MSCs can be modulated by exposure to different agents, giving promise to the field of tissue engineering [1, 7, 8, 20, 22, 23].

Adipose-derived MSCs have an inherent ability to locate damaged tissue. Their response to molecular signaling within the body has been demonstrated in studies using radionucleotide-tagged cells [22].

Regulatory Status

Adipose tissue is a human cell, tissue, and cellular and tissue-based product (HCT/P), which is defined in Title 21 of the Code of Federal Regulations Part 1271 (21CFR 1271.3(d)) as articles containing or consisting of human cells or tissues that are intended for implantation, transplantation, infusion, or transfer into a human recipient. Because of its unique nature, HCT/Ps have been regulated by the FDA through a tiered, risk-based approach. The FDA is authorized to apply the requirements in the Federal Food, Drug, and Cosmetics Act and the Public Health Service Act (PSA) to those products that meet the definition of drug, biologic, or device. Some HCT/Ps that meet specific criteria do not require premarket review and approval. In order to meet those criteria, specific registration, manufacturing, and reporting steps must be followed in order to prevent the introduction, transmission, and spread of communicable disease. The steps to qualify as an "exempt" HCT/P product are found in PSA Section 361.

The FDA released two draft guidance documents in 2014 that were finalized in 2017 that addressed specific definitions used to define a Section 361 "exempt" HCT/P product [24, 25]. The draft guidance documents further defined "same-day surgical procedure," "minimal manipulation," and "homologous use" with a specific mandate that a HCT/P that qualified as a Section 361 product must meet all three of these criteria.

To this day, there exists much contention between regenerative stem cell clinicians and the FDA regarding the definition of minimal manipulation, homologous use, and adipose tissue. The FDA contends that the use of adipose tissue via enzymatic digestion, ultrasonic cavitation, or other processing methods was considered more than minimal manipulation. Therefore, the process of SVF production was not considered to satisfy the three criteria of 21 CFR 1271 Section 361 HCT/P: homologous use, minimal manipulation, same-day surgical procedure. Many clinicians disagree with this interpretation as adipose tissue is defined not only as a structural support tissue but also as a metabolic endocrine tissue with endocrine and paracrine functions.

Continued widespread use of adipose as a HCT/P product that meets Section 361 exemption prompted the FDA to publish two additional guidance documents in 2017 [26, 27]. In line with the Twenty-First Century Cures Act passed in December 2016, the FDA published these documents as part of comprehensive FDA policy framework to address plans to support and expedite the development of regenerative products, including HCT/Ps. The Twenty-First Century Cures Act was designed to help accelerate medical product development and bring new innovations and advances to patients who need them faster and more efficiently and to simplify and streamline its application of the regulatory requirements for devices used in the recovery, isolation, and delivery of *regenerative medicine advanced therapies* (RMATs), including combination products [28, 29].

The FDA proclaimed its intent to apply a risk-based approach to enforcement of cell-based regenerative products through November 2020, taking into account how products are being administered as well as the diseases and conditions for which they intend to be used. That discretion will not be afforded to those that pose significant potential patient safety concerns.

There are currently several companies that have applied for consideration of their products as Investigational New Drug (IND). And several others are pursuing New Drug Applications (NDA) and Biologics License Applications (BLA).

Adipose-derived stem cells thus are considered and regulated as a drug, device, and/or biologic product. The FDA has clearly stated its position against adipose stem cell therapy meeting the necessary criteria for exemption of premarket review and approval and thus does not qualify for 361 exemption status. Even though clinicians continue to offer adipose stem cell therapy to their patients, there are potential repercussions. With new accelerated pathways for regenerative medicine companies to pursue FDA approval, it is almost certain that new and improved adipose therapies will evolve.

Basic Concerns and Contraindications

The clinical application of cell-based therapies is somewhat controversial. Considered experimental, the therapy is not FDA-approved as of 2019. These facts must be disclosed to all prospective patients. Potential patients should also be informed that their treatment might prevent them from participating in future clinical research studies.

Cell-based therapies are minimally invasive, relatively safe approaches to complex diseases, though the lack of conclusive evidence creates some questions as to their safety and efficacy. It is estimated, however, that hundreds of thousands of autologous stem cell treatments are done per year worldwide, with a paucity of reported complications.

There is some variation in the number of stem cells present in various donor sites and with donor age [30–32]. In general, the most efficient methods can isolate about 500,000–1,000,000 cells per gram of lipoaspirate tissue with a >80% viability. The number of viable cells required for treatment of a particular condition is unknown because there are insufficient data to establish a reliable dose versus effect relationship.

Angiogenesis and mitosis are effective outcomes of cell-based therapies, so there is a theoretical risk of tumorigenesis or increased growth of preexisting cancers. This result has not been seen clinically. Hernigou et al. reported no increased cancer risk in 1873 patients who were observed for an average of 12.5 years after treatment with autologous cell-based therapy using bone marrow–derived stromal progenitor cells [33]. Nevertheless, many physicians consider preexisting solid tumor disease a contraindication to stem cell therapy.

Contraindications for the use of autologous stem cell therapy in musculoskeletal medicine include retracted complete ligament or tendon tears and loose bodies in the articular space. In these cases, surgical therapy is warranted.

The use of stem cell therapy within the spine is nascent. Proper indications and contraindications will be developed as the therapy gains wider utilization, but it is clear that some findings within the spine would constitute a contraindication. These include spinal instability, disc extrusion, Modified Pfirrmann Grade VIII disc disease, critical spinal stenosis, and spinal infection.

Other conditions considered to be contraindications to autologous adipose stem cell therapy are preexisting local or systemic infection, severe cardiovascular disease, and blood dyscrasias.

Preoperative Considerations

Age, general health, nutritional status, and the availability of adipose tissue should be considered when evaluating a patient for autologous stem cell therapy. In patients with advanced age or nutritional or medical compromise, autologous therapy may not be the best option, and an allogeneic approach can be considered. Emaciated patients or high-performance athletes may not have an adequate volume of adipose tissue. In those cases, alternative treatments should be considered.

The health of each patient should be assessed preoperatively. Patients should be encouraged to stop smoking 4 to 6 weeks before treatment. Heavy alcohol consumption should be avoided. Nutrition should be optimized with clean, whole foods and nutritional supplementation.

NSAIDs should be avoided at least 2 weeks before and 4 weeks after autologous or allogeneic stem cell therapy. Steroid injections should be avoided for 4 weeks before and after treatment. Within the orthopedic literature, NSAIDs have been linked to impaired fracture healing and abnormal chondrocyte differentiation. Nonsteroidal anti-inflammatory medications and steroids have also been demonstrated altered stem cell (MSC) gene expression, decreased cell proliferation, and altered cell differentiation [34–36].

The patient's medical condition will determine the amount of adipose to be aspirated. Most systems utilize approximately 60 cc of adipose tissue to recover a therapeutic dose of SVF, which should contain 50 to 100 million stem cells. An adequate adipose harvest site must be selected. The abdomen is commonly used, but in some instances, one must resort to the flank or "love handles," the hips, or thighs. Careful examination of the area should include notation of any prior operative procedures that may have produced scar tissue within or near the lipoaspiration area. Topographical, superficial skin markings performed preoperatively with the patient standing may provide a useful guide during the procedure. Although this is not a cosmetic procedure, one should attempt to provide a symmetric and appealing outcome.

A proper procedure consent should be completed and signed by the patient on the day of the procedure, prior to any sedative medication, including the following points:

- Consent for tumescent anesthesia
- Consent for lipoaspiration
- Consent for reintroduction of the final product, whether that be a joint injection; a muscle, tendon, or ligament injection; or an intravascular or intrathecal injection
- Disclosure that the procedure is experimental
- Disclosure that the procedure is not FDA-approved
- Acknowledgment that a successful outcome is not guaranteed
- Disclosure that the treatment may eliminate the patient's candidacy for future clinical research studies

Assess whether the patient would like to "bank" or cryopreserve some cells. Several FDA-listed tissue banks will cryopreserve a patient's adipose-derived MSCs for a fee. Theoretically, the tissue that is stored will always be more youthful and beneficial than tissue available in the future. Most tissue banks require 60–100 mL of adipose to be shipped overnight. The adipose tissue is processed, and the cells are expanded and cryopreserved until future need. Currently, expanded cell products are considered highly processed and consequently are subject to the Public Health Safety Act, Section 351. For such tissue to be used in the United States, it would need to be licensed by the FDA as a biological drug [37].

Preoperative intravenous antibiotics may be considered, as well as an anxiolytic.

Equipment

Figure 5.1 shows some of the equipment needed for lipoaspiration:

- 14 g/25 cm garden spray infiltration cannula
- 3 mm/25 cm Mercedes cannula
- 60-cc syringe snap lock
- Syringe caddy
- 2-quart stainless steel bowl
- 60-cc Luer lock syringes × 4
- 60-cc Toomey syringe × 2
- #11 blade scalpel
- 10-cc syringe
- 18-gauge 1-inch needle
- 25-gauge 1.5-inch needle
- Sterile back cover drape
- Sterile half drape
- Sterile prep kit (povidone-iodine or Hibiclens®, Mölnlycke Health Care, Norcross, GA)

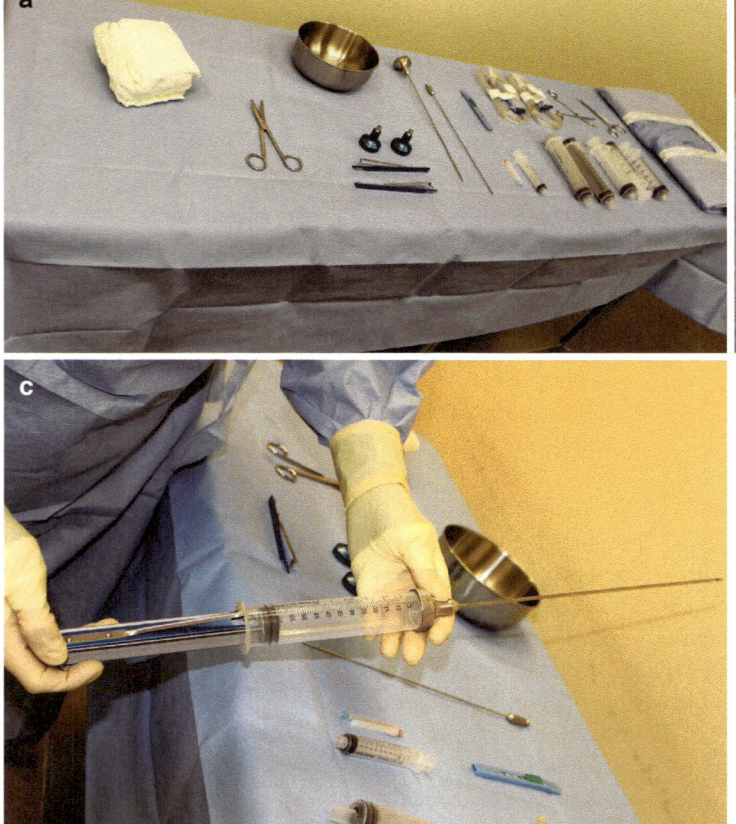

Fig. 5.1 (**a**) Back table setup for lipoaspiration procedure; (**b**) irrigation cannula; (**c**) lipoaspiration cannula with snap lock

- Sterile surgical marking pen

  Also to be used are several medications and some items of laboratory and tissue culture equipment:

- 0.9% sodium chloride IV solution (1000 mL)
- 8.4% sodium bicarbonate, 1 mEq/mL (50 mL vial)
- Lidocaine HCl 2% (50 mL vial)
- Epinephrine 1:1000 (30 mL vial)
- HEPA-filtered Class 100 laminar flow biological cabinet
- Centrifuge
- Dry block incubator or incubator shaker
- Disposable manual stem cell isolation kit

Technique of Lipoaspiration

Rodbell and James pioneered the initial techniques used to isolate cells from adipose tissue in the 1960s. The procedure has evolved to become a safe and minimally invasive procedure [38–40]. Today the isolation procedure includes the following steps:

- Tumescent liposuction, which finely minces tissue fragments (dependent on the size of the cannula)
- Washing to remove hematopoietic cells
- Enzyme or mechanical digestion
- Centrifugation to separate the SVF
- Isolating SVF with washing cells, centrifugation, and cell strainer
- Cells (SVF) prepared in the final solution

Preparation of Tumescent Anesthetic Fluid

The tumescent technique uses the standard anesthetic solution used for liposuction procedures. Tumescent fluid premixed on the day of the procedure is infiltrated into the subcutaneous tissue in order to anesthetize the procedure site locally. The amount of tumescent fluid used is calculated based upon the amount of adipose being harvested; it is limited by the maximum lidocaine dose based upon the patient's weight (4.5 mg/kg; 7 mg/kg when combined with epinephrine). The safe dosage for tumescent lidocaine was shown to be 35 mg/kg by Klein in 1990, and this has become standard of care for liposuction procedures [41].

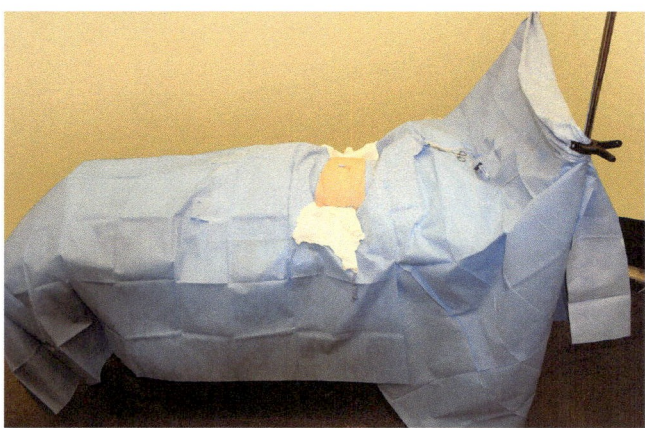

Fig. 5.2   Patient positioning and draping

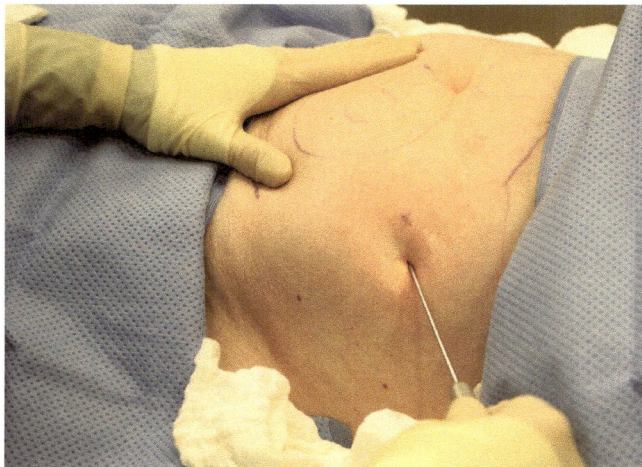

Fig. 5.3   Infiltration of tumescent fluid

- For harvesting small amounts of adipose tissue (i.e., 60–120 mL), a 0.1% tumescent solution may be utilized. Into a 1000-mL bag of 0.9% sodium chloride, introduce the following using the sterile technique:
  - 50 mL lidocaine 2%
  - 1 mL epinephrine 1:1000
  - 10 mL sodium bicarbonate 8.4%
- For harvesting large amounts of adipose tissue (>120 mL), a 0.05% tumescent solution can be utilized. Into a 1000-mL bag of 0.9% sodium chloride, introduce the following using the sterile technique:
  - 25 mL lidocaine 2%
  - 1 mL epinephrine 1:1000
  - 8 mL sodium bicarbonate 8.4%

The tumescent solution should be mixed on the same day as the procedure, and the epinephrine should be added immediately prior to use. The bag should be clearly identified and dated.

Infiltration of Tumescent Anesthetic Fluid

The patient is taken to the procedure suite and positioned supine for abdominal adipose harvesting or posterior or lateral decubitus for flank/hip adipose harvesting. Appropriate monitoring is placed. Sterile prep and drape is performed over the lipoaspiration site (Fig. 5.2). The port placement should be considered. If the abdomen is the harvest site, the port sites should be asymmetrically placed bilaterally at the anterior axillary line, at the level of the anterior iliac spine. Place a local anesthetic skin wheal at these sites. Using a #11 blade scalpel, make a 5-mm skin incision. The tumescent fluid is then infiltrated subcutaneously using the 14 gauge garden spray infiltrating cannula throughout the area of lipoaspiration. The tumescent fluid IV bag may be hung on an IV pole with pressure bag–assisted gravitational flow. As an alternative, the tumescent fluid may be delivered manually with a 60-cc syringe.

Tumescent fluid infiltration should be delivered slowly and evenly throughout the tissue. The irrigational cannula must remain parallel with the abdominal wall to avoid any unintentional transabdominal or peritoneal injury. Adequate infiltration is appreciated when the skin appears firm with turgor. There may be blanching, demarcating vasoconstriction associated with the epinephrine (Fig. 5.3).

Collection of Lipoaspirate

Once the tumescent fluid has been infiltrated, lipoaspiration is conducted with a 3 mm/25 cm Mercedes cannula attached to a 60-cc Toomey syringe, with a moderate amount of suction pressure. This is obtained by pulling back the syringe plunger after the cannula has been placed subcutaneously. Using a "snap lock" or "Johnnie snap" will support the plunger in this position while you work (Figs. 5.2 and 5.4).

The cannula is manipulated in a fanlike manner throughout the targeted tissue as the lipoaspirate is collected. The nondominant hand should be used as a guide to feel the tip of the cannula, ensuring that the cannula tip is not too superficial and does not extend beyond the intended treatment area (Fig. 5.4). Using this technique, deeper areas are aspirated first, followed by more shallow areas. Take care not to repeatedly course a specific area, as doing so may cause dimpling of the skin.

Continue to suction the aspirate until the syringe is full. Then place it upright in a syringe rack to allow the fat to rise above the supernatant fluid. Drain any supernatant fluid into a sterile stainless steel bowl and continue until the desired volume of fat has been harvested (Fig. 5.5).

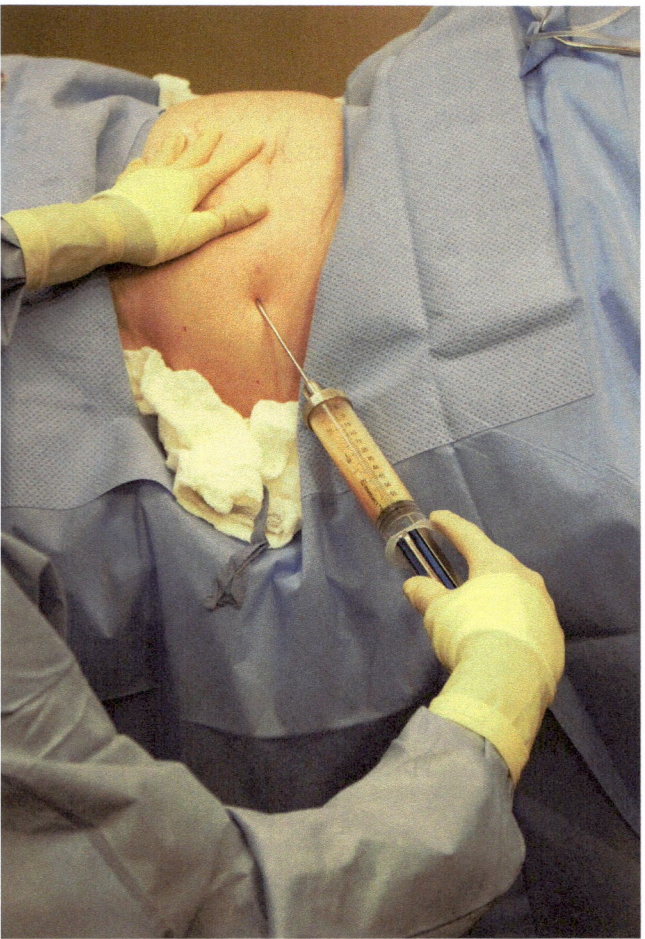

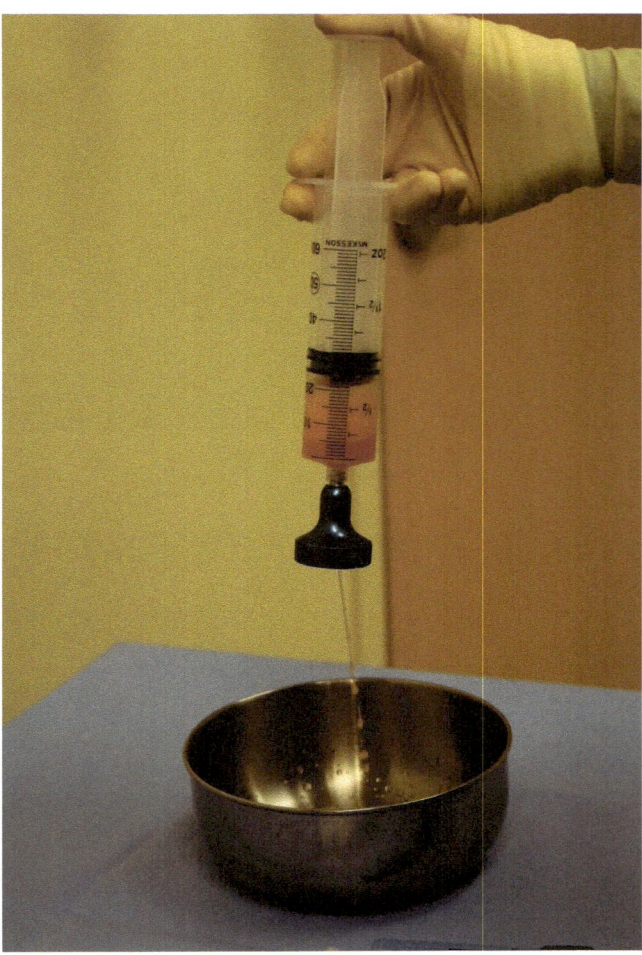

Fig. 5.4 Collection of lipoaspirate

Fig. 5.5 Separation of fat from supernatant fluid

Once collected, the harvested fat should be transferred to the processing area in a closed system. If using a syringe, cap the syringe for transport (Fig. 5.6).

Post-Procedure Care

- Gently express any excess tumescent fluid through the port sites.
- Close the port sites with steri-strips.
- Apply absorbent dressings over the port sites.
- Apply a compression garment. This can be a compression bodysuit for patients with thigh or hip lipoaspiration sites, but a simple abdominal binder will suffice for most patients. The patient should be instructed to wear the compression device continuously for the first 72 hours, and then daily for the next 3 to 4 days. Compression aids in hemostasis, improving post-procedure bruising and discomfort, and helps with post-procedure aesthetics.
- Transfer the patient to recovery and monitor vital signs.

Adipose Processing: Enzymatic and Mechanical

There are two generally accepted means for isolation of the SVF from adipose tissue: mechanical and enzymatic. Both methods are equally safe, but there are differences to be noted when choosing between them. Mechanical isolation is less costly and quicker to perform, but the end product will contain a higher concentration of blood mononuclear cells and fewer progenitor cells [42]. When contemplating using smaller quantities of adipose tissue for SVF extraction, the mechanical method may be considered. Enzymatic isolation, on the other hand, has been shown in studies to demonstrate a significantly greater efficiency in the separation process through a consistent and predictable digestion of the extracellular matrix (Table 5.3). For this reason, the authors advocate the enzymatic isolation method, as outlined here:

- Harvested adipose tissue should be processed in a clean setting. All specimens should be clearly marked with

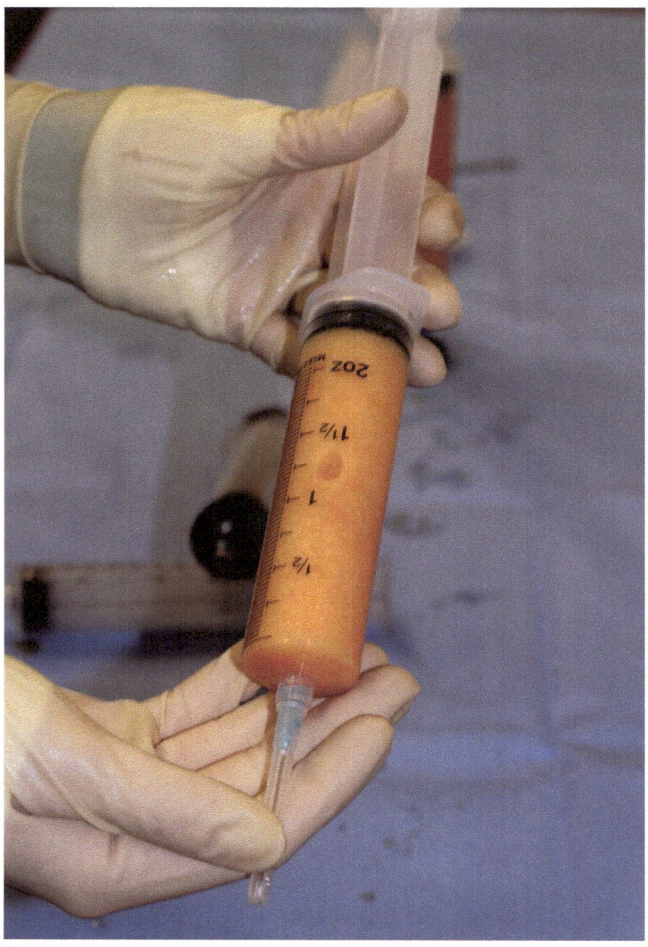

Fig. 5.6 Transport of the harvested adipose

Table 5.3 Comparison of mechanical vs. enzymatic isolation methods for extracting the SVF from harvested adipose tissue

	Mechanical isolation	Enzymatic isolation
Time to perform	15–30 minutes	2–3 hours
Cost	No added cost	$2–$5 per gram
Cell count (nucleated cells per cc of lipoaspirate)	1.0×10^4 to 2.4×10^5	1.0×10^5 to 1.3×10^6
Progenitor cell concentrations	Lower	Higher

patient identifiers. We recommend that all tissue handling outside of the sterile procedure suite occur under a Class 100 HEPA-filtered laminar flow biological cabinet using an aseptic technique (Fig. 5.7).

- Several companies offer proprietary formulas including protocol steps and unique digestive enzymes, which are packaged in disposable kits. The basic steps universally utilized to isolate adipose stem cells involve a cell wash and collagenase digestion, followed by centrifugal separation and filtration to isolate the single-cell SVF from the primary adipocytes.

The SVF is then resuspended in a carrier solution for final treatment. The carrier solutions include autologous platelet-rich plasma and preservative-free normal saline. Autologous platelet-rich plasma is the author's preferred carrier solution for musculoskeletal, intrathecal, or intravascular therapeutic applications. The total resuspension volume may range from 2 to 10 cc depending on the site of treatment.

Reintroduction of the Adipose Stem Cell Product

For musculoskeletal applications, the patient is transferred back to the clean procedure suite and positioned appropriately for the injection, with appropriate monitoring. The injection (whether intra-articular or soft tissue) should be done with direct visualization utilizing fluoroscopy or ultrasound. A 22-gauge or larger bore needle should be utilized to prevent shear force–induced cell rupture. Note that contrast material is cytotoxic and should not be used. Additionally, many local anesthetics are cytotoxic. One percent lidocaine is well tolerated.

For systemic applications, the resuspended stem cell solution will be injected into a peripheral vein through an IV catheter or needle. The injection should be done as an IV push slowly over 5–10 minutes. The patient's pulse, oxygen saturation, and blood pressure should be monitored before, during, and after the injection.

For intradiscal, facet joint, or sacroiliac joint injections, the patient is taken to the clean procedure suite and positioned prone with monitors applied. Fluoroscopy or ultrasound guidance should be used to confirm the accurate needle placement.

For intrathecal applications, the patient is taken to the clean procedure suite and positioned in a prone position with monitors applied. Sterile prep is carried out, and the patient is draped for a lumbar or cervical fluoroscopically guided intrathecal injection. A 22-gauge spinal or Tuohy needle may be used. Use minimal contrast, as larger amounts of contrast could be cytotoxic to the MSCs. Confirmation of the intrathecal location is demonstrated by CSF flow through the hub of the needle.

Most commercially available adipose harvesting and processing systems include a filtration step. All final products should be filtered through a 100-micron filter to prevent potential embolic events.

Post-Procedure Care and Potential Complications

- Neither NSAIDS nor steroids are recommended for 4 weeks after treatment [34–36].
- The anti-inflammatory properties of the treatment may result in positive effects within the first couple of weeks in

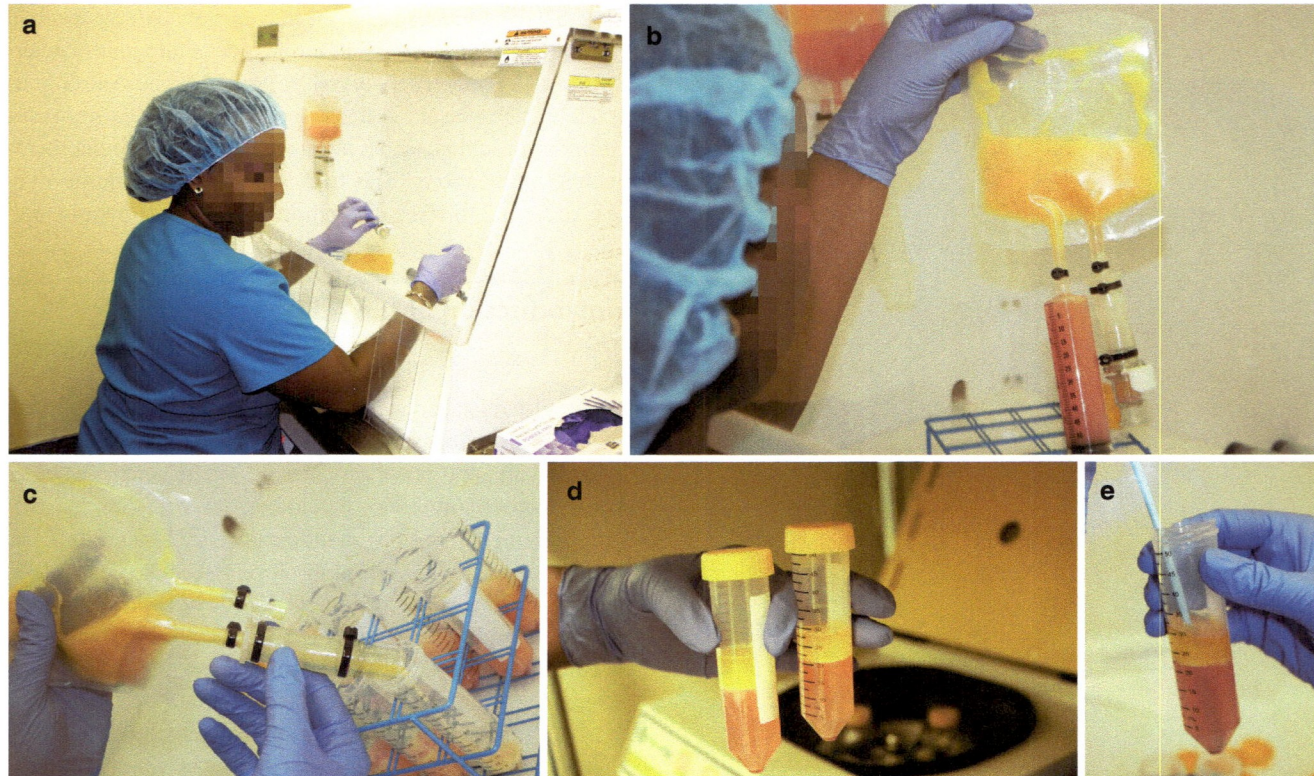

Fig. 5.7 (**a**) Processing of harvested adipose tissue under a laminar flow hood; (**b**) processing of harvested adipose tissue: washing the adipose; (**c**) processing of harvested adipose tissue: separating adipose into 50-ml conicals for centrifugation; (**d**) processing of harvested adipose tissue: after centrifugation the adipose has separated into the SVF at bottom and adipose at top; (**e**) processing of harvested adipose tissue: collecting the SVF pellet from the bottom of the conical

some cases, but the true therapeutic results may take 3 to 6 months to be realized.

- Normal light activity is recommended for the initial week after the procedure. A return to light exercise is recommended at 6 weeks after treatment.
- Many in the field believe that repeat treatments may be needed for many patients with severe local or systemic disease processes, though there is no research to support this idea. Autologous stem cell therapy may not offer a cure, but it certainly may offer a nonpharmacological treatment alternative.
- Although autologous adipose stem cell therapy is considered a safe same-day procedure, there are potential complications.
  - Infection due to poor sterile technique or contamination of the tissue product is possible. Fortunately, MSCs have demonstrated an antibiotic propensity to protect against this possibility.
  - Harvest site pain, soreness, or bruising may occur but is usually mild and can be treated with supportive therapy such as ice, acetaminophen, or analgesics. If symptoms persist, have the patient come in for a clinical evaluation.

- Injection site pain, soreness, or bruising is also usually mild and responsive to supportive care. If persistent, have the patient come in for a clinical evaluation.
- Skin dimpling or other cosmetic disfigurement is possible. It is always important to practice good techniques during lipoaspiration. Avoid excessive aspiration in any given area.
- Because you are using the patient's own tissue in this therapy, there is no risk of rejection, but if you are processing tissue samples from multiple patients on the same day, there is a risk of injecting the wrong sample into a patient. Always clearly label all specimens through the entire isolation process.

Clinical Pearls

- Standard universal precautions should be followed by all personnel with potential exposure to any patient tissue.
- There have been case reports of transient hypertension and tachycardia and/or symptoms of lightheadedness, flushing, or headache upon systemic intravascular

injections. Always monitor your patient and have oxygen and supportive medications available.

- Although it is highly unlikely that you will ever need it, have a crash cart and airway resuscitative equipment available. Many of your patients may have multiple comorbidities.
- Contrast, antibiotics, and many local anesthetics have been shown to be cytotoxic to mesenchymal stem cells. Use only 1–2% lidocaine, which has been shown not to be cytotoxic, and limit the amount of contrast injected when possible.
- Inject the final product with needles and catheters of 22-gauge or larger bore. This bore size does not disrupt the cell structure.
- Proprietary cell isolation techniques can provide safe, legal methods to consistently harvest approximately 50–100 million cells per 60–100 mL of adipose tissue, with reproducibility and validated analysis.
- Cell yield can be affected by several factors:
  - Surgical technique.
  - Location of fat.
  - Enzymatic digestion: Enzymatic digestion times and concentrations strongly modify the yield and viability of cells.
- Consider in advance the volume of injectate you will need for each area treated when performing your final resuspension. A small joint such as a finger or facet joint will only accommodate 1 to 2 mL of fluid, whereas a large joint such as a knee may require 6 to 8 mL or more.
- I frequently recommend an intravascular dose as well as an intra-articular dose in patients with osteoarthritis. Mesenchymal stem cells have demonstrated a unique homing ability. When introduced intravascularly, they make their way to the damaged tissue via cell signaling mechanisms.
- Although not endorsed by the FDA, there is a theoretical benefit for the use of intravascular stem cell treatments for autoimmune and inflammatory diseases, as well as for prevention and longevity.
- Do not advertise or make therapeutic claims with regard to this therapy. The FDA is hypervigilant regarding such public statements.

References

1. Murphy M, Moncivias K, Caplan AI. Mesenchymal stem cells: environmentally responsive for regenerative medicine. Exp Mol Med. 2013;45:e54. https://doi.org/10.1038/emm.2013.94.
2. Li H, Zimmerlin L, Marra KG, Donnenberg VS, Donnenberg AD, Rubin JP. Adipogenic potential of adipose stem cell subpopulations. Plast Reconstr Surg. 2011;128(3):663–72. Epub 2011/05/17.
3. Han J, Koh YJ, Moon HR, Ryoo HG, Cho CH, Kim I, Koh GY. Adipose tissue is an extramedullary reservoir for functional hematopoietic stem and progenitor cells. Blood. 2004;115(5):957–64.
4. Bourin P, et al. Stromal cells from the adipose tissue-derived stromal vascular fraction and culture expanded adipose tissue-derived stromal/stem cells: a joint statement of the International Federation for Adipose Therapeutics and Science (IFATS) and the International Society for Cellular Therapy (ISCT). Cytotherapy. 2013;15:641–8.
5. Traktuev DO, Merfeld-Clauss S, Li J, Kolonin M, Arap W, Pasgualini R, Johnstone BH, March KL. A population of CD-34 positive adipose stromal cells share pericyte and mesenchymal surface markers, reside in a periendothelial location, and stabilize endothelial networks. Circ Res. 2008;102(1):77–85.
6. Gimble, et al. Adipose derived stem cells for regenerative medicine. Circ Res. 2007;100:1249–60.
7. Caplan AI. Mesenchymal stem cells. J Orthop Res. 1991;9:641–50.
8. Caplan A, Mesenchymal DJE. Stem cells as trophic mediators. J Cell Biochem. 2006;98:1076–84.
9. Centeno CJ, Busse D, Kisiday J, Keohan C, Freeman M, Karli D. Increased knee cartilage volume in degenerative joint disease using percutaneously implanted, autologous mesenchymal stem cells. Pain Physician. 2008;11(3):343–53.
10. Emadedin M, Aghdami N, Taghiyar L, Fazeli R, Moghadasali R, Jahangir S, Farjad R, BaghabanEslaminejad M. Intra-articular injection of autologous mesenchymal stem cells in six patients with knee osteoarthritis. Arch Iranian Med. 2012;15(7):422–8.
11. Jo CH, Yg L, Shin WH, Kim H, Chai JW, Jeong EC, Kim EC, Kim JE, Shim H, Shin JS, Shin IS, Ra JC, Oh S, Yoon KS. Intra-articular injection of mesenchymal stem cells for the treatment of osteoarthritis of the knee: a proof-of-concept clinical trial. Stem Cells. 2014;32(5):1254–66.
12. Longo UG, Papapietro N, Petrillo S, Franceschetti E, Maffulli N, Denaro V. Mesenchymal stem cell for prevention and management of intervertebral disc degeneration. Stem Cell Int. 2012; https://doi.org/10.1155/2012/921053.
13. Sivakamansundari V, Lufkin T. Stemming the degeneration: IVD stem cells and regenerative therapy for degenerative disc disease. Adv Stem Cells. 2013; https://doi.org/10.5171/2013.724.
14. Hohaus C, Ganey TM, Minkus Y, Meisel HJ. Cell transplantation in lumbar spine disc degeneration disease. Eur Spine J. 2008;17(suppl 4):492.
15. Hiyama A, Mochida J, Iwashina T, Omi H, Watanabe T, Serigano K, Tamura F, Sakai D. Tranplantation of Mesenchymal stem cells in a canine disc degeneration model. J Orthop Res. 2008;26:589.
16. Henriksson HB, et al. Transplantation of human mesenchymal stem cells into intervertebral discs in a xenogeneic porcine model. Spine. 2009;34:141.
17. Sakai D, et al. Differentiation of mesenchymal stem cells transplanted to a rabbit degenerative disc model: potential and limitations for stem cell therapy in disc regeneration. Spine. 2005;30:2379.
18. Orozco L, et al. Intervertebral disc repair by autologous mesenchymal bone marrow cells: 2 pilot study. Transplantation. 2011;92(7):822–8.
19. Pettine KA, Murphy MB, Suzuki RK, Sand TT. Percutaneous injection of autologous bone marrow concentrate cells significantly reduces lumbar discogenic pain through 12 months. Stem Cells. 2015;33(1):146–56.
20. Mizuno H, Tobita M, Uysal AC. Concise review: adipose derived stem cells as a novel tool for future regenerative medicine. Stem Cells. 2012;30(5):804–10.
21. Folgiero V, Migliano E, Tedesco M, Iacovelli S, Bon G, Torre ML, Sacchi A, Marazzi M, Falcioni R. Purification and characterization of adipose derived stem cells from patients with lipoaspirate transplantation. Cell Transplant. 2010;19(10):1225–35.

22. Bailey AM, Kapur S, Katz AJ. Characterization of adipose derived stem cells: an update. Curr Stem Cell Res Ther. 2010;5(2):95–102.

23. Caplan AI, Correa D. The MSC: an injury drugstore. Cell Stem Cell. 2012;9:11–5.

24. FDA Draft Guidance (December, 2014). Minimal manipulation of human cells, tissues and cellular and tissue-based products.

25. FDA Draft Guidance for Industry (October, 2014). Same surgical procedure exemption under 21 CFR 1271.15(b): Questions and Answers Regarding the Scope of the Exception. http://www.fda.gov/BiologicsBloodVaccines/GuidanceComplianceRegulatoryInformation/Guidances/Tissue/ucm419911.htm.

26. FDA Draft Guidance for Industry and Drug Administration Staff (Nov 2017/Corrected Dec 2017). Regulatory considerations for human cells, tissue, and tissue based products minimal manipulation and homologous use. U.S. Department of Health and Human Services Food and Drug Administration Center for Biologics Evaluation and Research.

27. FDA Draft Guidance for Industry (Nov 2017). Evaluation of devices used with regenerative medicine advanced therapies. U.S. Department of Health and Human Services Food and Drug Administration Center for Biologics Evaluation and Research.

28. FDA Draft Guidance for Industry (Nov 2017). Expedited programs for regenerative medicine therapies for serious conditions. U.S. Department of Health and Human Services Food and Drug Administration Center for Biologics Evaluation and Research.

29. 21st Century Cures Act. US H.R. 6 114th congress, 2016. https://www.fda.gov/RegulatoryInformation/LawsEnforcedbyFDA/SignificantAmendmentstotheFDCAct/21stCenturyCuresAct/

30. Buschmann J, Gao S, Harter L, et al. Yield and proliferation rate of adipose-derived stromal cells as a function of age, body mass index and harvest site-increasing the yield by use of adherent and supernatant fractions. Cytotherapy. 2013;15(9):1098–105.

31. Jurgens WJ, Oedayrajsingh-Varma MJ, Helder MN, et al. Effect of tissue-harvest site on yield of stem cells derived from adipose tissue: implications for cell-based therapies. Cell Tissue Res. 2008;332:415–26.

32. Vilaboa SD, Navarro-Palou M, Llull R. Age influence on stromal vascular fraction cell yield obtained from human lipoaspirates. Cytotherapy. 2014;12:1092–7.

33. Hernigou P, Homma Y, Flouzat-Lachaniette CH, Poingnard A, Chevailler N, Rouard H. Cancer risk is not increased in patients treated for orthopedic diseases with autologous bone marrow cell concentrate. J Bone Joint Surg Am. 2013;95:2215–21.

34. Yshiya S, Dhawan A. Cartilage repair techniques in the knee: stem cell therapies. Curr Rev Musculoskelet Med. 2015:457–66.

35. Alaseem AM, Madiraju P, Aldebeyan SA, Noorwali, H, Antoniou J, Mwale F. Naproxen induces type X collagen expression in human bone marrow derived mesenchymal stem cells through up-regulation of 5-Lipoxygenase. Tissue Eng: Part A. 2015;21(1 and 2).

36. Fredriksson M, Li Y, Stalman A, Haldosen LA, Fellander-Tsai L. Doclofenac and triamcinolone acetonide impair tenocytic differentiation and promote adipocytic differentiation of mesenchymal stem cells. J Orthopaed Surg Res. 2013;8:30.

37. Halme DG, Kessler DA. FDA regulation of stem cell based therapies. N Engl J Med. 2006;355:1730–5.

38. Fournier PF. Reduction syringe lipo contouring. Dermatol Clin. 1990;8:539.

39. Hunsted JP. Tumescent and syringe liposculpture: a logical partnership. Aesth Plast Surg. 1995;19:321–33.

40. Housman TS, et al. The safety of liposuction: results of national survey. Dermatol Surg. 2002;08:971–8.

41. Klein J. Tumescent technique chronicles. Dermatol Surg. 1995;21:449–57.

42. Aronmitz JA, Lockhart RA, Hakakian CS. Mechanical verses enzymatic isolation of stromal vascular fraction cells from adipose tissue. Springerplus. 2005;4:713.

Intra-annular Fibrin Discseel®

6

Kevin Joseph Pauza, Maxim Moradian, and Gregory Lutz

Introduction

The Discseel® Procedure treats chronic low back and cervical discogenic pain, with or without extremity radiculopathy. The procedure is defined as a sequence of two steps: a diagnostic, nonprovocation annulogram, followed by intra-annular injection of nonautologous fibrin into every morphologically abnormal disc (torn disc) and into needle puncture holes created by the preceding diagnostic annulogram. Needle puncture holes are so imperceptibly small that some may believe this step of sealing needle puncture holes unnecessary, but highly favorable outcomes result by following this strict, pragmatic protocol. Prior attempts to regenerate discs by utilizing stem cells (mesenchymal precursor cells), PRP (platelet-rich plasma), or any biologic fail to reliably provide relief [1–10]. A prospective investigation pending publication demonstrates safety and statistically significant improvement of all 15 outcomes measured in 373 subjects at 24 months following the Discseel® Procedure [11].

Logic dictates that the efficacy of other intradiscal biologic treatments is compromised if those biologics leak from intervertebral discs, and conversely, efficacy should improve if that biologic remains within the disc, which is, after all, the intended site of action. Logic also dictates that any biologic failing to target pathology within the disc provides little to no value. In vivo investigations demonstrate that biologics, whether viscous or nonviscous, leak from degenerated, torn, or disrupted discs, even when encapsulated in hydrogels or other delivery systems. Annular tears obviously pose a problem and include all common disc pathologies. So any biologic not addressing tears serves little to no benefit. The need to seal tears easily explains the necessity of tissue adhesives such as fibrin. In a published in vivo investigation, all radiolabeled stem cells (MPCs) injected into intervertebral discs of rabbits leaked from those discs, negating any potential treatment efficacy [12]. More disconcerting, however, was the discovery that the radiolabeled cells, which leaked, migrated into adjacent bone and were found within new, exuberant osteophyte formations adjacent to the treated discs [13]. These osteophytes were readily evident through both radiographic and gross visual inspection. Seeing that stem cells were associated with new osteophyte (bone spur) formation changes everything. Because it is one thing to recognize that a specific stem cell treatment provides no patient benefit, yet it is an entirely different problem to recognize that injected stem cells meant to help may actually cause harm by noting new and potentially deleterious bone spurs in vertebral canals or foramen, which may already be compromised due to the nature of degenerated discs. The Discseel® Procedure does not cause injectate leakage because it utilizes FDA-approved fibrin, a soft, "disc-like consistency" bioadhesive made of two components meeting and coming together within the disc's annular layers. It is slightly stronger, yet equally soft, when compared with natural discs. Therefore, this biologic glue immediately seals, allowing no leakage. Even more important, fibrin instigates the new disc tissue growth [14, 15].

K. J. Pauza (✉)
Baylor Scott and White, Texas Spine and Joint Hospital, Tyler, TX, USA

Regenerative Sportscare Institute, Senior Physician, New York, NY, USA

M. Moradian
Interventional Physiatrist, iSCORE (Interventional Spine Care and Orthopedic Regenerative Experts, PC), Arcadia, CA, USA

G. Lutz
Hospital for Special Surgery, Regenerative Sports Care Institute, New York, NY, USA

Background

The Discseel Procedure uses an FDA-approved nonautologous fibrin as a tissue adhesive, which is FDA approved for multiple applications in the human body. Studies affirm fibrin's properties as a sealant, adhesive, anti-inflammatory, and chemotactic regenerative agent [14–16].

© Springer Nature Switzerland AG 2023
C. W Hunter et al. (eds.), *Regenerative Medicine*, https://doi.org/10.1007/978-3-030-75517-1_6

The efficacy realized that treating spines obviously depends, in a large part, on establishing an accurate pretreatment diagnosis. Many studies affirm that the ability to identify symptom etiology is not reliable or consistent when using MRIs, CTs, myelograms, or other common spine imaging modalities [16–25].

Investigations associated with the Discseel® Procedure strongly suggest that diagnosis made by annulograms, in conjunction with patient history, symptoms, and other findings, results in greater efficacy than any treatments relying on other diagnostic means, including those other treatments relying on provocation discography. Prior to these investigations relying on annnulograms, provocation discography was the gold standard and still may be optimal with those practitioners unskilled at performing nonprovocation annulograms, or without physicians available to perform nonprovocation annulograms. This is based on results realized when annulograms precede intra-annular fibrin injection.

Prior to the advent of the Discseel® Procedure, there were no treatments successfully treating annular tears. Knowing the annulus' morphology was not specifically necessary while employing those treatments injecting "something" into the nucleus pulposus. Injecting fibrin glue best addresses annular tears, and tears are best identified through annulograms because they evaluate every disc's annular morphology in the region of symptomology.

Discography & Annulograms

"Relatively-primitive" provocation discography tests were previously thought to improve diagnostic specificity when evaluating axial symptomology [25–67] in comparison to traditional imaging modalities, such as MRI or CT, and that is true, but now with the advent of annulograms, their sensitivity supersedes provocation discography. However, even though provocation discography was at one time considered the standard of care, it is important to note that no investigation directly correlates provocation discography results with a successful treatment outcome of any type. In comparison, the Discseel® Procedure directly associates positive annulograms with efficacy following intra-annular fibrin injection. Although discography was meant to establish symptom etiology, it lacks the ability to reliably evaluate annular tears residing within the outer portions of the 22–25 concentric layers of the discs' annular lamella [67].

Investigations by Caragee [49] suggested that provocation discography resulted in accelerated disc degeneration, disc herniation, loss of disc height, loss of MRI signal intensity, and the development of reactive end-plate changes when compared with matched controls. Those studies, however, are controversial and imply, but do not prove, a causal rela-

tionship between discography and disc degeneration. A potential benefit of annulograms proceeding sealing discs with fibrin is that the relationship between discography and premature disc degeneration may be mitigated for two reasons: first, because discography's iatrogenic disc damage does not necessarily apply to "low-pressure, nonprovocation" annulograms, with their comparatively low pressures and volumes. Second, fibrin immediately seals all intra-annular punctures created by performing the annulogram during the Discseel® Procedure. Together, these seem to mitigate concerns raised by Caragee's investigations [49], unless and until proven otherwise.

Discseel® Procedure vs. Intranuclear Fibrin

Injecting any mass into the disc's center nucleus pulposus is counterintuitive if one's treatment goal is to contain the nucleus by strengthening the outer annulus fibrosus. Interestingly ironic is that the other intra-discal treatments rely on injecting a mass into the center nucleus pulposus. That potentially and seemingly damaging technique is purportedly performed to regenerate nucleus pulposus cells. However, injecting any fluid or other mass-occupying substance into discs' center nucleus pulposus gel displaces that gel outward. This denotes Archimedes' property of displacement in fluid dynamics, and therefore the Discseel® Procedure intentionally avoids injecting fibrin into discs' centers. More preferably, fibrin injected intra-annularly creates a barrier maintaining existing nucleus pulposus within the discs' centers where it is needed. Both in vitro [70] and in vivo [11] investigations support the clinical utility of intra-discal, and more specifically, intra-annular, fibrin [11]. A randomized, blinded investigation comparing intradiscal, nonautologous fibrin vs. normal saline control disc injections demonstrated that statistically significant mechanical repair occurred along with improvement of the disc's biochemical milieu following the intradiscal fibrin treatment [68]. Disc pH increased to normal, and inflammatory constituents disappeared following intra-discal injection of fibrin but not following intra-discal injection of normal saline.

Early investigations of intradiscal fibrin demonstrated its value even before current refinements were incorporated into the Discseel® Procedure [11, 69–72]. Refinements incorporated to make the Discseel® Procedure include the following:

- Testing every disc in the region of symptoms with an annulogram and not relying on provocation discography
- Treating all discs in the region of symptoms that possess annular tears based on the annulogram and sealing needle puncture holes of every disc tested [11, 69–74]

This Discseel® Procedure results in statistically significant improvement in treated patients' pain, function, mental health, disability, and quality of life outcomes (Fig. 6.1) [11, 69–72]. Together, the combination of performing annulograms and injecting intra-annular fibrin at every location of disc annular tearing defines the Discseel® Procedure. Improved outcomes are realized with the aforementioned specific methodology [11].

Another benefit of annulograms over provocation discography is that they allow for the identification of otherwise radiographically imperceptible annular tears in otherwise seemingly normal discs, which might cause debilitating

symptoms. Detecting small peripheral annular tears may be necessary to adequately treat these tears. This process includes fastidiously performed annulograms done with contrast, allowing dynamic radiographic visualization of annular tears.

When annulograms are performed, contrast is visualized while flowing through annulus fibrosis defects (tears) while passing nociceptors, which may be sensitized by inflammatory constituents. This contrast flow mimics the flow of inflammatory mediators, or nucleus pulposus, which may often be "one in the same," traveling through annular tears. The annulogram allows dynamic fluoroscopic visualization

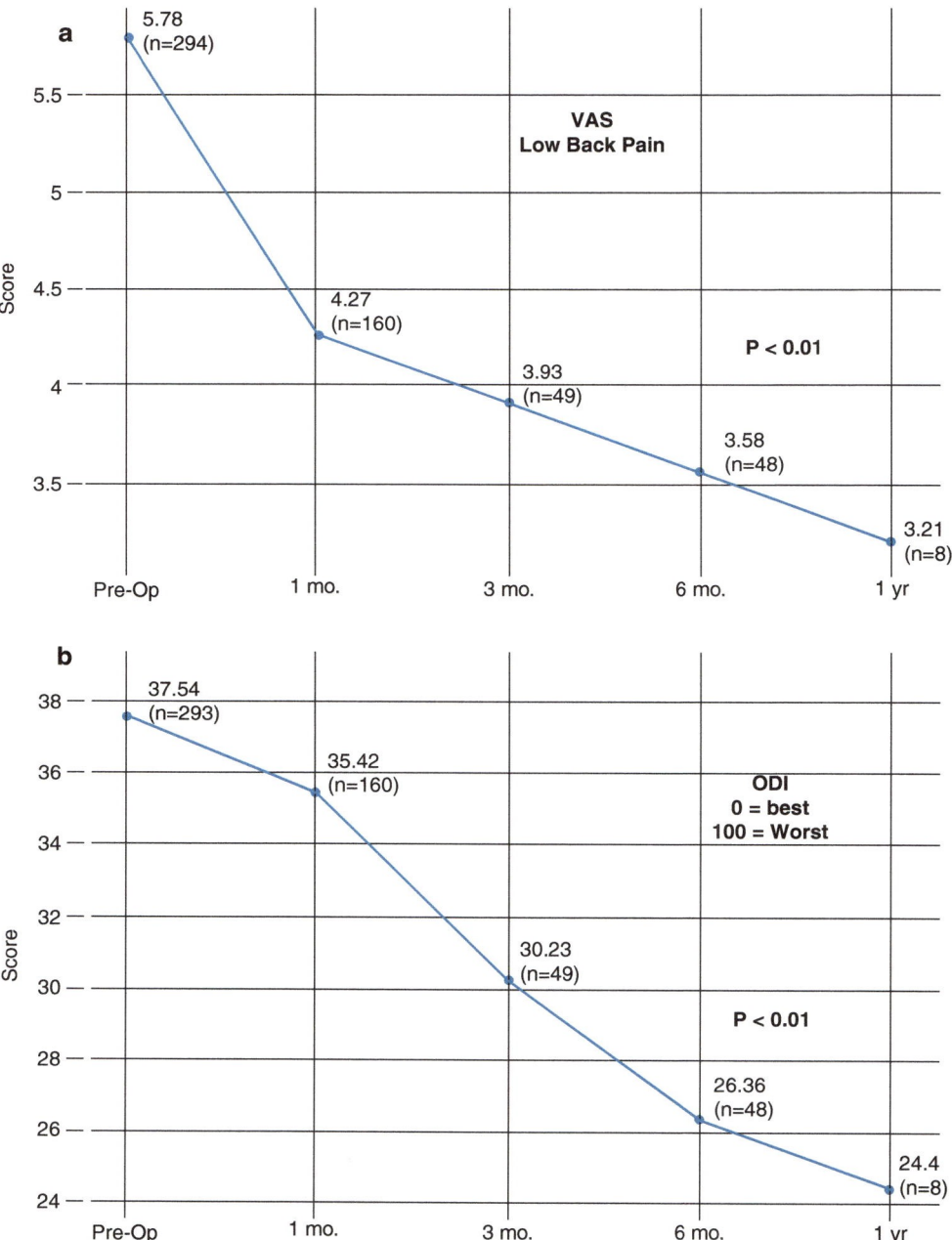

Fig. 6.1 (**a**) Visual Analog Scores (VAS) from a prospective investigation following lumbar Discseel® Procedure performed in subjects with chronic low back pain with and without lower extremity symptoms. (**b**) Oswestry Disability Index (ODI) from a prospective investigation following lumbar Discseel® Procedure performed in subjects with chronic low back pain with and without lower extremity symptoms

of contrast flow within the 22–25 annular layers and outward onto tissue and structures adjacent to torn discs. Post-annulogram computed tomography (CT) is unnecessary to identify annular tears because static and dynamic plain film fluoroscopy obtained during annulograms provides adequate and precise tear identification.

Microanatomy and Biochemistry

The greatest concentration of nociceptors resides within the posterior aspect of the annulus fibrosus. The second greatest concentration exists in the posterolateral annulus fibrosus, and the lowest concentration of nociceptors exists within the anterior annulus fibrosus [73–80].

There is an innate region of weakness of the intervertebral disc at its posterolateral portion, which unfortunately corresponds with the region having greatest density of nociceptors. Therefore, disc pain is predictably common when one recognizes that the dense concentration of nociceptors resides in the region most commonly associated with annulus fibrosis tears or failure. These annular tears are important because they allow the extravasation of nucleus pulposus within the disc in the region with the greatest concentration of pain-transmitting nociceptors. Interestingly, the body reacts to this leaked nucleus pulposus as a foreign sustenance, responding with inflammation and the autoimmune response, causing pain. Resultant inflammatory cytokines play a role in mediating discogenic low back pain and lumbar radiculopathy. Histochemical studies in human and animal intervertebral discs show that the nucleus pulposus in contact with torn annulus fibrosus instigates the formation of inflammatory and autoimmune constituents that includes the formation of peptides such as calcitonin gene-related peptide (CGRP), vasoactive intestinal peptide (VIP), and substance P, which heightens the sensitivity of the local nociceptive nerve fibers [81–85].

Indications

Patient Selection

Patients should have experienced chronic low back pain for 6 months or more in duration, with or without lower extremity symptoms (either radicular or nonradicular, somatic referred leg symptoms). Studies currently pending publication also demonstrate high safety and efficacy of the Discseel® Procedure in treating chronic cervical pain, with and without extremity radiculopathy. Patient screening should consider limiting treatment of patients with abnormal psychosocial factors.

Annulograms, like any test, should only be performed when their results directly affect the direction of the treatment algorithm. Annulograms, or the Discseel Procedure, may not typically be indicated for patients with acute or subacute LBP and/or leg pain because those symptoms will likely resolve spontaneously within six months. In comparison, those with chronic LBP are likely to experience pain persisting of at least five years duration following symptom onset. However, each situation deserves individual evaluation and consideration. Might an incapacitating acute injury merit treatment consideration, as routinely occurs with surgical discectomy?

In patients presenting with axial back pain with lumbar extension and relief with flexion, one should consider ruling out zygapophyseal joint etiology by performing diagnostic medial branch blocks or intra-articular zygapophyseal (Z-joint) injections with local anesthetic to determine whether the pain originates from the Z-joints. Extremity weakness including foot drop or reflex loss is not an exclusion criterion for the Discseel Procedure.

Blood thinners are relative contraindications, and patients should consult with the prescribing physician's office as well as the proceduralists prior to the procedure. Pre- and post-treatment blood coagulation lab values should be checked, and the risks and benefits discussed with each patient.

Prior Nonsurgical Treatment

The Discseel® Procedure pretreatment instructions do not require that patients undergo an epidural injection of corticosteroid because this injection is not site specific and will not corroborate the existence and exact location of a painful intervertebral disc and, although exceedingly low, the risks associated with epidural injections of corticosteroids are not zero [86].

Prior Spine Surgical Treatment

Prior spine surgery does not exclude patients, including laminectomy, laminotomy, discectomy, foraminotomy, anterior or poster interbody fusion with screws and rods, interspinous decompression devices, implanted spinal cord stimulators, or pumps. None excludes patients from undergoing the Discseel® Procedure. There are also some surgeries that increase stress and strain on the adjacent intervertebral discs, which may increase the need for the Discseel® Procedure [87–98].

Anatomic Exclusion Criteria

Exclusion criteria include severe vertebral canal or intervertebral foraminal stenosis, severe compression of the cauda equina or spinal cord at the level targeted for treatment, or upper motor neuron signs, or cauda equina syndrome. These apply only to the location or level being treated. Displacement of the dura, spinal nerves, or spinal cord is not an exclusion criterion. Preferable, but not required, is nonoccupied vertebral canal space. A nonsequestered, extruded disc herniation causing moderate to severe stenosis is a relative contraindication, not an absolute contraindication. Motor weakness, including foot drop, is not a Discseel® Procedure absolute contraindication. At this juncture, no scientific basis exists to mandate that segmental instability be considered a contraindication. It is possible that instability may be caused by ligamentous laxity (anterior longitudinal ligament and posterior longitudinal ligament), which may improve with disc tissue growth following the Discseel® Procedure.

Patients with severe spinal stenosis may elect surgical decompression prior to undergoing the Discseel® Procedure. Because surgical discectomies cause iatrogenic annular disruption, a post-surgical Discseel® Procedure may be a treatment option to address discs that have had a portion of the annulus surgically removed during discectomy. The various discectomy techniques all increase annulus fibrosus disruption resulting in the potential for recurrent disc herniation or accelerated degeneration at the level of the discectomy [97, 98].

In addition to the indication for the performance of the Discseel® Procedure after discectomy, it may be performed after interbody fusion on the adjacent segments to treat the annular tearing caused increased aberrant forces known to increase the likelihood of adjacent segment accelerated degeneration [87–96]. Annular fibrin can also be injected into a disc that has been "fused" by a spanning pedicle screw and rod construct because the fused disc may still leak inflammatory mediators and produce pain. The intervertebral disc can be tested with an annulogram and treated with fibrin, if torn and considered possibly symptomatic, even following fusion.

Absolute Contraindications

- The patient is unable or unwilling to consent to the procedure.
- The patient has evidence of untreated localized infection at the procedural site.
- The patient is pregnant.

Relative Contraindications

- The patient has a known allergy to any of the substances used for the injections.
- There is the presence of an active bleeding diathesis.
- The patient is currently on anticoagulants.
- There is a known systemic infection present.

- The patient has undergone a dental procedure one week prior to treatment or six weeks following the procedure.

Equipment

The Discseel® Procedure is performed in a room suitable for fluoroscopically guided aseptic procedures. A sterile surgical suite is not necessary. The room must be equipped with fluoroscopy (C-arm or two-plane image intensifier) and an x-ray compatible table. The room should also be equipped with minimally invasive cardiopulmonary monitoring equipment including an ECG, pulse oximeter, and blood pressure cuff. Supplemental O_2 should also be available.

Sterile skin preparation may be an iodine-based solution (e.g., povidone-iodine), or an alcohol-based antiseptic (e.g., chlorhexidine gluconate 0.5% in 70% alcohol), or a combination of the two.

The placement of sterile drapes or sheets achieves a sterile field and an aseptic region for the injection site. A two-needle technique may be used to test and treat the discs but is not a necessary component of the procedure. The single needle technique employs an 18-gauge 150-mm (6.0 inch) curved tip Tuohy needle. A 90-mm (3.5 inch) small-gauge (23–27 gauge) needle is utilized for anesthetizing the skin. Sterile gloves and standard radiation protection are mandatory for the proceduralists, and a sterile gown and mask are optional, based on physician preference.

A 10–20-ml syringe can be used to inject contrast, or the contrast may be dispensed into smaller volume syringes for easier contrast injection and attached to minimal volume, short extension tubing for precise annulogram injection control.

Intravenous cannula access is recommended for administering sedation and emergent medication or fluids for cardiovascular emergencies.

Staff

At least one assistant in the procedure room prepares the contrast and heats and prepares the prothrombin and fibrinogen in the procedure room. Care should be taken to dispense each into its correctly labeled respective syringe.

A second assistant or radiologic technologist may operate the fluoroscope.

Pre-Procedure

History and Physical Examination
An appropriate pre-procedure history is obtained, and a physical examination is performed to establish the patient's

suitability for the diagnostic annulogram and Discseel® Procedure. The patients should avoid any dental procedures one week prior to their treatment and six weeks following their Discseel® Procedure.

Informed Consent

The patient should be informed of all aspects of the procedure, the risks and benefits of the procedure, and suitable alternative options. The patient also consents to understand the definition of off-label use of an FDA-approved medication. Off-label use of FDA-approved medications is commonplace and a well-accepted practice in the fields of medicine and surgery. For consideration, epidural injections of corticosteroids are an off-label use of corticosteroids because corticosteroids have never been specifically approved by the FDA for their epidural placement.

Premedication

The patient should be given standard NPO orders with the time specifications in accordance with the institution if IV sedation is offered.

Antibiotics Antibiotic prophylaxis against discitis, including cephazolin 1 g, clindamycin 900 mg, or ciprofloxacin 400 mg, may be administered intravenously within 15–60 minutes before the procedure, but this is not a mandate and was not performed in investigations referenced. This is also per consensus guidelines previously adopted by recognized spine procedural societies.

If the patient is allergic to penicillin, an alternative is clindamycin IV 600—900 mg [51].
Anesthetic and Sedation   Local anesthetic (lidocaine 1.0—2.0%) is used for skin infiltration, and conscious sedation using the sedative medications of choice (i.e., midazolam/ fentanyl/ketamine) may be used for patient comfort.

Allergy

If the patient has a known allergy to contrast medium, they may be pretreated with H1 and H2 blocking medications and corticosteroids prior to the procedure. Another option is to utilize gadolinium in those patients with a known contrast allergy.

A patient's ability to tolerate the anxiety associated with any invasive test, especially disc access procedures, is variable. Because of this, careful administration of sedatives and opioid medications is essential, allowing the patient to remain awake enough to convey sensations and locations of pain. This will allow for more accurate localization of the patient's pain generator and can help to avoid injury to the adjacent spinal nerves during the procedure. A full provocative discography procedure with a complete record of the patient's response is unnecessary.

Technique

Positioning The patient is positioned prone on a procedure table.

Sterility The skin of the lumbar region and upper gluteal region is prepared for an aseptic procedure as discussed above. The operator and any personnel within the vicinity of the patient and fibrin mixing station should wear clean attire (scrubs suits, for example). Surgical caps and masks are recommended, but not mandatory.

If the operator performs the sterile skin preparation, they should don fresh gloves after the skin has been prepared and prior to inserting any needles.

To help minimize the chance of bacterial contamination to the needle and/or the disc, the needle should not be unnecessarily exposed to the atmosphere. Upon being withdrawn from its scabbard, it should be inserted without significant delay. Although not scientifically validated, to further minimize the likelihood of disc contamination from skin bacterial flora, one may puncture the skin with a sterile, larger gauge needle (14–18 g) at the skin entry point and direct a smaller needle that will be used to puncture the disc through the outer needle [98].

Selecting Disc Levels to Test and Treat

If technically feasible, test every disc in the region of the patient's symptoms. Typically three to four intervertebral discs will be tested. Higher segmental levels should be tested if they correlate with recognized radicular or somatic pain patterns as expressed via the patient's history.

Target Identification

An anteroposterior (AP) image of the lumbar spine is obtained, and the target disc is identified.

The disc may be approached from either side, but it is recommended to optimize the ease of access by approaching the disc from the side that is less encumbered by osteophytes and less narrowed due to scoliosis or fusion hardware. If necessary, the needle may be rotated gently in alternating directions to penetrate bone overgrowth that is hindering disc access. Testing and treating incompletely fused discs is appropriate because annular disc tissue may be intact, even in segments with implanted cages.

Once the disc approach side has been selected, the fluoroscope is tilted caudal or cranial (tilted to the feet or head, respectively) so that the X-ray beam passes parallel to the ring apophysis or the end plates of the vertebrae to maximize the radiographic height of the targeted disc.

The fluoroscope beam is rotated obliquely, allowing visualization of the target disc from the ipsilateral posterolateral oblique aspect. During traditional discography, the beam is rotated obliquely until the anterior aspect of the superior articular process (SAP) overlying the target disc lies parallel to the axial division of the anterior two-thirds and posterior one-third of the target disc. That view will allow the needle to be advanced parallel to the x-ray beam, directing the needle tip intentionally to the center of the nucleus pulposus as it passes across the anterior surface of the superior articular process. It is important to note that this discogram approach view differs from the Discseel® Procedure view described later insofar as the discogram view intentionally directs the needle tip trajectory to the center of the nucleus pulposus and the Discseel procedure view directs the needle to the posterior annulus. In comparison to a discogram, the Discseel™ Procedure's intent is to target the annulus fibrosus at its most posterior aspect of the intervertebral disc.

There are two differences in the Discseel® Procedure needle trajectory that allows for the needle tip to reach the posterior annulus fibrosis instead of the center nucleus pulposus.

One difference in the technique is to continue rotating the image intensifier to a more oblique position until the SAP is seen over the disc at the 1/3 posterior and 2/3 anterior junction (instead of at the 2/3 posterior and 1/3 anterior junction).

The second difference in technique is when the physician employs the common discogram imaging method previously described (SAP at the 1/3 posterior-2/3 anterior junction). Then, instead of marking and penetrating the skin at the typical location overlying the anterior portion of the SAP and inserting the needle directly along the pathway of the x-ray beam into the center of the nucleus pulposus, the physician marks and penetrates the skin over the radiographic anterior aspect of the disc. The needle then enters the skin at this slightly more anterior and lateral position and is directed posteromedially instead of parallel to the x-ray beam. Ideally, the needle will come in contact with the anterior portion of the SAP, so the appropriate needle tip depth can be determined without changing the position of the image intensifier. This will also ensure that the needle tip is in position to avoid injuring the descending spinal nerve that will be descending from the level above just anterior to the pathway of the needle. This needle trajectory will direct the needle tip to the desired target, the posterior aspect of the disc's annulus fibrosis.

Another precaution that can be taken to minimize the likelihood of the needle injuring a spinal nerve is to avoid injecting local anesthetic into the region of the disc or spinal nerves until needle advancement and maneuvering is complete.

The target point for puncture of the annulus fibrosus lies at the superoinferior midline of the target disc, just lateral to the lateral margin of the superior articular process.

At the L5-S1 level, the iliac crest may overlie the disc target in the posterolateral oblique view. Care should be taken to obtain a view such that the target point lies between the superior articular process of S1 medially and the iliac crest laterally.

If the iliac crest continues to overly the L5-S1 target, a skin puncture point could be placed over the location on the iliac crest closest to the target location. By the time the needle reaches the depth of the iliac crest, it may have traversed medially enough to bypass the bony crest itself. Alternatively, the needle entry point can be located over the point on the iliac crest closest to the target area, and the needle can be directed medially around the crest and back to the disc entry target.

In the posterolateral oblique view, a puncture point on the skin is selected, and a skin wheal is raised with local anesthesia (lidocaine 1% or 2%) using a 23–27-gauge skin needle.

Technique Needle Placement

New needles should be used for each disc injected to minimize infection likelihood. The skin overlying the target disc is marked and anesthetized with local anesthetic, but needle tract or disc region should not be anesthetized to avoid anesthetizing the descending spinal nerves. When performing the procedure, any new onset of leg pain reported by the patient should be noted to avoid injuring the descending ventral ramus with the needle or an errant injection. If the patient complains of paraesthesia or radicular pain, needle insertion should cease immediately and the needle is withdrawn slightly and redirected to avoid a nearby descending spinal nerve.

Additionally, because there are a limited number of nociceptors that exist along the needle trajectory from the skin to the disc, appropriate needle advancement should not cause undue patient discomfort. When the needle encounters the annulus, a firm and rubbery resistance is typically felt and the needle's progress is monitored by alternating between AP and lateral fluoroscopic projections.

When a single needle technique is used, a 22-gauge, 200-mm Tuohy needle is advanced carefully to the target within the annulus fibrosis at the posterior portion of the disc. Anteroposterior and lateral views confirm the correct needle tip position before a trace amount of nonionic radiopaque contrast is injected into the annulus fibrosis under dynamic fluoroscopic visualization (Fig. 6.2). Careful observation

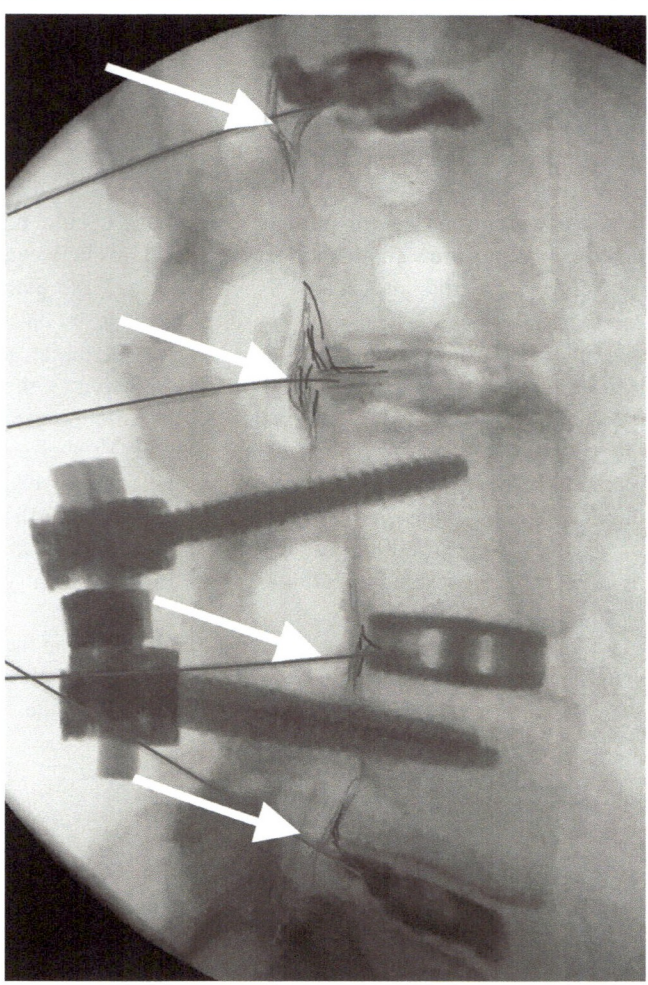

Fig. 6.2 Lateral fluoroscopic view showing the needles inserted into the lumbar intervertebral discs (white arrows) with contrast injected into the discs (white arrowheads)

allows visualization and documentation of contrast flow patterns within and outside the disc.

Following the annulogram, the connection tubing is disconnected and the apparatus combining the prothrombin and fibrinogen is connected to the needle hub. Using gentle pressure, the plunger is advanced. When injected, the prothrombin combines with the fibrinogen and aprotinin, producing fibrin as it is injected into the disc. Fibrin is slowly injected with gentle pressurization of the syringe. If resistance prevents the flow of fibrin, the needle's metal stylet is replaced to occupy the entire central portion of the needle. To assure that the stylet occupies the needle's entirety, rotate the stylet's notch until it rests entirely within the hub's groove. This assures that there is no fibrin or tissue obstruction or needle "kink" at the needle's most distal aspect that would impede flow. The physician cannot assume that the disc is entirely filled or sealed maximally due to the perception of complete resistance without first reinserting the stylet to

ensure that fibrin has not activated and obstructed the needle's lumen. Total fibrin volume injected per disc is highly variable and dependent on the disc's morphology but typically ranges from approximately 0.5 cc to 6.0 cc. Observation of the contrast departing the annular tears when injecting the fibrin indicates the presence of fibrin within those tears. Although the disc height often increases during fibrin injection, this is not necessarily the goal of the Discseel® Procedure. During a typical procedure in a patient with chronic low back pain, usually 3–4 discs are tested and treated.

Additionally, more cephalad discs may be tested and treated if the patient's symptoms and imaging studies indicate the need for this. If a morphologically normal disc is identified based on annulogram, the needle puncture site is sealed with fibrin to try to protect and preserve the integrity of that disc.

Post-Procedural Care

After needles are removed and the skin puncture points sterilely dressed, the patient is taken to recovery where cardiopulmonary monitoring is performed for approximately 30 min or longer if indicated. If the patient is stable at this point, they are discharged to a caregiver or a family member. Short-acting analgesics may be prescribed at this time. Patients are instructed not to drive on the day of their procedure and to expect increased discomfort for a few days to a few weeks. Intra-discal biologics that occupy the annular fissures instead of leaking can maintain the intervertebral disc height and can cause increased patient symptoms from the injected disc(s). These symptoms can be referred to as a full feeling, deep pressure, prolonged pain, or other post-procedure symptoms not commonly reported with other intradiscal procedures. Therefore prescriptions for pain medication to address this period of increased discomfort may be considered.

Conclusion

Being safe and efficacious, the Discseel® Procedure treats spine pathology with relative ease and high success. Because the Discseel® Procedure returns discs to their normal states, both mechanically and biochemically, it eliminates common and costly problems associated with all spine surgery, even minimally invasive spine surgery.

The Discseel® Procedure's ability to spare patients from needing additional spine surgery due to adjacent disc degeneration and "the domino effect" saves society pain, suffering, and billions of healthcare dollars.

The field of spine specialists is observing "decision-makers" evolve from old and contrarian spine treatments, including spine surgery, to new treatments restoring spines to their pre-injury state, benefiting patients and the healthcare system.

References

1. Richardson S, Kalamegam G, Pushparaj P, et al. Mesenchymal stem cells in regenerative medicine: focus on articular cartilage and intervertebral disc regeneration. Methods. 2015.
2. Chen WH, Lo WC, Lee JJ, et al. Tissue-engineered intervertebral disc and chondrogenesis using human nucleus pulposus regulated through TGF-beta1 in platelet-rich plasma. J Cell Physiol. 2006;209(3):744–54.
3. Gullung GB, Woodall JW, Tucci MA, et al. James J platelet-rich plasma effects on degenerative disc disease: analysis of histology and imaging in an animal model. Evid Based Spine Care J. 2011;2(4):13–8.
4. Obata S, Akeda K, Imanishi T, et al. Effect of autologous platelet-rich plasma-releasate on intervertebral disc degeneration in the rabbit annular puncture model: a preclinical study. Arthritis Res Ther. 2012;14(6).
5. Kim HJ, Yeom JS, Koh YG, et al. Anti-inflammatory effect of platelet-rich plasma on nucleus pulposus cells with response of TNF alpha and IL-1. J Orthop Res. 2014;32:551–6.
6. Sawamura K, Ikeda T, Nagae M, et al. Characterization of in vivo effects of platelet-rich plasma and biodegradable gelatin hydrogel microspheres on degenerated intervertebral discs. Tissue Eng Part A. 2009;15:3719–27.
7. Gullung G, Woodall J, Tucci M, et al. Platelet-rich plasma effects on degenerative disc disease: analysis of histology and imaging in an animal model. Evid Based Spine Care J. 2011;2:13–8.
8. Tuakli-Wosornu YA, Terry A, Boachie-Adjei K, Harrison JR, et al. Lumbar intradiscal platelet-rich plasma (PRP) injections: a prospective randomized double-blind study. PM and R. 2016;8(1):1–10.
9. Harrison JR, Herzog RJ, Lutz GE. Increased Nuclear T2 signal intensity following Intradiscal platelet rich plasma: a case report. Submitted to PM&R.
10. Levi D, Horn S, Tyszko S, et al. Intradiscal platelet-rich plasma injection for chronic discogenic low back pain: preliminary results from a prospective trial. Pain Med. 2015;0:1–13.
11. Amer. Acad. Ortho. Med. Annual Meeting. Evaluating the Safety and Efficacy of Fibrin to Treat Multi-level Chronic Discogenic Low Back Pain and associated Radiculopathy. 2020. Pending Publication, Spine Jour. 2021.
12. Vadala G, Sowa G, Hubert M, et al. Mesenchymal stem cells injection in degenerated intervertebral disc: cell leakage may induce osteophyte formation. J Tissue Eng Regen Med. 2012;6(5):348–55.
13. Li YY, Diao HJ, Chik TK, et al. Delivering mesenchymal stem cells in collagen microsphere carriers to rabbit degenerative disc: reduced risk of osteophyte formation. Tissue Eng. Part A. 2014;20(9–10).
14. Ahmed TA, Dare EV, Hincke M. Fibrin: a versatile scaffold for tissue engineering applications. Tissue Eng Part B Rev. 2008;14(2):199.
15. Colombini A, Ceriani C, Banfi G, et al. Fibrin in intervertebral disc tissue engineering. Tissue Eng Part B Rev. 2014;20(6):713–21.
16. Schek RM, Michalek AJ, Iatridis JC. Genipin-crosslinked fibrin hydrogels as a potential adhesive to augment intervertebral disc annulus repair. Eur Cell Mater. 2011;5:275.
17. Jansen M, Brant-Zawadzki M, Timo K, et al. Magnetic-resonance imaging of the spine in people without back pain. N Engl J Med. 1994;331:69–73.
18. Grubb SA, Lipscomb HJ, Guilford WB. The relative value of lumbar roentgenograms, metrizamide myelography, and discography in the assessment of patients with chronic low-back-syndrome. Spine. 1987;12:282–6.
19. Gaensler E. Nondegenerative diseases of the spine. In: Brant W, editor. Fundamentals of diagnostic radiology. Baltimore: Williams and Wilkins; 1999. p. 233–380.
20. Jarvik JG, Deyo RA. Diagnostic evaluation of low Back pain with emphasis on imaging. Ann Intern Med. 2002;137:586–97.
21. YoshidaH FA, TamaiK, et al. Diagnosis of symptomatic disc by magnetic resonance imaging: T2-weighted and gadolinium-DTPA-enhanced T1 weighted magnetic resonance imaging. J Spinal Disord Tech. 2002;15:193–8.
22. Toyone T, Takahashi K, Kitahara H, et al. Vertebral bone-marrow changes in degenerative lumbar disc disease: an MRI study of 74 patients with low back pain. J Bone Joint Surg Br. 1994;76:757–64.
23. Weishaupt D, Zanetti M, Hodler J, et al. Painful lumbar disk derangement: relevance of endplate abnormalities at MR imaging. Radiology. 2001;218:420–7.
24. Sandhu HS, Sanchez-Caso LP, Parvataneni HK, et al. Association between findings of provocative discography and vertebral endplate signal changes as seen on MRI. J Spinal Disord. 2000;13:438–43.
25. Braithwaite I, White J, Saifuddin A, et al. Vertebral end-plate (Modic) changes on lumbar spine MRI: correlation with pain reproduction at discography. Eur Spine J. 1998;7:363–8.
26. Vanharanta H, Sachs BL, Spivey MA, et al. The relationship of pain provocation to lumbar disc deterioration as seen by CT/discography. Spine. 1987;12:295–8.
27. Crock HV. Internal disc disruption: a challenge to disc prolapse. Spine. 1986;11:650–3.
28. Moneta GB, Videman T, Kaivanto K, et al. Reported pain during lumbar discography as a function of annular ruptures and disc degeneration. A re-analysis of 833 Discograms. Spine. 1994;1994(17):1968–74.
29. Collis JS, Gardner WJ. Lumbar discography—an analysis of 1,000 cases. J Neurosurg, 1962;19:452–461. 30. Erlacher PR: Nucleography. J Bone Joint Surg. 1952;34B:204–10.
30. Hsien-Wen S, Yu-Min C, Hsing-T'Ang K, et al. Lumbar discography: an experimental and clinical study. Chin Med J. 1964;83:521–30.
31. Nordlander S, Salen EF, Unander-Scharin L. Discography in low back pain and sciatica. Acta Orthop Scandinav. 1958;28:90–102.
32. Walk L. Clinical significance of discography. Acta Radiol. 1956;46:36–7.
33. Lindblom K. Diagnostic disc puncture of intervertebral disks in sciatica. Acta Orthop Scandinav. 1948;17:231–9.
34. Lindblom K. Technique and results in myelography and disc puncture. Acta Radiol. 1950;34:321–30.
35. Hirsch C. An attempt to diagnose the level of a disc lesion clinically by disc puncture. Acta Orthop Scandinav. 1949;18:132–40.
36. Lindblom K. Technique and results of diagnostic disc puncture and injection (discography) in the lumbar region. Acta Orthop Scandinav. 1951;20:315–26.
37. Lindblom K. Discography of dissecting transosseous ruptures of intervertebral discs in the lumbar region. Acta Radiologica. 1951. 36:13–16; Friedman J, Goldner MZ. Discography in evaluation of lumbar disc. Radiology. 1955;65:653–62.
38. Feinberg SB. The place of discography in radiology in 2,320 cases. AJR. 1964;92:1275–81.
39. Butt WP. Lumbar discography. J Can Assoc Radiol. 1963;14:172–81.
40. Gardner WJ, Wise RE, Hughes CR, et al. X-ray visualization of the intervertebral disk with a consideration of the morbidity of disk puncture. Arch Surg. 1952;64:355–64; Braithwaite I, White J, Saifuddin A, et al. Vertebral end-plate (Modic) changes on lumbar spine MRI: correlation with pain reproduction at discography. Eur Spine J. 1998;7:363–8.

41. Keck C. Discography: technique and interpretation. AMA Arch Surg. 1960;80:580–6.
42. Wilson DH, MacCarty WC. Discography: its role in the diagnosis of lumbar disc protrusion. J Neurosurg.
43. Holt EP. The question of lumbar diskography. J Bone Int Surg. 1968;50A:720–5.
44. Pauza KJ, Howell S, Dreyfuss, et al. NASS OUTSTANDING PAPER. Prospective double blind, placebo controlled study evaluating the efficacy of intradiscal electrothermal therapy for the treatment of chronic discogenic low back pain. Spine J. 2004;4(1):27–35.
45. Bogduk N, editor. Practice Guidelines for spinal diagnostic and treatment procedures. ISIS; 2004.
46. Carragee EJ, Tanner CM, Yang B, et al. False-positive findings on lumbar discography. Reliability of subjective concordance assessment during provocative disc injection. Spine. 1999;24(23):2542–7.
47. Carragee EJ, Tanner CM, Khurana S, et al. The rates of false-positive lumbar discography in select patients without low back symptoms. Spine. 2000;25:1373–81.
48. Carragee EJ, Alamin TF, Miller J, et al. Provocative discography in volunteer subjects with mild persistent low back pain. Spine J. 2002;2:25–34.
49. Carragee EJ, Chen Y, Tanner CM, et al. Can discography cause long-term back symptoms in previously asymptomatic subjects? Spine. 2000;25:1803–8.
50. Carragee EJ, Chen Y, Tanner CM, et al. Provocative discography in patients after limited lumbar discectomy: a controlled, randomized study of pain response in symptomatic and asymptomatic subjects. Spine. 2000;25:3065–71.
51. Derby R, Kim BJ, Lee SH, et al. Comparison of discographic findings in asymptomatic subject discs and the negative discs of chronic LBP patients: can discography distinguish asymptomatic discs among morphologically abnormal discs? Spine J. 2005;5:389–94.
52. McCutcheon ME. CT scanning of lumbar discography: a useful diagnostic adjunct. Spine. 1986;11:257–9.
53. Sachs BL, Vanharanta H, Spivey MA, Guyer RD, Videman T, Rashbaum RF, Johnson RG, Hochschuler SH, Mooney V. Dallas discogram description: a new classification of CT/discography in lowback disorders. Spine. 1987;12:287–94.
54. Aprill C, Bogduk N. High intensity zone: a diagnostic sign of painful lumbar disc on magnetic resonance imaging. Brit J Radiol. 1992;65:361–9.
55. Vanharanta H, Sachs BL, Spivey MA, et al. The relationship of pain provocation to lumbar disc deterioration as seen by CT/discography. Spine. 1987;12:295298.
56. Moneta GB, Videman T, Kaivanto K, et al. Reported pain during lumbar discography as a function of annular ruptures and disc degeneration. A re-analysis of 833 discograms. Spine. 1994;17:1968–74.
57. Polk HC, Christmas AB. Prophylactic antibiotics in surgery and surgical wound infections. Am Surg. 2000;80:105–11.
58. International Spine Intervention Society (ISIS), Bogduk B. Proposed discography standards. ISIS Newsletter, Vol. 2(1). Daly City, California: International Spinal Injection Society; 1994. p. 10–3.
59. Bogduk N, Aprill C, Derby R. Discography. In: White AH, editor. Spine care, Vol. 1. St Louis: Mosby; 1995. p. 219–38.
60. Bogduk N, Chr PK. International spine intervention society practice guidelines for spinal diagnostic and treatment procedures. Oxford Blackwell Science; 2003.
61. Fraser RD, Osti AL, Vernon-Roberts B. Discitis after discography. J Bone Joint Surg. 1987;69B:26–35.
62. Alamin T. The functional anesthetic discogram: comparison of the results of a novel technique to that of provocative discography in a group of patients with chronic low back pain. International Society for the Study of the Lumbar Spine. Abstracts. June, 2006:52–53.
63. Ohtori S, Kinoshita T, Yamashita M, et al. Results of surgery for discogenic low back pain: a randomized study using discography versus discoblock for diagnosis. Spine. 2009;34(13):1345–8.
64. Ren J, Zhang Y, Chee, et al. Effects of local anesthetic and nonionic contrast agents on bovine intervertebral disc cells cultured in alginate. Abstract. SAS. 2010.
65. Derby R, Lee SH, Kim BJ, et al. Pressure-controlled lumbar discography volunteers without low back symptoms. Pain Med. 2005;6:213–21.
66. Derby R, Howard MW, Grant JM, et al. The ability of pressure-controlled discography to predict surgical outcomes. Spine. 1999;24:346–71.
67. DePalma M, Lee J, Peterson L, et al. Are outer annular fissures stimulated during discography the source of discogenic low-back pain? An analysis of analgesic discography data. Pain Med. 2009;10:3.
68. Buser Z, Kuelling F, Liu J, et al. Biological and biomechanical effects of fibrin injection into porcine intervertebral discs. Spine. 2011;36(18).
69. Yin W, Pauza K, Olan W, et al. Symptomatic lumbar internal disc disruption: results of a prospective multicenter pilot study with 24 month follow-up. Pain Med. 2014;15(1).
70. Pauza K, Yin W, Olan W, et al. Biostat Biologix intradiscal fibrin sealant used for the treatment of chronic low back pain caused by lumbar internal disc disruption: results of a 12 month, prospective multi-center pilot study. Surgical Arthrodesis Society. Annuall Meeting. 2010.
71. Pauza K, Wright C, Fairbourn A. Treatment of annular tears and "leaky disc syndrome". Techn Reg Anesth Pain Manag. (1–2):45–9.
72. Pauza K. Intradiscal biologics. In: Gebhart GF, Schmidt RF, editors. Encyclopedia of pain. Philadelpia; 2013.
73. García-Cosamalón J, del Valle ME, Calavia MG, et al. Intervertebral disc, sensory nerves and neurotrophins: who is who in discogenic pain? J Anat. 2010;217(1):1–15.
74. Yoshizawa H, O'Brien JP, Thomas-Smith W, et al. The neuropathology of intervertebral discs removed for low-back pain. J Pathol. 1980;132:95–104.
75. Korkala O, Gronblad M, Liesi P, et al. Immunohistochemical demonstration of nociceptors in the ligamentous structures of the lumbar spine. Spine. 1985;10:156–7.
76. Bogduk N, Tynan W, Wilson S. The nerve supply to the human lumbar intervertebral discs. J Anat. 1981;132:39–56.
77. Bogduk N. The innervation of the lumbar spine. Spine. 1983;8:286–93.
78. Groen GJ, Baljet B, Drukker J. Nerves and nerve plexuses of the human vertebral column. Am J Anat. 1990;188:282–96.
79. Malinsky J. The ontogenetic development of nerve terminations in the intervertebral discs of man. Acta Anat. 1959;38:96–113.
80. Konttinen YT, Gronblad M, Antti-Poika I, et al. Neuroimmunohistochemical analysis of peridiscal nociceptive neural elements. Spine. 1990 15:383–6.
81. Peng B, Wu W, Li Z, Guo J, Wang X. Chemical radiculitis. Pain. 2007;127:11–6.
82. Olmarker K, Rydevik B, Nordborg C. Autologous nucleus pulposus induces neurophysiologic and histologic changes in porcine cauda equina nerve roots. Spine 1993;18(11):1425–32.
83. Saal JS. The role of inflammation in lumbar pain. Spine. 1995;20(16):1821–7.
84. Bobechko W, Hirsch C. Autoimmune response to nucleus pulposus in rabbit. J Bone Joint Surg. 1965;47B:3; Marshall L, Trethewie E, Curtain C. Chemical radiculitis. A clinical, physiological, and immunological study. Clin Ortho Relat Res, 1977. 11.129.

85. Ohtori S, Inoue G, Ito T. Tumor necrosis factor-immunoreactive cells and PGP 9.5-immunoreactive nerve fibers in vertebral endplates of patients with discogenic low.

86. Chou R, Hashimoto R, Friedly J, et al. Epidural corticosteroid injections for radiculopathy and spinal stenosis: A systematic review and meta-analysis. Ann Intern Med. 2015;163(5):373–81.

87. Ekman P, Möller H, Shalabi A, et al. A prospective randomised study on the long-term effect of lumbar fusion on adjacent disc degeneration. Eur Spine J. 2009;18(8):1175–86.

88. Harrop J, Youssef J, Maltenfort M, et al. Lumbar adjacent segment degeneration and disease after arthrodesis and total disc arthroplasty. Spine. 2008;33(15):1701–7.

89. Chen C, Cheng C, Liu C, et al. Stress analysis of the disc adjacent to interbody fusion in lumbar spine. Med Eng Phys. 2001;23(7):483–91.

90. Pezowicz C, Schechtman H, Robertson P, et al. Mechanisms of anular failure resulting from excessive intradiscal pressure: a microstructural-micromechanical investigation. Spine. 2006;31(25):2891–903.

91. Throckmorton T, Hilibrand A, Mencio G, et al. The impact of adjacent level disc degeneration on health status outcomes following lumbar fusion. Spine. 2003;28(22):2546–50.

92. Adams M, Freeman B, Morrison H, et al. Mechanical initiation of intervertebral disc degeneration. Spine. 2000;25(13):1625–36.

93. Zhang C, Berven S, Fortin M, et al. Adjacent segment degeneration versus disease after lumbar spine fusion for degenerative pathology: a systematic review with meta-analysis of the literature. Clin Spine Surg. 2016;29(1):21–9.

94. Lee C. Accelerated degeneration of the segment adjacent to the lumbar fusion. Spine. 1988;13(3):375–7.

95. Lee C. Accelerated degeneration of the segment adjacent to a lumbar fusion. Spine. 1988;3(13):375–7.

96. Sheng C, Cheng-Kung C-L, et al. Stress analysis of the disc adjacent to interbody fusion in lumbar spine. Med Eng Phy. 2001;23(7):483–91.

97. O'Connell G, Malhotra N, Vresilovic E, Elliott D. The effect of discectomy and the dependence on degeneration of human intervertebral disc strain in axial compression. Spine. 2011;72(2):181–204.

98. Schroeder J, Dettori J, Brodt E, et al. Disc degeneration after disc herniation: are we accelerating the process? Evidence Based Spine Care J. 2012;3(4):33–40.

Allograft Therapies in Regenerative Medicine

7

Tory L. McJunkin, Arianna Cook, and Edward L. Swing

Introduction

Overview

Regenerative medicine is a novel field based on the use of growth factors and cellular products, such as stem cells, to supplement native cells in restoring damaged tissue function [1]. These treatments are intended to slow the degeneration or reverse damage of tissues in various locations throughout the body. Regenerative therapies include autografts, which utilize the patient's own stem cells and growth factors, as well as allografts, which rely on isolating stem cells and growth factors from the tissues of donors [2]. Allograft therapies can often be less invasive and less expensive than similar autograft therapies and can offer a potentially fruitful approach for regenerative therapies [3].

Allografts are a source of mesenchymal stem cells (MSCs) [2]. Mesenchymal stem cells are found in many adult tissues and can be used for their regenerative properties. The cells are multipotent and can differentiate into various cell types, including osteoblasts, adipocytes, myocytes, and chondrocytes. They also have a high proliferative capacity, are able to self-renew, and possess mesodermal differentiation potential.

Although bone marrow has been one of the main autograft sources of MSCs, harvesting bone marrow is difficult and the differentiation potential and maximal life span of MSCs from bone marrow decline with increasing age of the patient from which they are taken [3]. Therefore, alternative allograft sources are being studied for use as regenerative therapies. In particular, amniotic tissue, umbilical cord blood (UCB), and Wharton's jelly have been identified as potential tissue sources for allograft therapies and will be the focus of this chapter.

Guidelines for Allograft Therapy

A number of considerations should guide the use of regenerative allograft therapies in clinical practice. These should include medical, ethical, and legal considerations.

Medical considerations should include the following:

- *Appropriate source of allograft tissues.* Allograft tissues should be obtained from a tissue bank that closely follows the Food and Drug Administration (FDA) requirements, such as screening donors for communicable diseases. Although the tissues used in regenerative allografts tend to have low immunogenicity, some applications may warrant further consideration of matching the donor with the patient.
- *Patient selection.* Allograft therapies have been identified as having possible value for treating a number of patient indications (see Table 7.1 for a list of potential indications). Only patients with an appropriate medical diagnosis should be treated with allograft therapies. Contraindications to injection therapies in general (e.g., coagulation disorders) and biologic therapies (e.g., infections) in particular should be ruled out.
- *Treatment delivery.* The delivery of allograft treatments is most commonly performed through injection. Depending on the anatomy of the targeted body part, imaging techniques may be appropriate to ensure the appropriate delivery of the allograft. This may include ultrasound or fluoroscopic guidance. See Fig. 7.1 for examples of operative imaging of intra-articular knee, intra-articular facet, and intradiscal injections of amniotic tissue solutions.

T. L. McJunkin (✉)
Arizona Pain Specialists, Scottsdale, AZ, USA
e-mail: DrMcJunkin@ArizonaPain.com

A. Cook
University of North Carolina, Department of Anesthesiology, Chapel Hill, NC, USA

E. L. Swing
Phoenix Children's, Department of Medical Education, Phoenix, AZ, USA

© Springer Nature Switzerland AG 2023
C. W Hunter et al. (eds.), *Regenerative Medicine*, https://doi.org/10.1007/978-3-030-75517-1_7

Table 7.1 Potential indications for regenerative allograft therapies identified by in vitro studies, animal models, and clinical research

Indication
Osteoarthritis
Rheumatoid arthritis
Degenerative disc disease
Skin burns
Skin transplantation
Wound healing
Tendinopathy
Ligament injury
Corneal lesions
Conjunctival lesions
Spinal cord injury
Nerve injury
Neurodegeneration
Myelodysplasia
Hematological malignancies
Ischemic stroke
Multiple sclerosis
Buerger's disease

Ethical/legal considerations should include the following:

- *Informed consent.* Patients should be informed of the risks associated with the particular type of allograft procedure. It is also critically important that patients understand the experimental status of allograft therapies. Claims regarding the efficacy should be carefully characterized to avoid overstating either the degree of benefits or the certainty of those benefits.
- *Tissue graft* vs. *drug characteristics.* Tissue grafts are exempt from FDA approval, but the FDA has clarified that under certain conditions, tissue therapies (whether allografts or autograft) may be considered a drug and thus be subject to the requirement of FDA approval. Specifically, the tissues must be minimally manipulated, meaning no in vitro expansion of cells or exposure to an agent that modifies the properties of these cells (anticoagulants are permitted) and have homologous use (meaning the repair, reconstruction, replacement, or supplementation of a recipient's cells or tissues with a tissue that performs the same basic function or functions in the recipient as in the donor).

Amniotic Tissue

The unique makeup of human amniotic tissues, specifically amniotic membrane and amniotic fluid, has fostered the use of these cells for tissue growth and healing. The amniotic membrane in vivo contains the developing fetus and is filled with amniotic fluid [4]. Amniotic fluid contains the various

cells derived from different parts of the developing fetus and surrounding tissue, including the amniotic membrane. The content of the amniotic fluid changes over the course of gestation. The amniotic fluid and membranes serve as protection for the embryo and fetus and provide support and nutrients during embryogenesis and fetal development.

Amniotic tissues have several properties that make them well suited for allograft treatments. Human amniotic epithelial cells are broadly multipotent and can differentiate into mesodermal and nonmesodermal lineages [4]. Along with multipotency, the cells have anti-inflammatory effects as well as a low degree of immunogenicity [5]. The cells do not produce acute rejection when placed into another patient and can improve allograft tolerance [5, 6]. Amniotic MSCs act as an immunosuppressant and have been shown to decrease graft-versus-host disease in an animal model [6].

The antimicrobial properties of amniotic tissues also make these cells suitable for implantation. Amnion has shown to have antibacterial properties in in vitro studies [7]. When exposed to human interleukin (IL)-1β, amniotic epithelial cells produce elevated levels of numerous natural antimicrobial proteins, including human beta-defensins and secretory leukocyte protease inhibitor [8].

Potential Indications

Amniotic fluid-derived cells have the possible capability to repair injured tissues. Though injections of amniotic fluid are experimental, multiple studies have examined the potential of these injections to treat multiple conditions, including connective tissue and degenerative changes of joints and other tissues [9, 10]. Besides stimulating new tissue growth, amniotic fluid can decrease joint pain in conditions such as osteoarthritis. Other possible uses include assistance with wound healing, knee arthritis, Achilles tendinopathy, and neuropathy.

The clinical use of amniotic membrane began over a century ago, with the first reports describing its ability to treat skin burns and wounds [11]. Further studies have been conducted to advance the use of amniotic membrane in surgery [12]. Amniotic membrane is being used in areas such as corneal and conjunctival surface reconstruction, open skin ulcers and traumatic wound treatment, and skin transplantation [13–15]. Like amniotic fluid, amniotic membrane has the potential to treat tendinopathies and neuropathies.

Experimental Clinical Application

Amniotic stem cells (ASCs) are being tested as new experimental treatments for a variety of different conditions.

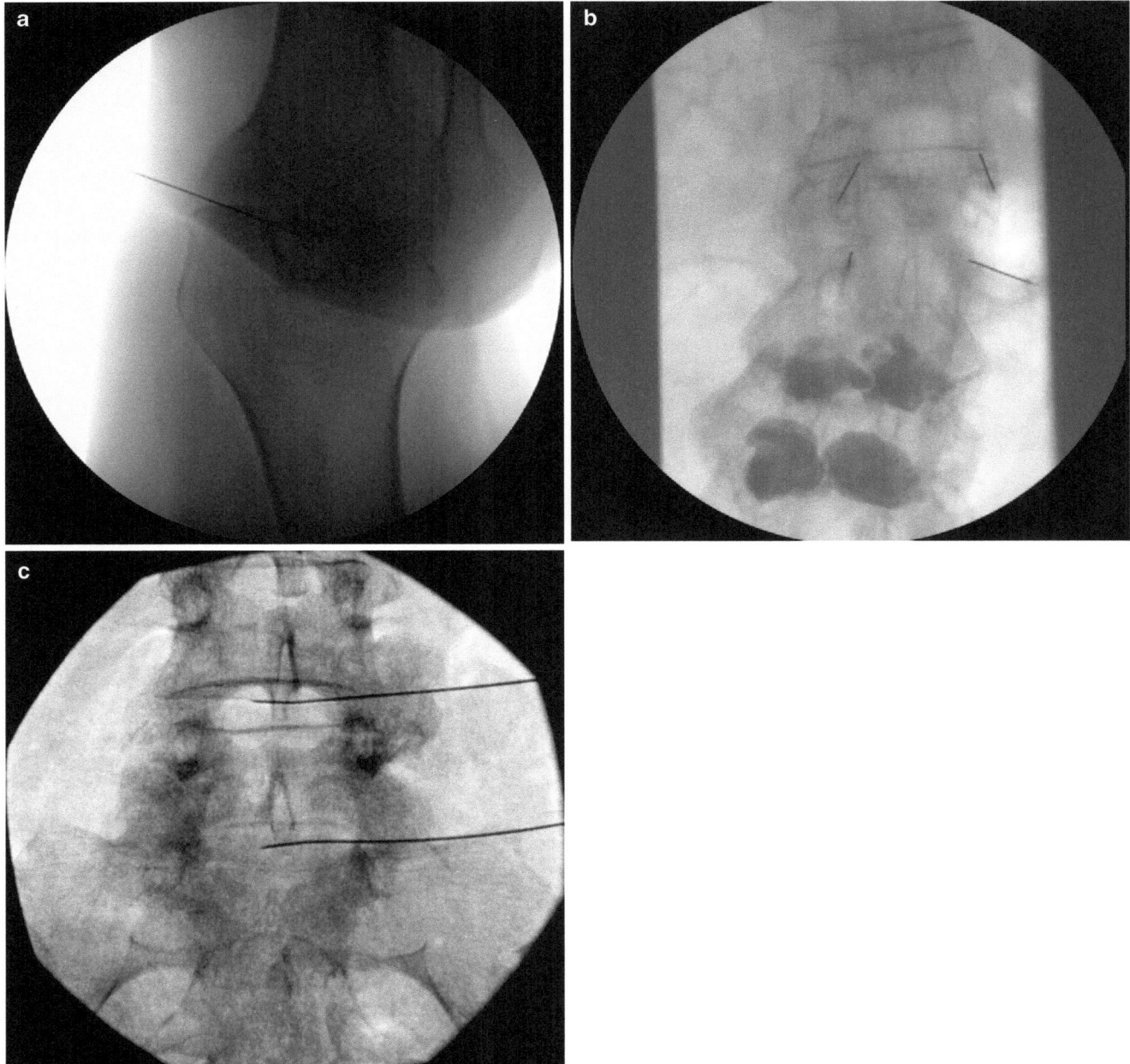

Fig. 7.1 Operative imaging of amniotic tissue injections for (**a**) knee osteoarthritis, (**b**) facet joint osteoarthritis, and (**c**) L4/5 and L5/S1 degenerative disc disease

Provided ASCs are not modified (e.g., in vitro expansion of the cells) or combined with products (other than anticoagulants), these cells are not considered a drug by the FDA and are thus not FDA-regulated [9]. However, in the absence of large-scale randomized clinical trials to test their efficacy, these treatments are not definitively proven to be effective for each of their specific uses and are typically not covered by insurance. Clinical evidence for amniotic tissue treatments for chronic pain conditions is limited to in vitro studies, in vivo animal models, case studies, and nonrandomized studies [16–21]. These studies, however, have shown the potential value of amniotic tissue to enhance healing and repair of the body. For example, Vines et al. tested intra-articular injection of amniotic suspension allografts in six patients with knee osteoarthritis [20]. Patients reported improvements in pain and activities of daily living at up to one year. In a rare randomized trial of amniotic tissues, Bhattacharya randomized 52 patients with knee osteoarthritis to receive intra-articular injections of either human amniotic fluid or a corticosteroid (triamcinolone). Those patients receiving human amniotic fluid injections showed greater improvements in pain (VAS) and walking distance at three- and six-month follow-up assessments [21].

There are multiple studies showing that amniotic tissues can help in healing skin wounds [22, 23]. Amniotic membrane naturally acts as a basement membrane that aids epithelial cell migration, stimulates epithelial differentiation, and produces growth factors that promote epithelization [22]. An in vitro study has shown that amniotic fluid stimulates reepithelialization in human skin wounds [23]. The reparative properties of the amniotic fluid were thought to be due to high concentrations of hyaluronic acid. When hyaluronic acid degraded, reepithelialization was diminished. Small, nonrandomized studies in humans have shown that treatment with amniotic membrane and fluid can promote leg ulcer healing [24, 25]. A case series of five patients with chronic nonhealing wounds showed that amniotic membrane accelerated wound healing and helped patients return to normal function [25].

Another field commonly using amniotic-derived tissues is ophthalmology. The promotion of reepithelialization, as well as the inhibition of angiogenesis and inflammation, makes amnion valuable for ocular use [9]. Amniotic fluid and membrane have been used to form a reconstructive scaffold after the removal of ocular surface lesions [22]. Amniotic membrane transplantation has been used with a variety of ocular surface procedures such as conjunctival and corneal surface reconstruction. Three eyes of three patients with up to 3 mm corneal perforations were treated with hyperdried cross-linked amniotic membrane [26]. Each patient successfully healed the corneal perforations within 28 days, and amniotic membrane was found to be an effective substrate for corneal repair [26]. Amniotic-derived cells have also been used ex vivo as a promoter of limbal and conjunctival stem cell regeneration [9, 22].

There is also clinical evidence that amniotic tissues can help with tendon and ligament repair. In an in vitro ovine model, amniotic membrane-derived cells had tenogenic differentiation potential and were capable of developing into three-dimensional tendon-like structures [27]. Amniotic tissue-derived injections were also shown to be effective at increasing the success of Achilles tendon repair in in vivo ovine and rat studies [28, 29]. Additionally, injection of MSCs from amnion was found to be effective in aiding in repair of equine tendon and ligament injuries [30]. Further studies have shown that the amniotic membrane products can be used for cartilage restoration in rat and sheep models [10].

Amniotic cells also have the potential to treat degenerative joint diseases. Injection of amniotic tissues could be a possible nonoperative treatment for osteoarthritis [10]. Amniotic fluid is thought to act as a homolog to synovial fluid, providing cushion and lubrication in joints [31]. In a rat model, an intra-articular injection of amniotic membrane decreased the number and severity of chondral lesions and reduced cartilage degeneration [32]. An interim analysis of a clinical registry study of 170 patients with knee osteoarthritis showed that amniotic fluid decreased pain and stiffness in knees that had been injected with it [31].

Evidence also suggests that amniotic tissue can help in treating spinal cord and nerve injuries. Amniotic-derived stem cells have the potential to release various neurotrophic factors that promote nerve regeneration [33]. Intramuscular injection of amniotic fluid-derived cells in a rat muscle denervation model produced preserved anterior horn cells and increased nerve myelination. An in vitro study found that amniotic-fluid-derived stem cells injected into acellular nerve allografts facilitated nerve regeneration following injury and demonstrated a more robust motor function recovery [34].

Concerns and Contraindications

Treatments with amniotic tissues such as amniotic fluid injections are still in the early phases of use. Although considered low-risk, various standard treatment options should be considered, particularly low-risk conservative treatments, before proceeding to amniotic tissue-based treatments [35]. Additional considerations exist when selecting appropriate patients and the mechanism of delivery of these therapies. Contraindications to amniotic tissue treatments are related to contraindications for any localized injection or treatment. These include systemic illness, coagulopathy, and site infection [36]. Currently, no allergic reactions to amniotic tissues have been reported.

Preoperative and Operative Considerations

It is important to understand the patient's condition and correctly diagnose the source of the problem. More conservative treatments should be tried first including but not limited to massage, chiropractic care, stretching, physical therapy, and nonexperimental injections [35]. When considering any injection, it is important to understand the anatomy and volume of the area of treatment [37]. For example, a knee may need a higher volume (2–4 mL) than a temporomandibular joint (0.5–1 mL). Standard sterile precautions should be followed as for any injection. Larger-gauge needles are typically used to direct amniotic tissue into the target area. The amniotic tissue can also be delivered along with perforating or needling the host tissue to stimulate tissue repair, promote circulation, and support healing. To ensure precise placement, many providers will use ultrasound or fluoroscopy to visualize and confirm the area of treatment. Imaging choice depends on provider imaging expertise, imaging preference, patient pathology, and the anatomy of the target area.

Umbilical Cord Blood

Since the first successful cord blood transplantation in 1988, umbilical cord blood has become a recognized stem cell source [38]. The umbilical cord is a cord of vascular channels between the developing fetus and the placenta [39]. The cord contains two umbilical arteries and one vein. The vein supplies the fetus with oxygenated, nutrient-rich blood, whereas the umbilical arteries carry away deoxygenated, nutrient-depleted blood back to the placenta. The amnion membrane and Wharton's jelly, a gelatinous matrix covering the umbilical cord, provide protection for the three vessels.

The umbilical cord can provide stem cells from the blood within the umbilical vessels as well as the walls of the vessels and from Wharton's jelly [39]. Cord blood can be harvested at birth using a sterile collection kit containing an anticoagulant. Cord blood samples can be collected in utero or ex utero. The process is noninvasive and has minimal to no danger to the mother or infant [38]. Once collected, cord blood units are transferred to designated laboratories where the stem cells are extracted [39].

Umbilical cord blood contains a variety of stem and progenitor cells, including MSCs and endothelial progenitor cells [38]. The MSCs of cord blood play a supportive role in hematopoiesis and can differentiate into a range of tissue lineages. The MSCs from umbilical cord blood also have immunoregulatory properties and are able to suppress T cell proliferation as well as T cell-mediated allogeneic responses.

Potential Indications

There are many potential clinical uses for umbilical cord blood MSCs in regenerative medicine. Cord blood has been used in hematopoietic stem cell transplantation [3]. Many studies also support its use for articular cartilage damage [40–42]. This includes the use of UCB MSC's for rheumatoid arthritis, osteoarthritis, and joint trauma.

Cord blood can also serve as a rich source of endothelial progenitor cells [38]. The cells have similar properties to embryonic angioblasts. The cells could be used to repair and regenerate vascular endothelium, such as in damage that occurs with ischemic disease. Furthermore, cord blood stem cells may be able to treat neurodegenerative disease and neuronal injuries.

Experimental Clinical Application

As early as 1939, umbilical cord blood has been used for its regenerative properties [43]. The first umbilical cord transplantation in a human was performed over 256 years ago in a child with Fanconi anemia. Today over 60,000 umbilical cord blood units have been stored for transplantation worldwide. Cord blood is composed of early stem cell progenitors and can give rise to multiple cell lineages [38]. Furthermore, lymphocyte populations of cord blood are composed of immunologically naïve cells (lacking class II human leukocyte antigen) that are typically not recognized by the host immune system. A meta-analysis of umbilical cord transplantation versus unrelated bone marrow transplantation showed a lower incidence of chronic graft-versus-host disease and no difference in acute graft-versus-host disease [44]. Umbilical cord blood transfusion has also been used to increase disease survival for myelodysplasia and other hematologic disorders [45]. A study analyzing 180 patients who underwent unrelated cord blood transplantation for myelodysplasia or secondary acute myelodysplastic leukemia revealed a 2-year disease-free survival and overall survival of 30% and 34%, respectively.

Studies have supported the use of cord blood in treating cartilage and joint damage. An in vitro study has shown that MSCs from ovine umbilical cord blood can be used to create cartilage [46]. The use of cord blood for painful joint diseases such as rheumatoid and psoriatic arthritis is being investigated. One study examined the use of human umbilical cord-derived MSCs as a rheumatoid arthritis therapy [47]. Mice with collagen-induced arthritis had significantly decreased severity of arthritis when treated with cord MSCs. The results were similar to the Etanercept-treated group. A case report evaluated a 56-year-old patient with standard treatment-resistant psoriatic arthritis and revealed the success of intravenous and intra-articular injections of cord blood stem cells [48]. The patient had a remission of symptoms and normalized inflammatory markers within the first 30 days following treatment.

Umbilical cord blood cells are a promising treatment for ischemic conditions as well. In a mouse model of myocardial ischemia, heart tissue had regenerated after cord blood-derived cardiac progenitor cells were transplanted [49]. The cells also integrated with local cells in heart tissue 14 days after transplantation. Thromboangiitis obliterans (Buerger's disease), a nonatherosclerotic, vaso-occlusive disease, currently has no curative medication or surgery, but cord blood MSCs have shown promising results [50]. Four men suffering from distal limb ischemia secondary to Buerger's disease were transplanted with cord MSCs and had significant relief of their ischemic rest pain. Necrotic skin lesions also healed within 4 weeks of transplantation, and follow-up angiography showed increased digital capillaries.

Besides ischemia in the heart and limbs, umbilical cord blood has the potential to improve functionality in ischemic stroke patients [51]. Ten adult patients suffering from a recent middle cerebral artery ischemic stroke were infused with umbilical cord blood 3–9 days after their stroke. Patients were assessed with modified Rankin Score (mRS) and National Institutes of Health Stroke Scale (NIHSS). The three-month assessment of patients revealed improvement by at least one grade in the mRS and at least four points in the NIHSS. In a rat model of acute ischemic stroke, intravenous administration of umbilical cord blood caused significant improvement in both behavioral and structural impairments [52]. Experimental findings also indicated that umbilical cord blood enhanced neurogenesis and suppressed inflammation in damaged areas leading to therapeutic protection and better functional recovery.

Concerns and Contraindications

Cord blood cells can be used to treat a wide variety of diseases, but there are several concerns for their use. These concerns include informed consent, legal implications, cost, and whether the cord blood comes from public banks or commercial banks [53]. As cord blood is still in its experimental phase, medical indications and claims of medical benefit are still being investigated. Standard treatment options should be explored before considering umbilical cord blood-based treatments. Contraindications for intraarticular injections of cord blood are similar to any localized injection and include systemic illness, coagulopathy, and site infection [36]. Contraindications to intravenous infusion include limited venous access in an injured, burned, or infected extremity. Although there are lower rates of graft-versus-host disease with a higher human leukocyte antigen (HLA) disparity, HLA matching should be considered for certain hematological conditions undergoing cord blood treatment.

Preoperative and Intraoperative Considerations

As with other allograft treatments, it is critical to correctly diagnose the patient's condition. More conservative treatments should be considered such as physical therapy and nonexperimental injections before proceeding to umbilical cord blood use. Considerations for umbilical cord blood injections are similar to those described for amniotic tissue injections. When umbilical cord blood stem cells are used for transplantation, the patients' lab values, especially their white blood cell production and platelet production, need to be closely monitored to ensure engraftment [54].

References

1. Toda A, Okabe M, Yoshida T, Nikaido T. The potential of amniotic membrane/amnion-derived cells for regeneration of various tissues. J Pharmacol Sci. 2007;105(3):215–28.

2. Kern S, Eichler H, Stoeve J, et al. Comparative analysis of mesenchymal stem cells from bone marrow, umbilical cord blood, or adipose tissue. Stem Cells. 2006;24(5):1294–301. https://doi.org/10.1634/stemcells.2005-0342.

3. Mahla RS. Stem cells applications in regenerative medicine and disease therapeutics. Inter J Cell Biol. 2016;2016:6940283. https://doi.org/10.1155/2016/6940283.

4. Loukogeogakis SP, De Coppi P. Concise review: amniotic fluid stem cells: the known, the unknown, and the potential regenerative medicine applications. Stem Cells. 2017;35(7):1663–73. https://doi.org/10.1002/stem.2553.

5. Anam K, Lazdun Y, Davis PM, et al. Amnion-derived multipotent progenitor cells support allograft tolerance and induction. Am J Transplant. 2013;13(6):1416–28. https://doi.org/10.1111/ajt.12252.

6. Yamahara K, Harada K, Ohshima M, et al. Comparison of angiogenic, cytoprotective, and immunosuppressive properties of human amnion- and chorion-derived mesenchymal stem cells. PLoS One. 2014;9(2):e88319. https://doi.org/10.1371/journal.pone.0088319.

7. Kjaergaard N, Hein M, Hyttel L, Helmig RB, et al. Antibacterial properties of human amnion and chorion in vitro. Eur J Obstet Gynecol Reprod Biol. 2001;94(2):224–9.

8. Stock SJ, Kelly RW, Riley SC, Calder AA. Natural antimicrobial production by the amnion. American J Obstet Gynecol. 2007;196(3):255.e1–6.

9. Larson A, Gallicchio VS. Amniotic derived stem cells: role and function in regenerative medicine. J Cell Sci Therapy. 2017;8:269. https://doi.org/10.4172/2157-7013.1000269.

10. Riboh JC, Saltzman BM, Yanke AB, Cole BJ. Human amniotic membrane-derived products in sports medicine: basic science, early results, and potential clinical applications. Am J Sports Med. 2016;44:2425–34.

11. Rennie K, Gruslin A, Hengstschläger M, et al. Applications of amniotic membrane and fluid in stem cell biology and regenerative medicine. Stem Cells Int. 2012;2012:721538. https://doi.org/10.1155/2012/721538.

12. Young RL, Cota J, Zund G, Mason BA, Wheeler JM. The use of an amniotic membrane graft to prevent postoperative adhesions. Fertil Steril. 1991;55:624–8.

13. Tseng SCG, Prabhasawat P, Lee SH. Amniotic membrane transplantation for conjunctival surface reconstruction. Am J Ophthalmol. 1997;124:765–74.

14. Lee SH, Tseng SCG. Amniotic membrane transplantation for persistent epithelial defects with ulceration. Am J Ophthalmol. 1997;123:303–12.

15. Mermet I, Pottier N, Santhillier JM, et al. Use of amniotic membrane transplantation in the treatment of venous leg ulcers. Wound Repair Regen. 2007;15:459–64.

16. Alviano F, Fossati V, Marchionni C, et al. Term amniotic membrane is a high throughput source for multipotent mesenchymal stem cells with the ability to differentiate into endothelial cells in vitro. BMC Dev Biol. 2007;7:11.

17. Liu PF, Guo L, Zhao DW, Zhang ZJ, Kang K, Zhu P, Yuan XL. Study of human acellular amniotic membrane loading bone marrow mesenchymal stem cells in repair of articular cartilage defects in rabbits. Genet Mol Res. 2014;13:7992–8001.

18. Chiang CY, Liu SA, Sheu ML, Chen FC, Chen CJ, Su HL, Pan CH. Feasibility of human amniotic fluid derived stem cells in alleviation of neuropathic pain in a chronic constrictive injury nerve model. PLoS One. 2016;11:e0159482.

19. Diaz-Prado S, Rendal-Vazquez E, Muinos-Lopez E, Hermida-Gomez T, et al. Potential use of the human amniotic membrane as a scaffold in human articular cartilage repair. Cell Tissue Bank. 2010;11:183–95.

20. Vines JB, Aliprantis AO, Gomoll AH, Farr J. Cryopreserved amniotic suspension for the treatment of knee osteoarthritis. J Knee Surg. 2016;29:443–50.

21. Bhattacharya N. Clinical use of amniotic fluid in osteoarthritis: a source of cell therapy. In: Bhattacharya N, Stubblefield P, editors. Regenerative medicine using pregnancy-specific biological substances. London: Springer; 2011. p. 395–403.

22. Mamede AC, Carvalho MJ, Abrantes AM, et al. Amniotic membrane: from structure and functions to clinical applications. Cell Tissue Res. 2012;349(2):447–58. https://doi.org/10.1007/s00441-012-1424-6.

23. Nyman E, Huss F, Nyman T, Junker J, Kratz G. Hyaluronic acid, an important factor in the wound healing properties of amniotic fluid: in vitro studies of re-epithelialisation in human skin wounds. J Plast Surg Hand Surg. 2013;47(2):89–92. https://doi.org/10.3109/2000656X.2012.733169.

24. Mermet I, Pottier N, Sainthillier JM, et al. Use of amniotic membrane transplantation in the treatment of venous leg ulcers. Wound Repair Reg. 2007;15(4):459–64.

25. Swan J. Use of cryopreserved, particulate human amniotic membrane and umbilical cord (AM/UC) tissue: a case series study for Application in the healing of chronic wounds. Surg Technol Inter. 2014;25:73–8.

26. Kitagawa K, Okabe M, Yanagisawa S, Zhang XY, et al. Use of a hyperdried cross-linked amniotic membrane as initial therapy for corneal perforations. 2011;55(1):16–21. https://doi.org/10.1007/s10384-010-0903-0.

27. Barboni B, Curini V, Russo V, et al. Indirect co-culture with tendons or tenocytes can program amniotic epithelial cells towards stepwise Tenogenic differentiation. PLoS One. 2012;7(2):e30974. https://doi.org/10.1371/journal.pone.0030974.

28. Barboni B, Russo V, Curini V, et al. Achilles tendon regeneration can be improved by amniotic epithelial cell Allotransplantation. Cell Transplant. 2012;21(11):2377–95. https://doi.org/10.3727/096368912X638892.

29. Philip J, Hackl F, Canseco JA, et al. Amnion-derived multipotent progenitor cells improve Achilles tendon repair in rats. Eplasty. 2013;13:e31.

30. Lange-Consiglio A, Tassan S, Corradetti B, et al. Investigating the efficacy of amnion-derived mesenchymal stromal cells in equine tendon and ligament injuries. 2013;15(8):1011–20. https://doi.org/10.1016/j.jcyt.2013.03.002.

31. Demesmin D, Beall D, Nalamachu S, et al. Amniotic fluid as a homologue to synovial fluid: interim analysis of prospective, multi-center outcome observational cohort registry of amniotic fluid treatment for osteoarthritis of the knee. Poster Presented at the 2015 American Academy of Pain Medicine Annual Meeting. Accessed at http://www.painmed.org/2015posters/abstract-lb004/.

32. Willett NJ, Thote T, Lin AS, et al. Intra-articular injection of micronized dehydrated human amnion/chorion membrane attenuates osteoarthritis development. Arthritis Res Ther. 2014;16(1):R47.

33. Chen CJ, Cheng FC, Su HL, et al. Improved neurological outcome by intramuscular injection of human amniotic fluid derived stem cells in a muscle denervation model. PLoS One. 2015;10(5):e0124624. https://doi.org/10.1371/journal.pone.0124624.

34. Ma X. Regeneration of large-gap peripheral nerve injuries using acellular nerve allografts plus Amniotic Fluid Derived Stem Cells (AFS). Pathology. 2016;30(1).

35. Werber B. Amniotic tissues for the treatment of chronic plantar Fasciosis and Achilles tendinosis. J Sports Med. 2015; 2015:6.

36. Cardone DA, Tallia AF. Joint and soft tissue injection. Am Fam Physician. 2002;66(2):283–6.

37. Diwan S, Deer T. Advanced procedures for pain management. New York: Springer; 2018.

38. Seres KB, Hollands P. Cord blood: the future of regenerative medicine? Reprod Biomed Online. 2010;20(1):98–102. https://doi.org/10.1016/j.rbmo.2009.10.012.39.

39. Forraz N, McGuckin CP. The umbilical cord: a rich and ethical stem cell source to advance regenerative medicine. Cell Proliferat. 44(Suppl 1):60–9. https://doi.org/10.1111/j.1365-2184.2010.00729.x.

40. Fuchs JR, Hannouce D, Terada S, Zand S, Vacanti JP, Fauza DO. Cartilage engineering from ovine umbilical cord blood mesenchymal progenitor cells. Stem Cells. 2005;23:958–64.

41. Chung JY, Song M, Ha CW, Kim JA, Lee CH, Park YB. Comparison of articular cartilage repair with different hydrogel-human umbilical cord blood-derived mesenchymal stem cell composites in a rat model. Stem Cell Res Ther. 2014;5:39.

42. Ha CW, Park YB, Chung JY, Park YG. Cartilage repair using composites of human umbilical cord blood-derived mesenchymal stem cells and hyaluronic acid hydrogel in a minipig model. Stem Cells Transl Med. 2015;4:1044–51.

43. Ballen KK, Gluckman E, Broxmeyer HE. Umbilical cord blood transplantation: the first 25 years and beyond. Blood. 2013;122(4):491–8. https://doi.org/10.1182/blood-2013-02-453175.

44. Hwang WY, Samuel M, Tan D, Koh LP, Lim W, Linn YC. A meta-analysis of unrelated donor umbilical cord blood transplantation versus unrelated donor bone marrow transplantation in adult and pediatric patients. Biol Bone Marrow Transplant J Am Soc Blood Marrow Transplant. 2007;13(4):444–53.

45. Robin M, Sanz GF, Ionescu I, et al. Unrelated cord blood transplantation in adults with myelodysplasia or secondary acute Myeloblastic Leukemia: a survey on behalf of Eurocord and CLWP of EBMT. Leukemia. 2011;25(1):75–81. https://doi.org/10.1038/leu.2010.219.

46. Fuchs JR, Hannouche D, Terada S, Zand S, Vacanti JP, Fauza DO. Cartilage engineering from ovine umbilical cord blood mesenchymal progenitor cells. Stem Cells. 2005;23(7):958–64.

47. Shin T-H, Kim H-S, Kang T-W, et al. Human umbilical cord blood-stem cells direct macrophage polarization and block inflammation activation to alleviate rheumatoid arthritis. Cell Death Dis. 2016;7(12):e2524. https://doi.org/10.1038/cddis.2016.442.

48. Coutts M, Soriano R, Naidoo R, Torfi H. Umbilical cord blood stem cell treatment for a patient with psoriatic arthritis. World J Stem Cells. 2017;9(12):235–40. https://doi.org/10.4252/wjsc.v9.i12.235.

49. Pham TB, Nguyen TT, Bui AV, Pham HT, et al. Evaluation of treatment efficacy of umbilical cord blood-derived mesenchymal stem cell-differentiated cardiac progenitor cells in a myocardial injury mouse model. Springer Nature Singapore; 2018.

50. Kim SW, Han H, Chae GT, Lee SH, et al. Successful stem cell therapy using umbilical cord-derived multipotent stem cells for Buerger's disease and ischemic limb disease animal model. Stem Cells. 2006;24(6):1620–6.

51. Laskowitz DT, Bennett ER, Durham RJ, et al. Allogeneic umbilical cord blood infusion for adults with ischemic stroke: clinical outcomes from a phase 1 safety study. Stem Cells Transl Med. 2018; https://doi.org/10.1002/sctm.18-0008.

52. Yoo J, Kim HS, Seo JJ, et al. Therapeutic effects of umbilical cord blood plasma in a rat model of acute ischemic stroke. Oncotarget. 2016;7(48):79131–40. https://doi.org/10.18632/oncotarget.12998.

53. Petrini C. Umbilical cord blood collection, storage and use: ethical issues. Blood Transfus. 2010;8(3):139–48. https://doi.org/10.2450/2010.0152-09.

54. Moise KJ. Umbilical cord stem cells. Obstetr Gynecol. 2005;106(6):1393–407.

Adult Mesenchymal Stem Cell Collection and Banking

Roy R. Liu and Houman Danesh

Stem Cell Applications in Regenerative Medicine

Regenerative cellular therapies are thought of by most as "experimental" therapies. However, its use is well established in the area of tissue and bone marrow transplantation, as well as in assisted reproductive technologies. Whereas these applications utilize hematopoietic and embryonic stem cells, respectively, mesenchymal stem cells (MSCs) are the most useful option for cell-based treatments of traumatic and degenerative bone, joint, and cartilaginous diseases. They have also been used in the treatment of other diseases, such as graft-versus-host-disease (GVHD), multiple sclerosis, Crohn's disease, and systemic lupus erythematosus (SLE), albeit with mixed results [1]. Mesenchymal stem cells have the ability to differentiate into cells of mesodermal origin, including osteoblasts, chondrocytes, and adipocytes. They can be differentiated from hematopoietic stem cells (HSCs) by the expression of cell surface antigens CD73, CD90, and CD105 and the lack of expression of CD14, CD20, CD34, and CD45. While there is no uniform agreement on how MSCs promote tissue healing, a large component of its efficacy may be related to immunomodulation. They have been shown to promote the secretion of various cytokines (e.g., transforming growth factor beta, vascular endothelial growth factor, epidermal growth factor) that stimulate local tissue repair [2] and suppress proliferation of T-cells and monocytes, thereby suppressing inflammation and apoptosis [3].

R. R. Liu
Icahn School of Medicine at Mount Sinai, Department of Anesthesiology, Perioperative and Paine Medicine, New York, NY, USA

H. Danesh (✉)
The Mount Sinai Hospital, Department of Anesthesiology, New York, NY, USA
e-mail: houman.danesh@mountsinai.org

Mesenchymal Stem Cell Sources

Mesenchymal stem cells are found in almost all organs and tissues after birth. Some sources, such as peripheral blood, contain so few MSCs that they are not clinically useful. Other tissue sources require invasive techniques to harvest. The first MSCs described in the literature were isolated from bone marrow stroma and are considered the "standard" source against which others are compared. The following discussion will focus on practical sources of MSCs that may be obtained and used for cell-based therapies.

Bone Marrow

Hematopoietic bone marrow-derived stem cells (HSCs) are the earliest type detected and the most well-studied. Hematopoietic stem cells have been used successfully to treat cancers of the blood and bone marrow. Bone marrow also contains populations of mesenchymal stem cells (MSCs), which are adult stem cells but have the potential to differentiate into multiple nonhematopoietic cell lineages (Fig. 8.1). There are several disadvantages of utilizing bone marrow-derived stem cells. Obtaining bone marrow aspirate is invasive, even at relatively accessible sites such as the iliac crest (Fig. 8.2). The necessity of placing patients under deep sedation or general anesthesia increases the cost of collection and introduces logistic inconveniences to the patient (e.g., preprocedural fasting, need for intravenous access, and need for an escort). Mesenchymal stem cells are relatively rare in bone marrow compared with other cell populations. It is estimated that they comprise at most 0.02% of the bone marrow stromal cell population [4]. Furthermore, there appears to be an age-related decline in the number and functionality of bone marrow-derived stem cells [5]. The invasiveness and inconvenience of collecting MSCs from bone marrow have prompted a search for alternative sites of cell harvest in recent years.

C. W Hunter et al. (eds.), *Regenerative Medicine*, https://doi.org/10.1007/978-3-030-75517-1_8

Fig. 8.1 Mesenchymal stem cell differentiation lineages

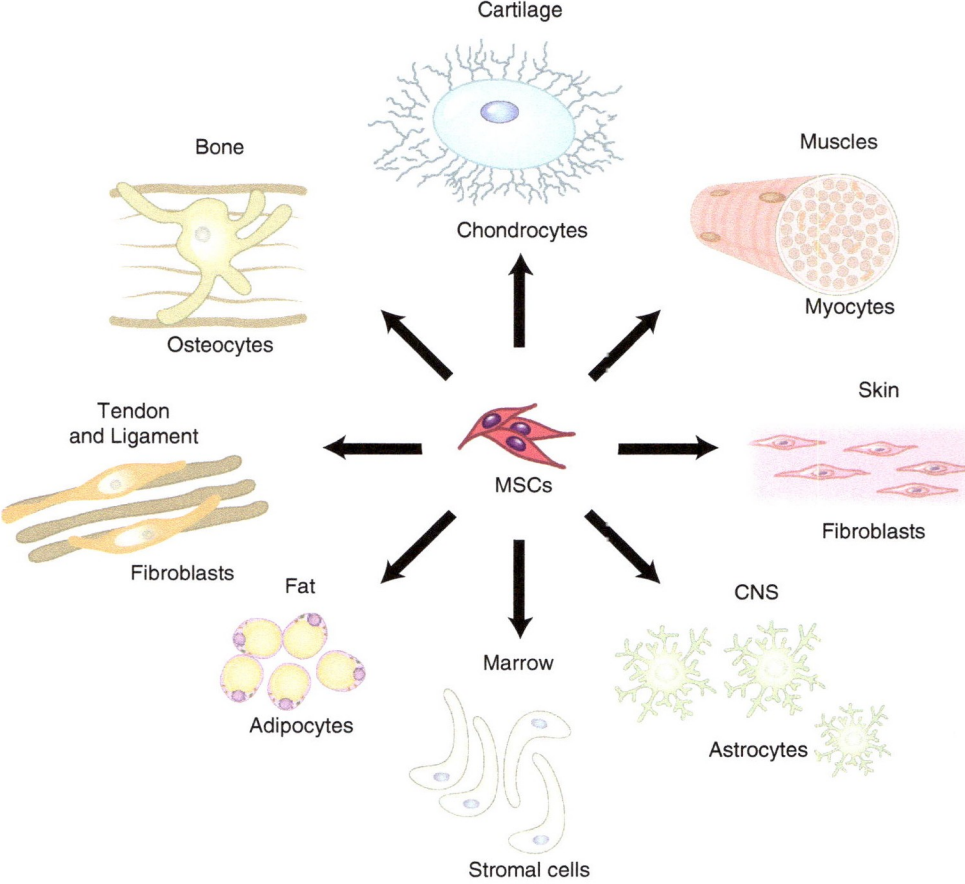

Fig. 8.2 Bone marrow aspiration from the iliac crest. (©Mayo Foundation for Medical Education and Research; used with permission)

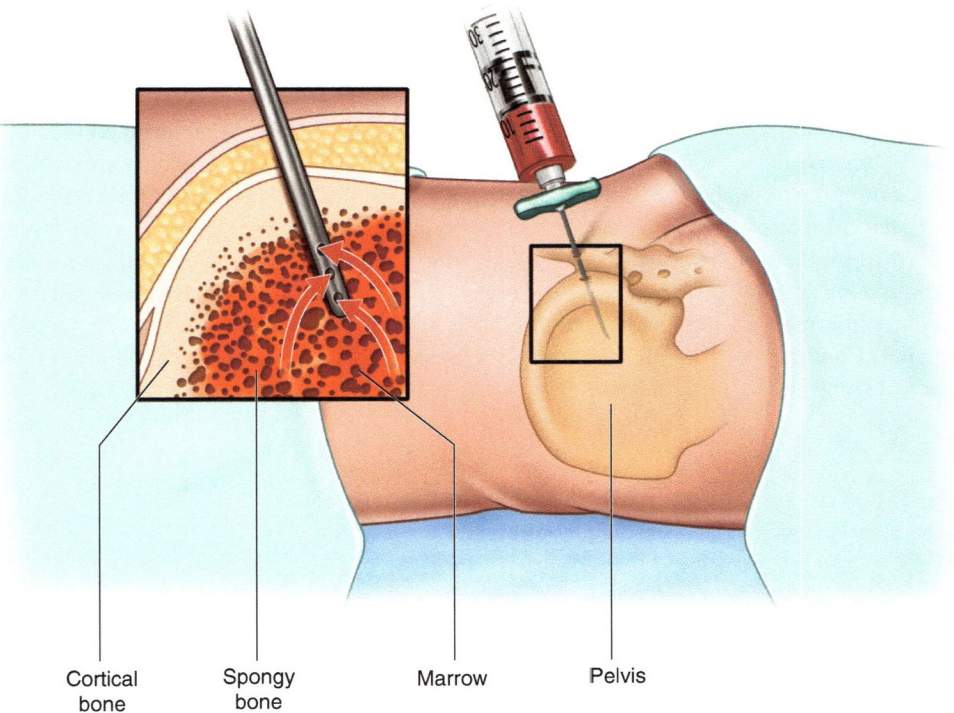

Umbilical Cord Blood and Tissue

Traditionally, umbilical cord blood has been harvested and banked to treat hematologic malignancies and diseases. Only recently has cord blood as well as cord tissue been utilized for regenerative medicine applications. Because it is essentially a byproduct of delivery that would otherwise be discarded as medical waste, collection is simple, at no discomfort to the patient, and not subject to ethical controversy (Fig. 8.3).

However, because collection is only possible around the time of delivery, banking time (and therefore cost) for umbilical cord blood and tissue is higher than for other tissues. Another downside of cord blood and tissue is the relative low density of MSCs. There is a wide range of published estimates, but even the highest estimates do not allow for most clinical applications without first expanding the cells in culture. This leads to additional regulatory concerns, as will be discussed later in this chapter.

Fig. 8.3 Umbilical cord blood/tissue collection of mesenchymal stem cells. (From: Van Pham P, Truong NC, Le PT, Tran TD, Vu NB, Bui KH, Phan NK. Isolation and proliferation of umbilical cord tissue derived mesenchymal stem cells for clinical applications. Cell Tissue Bank. 2016;17(2):289–302. https://doi.org/10.1007/s10561-015-9541-6. Epub 2015 Dec 17; used with permission)

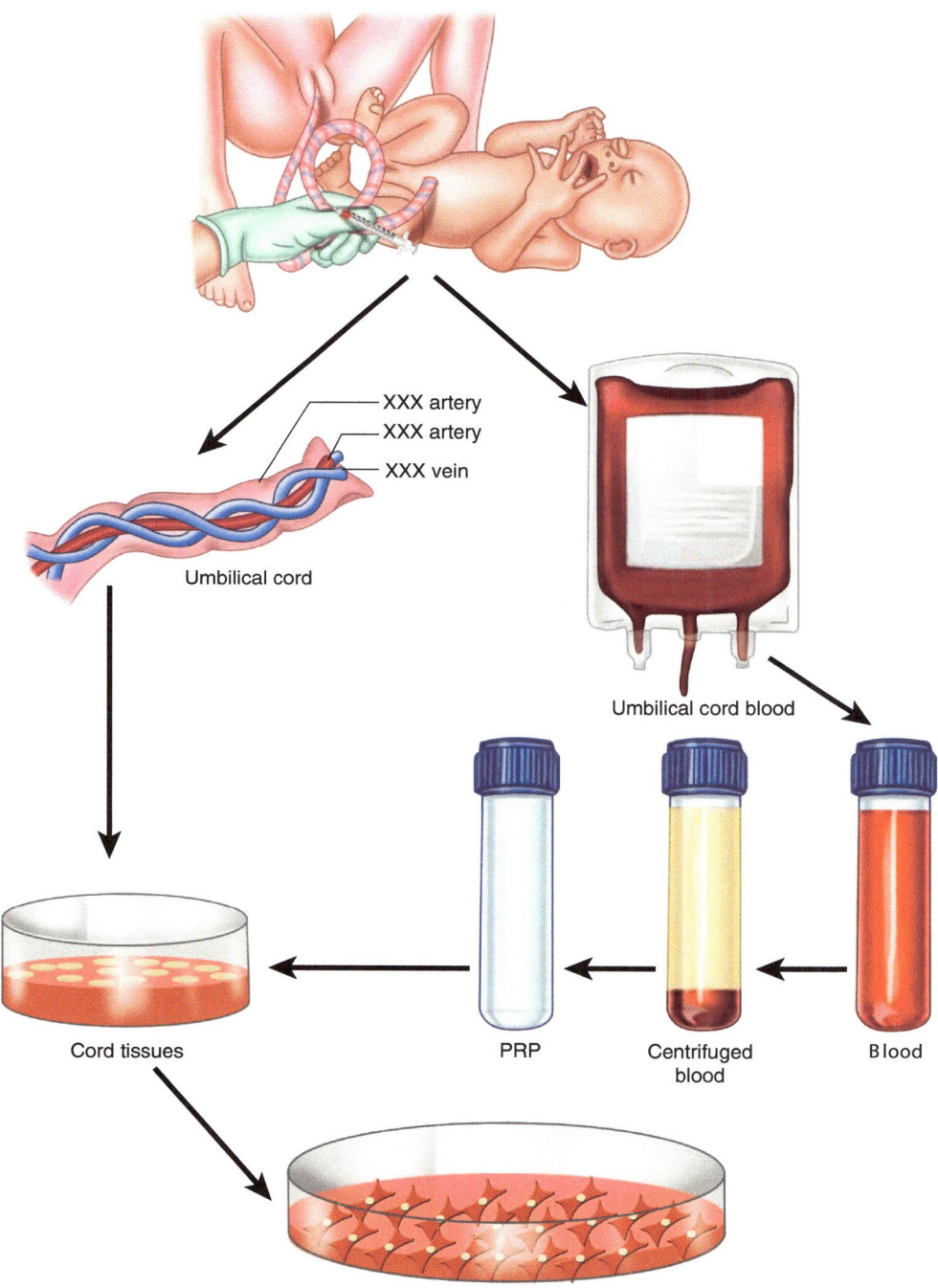

XXX artery
XXX artery
XXX vein

Umbilical cord

Umbilical cord blood

Cord tissues

PRP

Centrifuged blood

Blood

Mesenchymal stem cells

Adipose Tissue

Adipose tissue is an abundant and constant source of MSCs that holds a lot of promise in regenerative medicine. They have been shown to have similar differentiation potential to MSCs from other tissues, such as cord tissue [6]. Compared with bone marrow, adipose tissue represents a much more easily accessible and lower risk source of stem cells. In one study, the outer thigh was shown to have a higher concentration of MSCs compared with the abdomen, waist, or inner knee [7]. It is also believed to contain up to 500 times the density of MSCs compared with an equivalent volume of bone marrow [8]. It is feasible to collect tens to hundreds of millions of MSCs for immediate clinical application without in vitro expansion, as in the case of cord tissue. Adipose tissue MSCs can be harvested by various methods, such as needle aspiration under local anesthesia, or collected as a byproduct of surgical liposuction or abdominoplasty (Fig. 8.4).

Table 8.1 shows a summary of the advantages and disadvantages of bone marrow, umbilical cord blood and tissue, and adipose tissue as sources for MSCs.

Factors Impacting Stem Cell Effectiveness

There are multiple factors impacting the clinical usefulness of harvest stem cells for application in regenerative medicine.

Donor Age and Site of Collection

The use of autologous stem cells is preferred to allogeneic sources due to the low risk of alloimmunization and logistical simplicity. However, just as there is a linear relationship between the prevalence of musculoskeletal diseases and advancing age, there appears to be an inverse relationship between stem cell quantity and quality and age at harvest [9]. Stolzing et al. showed a reduction in the frequency of mesenchymal colony-forming cells (a marker of proliferation rate) and higher degrees of oxidative damage in bone marrow-derived MSCs of adults vs. children [10]. Studies on adipose tissue MSCs have shown mixed results, with some showing similar age-related decreases in number and proliferative potential as with bone marrow-derived stem cells [11], while others showing no significant effects on osteogenic differentiation potential [12, 13]. Combined with relative ease of collection, relative abundance of harvest sites, and significantly higher density of stem cells, adipose tissue compares favorably to bone marrow as the more practical source of MSCs for orthopedic regenerative applications.

Effects of Storage

Autologous use of MSCs for regenerative therapies is preferable to allogeneic use due to the absence of rejection risk. Unfortunately, the elderly medical population, who represent major targets for such therapies, are poor sources of high-quality MSCs. One theoretical solution to this conundrum is

Fig. 8.4 Lipoaspiration

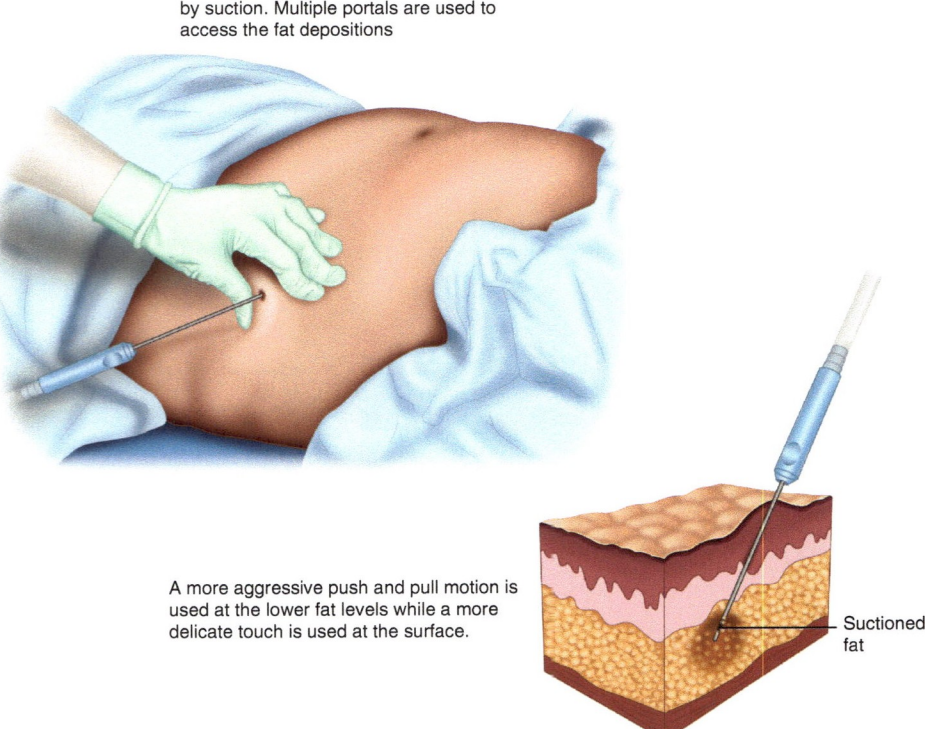

This subcutaneous fat is removed by suction. Multiple portals are used to access the fat depositions

A more aggressive push and pull motion is used at the lower fat levels while a more delicate touch is used at the surface.

Suctioned fat

Table 8.1 Advantages and disadvantages of various tissue sources of mesenchymal stem cells

	Pros	Cons
Bone marrow	Most well-studied	Invasive and uncomfortable collection process Yield highly dependent on collection methodology Relatively small population of MSCs compared with HSCs Limited applicability without ex-vivo expansion
Umbilical cord blood/ tissue	Noninvasive collection from the waste product of birth	One-time collection opportunity (for each live birth) Limited applicability without ex-vivo expansion
Adipose tissue	High MSC yield Less invasive collection than bone marrow Abundance of accessible sites on the body	Lack of established safety track record

to collect and cryopreserve the necessary cells from individuals who anticipate future needs while they are younger. For this to be a viable solution, the deleterious effects of cryopreservation on the viability and osteogenic potential of stored MSCs have to be significantly less than the effects of aging. Multiple authors have demonstrated that the retention of proliferation and differentiation potential in vitro of cryopreserved adipose-derived MSCs is comparable to freshly collected MSCs [14, 15].

Stem Cell Yields and Expansion

The optimal number of MSCs needed for most therapeutic applications is unclear. From existing literature, at least 1×10^6 MSCs/kg MSCs are needed [16, 17]. This means that for the average adult patient, 70–80 million cells are needed. Although bone marrow is still considered the "gold standard" source of stem cells, only 0.001–0.02% of cells isolated from bone marrow aspirate are MSCs [18, 19]. Assuming optimal bone marrow aspiration technique (e.g., minimal aspiration of peripheral blood) and optimal yield of MSCs within that sample, only about 3.6×10^5 MSCs can be isolated from a typical aspirate volume. Umbilical cord and tissue also have relatively low MSC yield. This necessitates the ex vivo expansion of MSC collections from the above sources prior to introduction into patients. In contrast, adipose tissue has up to 500 times the number of MSCs per equivalent volume of bone marrow aspirate, and it is possible to harvest enough MSCs for immediate application. For this reason, experts believe that adipose-tissue mesenchymal stem cells have become the new "gold standard" for musculoskeletal applications of cellular-based regenerative therapies.

Mesenchymal Stem Cell Collection and Banking

Bone Marrow

The iliac crest is the most common site of access for the extraction of bone marrow aspirate concentrate (BMAC). There are various commercially available bone marrow aspiration kits that can be used. After sterile preparation and draping of the surgical field, the anterior superior iliac spine (supine positioning) or posterior superior iliac spine (prone positioning) is palpated. The skin, soft tissue tract, and periosteum are then anesthetized with 1% lidocaine in a trajectory that is perpendicular to the ASIS/PSIS. A trocar is then introduced percutaneously along the anesthetized tract until contact with the cortical bone. Subsequently, a power drill or manual drill is used to advance the trocar into the medullary cavity of the iliac crest. After heparinization of the collection syringe, about 60 mL of bone marrow is aspirated. The bone marrow aspirate then undergoes processing by passage through a mesh filter, followed by density gradient centrifugation to isolate the bone marrow aspirate. The final yield is approximately 6 mL or 10% of the originally aspirated volume.

Umbilical Cord Blood and Tissue

Collection of cord blood is usually performed by accessing the umbilical vein after clamping of the umbilical cord and can be collected directly from the placenta after it has been delivered. A full-term birth yields roughly 75 mL of cord blood and up to 1.1×10^9 total nucleated cells [20]. Cord blood has a very high erythrocyte concentration, so collections are usually RBC-depleted prior to banking. There are multiple methods to isolate the buffy coat (containing the MSCs), including density gradient centrifugation, hydroxyl-ethyl starch sedimentation, and several commercially automated processes (AutoXpress by Cesca Therapeutics, Sepax™ by Biosafe, PrepaCyte®-CB Processing System by BioE, and Cord Blood 2.0™ by Americord) [21]. The final collection is then aliquoted into different compartments to allow for multiple future uses prior to cryopreservation.

Umbilical cord tissue is also collected shortly after the time of birth. A length of the umbilical cord is cut with sterile instruments and placed in a buffer solution for transport to the processing facility. It subsequently undergoes several rounds of washings with sterile saline and 70% ethanol before being cut into smaller segments and frozen at −180 C. When it is needed for use, the aliquoted tissue is thawed, washed, and resuspended. It is then placed in growth media for colony expansion. The average length of the umbilical cord is approximately 30 to 50 cm, and there are a wide range of MSC yields as described in the literature. One study quotes an unexpanded yield of roughly 0.65×10^6 cells/cm of tissue [22]. While this is more than the number of MSCs isolated from umbilical cord blood, both are too low to be useful for immediate clinical application and need to be first expanded in culture.

Adipose Tissue Stem Cells

Adipose tissue can be harvested as part of a different operation or specifically for MSC isolation. Examples of the former include collecting liquid fat during liposuction and solid fat during abdominoplasty. Solid fat requires one additional step in its processing as it is mechanically broken down into smaller pieces to expand its surface area in order to facilitate subsequent chemical breakdown by collagenase. According to one published protocol, the adipose preparation is then diluted by Dulbecco's phosphate-buffered saline (DPBS), washed by centrifugation, and mixed with collagenase to digest any solid matter. Dulbecco's modified Eagle media (DMEM) is added after some time to halt enzymatic digestion, and the resulting liquid mixture is centrifuged again to isolate the stromal vascular fraction (SVF) containing the MSCs [23]. The process of breaking down adipose tissue to obtaining the SVF takes less than 30 minutes with several commercially available isolation systems [24, 25]. However, isolating MSCs from the SVF currently takes up to 24 hours, possibly precluding the possibility of a same-day harvest-to-introduction model. Lastly, if the isolated SVF is not for immediate MSC isolation, it is resuspended in a combination of culture medium, fetal calf serum, and a cryoprotective solution such as dimethyl sulfoxide (DMSO) and then aliquoted into tubes and cryopreserved at -180 C for future use [26]. There has been debate as to whether protocols like the one just described adhere to current good manufacturing practice (CGMP), as the use of DMSO has been found to be cytotoxic and the use of animal serum potentially exposes humans to viral and prion disease transmission. While no official CGMP guidelines have been published pertaining to adipocyte-derived stem cells, an investigation into using minimal concentrations of DMSO and allogeneic human serum has yielded promising results [27].

Rules and Regulations

On July 1, 1902, the United States Congress passed the Biologics Control Act (BCA) in response to the deaths of children who contracted tetanus from contaminated diphtheria vaccines. This became the first federal law requiring manufacturers of biologic products to meet minimum safety guidelines. This law paved the way for several others that

exist today, including the Federal Food, Drug, and Cosmetic Act (FDCA) of 1938. In 1944, the Public Health Service Act (PHSA) was enacted to consolidate and codify previous laws (including the BCA) on biologic products, adding safety measures to the licensing and production of biologic products. The National Institutes of Health initially policed the new regulations as outlined in the PHSA, but the FDA assumed this responsibility in 1972 and has continued to do so through to the present day.

The FDA regulates stem cell-based products under multiple avenues of regulatory authority, including the PHSA and FDCA. They are classified as part of a broader category of human cells, tissues, and cellular- and tissue-based products, from here on referred to as HCT/Ps. They are defined under the Code of Federal Regulations (21 C.F.R.1271) as "articles containing or consisting of human cells or tissues that are intended for implantation, transplantation, infusion, or transfer into a human recipient" [28]. Therapies using autologous cells and tissues were exempt from this definition until 2006, when the FDA changed the wording from "into *another* human recipient' to "into *a* human recipient." The FDA has adopted a tier-based approach to regulate HCT/Ps based on their inherited degree of risk to patients. Low-risk HCT/Ps need to meet the criteria of being (a) minimally manipulated, (b) intended for homologous use, (c) *not* be combined with other reagents, (d) not have a systemic effect and is not dependent upon the metabolic activity of living cells for its primary function, and (e) for autologous or allogeneic use in close blood relatives. The HCT/Ps that meet the above criteria are regulated solely under section 361 of the PHSA, which includes guidance documents on preventing the introduction, transmission, and spread of communicable disease.

Any HCT/P that does not meet the above criteria is defined as a "biologic" and regulated under section 351 of the PHSA, subject to the same regulatory process as more traditional new investigational drugs. Their use requires submission and approval of a Biologics License Application (BLA), which involves preclinical followed by clinical studies and an Investigational New Drug (IND) Application review. This translates to a lengthy development-to-market time and high development costs, which are prohibitive to most regenerative medicine practitioners and clinics that offer stem cell-based therapies. Table 8.2 summarizes the key differences between "351" and "361" HCT/Ps.

Despite the specific definitions set forth to differentiate high- vs. low-risk therapies, controversy still exists over language in the statutes. For example, "minimal manipulation" was defined as processes that did not fundamentally change the composition of the collected tissue. The FDA has listed examples of "minimal manipulation" as including cutting, grinding, sterilizing, and density-gradient separation, centrifugation, cell isolation, and cryopreservation. However,

Table 8.2 Comparison of PHSA "361" vs. "351'"HCT/Ps

Low-risk HCT/Ps	High-risk HCT/Ps
Exempt from PHSA Section 351 regulations	Regulated under PHSA Section 351
NDA not required	NDA required for premarket approval (increased product conception to market time)
Manufactured under good tissue practices (GTPs)	Manufactured under GTPs and good manufacturing practices (GMPs)
High practicality	Low practicality
Low cost	High cost
Low therapeutic ceiling	High therapeutic ceiling

using an enzymatic process to isolate a specific cell population constitutes more than minimal manipulation. Enzymatic digestion of cord tissue prior to banking is a standard protocol of many tissue banks, and this practice seems to violate the FDA's definition of minimal manipulation, but specific regulation of cord and adipose tissue is not well-established and the FDA has not explicitly banned this practice. Another example of ambiguity in the current definitions revolves around using stem cells for homologous function. This means that the stem cells used in therapy need to function as they would otherwise function in the body. This becomes problematic for cord tissue-derived MSCs, as the function of these cells is lost at birth. Thus, definitions for the use of nonumbilical cord blood-derived stem cells need further clarification so as to ensure that practitioners are compliant with federal regulations.

Since the 1990s, the FDA has shown interest in expanding its regulatory influences over cellular medicine therapies, and in 2008, it backed up this rhetoric when it filed litigation against Regenerative Sciences LLC [29]. This stem cell clinic's patented Regenexx-C procedure involved taking blood and bone marrow samples from patients, expanding them in culture, and returning them to the patients via injection to the site of injury some weeks later. The FDA notified the company that this process of cell harvest and expansion was outside the definition of minimal manipulation. Thus, in their view, Regenexx was offering a "biologic" product (as defined by the PHSA) to patients and subject to a premarket approval process. Regenerative sciences made a counterclaim that its methodology for stem cell harvest and processing fell within the scope of standard medical procedure, and therefore no new drug was being introduced to patients. After a six-year legal battle, the United States Court of Appeals for the District of Columbia ruled in favor of the FDA.

Current and Future Directions

It is estimated that over 50 million Americans (or 1 in 5 adults) suffer from osteoarthritis [30], and it is one of the most common complaints seen in pain management. The US

government's clinical trials database (www.clinicaltrials.org) reveals 87 ongoing or completed clinical trials utilizing stem cells derived from bone marrow, cord blood, cord tissue, and adipose tissue to replace worn-down cartilage and slow down the progression of arthritis. This represents one of the most active areas of research in regenerative medicine.

There is a general perception that the FDA approval process for new drugs is slow and expensive. On average, it takes more than a decade and hundreds of millions of dollars to get from drug conception to market in the United States [31]. With massive support from pharmaceutical manufacturers, Congress enacted the Twenty-First Century Cures Act in 2016 in an effort to streamline the new drug approval process and make promising treatments accessible to the general public faster. One provision of particular interest to the field of regenerative medicine gives the FDA authority to give certain biologics products "Regenerative Medicine Advanced Therapy" or RMAT designation. A drug is eligible for RMAT designation if the following criteria are met:

(i) The drug is a regenerative medicine therapy, which is defined as a cell therapy, therapeutic tissue engineering product, human cell and tissue product, or any combination product using such therapies or products, except for those regulated solely under Section 361 of the Public Health Service Act and part 1271 of Title 21, Code of Federal Regulations.

(ii) The drug is intended to treat, modify, reverse, or cure a serious or life-threatening disease or condition.

(iii) Preliminary clinical evidence indicates that the drug has the potential to address unmet medical needs for such disease or condition [32].

As of 2018, 20 products have been granted RMAT designation. Although most of the conditions pain management physicians see patients for are not serious or life-threatening, one biologic product that did receive RMAT designation is MiMedx's Group's AmnioFix® Injectable, a micronized amniotic tissue product aimed at treating osteoarthritis of the knee. There is also speculation within the industry that Tigenix will apply for an RMAT designation for Cx601, a local administration of expanded adipose-derived stem cells (eASCs) [33].

References

1. Kim N, Cho S-G. New strategies for overcoming limitations of mesenchymal stem cell-based immune modulation. Inter J Stem Cells. 2015;8:54–68.
2. Freitag J, Bates D, Boyd R, Shah K, Barnard A, Huguenin L, Tenen A. Mesenchymal stem cell therapy in the treatment of osteoarthritis: reparative pathways, safety and efficacy—a review. BMC Musculoskelet Disord. 2016; https://doi.org/10.1186/s12891-016-1085-9.

3. Iyer SS, Rojas M. Anti-inflammatory effects of mesenchymal stem cells: novel concept for future therapies. Expert Opin Biol Ther. 2008;8:569–81.
4. Thirumala S, Goebel WS, Woods EJ. Clinical grade adult stem cell banking. Organogenesis. 2009;5:143–54.
5. Mueller SM, Glowacki J. Age-related decline in the osteogenic potential of human bone marrow cells cultured in three-dimensional collagen sponges. J Cell Biochem. 2001;82:583–90.
6. Daher SR, Johnstone BH, Phinney DG, March KL. Adipose stromal/stem cells: basic and translational advances: the IFATS collection. Stem Cells. 2008;26:2664–5.
7. Tsekouras A, Mantas D, Tsilimigras ID, Moris D, Kontos M, Zografos CG. Comparison of the viability and yield of adipose-derived stem cells (ASCs) from different donor areas. In Vivo. 2017;31:1229–34.
8. Harris D. Stem cell banking for regenerative and personalized medicine. Biomedicine. 2014;2:50–79.
9. Ganguly P, El-Jawhari JJ, Giannoudis PV, Burska AN, Ponchel F, Jones EA. Age related changes in bone marrow mesenchymal stromal cells: a potential impact on osteoporosis and osteoarthritis development. Cell Transplant. 2017; https://doi.org/10.3727/09636 8917x694651.
10. Stolzing A, Jones E, Mcgonagle D, Scutt A. Age-related changes in human bone marrow-derived mesenchymal stem cells: consequences for cell therapies. Mech Ageing Dev. 2008;129:163–73.
11. Zhu M, Kohan E, Bradley J, Hedrick M, Benhaim P, Zuk P. The effect of age on osteogenic, adipogenic and proliferative potential of female adipose-derived stem cells. J Tissue Eng Regen Med. 2009;3:290–301.
12. Khan W, Adesida A, Tew S, Andrew J, Hardingham T. The epitope characterisation and the osteogenic differentiation potential of human fat pad-derived stem cells is maintained with ageing in later life. Injury. 2009;40:150–7.
13. Dufrane D. Impact of age on human adipose stem cells for bone tissue engineering. Cell Transplant. 2017; https://doi.org/10.3727/0 96368917x694796.
14. Choudhery MS, Badowski M, Muise A, Pierce J, Harris DT. Cryopreservation of whole adipose tissue for future use in regenerative medicine. J Surg Res. 2014;187:24–35.
15. Roato I, Alotto D, Belisario DC, Casarin S, Fumagalli M, Cambieri I, Piana R, Stella M, Ferracini R, Castagnoli C. Adipose derived-mesenchymal stem cells viability and differentiating features for orthopaedic reparative applications: banking of adipose tissue. Stem Cells Int. 2016;2016:1–11.
16. Jung S, Panchalingam KM, Rosenberg L, Behie LA. Ex vivo expansion of human mesenchymal stem cells in defined serum-free media. Stem Cells Int. 2012;2012:1–21.
17. Mcdaniel JS, Antebi B, Pilia M, Hurtgen BJ, Belenkiy S, Necsoiu C, Cancio LC, Rathbone CR, Batchinsky AI. Quantitative assessment of optimal bone marrow site for the isolation of porcine mesenchymal stem cells. Stem Cells Int. 2017;2017:1–10.
18. Pittenger MF. Multilineage potential of adult human mesenchymal stem cells. Science. 1999;284:143–7.
19. Dragoo JL, Debaun MR. Stem cell yield after bone marrow concentration. Orthop J Sports Med. 2017; https://doi.org/10.1177/232596 7117s00445.
20. Harris. 2014.
21. Cord Blood Processing—AutoXpress® (AXP), Sepax, PrepaCyte®, Cord Blood 2.0. (2018, August 28). Retrieved 28 Aug 2018, from https://bioinformant.com/trending-now-fully-automated-umbilical-cord-blood-processing/#processing.
22. Tsagias N, Koliakos I, Karagiannis V, Eleftheriadou M, Koliakos GG. Isolation of mesenchymal stem cells using the total length of umbilical cord for transplantation purposes. Transfus Med. 2011;21:253–61.
23. Schneider S, Unger M, Griensven MV, Balmayor ER. Adipose-derived mesenchymal stem cells from liposuction and resected fat are feasible sources for regenerative medicine. Eur J Med Res. 2017; https://doi.org/10.1186/s40001-017-0258-9.
24. Oberbauer E, Steffenhagen C, Wurzer C, Gabriel C, Redl H, Wolbank S. Enzymatic and non-enzymatic isolation systems for adipose tissue-derived cells: current state of the art. Cell Regenerat. 2015; https://doi.org/10.1186/s13619-015-0020-0.
25. Raposio E, Simonacci F, Perrotta RE. Adipose-derived stem cells: comparison between two methods of isolation for clinical applications. Anna Med Surg. 2017;20:87–91.
26. Schneider S, Unger M, Griensven MV, Balmayor ER. Adipose-derived mesenchymal stem cells from liposuction and resected fat are feasible sources for regenerative medicine. Eur J Med Res. 2017; https://doi.org/10.1186/s40001-017-0258-9.
27. Thirumala S, Goebel WS, Woods EJ. Clinical grade adult stem cell banking. Organogenesis. 2009;5:143–54.
28. CFR—Code of Federal Regulations Title 21. In: accessdata.fda. gov. https://www.accessdata.fda.gov/scripts/cdrh/cfdocs/cfcfr/CFRSearch.cfm?CFRPart=1271. Accessed 6 Oct 2018.
29. United States v. Regenerative Sciences, LLC, 741 F.3d 1314 (D.C. Cir. 2014).
30. www.arthritis.org. Available online: http://www.arthritis.org. Accessed on 1 Oct 2018.
31. Paul SM, Mytelka DS, Dunwiddie CT, Persinger CC, Munos BH, Lindborg SR, Schacht AL. How to improve RandD productivity: the pharmaceutical industrys grand challenge. Nat Rev Drug Discov. 2010;9:203–14.
32. Center for Biologics Evaluation and Research. (n.d.) Cellular and gene therapy products—Regenerative medicine advanced therapy designation. Retrieved 11 Sept 2018, from https://www.fda.gov/BiologicsBloodVaccines/CellularGeneTherapyProducts/ucm537670.htm.
33. Center for Biologics Evaluation and Research. (n.d.) Cellular and gene therapy products—Regenerative medicine advanced therapy designation. Retrieved 11 Sept 2018, from https://www.fda.gov/BiologicsBloodVaccines/CellularGeneTherapyProducts/ucm537670.htm.

Exosomes

9

Timothy Ganey, H. Thomas Temple, and Corey W Hunter

Introduction

A discussion of exosomes in the context of regenerative medicine becomes challenging to entail without a context that enriches the story from where they emerge, how they are processed, what parts they might play, and what regulates the ecology that prioritizes their functional role in the nanosphere of being engendered in a living organism. Also evident from the work in exosomes is that cell products are inextricable derivatives of cell activity, cycle modulation, and reaction to metabolic shifts to the organism that result in systemic reaction. Over the past 30–35 years, the pendulum of cell biology shifted therapeutic strategy with the potential for stem cell therapy to afford a direct asset to clinical applications. Stem cell biology is a fascinating field of science that overlays traditional developmental biology, cutting edge genetics, but to date has relied on an intellectual extrapolation of utility allowing one to visualize new organs generated with personalized stem cells, or complete rejuvenation devoid of the footprints of time. While such enthusiasm that has enabled excitement can be appreciated, it is equally important to respect the complexity of stem cells and to appreciate the inherent challenges of managing the magic.

The route of scientific development for the "bench to bedside" generally is sculpted through discovery of an unknown agent or idea, careful study in the laboratory conditions in defined conditions, and then scaling in animal models to understand systemic regulation before human testing for either safety or efficacy. What emerges from that scientific stew are clear understandings that the genetics of cell development and systemic expression are intricate and convoluted, that successful results observed in lab animals (i.e., mice) do not always replicate in humans or predict similar outcomes, that transplanted cells do not always function in the expected manner – sometimes, they develop into the wrong cells, or that freshly transplanted cells are outright rejected by our immune system.

As of 2019, cord blood therapies are the only FDA-approved, stem-cell based therapies in the United States, and they are limited to treating patients with blood disorders. In contrast, quasi-legal, stem cell clinics have been very quick to manifest a potential use across America with stem cell clinics number in the 700+ offering non-FDA approved therapies in the United States. In a final point to introducing exosomes, Arnold Caplan, perhaps the most known individual in the field of stem cell use and credited for coining the term mesenchymal stem cell (MSC), now proposes that it is not the cell but the paracrine signaling potential of perivascular pericytes that is responsible for regenerative effect. Those signaling molecules, extracellular vesicles including macrovesicles, microvesicles, and exosomes, represent a large factor in the activity that is likely an intermediary in never-ending chain of future elucidation.

Cell-based therapies and the field of regenerative medicine for some time have been heralded as the coming pillar of medical care. In the course of time over which the trumpet has sounded, triumphs beyond blood diseases have been slow to arrive. With the passing of the twenty-first Century Cures Act, driven in large part by a desire to broaden the clinical reach of stem cells, the therapeutic basis for the cellular products has accented debate. Such debate stems not only from clinical consideration, beneficial use, or efficacy, but enduring topics such as cell source, cell expansion, and even ethical considerations originally associated with using pluripotent embryonic cells remain harbored in regenerative medicine discussions. With that foundation, it is not surprising that most clinical trials utilize multipotent stem cells

T. Ganey (✉)
Vivex Biomedical, Inc., Miami, FL, USA
e-mail: tim@bonepharm.com

H. T. Temple
Department of Orthopaedic Surgery, HCA Healthcare Inc., Miami, FL, USA

C. W Hunter
Ainsworth Institute of Pain Management, New York, NY, USA

Physical Medicine & Rehabilitation, Icahn School of Medicine at Mount Sinai Hospital, New York, NY, USA

© Springer Nature Switzerland AG 2023
C. W Hunter et al. (eds.), *Regenerative Medicine*, https://doi.org/10.1007/978-3-030-75517-1_9

anchoring performance expectations in the echo that in vitro experiments will mirror the many facets of living systems. The plant ecologist Frank Egler, aside from a life work in entomology and contributions to Rachel Carlson's *Silent Spring*, remains recognized for his quip, "Nature is not more complex than we think, but more complex than we can think" [1].

The topic of exosomes, or exosome use for therapeutic care, is an emerging dynamic of past quarter century of cell-based strategies. In the confidence that regenerative medicine can align illness and refresh health, support for nearly every cell-based therapy as useful in some application has been suggested; paracrine factors from those cells offer imposing potential to manipulate the margins of useful care (Fig. 9.1). With the likelihood of exosomes being used as extensions of cell approaches, sourcing derived from either autologous marrow or fat, or allogeneic cells harvested from a young, healthy, master donor, or cells selected, separated, identified, and expanded for off-the-shelf convenience are first choice considerations. Like issues with stem cell clinic appropriation, and despite an underlying intention to revive regenerative potential by imbuing new life force, the more resulting reality has accepted a strategy overhauled if not forged by the fire of policy. Noted earlier, with a single exception to cord blood cell approval via FDA oversight, ungoverned access has led outcome perceptions that trump realistic expectations treatment might afford, and in the process have formulated a contradictory model for health care where the patient assumes the cost in a system that is decidedly tiered in third-party reimbursement. That market in the United States alone and separate from medical tourism approaches 300 M USD in revenue and remains shadowed by several billion-dollar global market that addresses cell acquisition, cell production in both expansion and subcul-

ture, and the cryoprotection of cells and cell products [2]. The use of exosomes is expected to parallel that utility, and applications of immune-oncology drive forge an entire exosome entity as an application.

Policy protects untoward use; no debate needed. But in scope, ideas, acceptance, and the stringent oversight designed to sieve suitable therapeutic use from risky and irresponsible marketing might also stunt inherent potential for advancing understanding. In particular, product commercialization in regenerative medicine has largely been drawn upon HCTP regulations that require less FDA approval. One downside of this direction of development is that without claims possibly gained through a Biologics License Application (BLA), many treatments have been sustained by inference-based clinical anecdotes without detailing or even attempting to show how or why they work.

In most cases, this has resulted in a dearth of evidence to adequately assess the clinical use of cell products, not for the lack of interest in outcomes, but for the economics required to conduct trials under tight inclusion and exclusion treatment. Such protocols have defined the approval process but may inadequately mimic the larger population. There are several reasons that make this a critical consideration. FDA and the scientific process have long based biologic success from the expectation that a proof of mechanism can be defined. While a single variable within the architecture of defined laboratory testing could possibly predict a clinical response seasoned with "p-values," authenticating complex biology that includes the dynamic of cell composition and performance would require a seer of nearly infinitesimal clairvoyance.

Choices that have been made seem purposeful; autologous stem cells offer the advantage of being immunologically matched but potentially limited in potency if harvested from older patients. Autologous stem cell approaches also

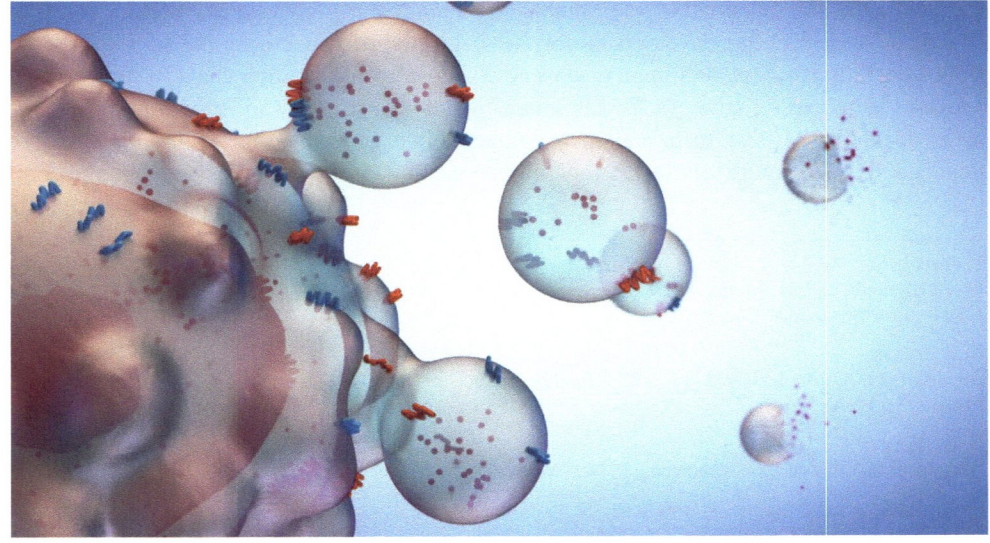

Fig. 9.1 Exosomes are a form of extracellular vesicle (EV) released from most kinds of prokaryotic and eukaryotic cells. Once thought to be mere detritus from cellular metabolism, they are now linked with many critical forms of cell signaling and immune function and play a vital role in a host of diseases, particularly cancer, where they may act to aid metastasis and thwart anticancer therapies. Graphic by Jason Drees

have the disadvantage of being difficult to standardize and scale, with quality assurance infrastructure hurdles that may limit widespread use by cost as much as efficacy. In disease modifying therapies (DMT) such as those employed for treating multiple sclerosis (MS), stem cell therapy has offered an effective regenerative action that overcomes limitations of currently available therapies. Now available all over the world, the cost of treatment varies in every country as per its own regulations and policies. Direct and indirect health care costs range from US$8528 to $54,244 per patient per year in the United States, with MS ranking second only to congestive heart failure in terms of price when compared with other chronic conditions [3, 4].

Because primitive cells have relatively few surface antigens and do not trigger acute immune responses, allogeneic transplants provide an attractive alternative, with the presumed benefit of being sourced from younger, healthier donors. Improved manufacturing margins also make allogeneic approaches an appealing model from a commercial perspective. As allogeneic cells engraft and differentiate, however, they can begin to express surface markers that are recognized by the immune system. So, while allogeneic stem cells may provide a powerful, short-term paracrine signal that may avoid triggering significant acute immune responses, long-term effects have been insufficiently studied to know whether either function or transparency will diminish over time.

Among the most avid customers for stem cell and platelet treatments are people with arthritis or sports injuries, those who find that mainstream medicine does not provide lasting relief. More than 30 million people in the United States suffer from osteoarthritis, including 14 million with bad knees. More than 700,000 knees are replaced in the United States each year, and more than half a million hips, according to the Arthritis Foundation. It is not surprising that a population of magnitude and prevalence will not seek out remedy for debilitating function.

In this regard, patients rely on testimonials and other informal evidence. But experts return to the admonition of use, cautioning that word-of-mouth experiences are not a substitute for rigorous studies; and in the context of containing policy, there are few abstaining from opinion in this forum. In many regards, the tide of caution protects the economics of expectations and actuarial economics of healthcare. A recent quote from the Center for Medical Technology Policy offers, *The power of anecdotes is just amazing when it just catches on.* As a nonprofit research group funded by pharmaceutical and reimbursement agencies, it is important to resist taking exception to comments such as, *This is how snake oil has been sold for generations* [5]. Perhaps there is enormous hubris in all technology and only with better understanding do we accept the new and notice the naïve we

have accented as state of the art. In a cynical context, an argument could be made that the body of knowledge forwarded from the past as innovation has been reluctant to separate the methods used to define it. Carried to the extreme analogy, extrapolating the way medical specialties have traditionally arisen, advances in technology support the expansion of knowledge in care delivery. For example, Röntgen's discovery of X-rays in 1895 provided an invention so remarkable that many did not believe the first reports of its use, referred to it mockingly as Dr. Röntgen's "alleged discovery of how to photograph the invisible" [6]. First radiologists, and now interventional radiologists, offer the intercept of acceptance of discovery that has evolved as imaging and medical intervention. More cautionary technologies bear merit as well. António Egas Moniz, a Portuguese neurologist, performed the first modern lobotomy, severing neural connections in the brain's frontal cortex to treat delusional or violent patients. Despite a lack of training as a surgeon, Moniz was awarded the Nobel Prize in Medicine or Physiology in 1948 for his invention [7]. A sequel society to lobotomists thankfully collapsed. And while the currency of creativity is rarely recognized at the outset, scientific process and hypotheses spring forward.

A similar revolution of understanding has emerged that has captured the interest of scientists, clinicians, industry, patients, and conjugally regulatory and reimbursement specialties as well. Unmasking the marvel of cell therapy has given paracrine function a bright introduction as a mechanism for cell therapy, complementing that measure of success with a cell communication language. Placed in the context of what is known about stem cells, or about extensive potential, where do exosomes fit in. They certainly are not tucked into a remote vacuum of the biologic process, and as conjugate as gear teeth auger biologic process in predictable patterns that in turn will fail to acknowledge or even be aware of the next truth to be elucidated.

Three basic premises are accepted in defining a stem cell. First is self-renewal: stem cells divide to produce identical daughter cells and thereby maintain the stem cell population. Second, stem cells divide asymmetrically to yield an identical cell and a daughter cell that acquires specific morphology, phenotype, and physiological properties that categorize it as a cell belonging to a particular tissue. The third property of stem cells is that they may renew the tissues that they populate. This asset in the evolution of understanding stem cells offers growing support that phenotypic expression can occur via lateral genetic transfer in secretion and inclusion of liposomal inclusions. And that is where exosome biology has served as a translator of potential that accommodates recognition, accentuates regeneration, and accounts potential for restorative potential for diminishing tissue performance.

Indications

Under current regulations, FDA has defined therapeutic claims based on the use of exosomes to be regulated as a biologic [8]. Clearly, without an overture to remove exosomes and given their role in cell-cell communication, there has been exosome transplantation since the advent of allograft tissue use. It is similarly apparent that the ubiquitous literature on serum-supported cultures has also contributed to elaboration of cell roles in regeneration as it is known that content, activity, and cell reaction are all governed by exosome content in serum and that serum content has been developed as well as a diagnostic tool [9].

Indications for use transcend musculoskeletal as would be expected, given their universal tissue presence in organ, tissue, cells, blood, and lymphatic location. At the time of this publication, there are more than 70 current studies ongoing in the United States on a variety of indications that span a range of sources of exosomes and tissue types to be treated. It is not surprising that exosomes derived from plant material are similarly being investigated for therapeutic, nutritional, or cosmetic uses [10]. A recently published review addresses the endogeneity and heterogeneity of exosomes affording extensive and unique advantages in the field of disease, diagnosis, and treatment. That offered, however, the authors hasten clarity on understanding the storage stability, low yield, low purity, and weak targeting of exosomes that limit current clinical applications [11].

Exosomes have a wide utility due to the general ease of use – this ranges from hair loss and general cosmetic use to osteoarthritis and chronic obstructive pulmonary disease (COPD). It should be noted that many of the "uses" of exosomes are still under investigation in the United States and/or considered "experimental" at the time of this publication (i.e., cancer, chronic kidney disease, bronchopulmonary dysplasia, genetic disorders, etc.).

Many of the popular musculoskeletal uses of exosomes do not have a great deal of data. For example, exosomes are commonly offered for osteoarthritis (OA) . While this is a generally benign diagnosis for one to attempt regenerative therapy, there is very little data to support the use of exosomes. In fact, exosomes are an intercellular communication mediator that is known to contribute to and maintain OA. There is some animal research to suggest that exosomes may slow the progression of early osteoarthritis and prevent severe knee articular cartilage damage; however, nothing has been formally studied in humans to show this benefit. As such, the authors do not recommend using exosomes for the treatment of joint OA at this time.

Based on the currently available data, the following is a list of possible indications with which exosomes could be potentially considered:

- Rotator cuff tendinopathy – specifically the supraspinatus (based on animal studies using tendon-derived exosomes)
- Osteoporotic fractures via upregulation of osteoblast genes (based on animal studies)
- Achilles tendonitis or tendinosis (based on animal studies)
- Decrease muscle atrophy and improve muscle regeneration associated with rotator cuff tears (based on animal studies using adipose-derived exosomes)
- Muscle strain (based on animal studies in the tibialis anterior using exosomes derived from platelet-rich plasma and mesenchymal stem cells)
- Skeletal muscle atrophy and/or defect (based on animal studies using exosomes derived from mesenchymal stem cells)

Please make note that the evidence for the above indications is basic science level and only studied on laboratory animals – not in humans. As such, caution should be taken to utilize exosome therapy within the confines of federal and state guidelines.

Future Perspectives

Data suggests that EVs and exosomes may have clinical applications. EVs and exosomes

have many characteristics that make them ideal drug-delivery vehicles: they can contain both proteins and genetic material, are well tolerated in the body as demonstrated by their presence in all biological fluids, and are able to cross the plasma membrane to release their contents within target cells. Because of the lack of size limitations, the intrinsic ability to target tissues might be modified to enhance cell-type specific targeting. Commercial stem cell and viable allograft products today contain exosomes. Given the challenge of separating them from cell preparations, they are an endemic component of tissue transfer products. It is also known that donor material from tissue banks is laden with exosomes and microRNA material as well. It is not surprising to understand that transfer of allograft material, sourced from non-self donor, constitutes an exosome-based lateral gene transfer. A similarly acceptable premise given the demonstration of exosome transfer is that tissue pairing might be used to define the direction if not the dimension of differentiation and phenotypic adoption. As immunogenicity is low, thus facilitating the use of allogenic donors, isolated cells might be expanded, manipulated, and modified to affect specific EVs for defined targets [12].

This hypothesis opens novel therapeutic opportunities, given that the harvested EVs can also be modified by the addition of mRNA or miRNA, or by loading with therapeutic drugs [13, 14]. Several considerations for exosome delivery have been proposed, either in vivo during their biogenesis or in vitro after their purification although the in vivo drug loading of both EVs and exosomes requires an in-depth understanding of their biogenesis [15]. The final step is that administration of EVs depends on the target disease; EVs administered intravenously or intranasally cross the blood-brain barrier and can deliver the cargo directly into the brain [16]. In vivo studies have already demonstrated the beneficial effects of intravenously injected exosomes in tissue repair [17].

In conclusion, EVs and exosomes released from stem cells mimic the effect of the cells, suggesting a potentially valuable role in regenerative medicine. It is however essential to understand the mechanisms responsible for microparticle release and the selective enrichment of paracrine factors and RNAs. Most cell types have been shown to release exosomes into the extracellular environment. Possible advancements over natural exchange might offer the inclusion of proteins and nucleic acids within EVs that would further protect them from the extracellular environment during early phases of inflammation.

Exosomes demonstrate several possible advantages over stem cells in terms of their use in regenerative medicine. Importantly, they are more stable and induce stronger signaling and are produced in higher concentrations than stem cells. Additionally, exosomes as delivery vehicles possess intrinsic homing abilities relative to other synthetic particles, thus avoiding unwanted accumulation in organs other than the target tissue. They demonstrate no inherent toxicity, are not associated with any long-term maldifferentiation, and carry little risk of immune rejection following in vivo allogenic administration [18]. Progenerate effects mediated by EVs might gain imposing potential introducing drugs, in addition to extant miRNA and autologous cargo involved in healing and regeneration. Further studies are needed to further clarify the stimuli and pathways regulating the assembly of bioactive molecules within vesicles. Understanding the subtleties of generative response tolls a language that baffles the borders of collective thought. Taking for granted a single signal–single response is insufficient strategy to translate the next iteration of biology.

In the understanding of the communication, do the signals triggering their release adhere to a more complex physics of membrane charge, juggling anionic and cationic balance as the fine-tuning mechanism? Using the empiric of 3000 exosomes of 100 nanometer from each cell, a stimulated exchange of 31,400 nanometers could be considered. Using a 30-micron cell as a nominal size (and with the shape simplicity of a sphere considered, 15,000-nm radius), a surface of 2.82 billion square nanometers emerges as nearly 0.11% of surface area of a cell. Although this seems an inconsequential contribution, novel work underway examining the heterogeneous distribution of lipids has begun to calibrate the impact of membrane charges on regulating the association of proteins with the plasma membrane [19]. Charged lipids are asymmetrically distributed between the two leaflets of the plasma membrane, resulting in the inner leaflet being negatively charged and a surface potential that attracts and binds positively charged ions, proteins, and peptide motifs. These interactions not only create a transmembrane potential but also facilitate formation of charged membrane domains. Other investigations have evaluated cell protectants, charged coatings, as a means of integrating charge variation as a direct correlate of exosome size and charge [20]. The context of charge, time varying change, cargo deployment, and physics attends a unique conjunction of biology that shifts the paradox of paracrine to a much deeper entrenchment that is based on voltage and physiology.

The detail needed to understand more fully remains staggering, but the thrust and vector of regenerative care continue to pursue health that even if asymptotic is directional to provide a higher level of health and healthcare.

Microanatomy and Biochemistry

Cell-to-cell communication has been known for some time to coordinate development among different cell types within adult tissues. Cells have been known to communicate via secreted molecules and by cell surface molecules, or by direct cell-to-cell contact by specialized molecules in connected channels that syncopate both charge and attachment to vary membrane current, protect the phenotypic physiology needed to buffer complex conditions, and facilitate an exchange of cell activity and cycle in a reciprocity that is both restorative and resolvable [21]. These molecules control cell migration, proliferation, differentiation, and apoptosis with inherent variation to tissue type but sustain an emphasis where sustainable biologic resolution does not sacrifice the integral nature of the organ to preserve the individual cellular makeup. Recognizing biologic integration aside from the immediate adjacency to cells led to the discovery of the extracellular matrix (ECM) interaction and its mediation by a broad spectrum of receptors including syndecans, dystroglycans, and integrins [22]. This understanding occurred as a parallel ascent of discovery, wherein research and innovation uncovered still another codex of biologic communication that emerged in the broad description as paracrine was further identified as extracellular vesicles.

Microvesicles released by cells represent the evolving mediator of cell-to-cell communication and are also an integral part of the intercellular microenvironment [23–25]. This new scenario accentuates an understanding of signal and molecule transfers between cells, not only locally but more importantly over distances connecting organs as much as cells. The presence of microvesicles in the extracellular space was initially reported by Chargaff and West, as a precipitable factor in platelet-free plasma and then again in 1960 [26, 27]. For many years microvesicles were considered to be inert cellular debris until De Broe et al. suggested that microvesicles released from human cells may result from a specific process [28]. It is now accepted that most cell types release microvesicles (e.g., epithelial [29], fibroblast [30], hematopoietic [31], immune [32], tumor [33]), and recent studies indicate that these vesicles may have crucial roles in both physiological and pathophysiological processes in the emerging biology of stem cells [34].

Paracrine Expression and Phenotypic Modulation

Within the metric of a nano dimension, exosome biology illuminates a bright potential; − latent context of ever smaller indivisibles that might be used to govern fate, or at least help conquer control of errant processes. As an extension of the reductionist empiric to think of "Biology, as the most lawless of the sciences" according to Mukherjee [35], enabling a potential that continues to contend for conscripted understanding does not seem a reach but again contends for the visage of technology so impressive that it seems nearly magic.

Research follows the imagination and fortunately the original illumination affords many a better view. Over the past 5 years, the focus on exosomes as sources of gain of function biologics has soared. Publications in this field have reached nearly 1000 manuscripts per year and continue to grow as more information emerges and more interests attend better understanding. In the context of the chapter heading, a short description of exosomes will introduce activity, modulation potential, and give a few examples currently being

developed based on both in vitro and in vivo observations. Cardiovascular, liver, neuroscience, orthopedics, and intervertebral disc applications span the breadth of interest and broad applicability. Secretion of cytoplasmic components in encapsulated membrane vesicles is emerging as acknowledged pathway of cell-cell communication. Bearing the heritage of phylogenetic conservation tracing back to bacteria, almost all living cells secrete membrane vesicles with functions varying from defense against viral attack, exchange of genetic material, and transfer of neurotransmitters or cytokines to more capricious characterizations as phenotypic inflection and durable maintenance of physiology [36, 37]. Lane links several existing treatises as a cogent argument for endosymbiosis and eukaryotic existence. His contention of lateral transfer, amoebic fusion, and nuclear membrane chimerism to accent energy support for cytoskeleton elaboration offer fascinating contexts for regenerative medicine. Given the understanding of cell metabolics, he cleverly demonstrates purpose in attachment and conveyance as more valuable than duplication of DNA and cell division. The thesis at every step is about proton transfer and avoiding the efficiency of entropy.

Although all cells sustain the function, mammalian cells export three types of microvesicles (MV): exosomes, microparticles, and apoptotic bodies, each originating from different subcellular compartments (Fig. 9.2). Exosomes differ in the presence of certain markers on the their surface - exosomes contain proteins such as CD9, CD63, Alix, flotillin-1, Tsg101 and clathrin and low amounts of phosphatidylserine, while microvesicles expose high amounts of phosphatidylserine, and feature proteins associated with lipid rafts and are enriched in cholesterol, sphingomyelin and ceramides [38, 39]. Both exosomes and microparticles contain coding and non-coding RNA (microRNA) in addition to proteins and lipids with biological functions.

Apoptotic bodies originate in membrane blebbings produced by the cells undergoing cell death and are much larger

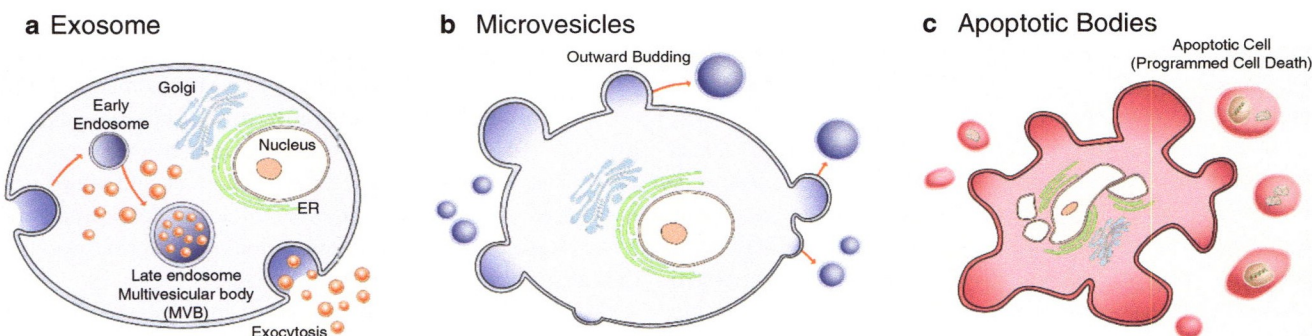

Fig. 9.2 Schematic representation of extracellular vesicles. From: Gurunathan et al. [61]. This is an open access article distributed under the Creative Commons Attribution License which permits unrestricted use, distribution, and reproduction in any medium, provided the original work is properly cited (CC BY 4.0)

than exosomes and microvesicles (500–2000 nm in diameter). They contain cytoplasm with packed organelles and may contain nuclear fragments, essentially strategic packaging to prevent leakage of cell content and avoid inflammation. Understanding the structure and some of the biochemistry is instructive, but grasping the biologic impetus for release and harnessing the technology to predictably replicate or simulate cell strategy are the technologies that beckon.

Early work focused on diagnostic potential as a means of predictive analytics to align membrane presence with specific malady. Alterations in count and conditions lead to several observations of vesiculation and conditions causal to alterations of the MV content responsive to shear stress and even storage conditions [40, 41].

Exosomes are small vesicles (30–150 nm) containing sophisticated RNA and protein cargos (Fig. 9.3). Secreted by all cell types in culture, they are found naturally in body fluids, including blood, saliva, urine, CSF, amniotic fluid, and breast milk. MSCs also synthesize and secrete functional exosomes that are cholesterol-rich phospholipid vesicles. Steering within the dimensions of regenerative medicine in the context of extending mechanism to the use of clinical use of stem cell products and outlining where mesenchymal stem cells have been considered for therapeutic intervention provides a somewhat restrictive discussion that is manageable and applicable to wider evaluation.

Arnold Caplan is credited with establishing a large presence of mesenchymal stem cell awareness and he is recognized internationally for his contributions to the science of stem cells. Coining the term MSC, an entire sector of regen-

erative biology was borne on the premise that specific cells are endowed with pluripotency and remain responsive to injury to the extent of modeling to mend. In works too numerous to cite, he opened many minds to the potential for cell-based therapy to auger an evolution of care in regenerative applications [42–45]. The fruition of thought has been self-evident in a very public and progressive understanding of the field based on emerging biology – developing first as cell registration or identity as the guide to activity and evolving as a description of *Medicinal Signaling Stores* (MSC remains the same acronym) as drug stores for sites of injury or inflammation [46]. In essence, perivascular cells have assumed an identity responsive as site-regulated, multi-drug delivery vehicles. MSCs in the activation and transitions to bone are guiding not only the osteogenic but trophic and immunomodulatory functions as well – nominally functional in resonating potency, but no longer stromal cells. These cells are responding to small packets of RNA packaged in lysosomal inclusions known as exosomes. An awareness that cell-to-cell transport will stimulate adoption in record time developed a basis seeded in a foundation of biologic regeneration – long ago accepted as a strategic initiative to eclipse device-based replacement. Occurring in balance and recognizing Janus-like dual functionality of TGF-β (pro-differentiation, growth-promoting, wound repair contrasted to anti-proliferation and pro-cell death), a foreshadowed context and concentration dependency served a concept that diverse functions impacting multiple cell types including immune function occur with regularity. Recent advances in understanding SMAD activities in the nucleus are beginning to reveal the molecular basis of the pleiotropic function of TGF-β and its ability to both promote and buffer functions (Fig. 9.4) [47].

Although the transition of attributing cell therapy mechanism to exosomal transfer of potency is in its infancy with regard to therapeutic translation, the role of MiRNAs in post-transcriptional regulation of gene expression has been implicated in the pathogenesis of disease for some time [48, 49]. Bone marrow-derived MSCs release exosomes that can promote breast cancer cell dormancy in a metastatic niche or protect the vascularization and suppress risk [50]. In keeping with the likelihood that cells considered as sources for cell therapy will be adopted similarly for exosome consideration, adipose MSCs also have been shown to secrete exosomes and microvesicles, and in turn regulate angiogenic potential of MSCs [51]. Moreover, injection of exosomes from MSCs into stroke rats has similarly been shown to relieve symptoms by promoting angiogenesis, neurite remodeling, and neurogenesis [52]. The precise molecular mechanics for their secretion and uptake, as well as their composition, "cargo," and resulting functions, is incompletely understood and has only begun to be unraveled. Exosomes are viewed as specifically secreted vesicles that enable intercellular com-

Fig. 9.3 Exosomes are small vesicles (30–150 nm) containing sophisticated RNA and protein cargos (Fig. 9.2). Secreted by all cell types in culture, they are found naturally in body fluids, including blood, saliva, urine, cerebrospinal fluid, amniotic fluid, and breast milk. Note legend and scaling in this magnification. Image courtesy of Renaud Sicard

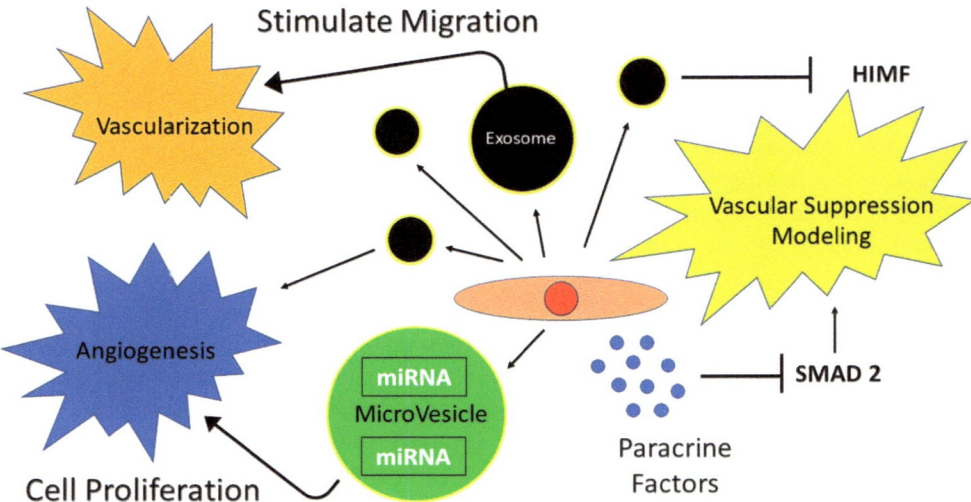

Fig. 9.4 Effects of MSCs on vasculature. MSCs secrete exosomes that facilitate endothelial cell migration, and thus contribute to vascularization. Angiogenic action of MSCs is mediated by MSCs-secreting EVs and MVs. MSCs generate exosomes and paracrine factors to inhibit HIMF and Smad2, exerting an effect as antivascular remodeling; in essence vascular suppression, effect, and balance wholescale ramification of endothelial sprouting (MSC) – Mesenchymal Stem Cell; (MV) – MicroVesicles; (HIMF) – Hypoxia-Induced Mitogenic Factor; (Smad-2)- signal transduction factor for the (TGF-B) –transforming growth factor beta

Fig. 9.5 Mechanisms whereby EVs achieve biologic effects. EVs activate cell signaling by physical ligand-receptor interactions, or by fusing with their target cells and transferring their contents. They may also be endocytosed by the target cells or may release their contents into the extracellular space. Dotted line depicts and delineates three mechanisms of exosome-cell effects

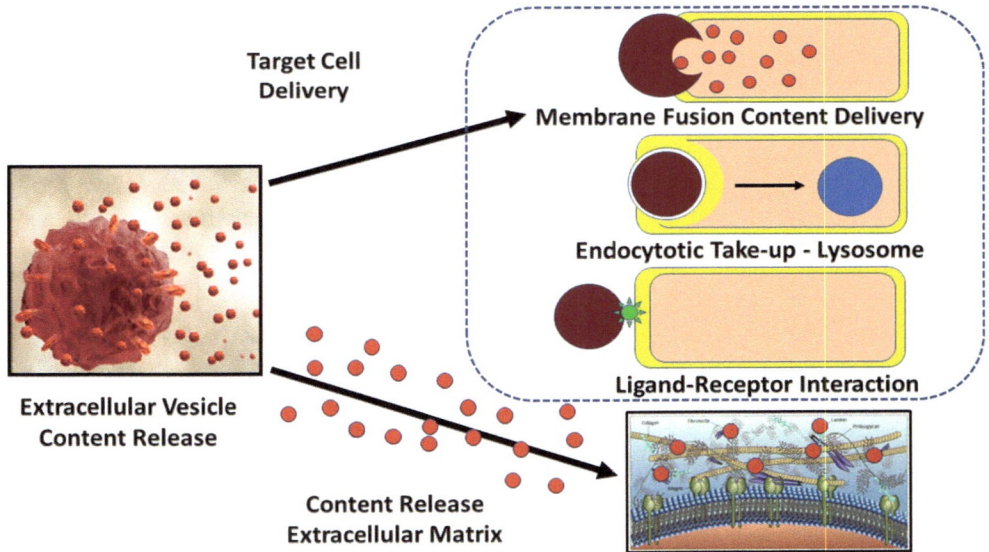

munication and have become the focus of exponentially growing interest, both to study their functions and to understand ways to use them in the development of minimally invasive diagnostics (Fig. 9.5).

It is indisputable that embryology is little more than an orchestrated wave of cell expansion, organ differentiation, coalescing form that inevitably gives way to loss of function, deterioration or disease, and extinction of body shape that had provided the resonance we recognize as individual identity. The model for regeneration has been to auger the echo of development as a reset that will allow rejuvenated potential to use tissue morphology as the cornerstone and scaffold to reflect both size and anatomical appropriate anatomy. Some of the first cells to be used as an "adult stem cell" have taken advantage of the isolation and awareness of adipose, umbilical cord, and bone marrow stem cells and the possibility of banking for future autologous use or for allogeneic applications, given the low risk of immunologic rejection and the relatively "naïve" state of development.

Recent attention has focused on the capacity of EVs to alter the phenotype of neighboring cells to make them resemble EV-producing cells. Stem cells are an abundant source of EVs, and the interaction between stem cells and the microenvironment (i.e., stem cell niche) plays a critical role in deter-

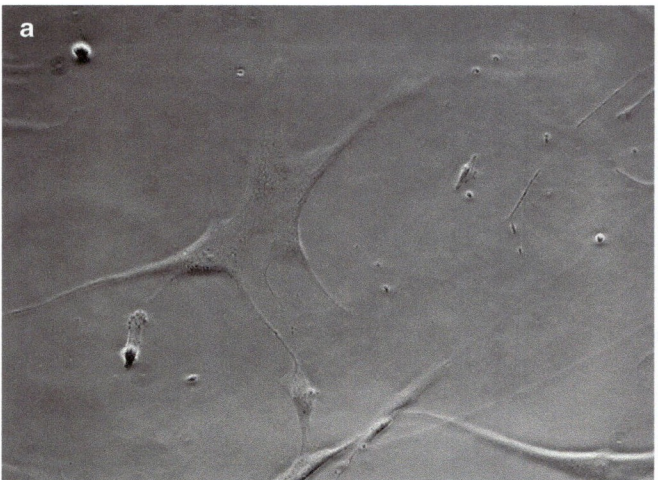 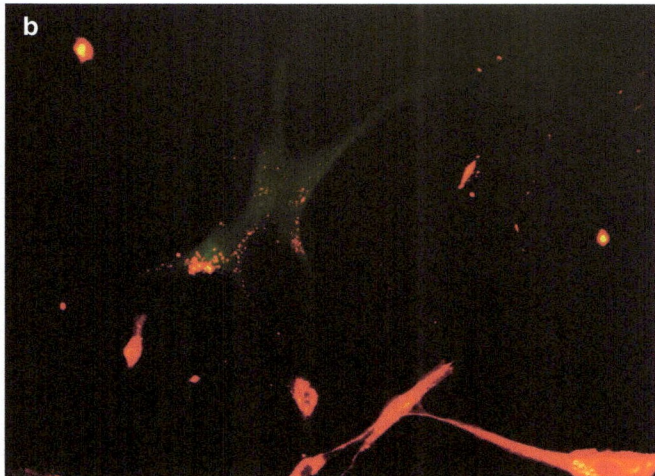

Fig. 9.6 Inter-cellular material exchange. (**a**) Bone marrow isolated cells and commercially available mesenchymal stem cells were co-cultured for 3 days. Bone Marrow cells were stained with DIL (fluorescent lipophilic cationic indocarbocyanine dye) and MSCs were stained with CFDA (Carboxyfluorescein diacetate) to highlight actin filaments.

(**b**) Exchange between cells of different origin apparently shares exosome and cellular material is evident under fluorescent illumination demonstrating the transfer of material from the bone marrow lineage to the mesenchymal cell lineage

mining stem cell phenotype. The stem cell niche hypothesis predicts that stem cell number is limited by the availability of niches releasing the necessary signals for self-renewal and survival, and the niche thus provides a mechanism for controlling and limiting stem cell numbers. EVs may play a fundamental role in this context by transferring genetic information between cells. EVs can transfer mRNA and microRNA to target cells, both of which may be involved in the change in target-cell phenotype toward that of EV-producing cells. The exchange of genetic information may be bidirectional, and EV-mediated transfer of genetic information after tissue damage may reprogram stem cells to acquire phenotypic features of the injured tissue cells (Fig. 9.6). In addition, stem cell–derived EVs may induce the de-differentiation of cells that survive injury by promoting their reentry into the cell cycle and subsequently increasing the possibility of tissue regeneration. The functions of extracellular vesicles depend on the phenotype of their parental cells. Although their cargo reflects the cell from which they were released, selective enrichment of specific molecules has been shown to occur [53]. Ongoing work has shown that both somatic and tumor cells are capable of exchanging exosome material [54].

EVs influence target-cell behavior in several ways. They can act as a signaling complex, transfer membrane receptors between cells, deliver proteins to target cells, and modify the receiving cells by horizontal transfer of genetic information. Membrane vesicles act as signaling complexes during development. EVs play an important role in developmental signaling and morphogenesis in multicellular organisms [55]. The formation of morphogen gradients is essential for tissue patterning, and morphogens are generally released from pro-

ducing cells and spread through adjacent tissues. For example, some cells express developmental gradients during tissue differentiation by secreting specific proteins such as Hedgehog, Wingless, or Decapentaplegic [56]. Morphogens tightly associated with the cell membrane are released via morphogen-enriched vesicles, thus creating a morphogen gradient. This functional facet of vesicle and exosome technology in regenerative medicine is an exceptional and intriguing context to consider. Can an acellular product complement, or even replace, the use of cells?

Work in this area is just emerging but direct experience makes one caution the classic transitive assurance that conditions and outcomes are identical. Biological nanoparticles are such a rapidly growing area of research in the life sciences and nanomedical field that in vitro measurement of multiple physical parameters such as size, concentration, surface charge, and phenotype characteristics has developed technology to match the interest. Particle Metrix ZetaView® (Particle Metrix, Inning, Germany) was used to for isolating exosomes from bone marrow and to quantify as approximately 3000 per cell. Given that calculation, acellular tests were run for osteoinductivty using a standard alkaline phosphatase assay for OI. Biologic activity at dosages in the 10 billion exosomes was as bioactive as 50 ng of BMP-4. Considering the number of cells required to yield a similar exosome count, three million cells would be needed. What makes the story compelling is that exomes are thermostable and retain potential during storage. Noting briefly the stem cell market valuation and the dependence on cryoprotection, such possibility is appealing and economically attractive to broader distribution with less cost.

Basic Concerns and Contraindications

Given the prevalence of exosomes, it should not be surprising that tumor cells can also secrete a large number of exosomes, and the specific antigens on their surface can reflect the nature of donor cells [57, 58]. Therefore, tumor exosomes have attracted great attention in cancer research. Tumor exosomes not only play an important role in the process of tumor growth, metastasis, and immune regulation but can also monitor the development of diseases and serve as diagnostic markers for diseases [59]. Not intending to castigate the use of exosomes to risk, the message is to underscore the need for critical understanding of the diagnostic potential enabled by identifying. As exosome therapeutics has yet to emerge from clinical trials, as an immature field it will need time for the technology approval to balance the biologic potential. As such, the use of exosomes should be limited to research and clinical trials.

Equipment, Techniques, Extraction, and Purification

Given the surging in-depth studies of exosomes, potential application value has been continuously tapped and unique and diverse technologies have been applied to amplification, identification, collection, and delivery possibility, all critical to the regulatory approval where purity, stability, potency, and predictability are critically assessed. Reproducible isolation and enrichment of exosomes will help assess their biological functions and elucidate what can be predictably and safely administered. However, as exosomes are heterogeneous in size, content, function, and source, the metrics mentioned are challenging if not difficult to develop [60].

Currently, the science of efficiently enriching exosomes is a major issue and one which is crucial for downstream analysis of exosomes. For different purposes and applications, different isolation methods are selected, among which ultracentrifugation, size-based isolation techniques, polymer precipitation, and immunoaffinity capture techniques are more commonly used. The separate discussion of these techniques strays from the topic intended. A process hastened beyond understanding carries risk, and exosomes are in many regards able to laterally shift genetic material that alters phenotype. While exosomes have always been an extant asset of biological systems, modification and metrics of identified functions are incompletely understood. A future of designer exosomes might one day capitalize on unique therapeutic specificity. For now, the narrow province of understanding is not nested in the broad universe of anticipated uses. As information is gathered, and proofs accepted, exosome therapeutics will hopefully be a mainstay of interventional medicine.

Clinical Pearls
- There is little-to-no data currently available on the safe and effective use of exosomes for musculoskeletal pathologies.
- Due to the specificity with which they are derived/extracted (i.e., adipose-derived, tendon-derived, etc.), exosomes could, in theory, provide more tissue-specific healing and potentially lead to greater improvements.
- As this aims to be an "off-the-shelf"-type treatment similar to allograft or corticosteroids, harvesting and in-office processing would not be a requirement.

References

1. Carson R, Darling L, Darling L. Silent Spring. Boston: Houghton Mifflin; 1962.
2. Grand View Research, Stem Cell Market Size Analysis Report By Product (Adult, hESC, Induced Pluripotent), By Application (Regenerative Medicine, Drug Discovery & Development), By Technology, By Therapy, And Segment Forecasts, 2019–2025, April 2019.
3. Shroff G. A review on stem cell therapy for multiple sclerosis: special focus on human embryonic stem cells. Stem Cells Cloning. 2018;11:1–11.
4. Multiple sclerosis by the numbers: facts, statistics, and you. Accessed 20 Sept 2017. Available from: http://www.healthline.com/health/multiple-sclerosis/facts-statistics-infographic.
5. Grady D, Abelson R. Stem cell treatments flourish with little evidence that they work. NYTimes. 13 May 2019.
6. Nitske RW. The Life of W. C. Röntgen, Discoverer of the X-Ray. Tucson: University of Arizona Press; 1971.
7. Dittrich L. Patient H.M.: A story of memory, madness and family secrets. New York: Random House Audio; 2016.
8. https://www.fda.gov/vaccines-blood-biologics/consumers-biologics/important-patient-and-consumer-information-about-regenerative-medicine-therapies
9. Jung HH, Kim JY, Lim JE, et al. Cytokine profiling in serum-derived exosomes isolated by different methods. Sci Rep. 2020;10:14069. https://doi.org/10.1038/s41598-020-70584-z.
10. Dad HA, Gu T-W, Zhu A-Q, Huang L-Q, Peng L-H. Plant exosome-like nanovesicles: emerging therapeutics and drug delivery nanoplatforms. Molecular Therapy. 2021;29(1):13–31, ISSN 1525–0016,. https://doi.org/10.1016/j.ymthe.2020.11.030.
11. Zhang Y, Bi J, Huang J, Tang Y, Du S, Li P. Exosome: a review of its classification, isolation techniques, storage, diagnostic and targeted therapy applications. Int J Nanomed. 2020;15:6917–34. Published 2020 Sep 22. https://doi.org/10.2147/IJN.S264498.
12. Uccelli A, Moretta L, Pistoia V. Mesenchymal stem cells in health and disease. Nat Rev Immunol. 2008;8:726–36.
13. Alvarez-Erviti L, Seow Y, Yin H, Betts C, Lakhal S, Wood MJ. Delivery of siRNA to the mouse brain by systemic injection of targeted exosomes. Nat Biotechnol. 2011;29:341–5.
14. Zhuang X, Xiang X, Grizzle W, Sun D, Zhang S, Axtell RC, Ju S, Mu J, Zhang L, Steinman L, Miller D, Zhang HG. Treatment of brain inflammatory diseases by delivering exosome encapsulated anti-inflammatory drugs from the nasal region to the brain. Mol Ther. 2011;19:1769–79.

15. Lai RC, Yeo RW, Tan KH, Lim SK. Exosomes for drug delivery – a novel application for the mesenchymal stem cell. Biotechnol Adv. 2013;31:543–51.

16. Alvarez-Erviti L, Seow Y, Yin H, Betts C, Lakhal S, Wood MJ. Delivery of siRNA to the mouse brain by systemic injection of targeted exosomes. Nat Biotechnol. 2011;29:341–5.

17. Baglio SR, Pegtel DM, Baldini N. Mesenchymal stem cell secreted vesicles provide novel opportunities in (stem) cell-free therapy. Front Physiol. 2012;3:359.

18. Bruno S, Grange C, Deregibus MC, Calogero RA, Saviozzi S, Collino F, Morando L, Busca A, Falda M, Bussolati B, Tetta C, Camussi G. Mesenchymal stem cell-derived microvesicles protect against acute tubular injury. J Am Soc Nephrol. 2009;20:1053–67.

19. Ma Y, Poole K, Goyette J, Gaus K. Introducing membrane charge and membrane potential to T cell signaling. Front Immunol. 2017;8:1513. https://doi.org/10.3389/fimmu.2017.01513.

20. Lim, et al. Development of a robust pH-sensitive polyelectrolyte ionomer complex for anticancer nanocarriers. Int J Nanomedicine. 2016;11:703–13.

21. Martin PE, Evans WH. Incorporation of connexins into plasma membranes and gap junctions. Cardiovasc Res. 2004;62:378–87.

22. Hynes RO. Integrins: bidirectional, allosteric signaling machines. Cell. 2002;110:673–87.

23. Cocucci E, Racchetti G, Meldolesi J. Shedding microvesicles: artefacts no more. Trends Cell Biol. 2009;19:43–51.

24. Majka M, Janowska-Wieczorek A, Ratajczak J, Ehrenman K, Pietrzkowski Z, Kowalska MA, Gewirtz AM, Emerson SG, Ratajczak MZ. Numerous growth factors, cytokines, and chemokines are secreted by human CD34(_) cells, myeloblasts, erythroblasts, and megakaryoblasts and regulate normal hematopoiesis in an autocrine/paracrine manner. Blood. 2001;97:3075–85.

25. Ratajczak J, Wysoczynski M, Hayek F, Janowska-Wieczorek A, Ratajczak MZ. Membrane-derived microvesicles: important and underappreciated mediators of cell-to-cell communication. Leukemia. 2006;20:1487–95.

26. Chargaff E, West R. The biological significance of the thromboplastic protein of blood. J Biol Chem. 1946;166:189–97.

27. Behnke O. Electron microscopical observations on the surface coating of human blood platelets. J Ultrastruct Res. 1968;24:51–69.

28. De Broe ME, Wieme RJ, Logghe GN, Roels F. Spontaneous sheddingof plasma membrane fragments by human cells in vivo and in vitro. Clin Chim Acta. 1977;81:237–45.

29. Lee H, Zhang D, Wu J, Otterbein LE, Jin Y. Lung epithelial cell-derived microvesicles regulate macrophage migration via MicroRNA-17/221-induced integrin β_1 recycling. J immunol (Baltimore, Md. : 1950). 2017;199(4):1453–64.

30. Hu L, Wang J, Zhou X, Xiong Z, Zhao J, Yu R, et al. Exosomes derived from human adipose mensenchymal stem cells accelerates cutaneous wound healing via optimizing the characteristics of fibroblasts. Sci Rep. 2016;6:32993. https://doi.org/10.1038/srep32993.

31. Butler JT, Abdelhamed S, Kurre P. Extracellular vesicles in the hematopoietic microenvironment. Haematologica 2018. 2018;103(3):382–94.

32. Wu R, Gao W, Yao K, Ge J. Roles of exosomes derived from immune cells in cardiovascular diseases. Front Immunol. 2019;10:648.

33. Kuzet SE, Gaggioli C. Fibroblast activation in cancer: when seed fertilizes soil. Cell Tissue Res. 2016;365:607–19.

34. Malda J, Boere J, van de Lest CH, van Weeren PR, Wauben MH. Extracellular vesicles - new tool for joint repair and regeneration. Nat Rev Rheumatol. 2016;12:243–9. https://doi.org/10.1038/nrrheum.2015.170.

35. Mukherjee S. The gene: an intimate history. London: Bodley Head; 2016.

36. Wang Y, Chen LM, Liu ML. Microvesicles and diabetic complications – novel mediators, potential biomarkers and therapeutic targets. Acta Pharmacol Sin. 2014;35(4):433–43.

37. Lane N. The vital question: energy, evolution, and the origins of complex life. New York: WW Norton and CO; 2015.

38. Elmore S. Apoptosis: a review of programmed cell death. Toxicol Pathol. 2007;35(4):495–516.

39. Rosca A, Rayia DMA, Tutuianu R. Emerging role of stem cells - derived exosomes as valuable tools for cardiovascular therapy. Curr Stem Cell Res Ther. 2017;12:134–8.

40. Miyazaki Y, Nomura S, Miyake T, Kagawa H, Kitada C, Taniguchi H, Komiyama Y, Fujimura Y, Ikeda Y, Fukuhara S. High shear stress can initiate both platelet aggregation and shedding of procoagulant containing microparticles. Blood. 1996;88(9):3456–64.

41. Bode AP, Orton SM, Frye MJ, Udis BJ. Vesiculation of platelets during in vitro aging. Blood. 1991;77(4):887–95.

42. Caplan AI. Mesenchymal stem cells. J Orthop Res. 1991;9(5):641–50.

43. Bruder SP, Fink DJ, Caplan AI. Mesenchymal stem cells in bone development, bone repair, and skeletal regeneration therapy. J Cell Biochem. 1994;56(3):283–94.

44. Caplan A, Correa D. The MSC: An injury drugstore. Cell Stem Cell. 2011;9(1):11–5.

45. Caplan AI. MSCs: the sentinel and safe-guards of injury. J Cell Physiol. 2016;231(7):1413–6. Epub 2015 Nov 26

46. Caplan AI. Mesenchymal stem cells: time to change the name! Stem Cells Transl Med. 2017;6:1445–51.

47. Li MO, Flavell RA. TGF-beta: A master of all T cell trades. Cell. 2008;134(3):392–404.

48. Trams EG, Lauter CJ, Salem N Jr, Heine U. Exfoliation of membrane ecto-enzymes in the form of micro-vesicles. Biochim Biophys Acta. 1981;645(1):63–70.

49. Lee RC, Feinbaum RL, Ambros V. The C. elegans heterochronic gene lin-4 encodes small RNAs with antisense complementarity to lin-14. Cell. 1993;75:843–54.

50. Ono M, Kosaka N, Tominaga N, Yoshioka Y, Takeshita F, Takahashi RU, et al. Exosomes from bone marrow mesenchymal stem cells contain a microRNA that promotes dormancy in metastatic breast cancer cells. Sci Signal. 2014;7:ra63. https://doi.org/10.1126/scisignal.2005231.

51. Lopatina T, Bruno S, Tetta C, Kalinina N, Porta M, Camussi G. Platelet-derived growth factor regulates the secretion of extracellular vesicles by adipose mesenchymal stem cells and enhances their angiogenic potential. Cell Commun Signal. 2014;12:26.

52. Xin H, Li Y, Cui Y, Yang JJ, Zhang ZG, Chopp M. Systemic administration of exosomes released from mesenchymal stromal cells promote functional recovery and neurovascular plasticity after stroke in rats. J Cereb Blood Flow Metab. 2013;33:1711–5. https://doi.org/10.1038/jcbfm.2013.152.

53. Li CC, Eaton SA, Young PE, Lee M, Shuttleworth R, Humphreys DT, Grau GE, Combes V, Bebawy M, Gong J, Brammah S, Buckland ME, Suter CM. Glioma microvesicles carry selectively packaged coding and non-coding RNAs which alter gene expression in recipient cells. RNA Biol. 2013;10:1333–44.

54. Weston WW, Ganey T, Temple HT. The relationship between exosomes and cancer: implications for diagnostics and therapeutics. BioDrugs. 2019;33(2):137–58. https://doi.org/10.1007/s40259-019-00338-5. Review

55. Greco V, Hannus M, Eaton S. Argosomes: a potential vehicle for the spread of morphogens through epithelia. Cell. 2011;06:633–45.

56. Lakkaraju A, Rodriguez-Boulan E. Itinerant exosomes: emerging roles in cell and tissue polarity. Trends Cell Biol. 2008;18:199–209.

57. Bernardi S, Foroni C, Zanaglio C, et al. Feasibility of tumor-derived exosome enrichment in the onco-hematology leukemic model of chronic myeloid leukemia. Int J Mol Med. 2019;44(6):2133–44. https://doi.org/10.3892/ijmm.2019.4372.

58. Sanderson RD, Bandari SK, Vlodavsky I. Proteases and glycosidases on the surface of exosomes: newly discovered mechanisms

for extracellular remodeling. Matrix Biol. 2019;75–76:160–9. https://doi.org/10.1016/j.matbio.2017.10.007.

59. Zlotogorski A, Vered M, Chaushu G, Dayan D. Exosomes isolated from saliva of cancer patients differ from those of healthy individuals. Oral Oncol. 2013;49:S70–1. https://doi.org/10.1016/j.oraloncology.2013.03.185.

60. Kalluri R. LeBleu VS.:review the biology, function, and biomedical applications of exosomes. Science. 2020;7:367(6478).

61. Gurunathan S, Kang M-H, Jeyaraj M, Qasim M, Kim J-H. Review of the isolation, characterization, biological function, and multifarious therapeutic approaches of exosomes. Cells. 2019;8:307.

Part II

Imaging, Guidance, and Planning

Basics of Ultrasound

Anthony Tran and Amitabh Gulati

Introduction

The scientific origin behind medical ultrasound can be attributed to the year of 1793 when Lazzaro Spallanzani observed that bats were able to navigate in complete darkness while blindfolded but were unable to do so with waxy ear plugs [1]. Thus, he concluded that the ability to hear was critical for bats to navigate. More than seven decades later, a seemingly unrelated discovery was made by Pierre and Jacques Curie termed the "piezoelectric effect", a physical property of certain solids (e.g., crystals) that enable the conversion of mechanical energy to electrical energy in a reversible manner [2]. The application of the piezoelectric effect with the concept of sound as a navigation tool took shape during World War I (1914–1918), in efforts to detect submarines [3]. Piezoelectric-compatible crystals were exploited to create a transducer able to receive electrical energy from the ship's engine and in-turn produce mechanical energy in the form of sound waves. These sound waves could be emitted underwater, reflect off surfaces, and bounce back to a receiver on the ship. The returning sound waves could then be reconverted to electrical energy and the amount of time required to receive the echo could be used to calculate the distance between the ship and an object. It was not until nearly three decades after the start of World War I, did the first documented application of sound for human medical diagnosis emerge. Adopting similar ideas from ship navigation, Austrian neurologist (Karl Dussik) attempted to detect brain tumors within a human skull by directing sound waves through patients' heads and analyzing the echos [4]. In the decades following Dr. Dussik's efforts, technological advancements to be further discussed in this chapter have enabled the routine use of sound for diagnostic and image guidance purposes in clinical medicine.

Contemporary utilization of sound for medical diagnosis and procedural guidance utilizes the general concept of a transducer to emit sound, but also to capture reflected sound signals and funnel information to a central processing unit (CPU) for image generation. In order to understand how an image is generated, the basic physical properties of sound need to be introduced.

Sound is described as mechanical energy that propagates longitudinally via compressions and expansions of molecules within a given medium [5]. Sound can be illustrated using a sinusoidal curve to illustrate the cyclic fluctuation of pressure with time (Fig. 10.1). A sound wave has various descriptors including volume (amplitude), distance between two adjacent points in a wave (wavelength), and number of repeated wave cycles that occur in 1 second (frequency) in unit Hertz (Hz). The human hearing range for sound is described to be between 20 Hz and 20,000 Hz [6]. An ultrasound, therefore, is sound beyond the audible range of humans. In the context of ultrasound image generation, the amount of detail within an image (resolution) and the depth of the scanning area achieved (penetration) are interrelated

A. Tran
New York-Presbyterian/Columbia and Cornell, Department of Rehabilitation Medicine, New York, NY, USA

A. Gulati (✉)
Memorial Sloan Kettering Cancer Center, Department of Anesthesiology and Critical Care, New York, NY, USA
e-mail: gulatia@mskcc.org

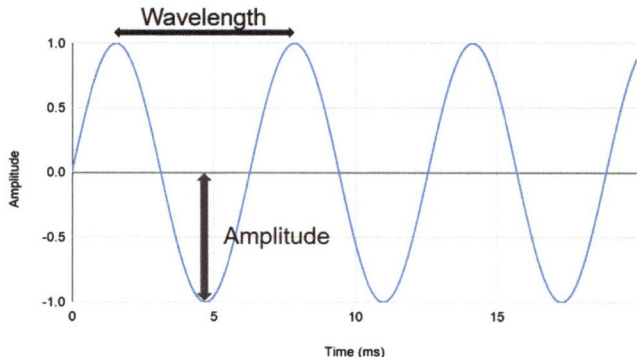

Fig. 10.1 Sound wave characteristics. Wavelength (λ) is the distance between two adjacent points in a wave. Amplitude is the distance from origin to crest/trough

C. W Hunter et al. (eds.), *Regenerative Medicine*, https://doi.org/10.1007/978-3-030-75517-1_10

to ultrasound frequency. While frequency is directly related to resolution, it is inversely related to penetration. That is, higher ultrasound frequencies produce higher resolution images with lower depths of scanning. The behavior of sound within the human body is dependent on the tissue of which the sound is propagating. As sound propagates through tissue, there is gradual loss of energy (attenuation) and some of the sound waves bounce back (reflection) or deflect obliquely while passing from one tissue type to another (refraction). There are various tissue types within a human body and the degree of resistance of an ultrasound beam through a particular tissue type (acoustic impedance) is illustrated in

Table 10.1 [7, 8]. Generally, the larger the difference between acoustic impedance between two tissue types, the increased proportion of reflected sound waves which correlate with signal intensity. Typically, ultrasound images are depicted in a gray scale continuum with brighter shades of gray indicating higher intensity and darker shades indicating lower intensity of signal. Therefore, a medical ultrasound image can be thought of as a compilation of reflected ultrasound waves of varying signal intensities from tissues and structures within the body. Let us explore this idea further by evaluating an ultrasound machine.

Equipment Ultrasound machines come in many different shapes and sizes. Fortunately, there are commonalities with ultrasound machines that often remain consistent regardless of make. An ultrasound transducer enables real-time interrogation of a field of interest using emitted and receiving reflected sound waves (Fig. 10.2a). The recovered signal from the transducer is processed at a CPU (Fig. 10.2b) for image generation. The image can be displayed on a monitor (Fig. 10.2c) and/or sent to a storage device where a keyboard (Fig. 10.2d) is helpful to input information detailing a study.

Table 10.1 Acoustic Impedance based on material type

Material	Acoustic Impedance (Mrayl)
Air	0.0004
Water	1.48
Fat	1.34–1.38
Liver	1.65
Blood	1.61–1.65
Muscle	1.62–1.71
Skull bone	6.0–7.8

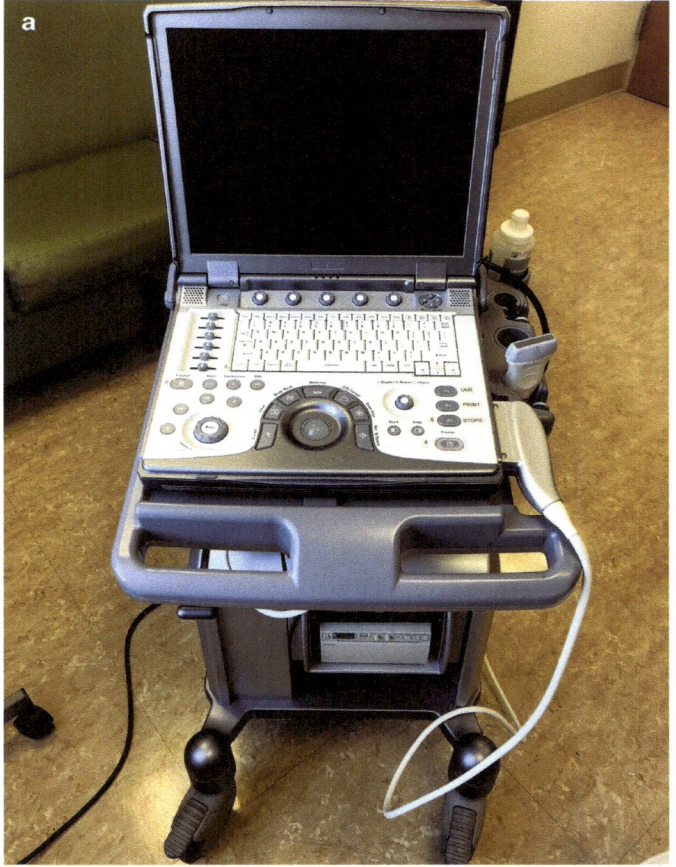 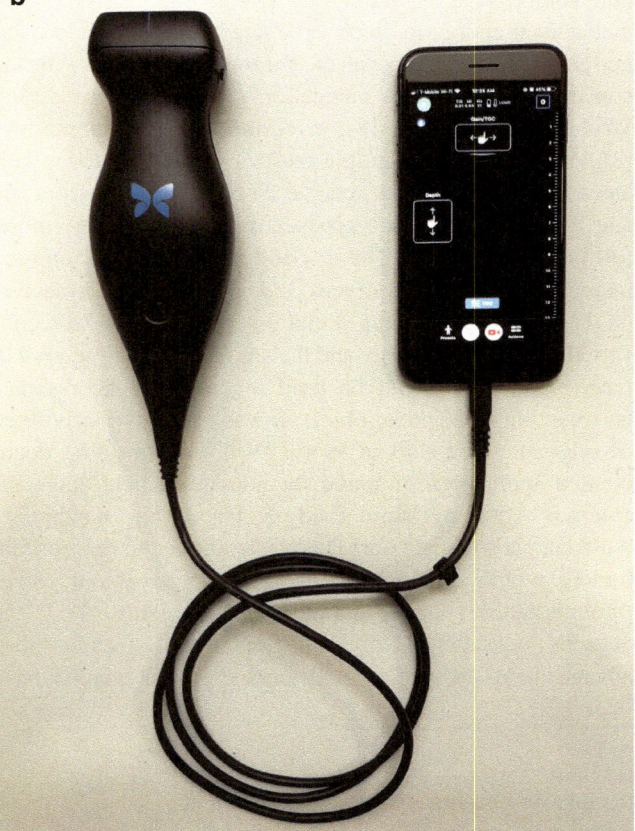

Fig. 10.2 Ultrasound machine components may differ in shape and size, a comparison of major functional components between two different manufacturers. Transducer (**a**). CPU (**b**). Monitor (**c**). Keyboard (**d**)

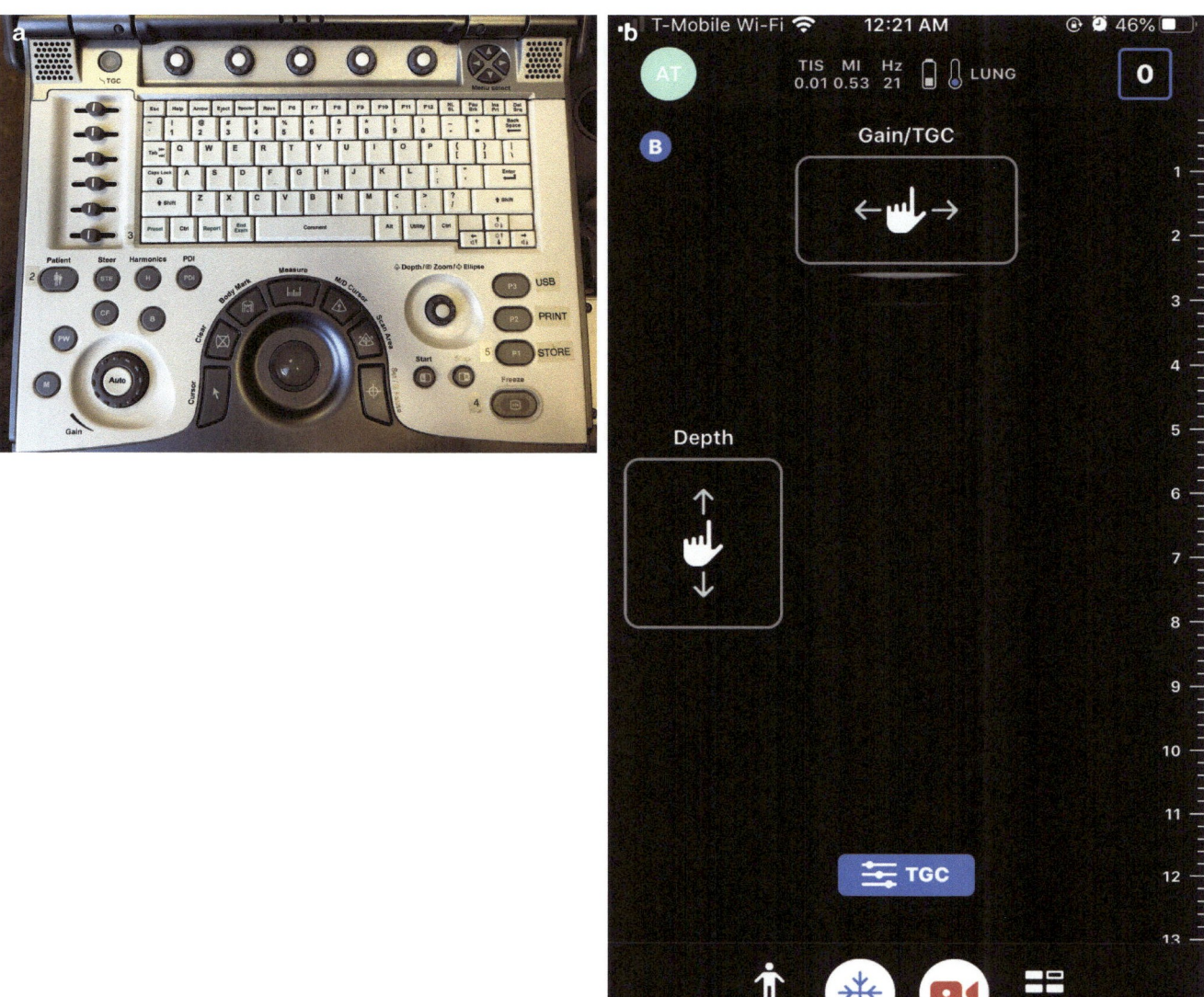

Fig. 10.3 Basic ultrasound knob comparison between two different manufacturers. Power button (**a**). Depth control (**b**). Gain control (**c**). Freeze button (**d**)

In regards to the control panel of an ultrasound machine (Fig. 10.3), there are numerous knobs. The knobs that are typically reproduced from machine to machine include a power button (toggle on/off) (Fig. 10.3a), depth (to adjust shallowness/deepness of an image's field of vision) (Fig. 10.3b), gain (to adjust darkness/brightness of sound signal) (Fig. 10.3c), and freeze (to hold an image at a current selected frame) (Fig. 10.3d). Now that a basic layout of an ultrasound machine has been established, let us learn how to handle a transducer and use the machine.

Transducers The basic types of transducers are linear (Fig. 10.4a), curvilinear (Fig. 10.4b), and phased. Transducer types vary in frequency and footprint, defined as the area of a transducer to which ultrasound rays are emitted. Each transducer has an indicator (a raised bump) on one side and by convention, the indicator is oriented to reflect the left side of the monitor. The upper limit of frequencies that can be achieved by any one transducer depends on the type, and may even vary within a particular type depending on the manufacturer. Generally, the order of relatively high to low frequencies based on type is considered to be: Linear > Curvilinear > Phased. Typically, the relatively high frequency transducers (e.g., linear) are used for superficial structures (e.g., bone, ligaments, and muscle) due to better resolution at shallower depths (poor penetrance). The curvi-

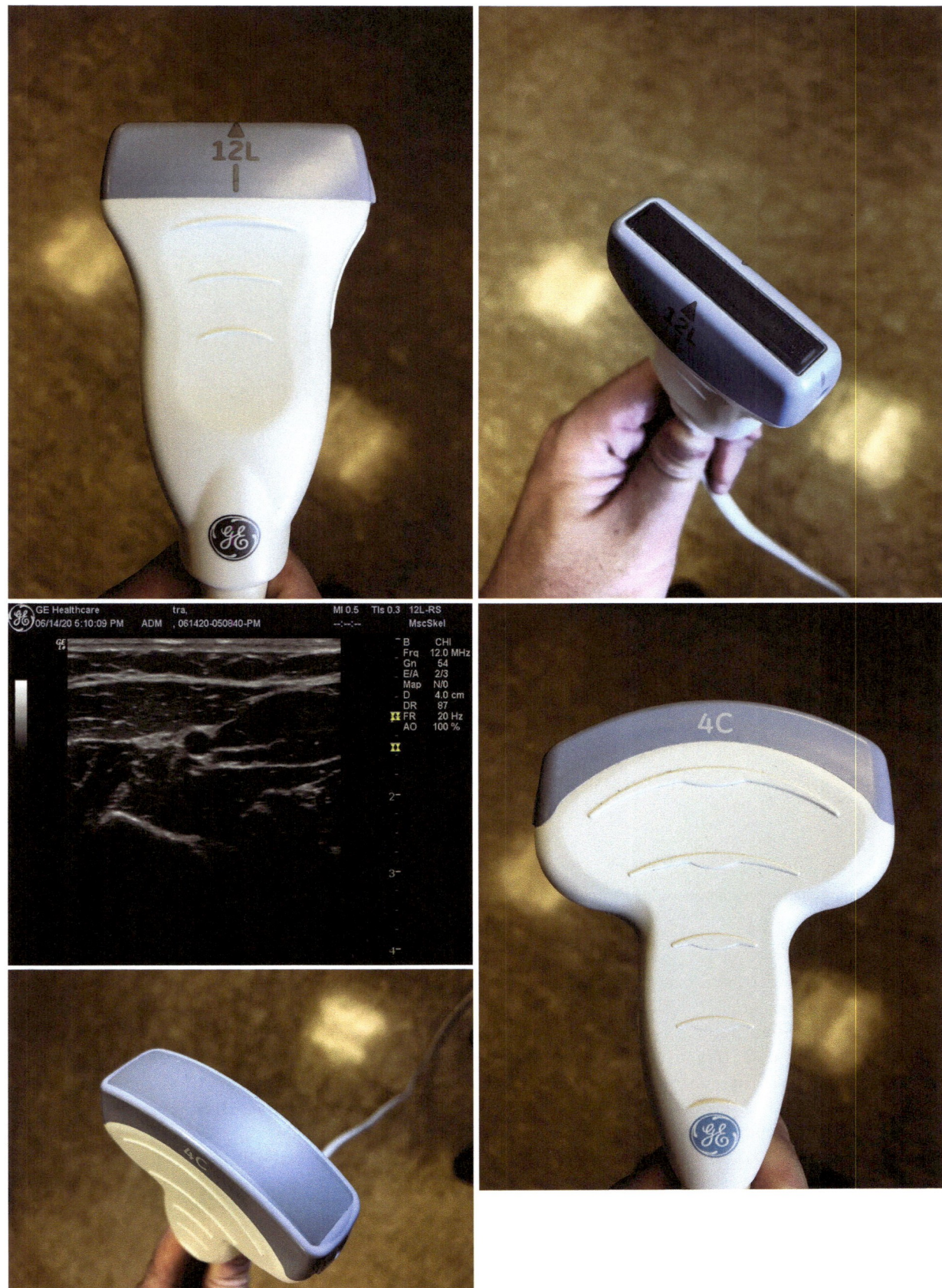

Fig. 10.4 Linear transducer (**a**) with view of footprint along with accompanying field of vision of a left forearm. Curvilinear transducer (**b**) with view of footprint along with accompanying field of vision of a left forearm

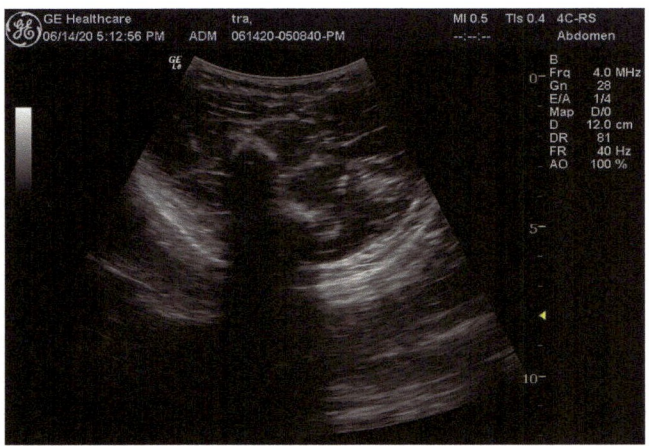

Fig. 10.4 (continued)

linear probes emit ultrasound at a relatively lower frequency allowing for deeper penetration and a wider depth of field (given the shape of its footprint). As a result, the curvilinear transducer is the preferred type for intra-abdominal structures. Phased transducers have a compact footprint enabling visualization from a narrow vantage point, such as in-between ribs to evaluate the thoracic cavity or via a subxiphoid view for cardiac examination (echocardiogram).

Medical ultrasound is highly user-dependent; a number of transducer motions have been described [9] and are demonstrated in Fig. 10.5.

Settings & Image Adjustment Once an ultrasound image is generated and displayed on the machine's monitor, there are a variety of modes that can be selected from. Three modes will be discussed here: B-mode, M-mode, and Color doppler (Fig. 10.6). B-mode, also known as "Brightness" mode, is a 2D gray scale image that correlates intensities of returning echoes with degrees of brightness, the higher the intensity the brighter the signal. This is often the default ultrasound mode. M-mode, also known as "Motion" mode, provides axial and temporal resolution of an ultrasound image. This mode is able to follow points that are longitudinally adjacent and record the position of these points with successive images over time. This mode is often used in echocardiograms, to evaluate the motion of heart valves over time. Lastly, Color doppler provides information regarding velocity of an area of interest. Often, color doppler is utilized to assess the speed and direction of blood flow within blood vessels and within the heart. Color doppler incorporates "doppler shift", a change in frequency due to movement of a reflector away or towards a transducer. A positive doppler shift denotes an increase in sound frequency due to an object moving towards the transducer, whereas a negative doppler shift denotes a decrease in sound wave frequency due to an object moving away. By convention, red on ultrasound imaging is movement towards the transducer (positive doppler shift) and blue is movement away from the transducer (negative doppler shift).

Ultrasound images are subject to a number of artifacts [10] that can be misleading for an image interpreter. The artifacts that will be highlighted here include: acoustic shadowing, posterior enhancement, anisotropy, and reverberations and mirror image. Acoustic shadowing is a poor signal beyond a structure that is strongly reflective or absorptive of the ultrasound beam (e.g., gallstones within a gallbladder or bone) (Fig. 10.7a). Posterior enhancement is an increased signal intensity deep to structures that transmit ultrasound beams well (e.g., fluid-filled structures) (Fig. 10.7b). Anisotropy is a reflection of ultrasound beam not directly back at the transducer, often seen when scanning structures with many fibrils (e.g., tendons, muscles) and may appear erroneously hypoechoic. Reverberations are ultrasound waves that bounce back and forth from two strong parallel reflectors within the body (Fig. 10.7c). A mirror image artifact is an exact copy of an image from a highly reflective surface.

Ultrasound imaging enables the benefit of real-time evaluation of structures within the human body, void of any ionizing radiation. The risks associated with ultrasound are minimal but include the thermal heating of tissues [11]. The long-term consequences of thermal heating by ultrasound have yet to be fully elucidated or proven to be significant in an adult human. Potential neonatal effects with ultrasound imaging in the obstetric setting have remained inconsistent [12]. Nonetheless, the ultrasonographer in an effort to achieve a clinically relevant ultrasound image should strive to minimize the exposure time to ultrasound waves for all associated participants.

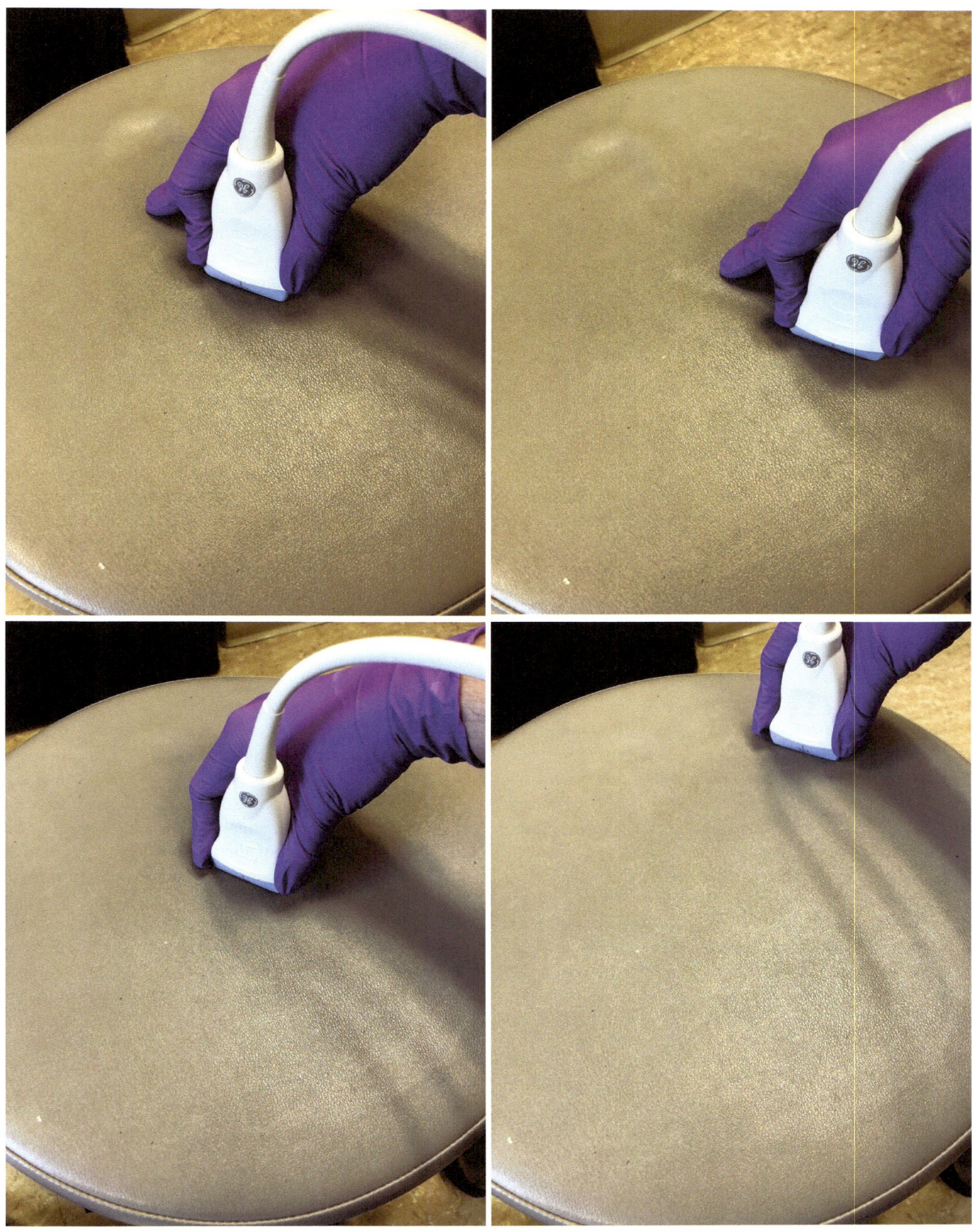

Fig. 10.5 (**a**) Sliding: probe motion side to side in the long axis; (**b**) Sweeping: probe motion forward and backward along the short axis; (**c**) Rotating: circular motion of probe; (**d**) Rocking: angling probe towards or away from the indicator; (**e**) Fanning: short-axis tilting of probe; (**f**) Compressing: downward pressure along probe, often to evaluate tissue/structural changes (e.g., delineating a compressible vein vs. an incompressible artery)

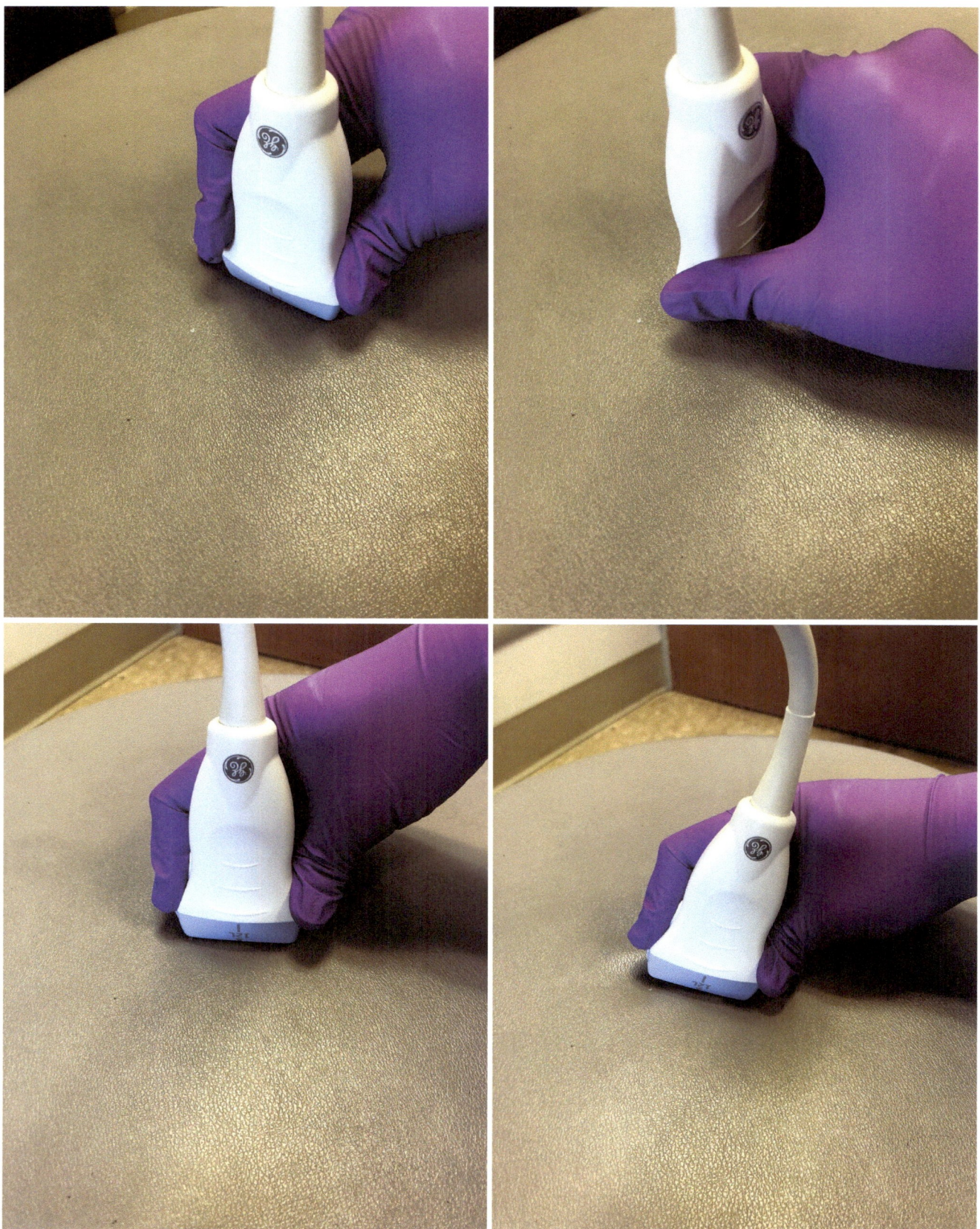

Fig. 10.5 (continued)

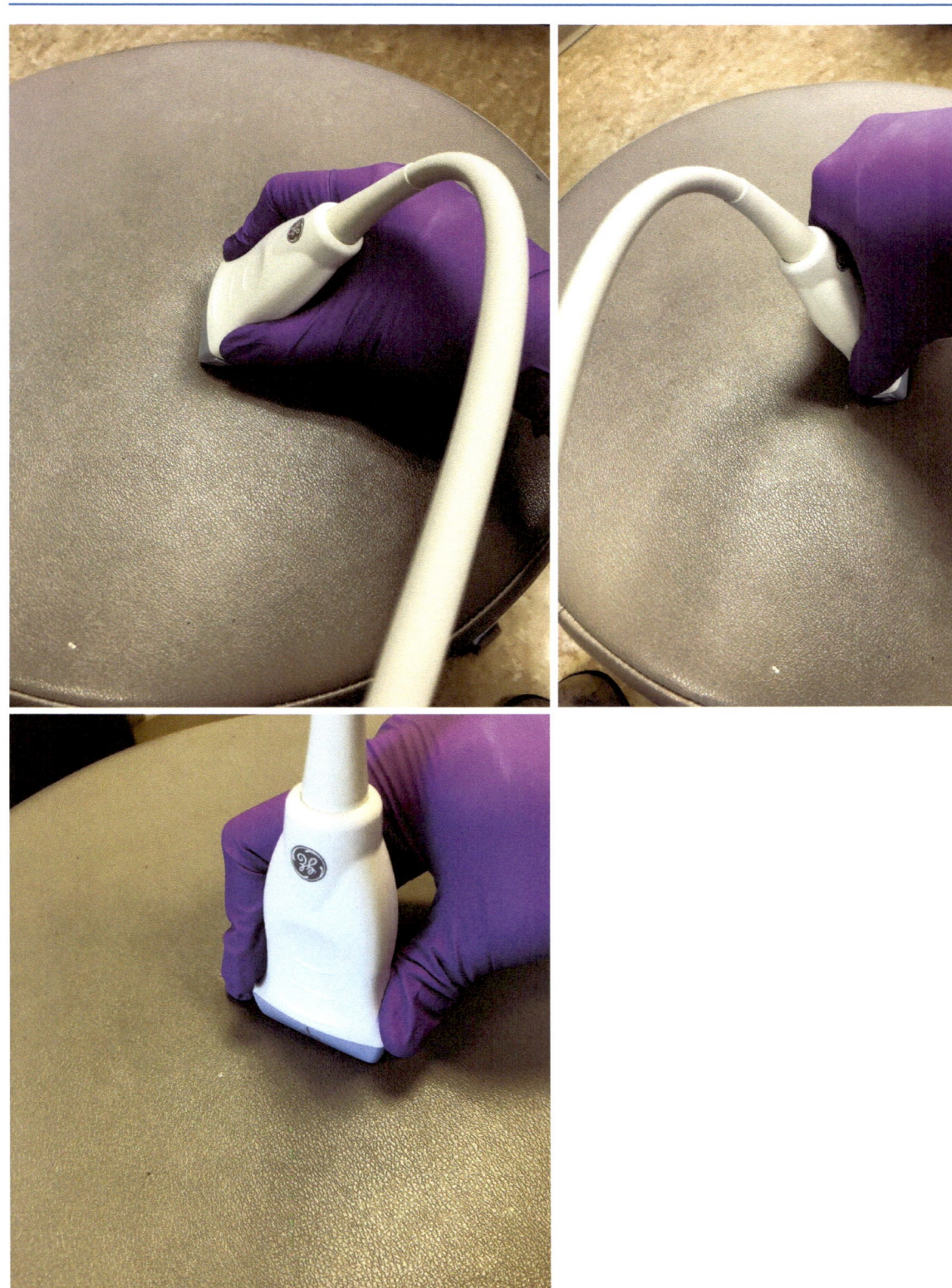

Fig. 10.5 (continued)

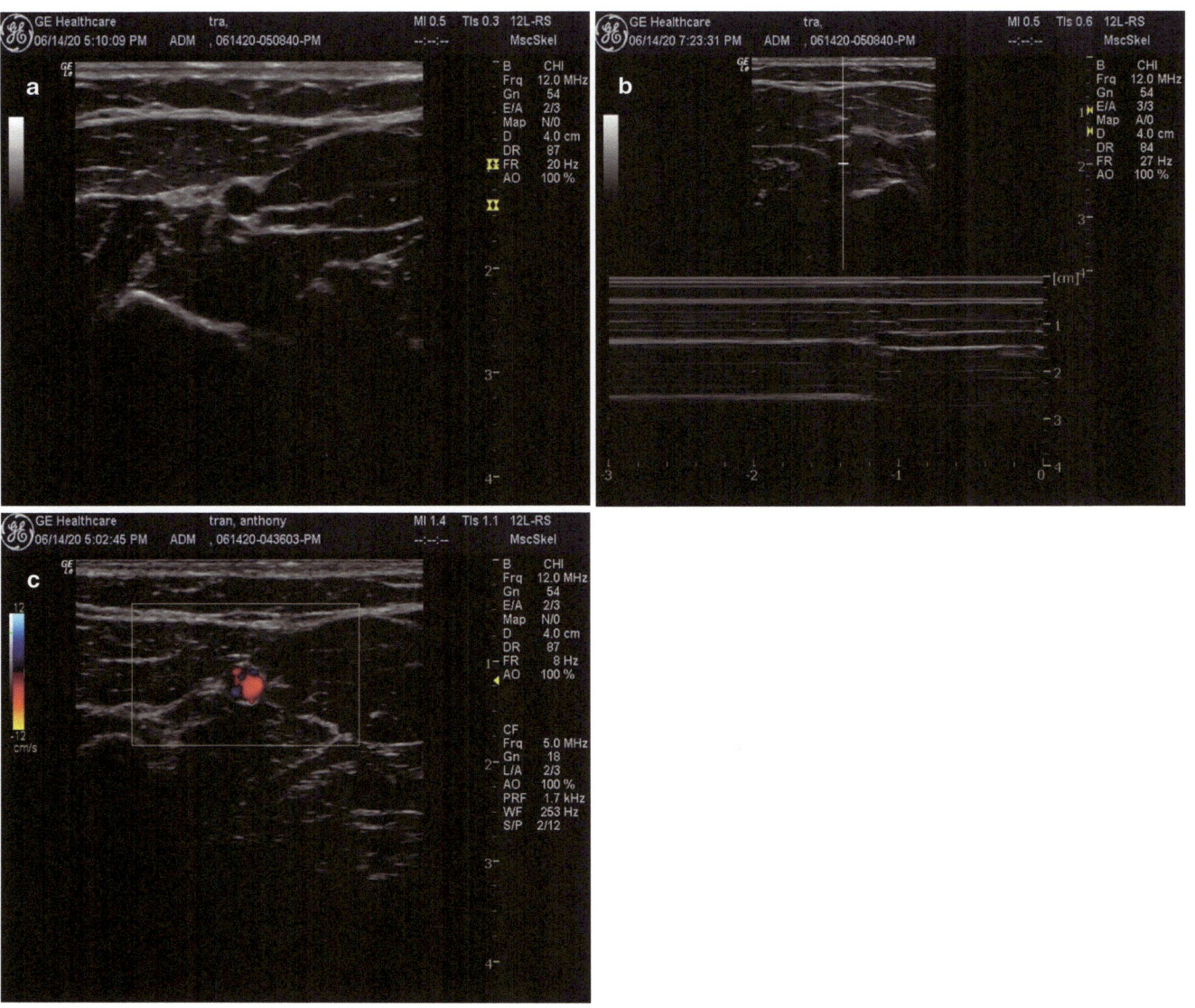

Fig. 10.6 Ultrasound images demonstrating various mode types. B-mode (**a**). M-mode (**b**). Color Doppler (**c**)

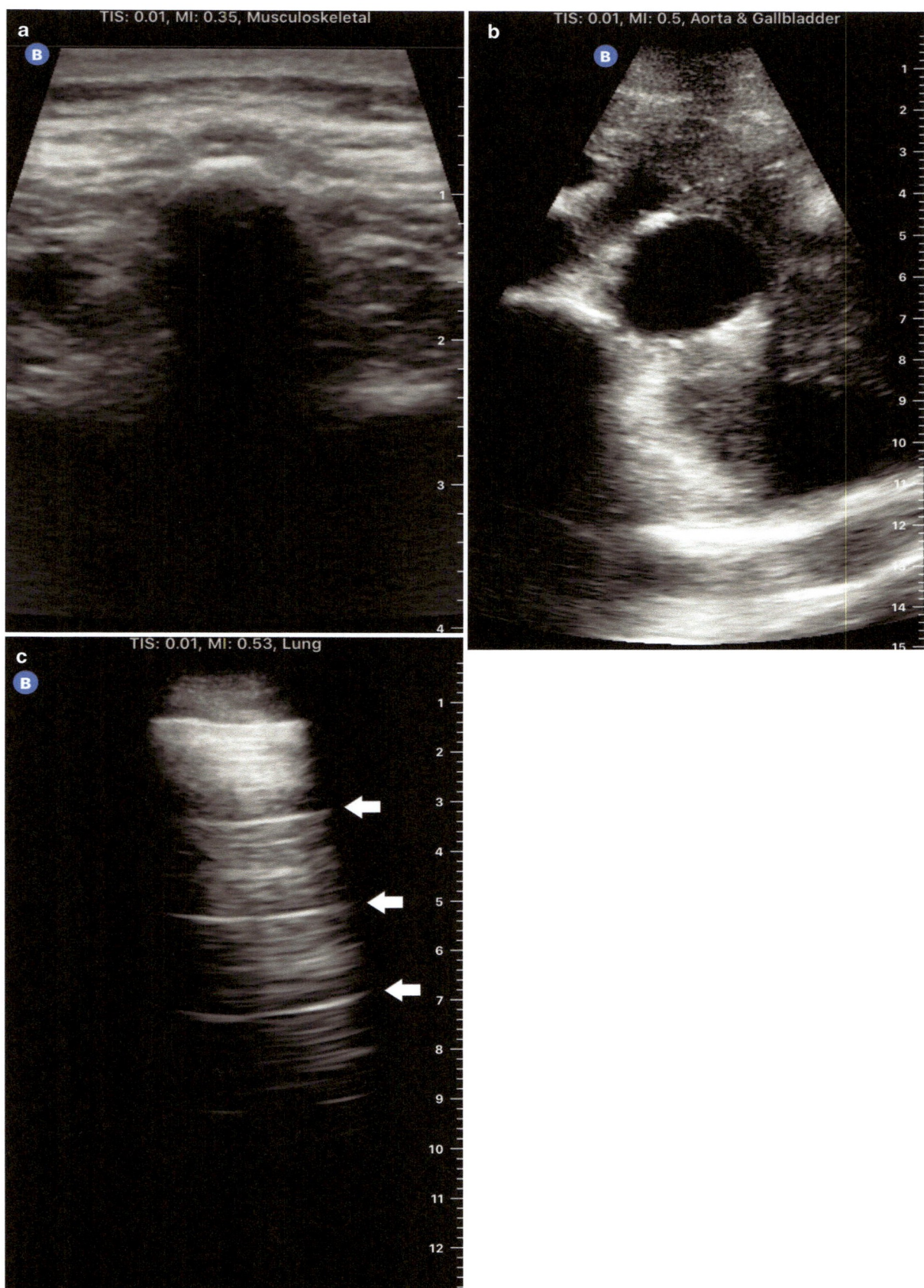

Fig. 10.7 Ultrasound images demonstrating various artifacts. Axial view of a lumbar spinous process with acoustic shadowing (**a**). Axial view of an abdominal aorta with posterior enhancement (**b**). Right lung with reverberation artifact (**c**, arrows)

References

1. Spallanzani L. Letters on the suspicion of a new sense in bats. Torino (Turin), Italy: Royal Press; 1794.
2. Curie J, Curie P. Development, via compression, of electric polarization in hemihedral crystals with inclined faces. Bulletin de la Societe de Minerologique de France. 1880;3:90–3.
3. "World War I: 1914–1918." Discovery of Sound in the Sea, 9 July 2017, dosits.org/people-and-sound/history-of-underwater-acoustics/world-war-i-1914-1918/.
4. Dussik KT. On the possibility of using ultrasound waves as a diagnostic aid. Neurol Psychiat. 1942;174:153–68.
5. Bamber JC, ter Haar GR. Physical principles of medical ultrasonics. Chichester: Wiley; 2004.
6. Purves D, Augustine GJ, Fitzpatrick D, et al., editors. Neuroscience. 2nd ed. Sunderland (MA): Sinauer Associates; 2001.
7. Shung KK. Diagnostic ultrasound: imaging and blood flow measurements. Boca Raton, FL: CRC Press; 2006.
8. Zagzebski JA. Essentials of ultrasound physics. St. Louis, MO: Mosby; 1996.
9. Bahner DP, Blickendorf JM, Bockbrader M, Adkins E, Vira A, Boulger C, Panchal AR. Language of transducer manipulation. J Ultrasound Med. 2016;35:183–8.
10. Taljanovic MS, Melville DM, Scalcione LR, Gimber LH, Lorenz EJ, Witte RS. Artifacts in musculoskeletal ultrasonography. Semin Musculoskelet Radiol. 2014;18(1):3–11.
11. Center for Devices and Radiological Health. "Ultrasound Imaging." U.S. Food and Drug Administration, FDA, www.fda.gov/radiation-emitting-products/medical-imaging/ultrasound-imaging.
12. Shankar H, Paul SP. Potential adverse ultrasound-related biological effects: a critical review. Anesthesiology. 2011;115(5):1109–24.

Ultrasound: In-Plane and Out-of-Plane

11

Alexander Bautista, Clairese M. Webb,
and George C. Chang Chien

Introduction

Although the ultrasound transducer emits a three-dimensional beam, the beam is only 1–3 mm in thickness. For this reason, the beam can be regarded as a mere 2-dimensional plane. Directing a needle to a specific target and keeping the needle in view ais one of the many challenges that are faced by operators doing ultrasound-guided injections. In-plane and out-of-plane techniques are the two approaches utilized to perform the injections. Recognition of the needle and its tip as it appears on ultrasound images is critical to be able to do a successful injection without damaging the surrounding structures or unintended intravascular injections.

In-Plane Approach

The In-plane technique is the more commonly employed technique between the two. This approach allows the needle and transducer in the same plane as the ultrasound beam. The needle insertion is advanced from the side of the transducer (see Fig. 11.1). The needle should be visible at all times during advancement.

Technique

After identification of the target and other pertinent anatomical structures, the target should be focused on the screen to allow the visualization of the needle trajectory (see Fig. 11.2). The needle path should be scanned to look for any sensitive structures that may be traversed along the path of the needle. The proposed target should be at the center or lower third of the screen opposite to where the needle should be entering. This would allow keeping track of the deep and important structures along the needle path. It is important to note that the needle is easier to visualize under ultrasound if the shaft of the needle is more perpendicular to the ultrasound beam. It is very important to stabilize the transducer by bracing the operator's hand against the patient's body; this will prevent the operator to lose the optimal view of the target and the needle path. Oftentimes, after putting local anesthetic on the needle entry point, the planned needle path can be anesthetized too.

Novice practitioners may have some difficulty in keeping the needle and its tip on focus. The most common reason for this is that the needle does not enter the skin perfectly aligned with the ultrasound transducer. It is not safe to advance the needle without being able to visualize during advancement. This may lead to unintended injury to structures forfeiting the mere purpose of using the ultrasound. If this happens,

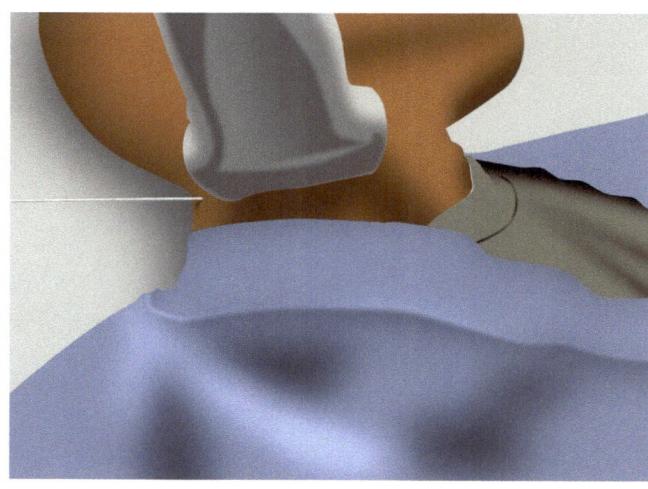

Fig. 11.1 In-Plane technique

A. Bautista (✉)
University of Louisville, Department of Anesthesiology and Perioperative Medicine, Louisville, KY, USA

C. M. Webb
University of Oklahoma Health Sciences Center, Department of Anesthesiology and Pain Medicine, Edmond, OK, USA

G. C. Chang Chien
GCC Institute, Department of Musculoskeletal Medicine and Medical Aesthetics, Newport Beach, CA, USA

attempt to visualize the needle by toggling the probe. However, by doing this, it is also possible to lose the target as well. If the attempt in visualizing the needle and target on the same plane remains to be futile, it is imperative to keep the target in view and reinsert the needle.

The advantages of the In-Plane technique are that safety is maximized. By visualizing the entire path of the needle, and tracing it as you progress the needle tip, adjustments can be made to avoid any sensitive structures. The disadvantages of the In-Plane technique include the necessity for a flat, planar approach to visualize the needle. This leads to a greater distance of tissue that is penetrated by the needle not only causing discomfort in the patient but also increases the potentially sensitive structures that need to be accounted for. Some areas of the body such as the anterior neck, or superficial structures such as ligaments, small joints, and tendons are not amenable to In-plane injection due to body habitus or lack of a safe needle entry point. Additionally, artifacts such as reverberation, and shadowing can occur due to the interactions between the ultrasound beams and the needle.

A Reverberation artifact occurs when an ultrasound beam encounters two strong parallel reflectors. When the ultrasound beam reflects back and forth between the reflectors, the ultrasound transducer interprets the sound waves returning as deeper structures. (see Fig. 11.3).

A shadow artifact, or acoustic shadowing, occurs signal void behind structures that strongly absorb or reflect ultrasound energy. This shadow artifact occurs deep to the needle shaft due to the strongly reflective nature of the metal.

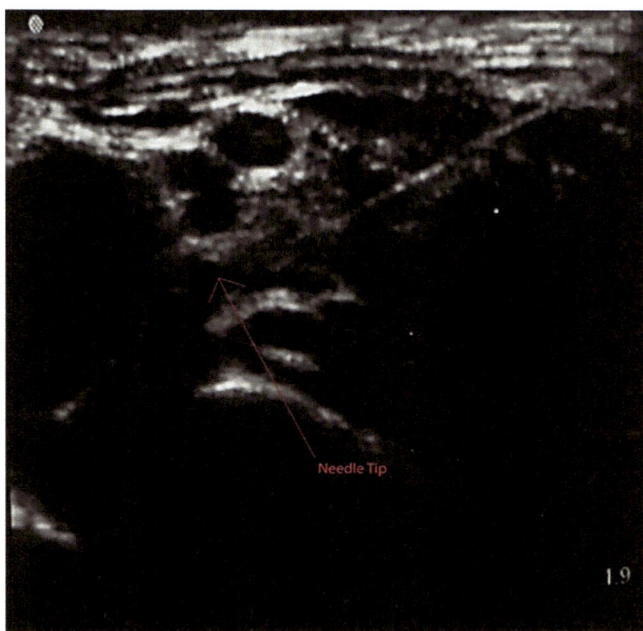

Fig. 11.2 Supraclavicular nerve block using in-plane ultrasound technique displaying the needle traversing the nerve bundle

Out-of-Plane Approach

The Out-of-Plane approach is technically more challenging than the In-plane approach. This requires more familiarity with the structures and comfort in handling the probe and needle. The needle for the out-of-plane approach enters the skin away from the probe and is directed perpendicular to the place of the ultrasound beam, or near-parallel to the beam as it projects from the probe (see Fig. 11.4). By using this tech-

Fig. 11.3 Reverberation artifact

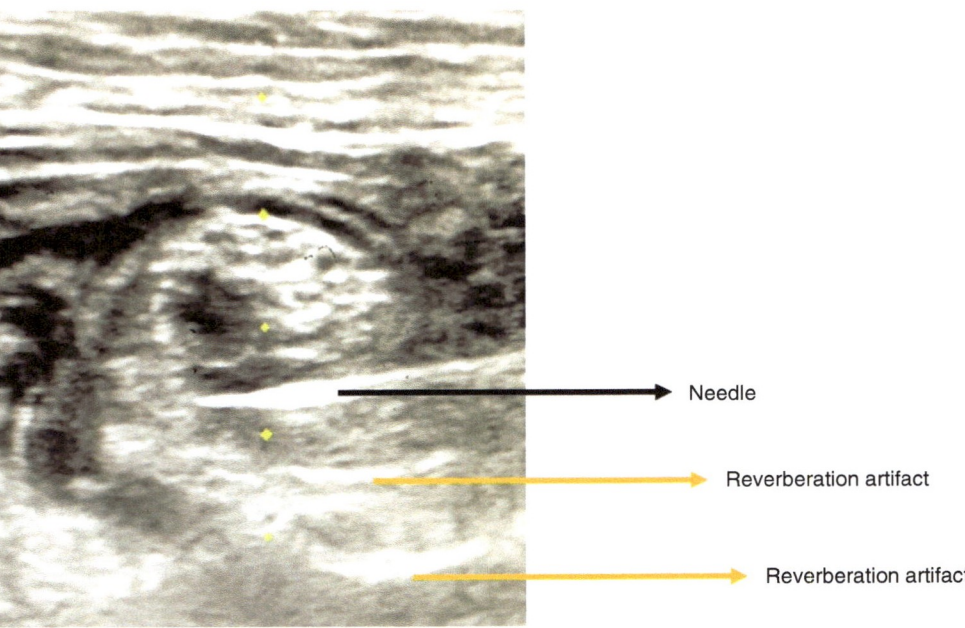

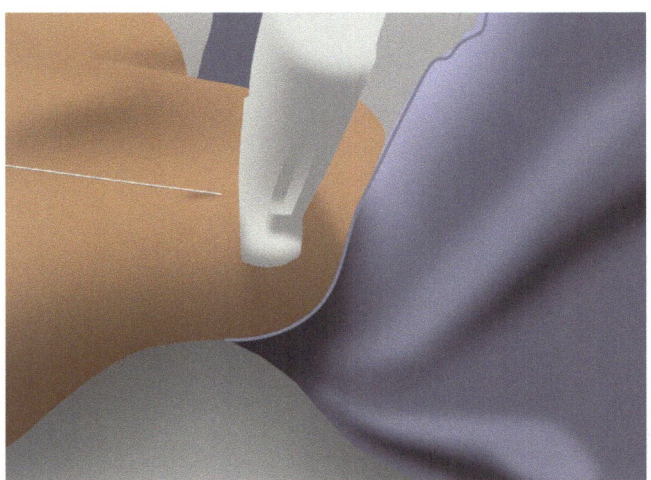

Fig. 11.4 Out-Of-Plane Technique

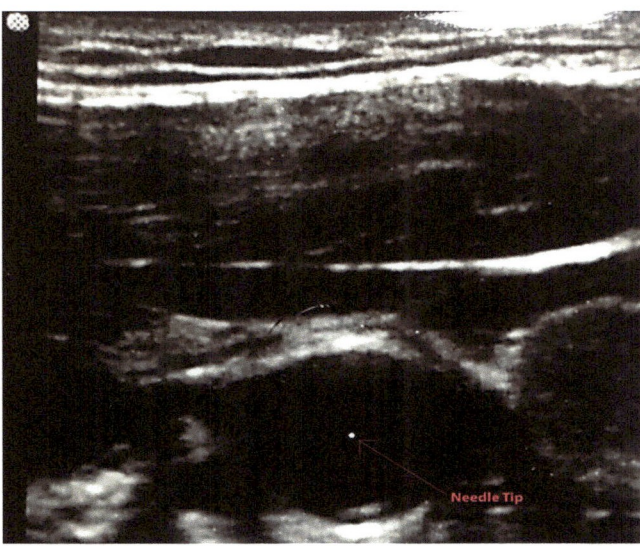

Fig. 11.5 Cannulation of internal jugular vein under out-of-plane ultrasound guidance displaying the needle tip within the vessel lumen

nique, it allows the sonographer to visualize the structures on either side of the target. In comparison to the in-plane technique, the out-of-plane approach requires a shorter path for the needle to travel, hence less discomfort for the patient. However, the downside for this technique is the inability to visualize the needle during advancement. The needle position can only be inferred as tissue movement is observed and/or by hydro-dissection of the structures.

Technique

The target is identified, preferably centered in the monitor. The needle entry point is situated very close to the transducer. After penetrating the skin, the needle is advanced almost straight down directed to the target. The needle is then kept track by observed tissue movement, using hydro-dissection and/or observing the needle tip cross the place of the ultrasound beam. It is important to note that while advancing the needle using the out-of-plane technique, it is imperative to observe the image for signs that the needle has crossed the ultrasound beam. The practitioner should understand that once the needle has crossed the place of the ultrasound beam, it will produce a small hyperechoic dot (see Fig. 11.5). If the needle continues to be advanced, it will not change the appearance on the screen and the dot will still be on the same position. This can lead to injury to structures on the other side of the beam. Tilting the probe can help with tracking of the needle and this allows needle tip visualization until it reaches the target. The tip is located as it crosses the beam plane.

The Oblique approach is a combination of the In-plane and Out-of-Plane approach. The Oblique approach is similar to the Out-of-Plane approach as the needle shaft is only partially visualized along its trajectory. It allows for minimizing the traversed tissue while allowing for more play in the geometry of the entry point and trajectory for the needle. It is used to maintain the ultrasound view in a more traditional long- or short-axis to the target. (see Fig. 11.6).

Fig. 11.6 Oblique Technique. Note how the shaft of the needle crosses the beam of the ultrasound probe in out-of-plane manner

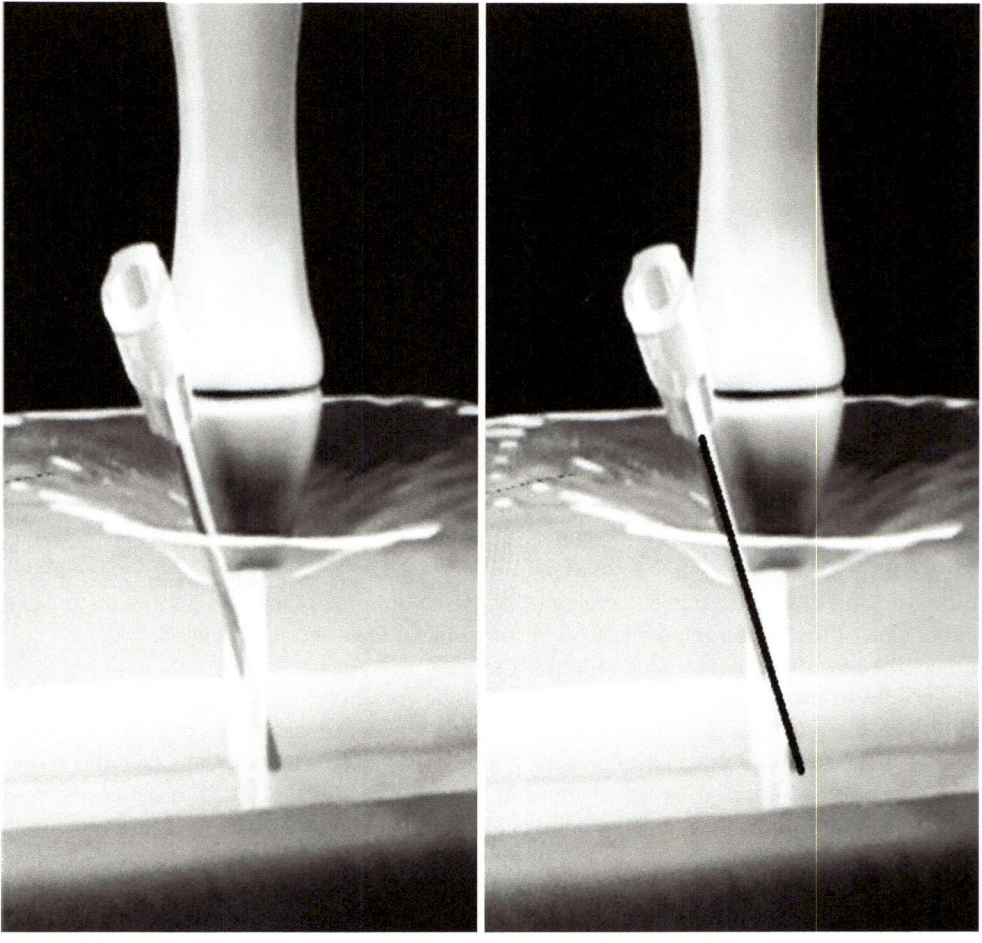

Suggested Reading

1. Niazi PA, Chan V. Ultrasound-guided anesthesia performance in the early learning period. Reg Anesth Pain Med. 2012;37:51–4.
2. Chin K, Perlas A, Chan V, Brull R. Needle visualization in ultrasound-guided regional anesthesia: challenges and solutions. Reg Anesth Pain Med. 2008;33:532–44.

Probe Selection

12

Elise M. Itano and George C. Chang Chien

Probe Selection

Proper selection of transducer frequency is an essential concept for attaining optimal image resolution in diagnostic and procedural ultrasound. The frequency and wavelength of soundwaves are inversely related. Thus the higher the frequency, the shorter the wavelength, and vice versa (Fig. 12.1). As a sound wave propagates through tissue, the sound energy is converted to other types of energy, heat for example, through absorption. The combined effect of energy absorption and scattering is called attenuation. Ultrasonic attenuation is the decay rate of the wave as it travels through a medium. As a wave is attenuated and decays, the amplitude decreases while the frequency remains constant. Higher frequency wavelengths have greater attenuation, and thus, do not penetrate as deeply into tissues. However, because of the short wavelength, and more waves in a given distance, there is better discrimination between two separate structures parallel to the plane of wave propagation [1–3]. This is called axial resolution.

Probes are commonly described by the size and shape of the transducer footprint. There are two basic types of transducers used in musculoskeletal and spine ultrasound: straight array linear and curvilinear. Linear probes are generally high frequency and therefore are better for imaging superficial structures. Curvilinear probes typically have a wider footprint and are generally low frequency, more suited for viewing deeper structures of the body. Ideally, the ultrasonographer will choose the probe with the highest possible frequency with a depth that penetrates far enough to visualize the target.

E. M. Itano (✉)
Boulder Medical Center, Interventional Sports and Spine, Department of Physical Medicine and Rehabilitation, Louisville, CO, USA

G. C. Chang Chien
GCC Institute, Department of Musculoskeletal Medicine and Medical Aesthetics, Newport Beach, CA, USA

Straight Linear Array Probe

The straight linear array probe is best used when scanning superficial structures, usually to depths up to 6 cm, for example when performing stellate ganglion or peripheral nerve blocks (Fig. 12.2a) [3]. There are a variety of linear array transducers, including large size (image width of >40 mm), medium size (<40 mm), and small field-of-view (hockey stick) probes [4]. Linear transducers generally have high (8–20 MHz) to medium (6–10 MHz) frequency waves which have less penetration compared to low-frequency waves; however, produce better axial resolution. The piezoelectric crystals are linearly aligned on the flat, rectangular face and produce sound waves in a straight line [1–4]. The waves are also received in a linear fashion on the probe and a rectangular-shaped image is produced (Fig. 12.2b).

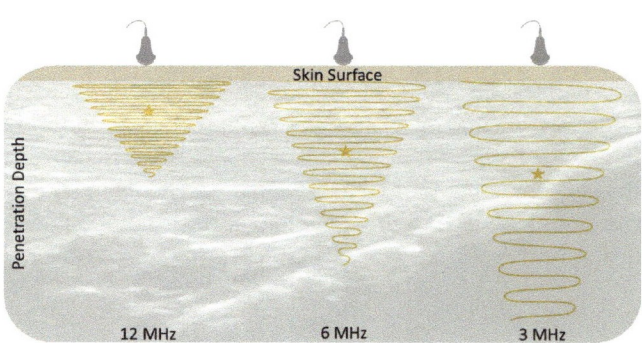

Fig. 12.1 Frequency and wavelength of ultrasound are inversely related. Frequency is the number of times the wave is repeated in one second (1/s = 1 Hz). Wavelength is the distance between successive crests, or one cycle. Properties of a high-frequency ultrasound include better resolution, greater attenuation, and less penetration. In comparison, properties of a low-frequency ultrasound include worse resolution, less attenuation, and deeper penetration

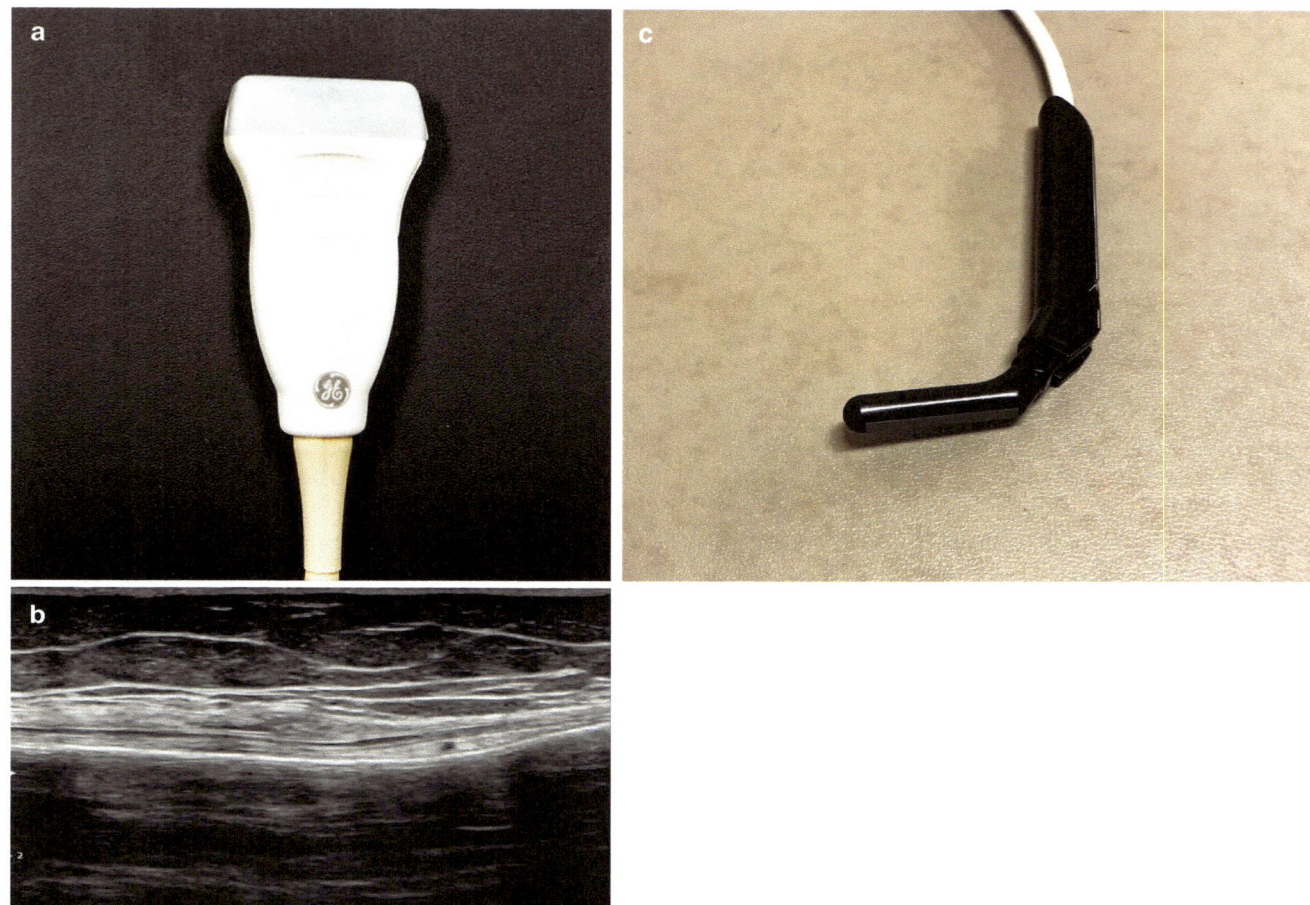

Fig. 12.2 (a) Straight linear array probe. (b) Genicular nerve scan using the straight linear array probe. Note the rectangular-shaped image. (c) Hockey-stick probe

Hockey-stick transducers are best used for imaging small superficial structures, especially irregular bony landmarks where the skin surface does not allow adequate contact with larger probes, for example viewing the anterior talofibular ligament (ATFL) at the ankle or the A1 pulley of the flexor digitorum in the hands (Fig. 12.2c). Because of the very small Field of view (FOV), however, there is some compromise viewing the complete structure and adjacent anatomy.

Curvilinear Array Probe

The curvilinear array probe, or convex probe, is best used when viewing deeper structures in the body such as the sciatic nerve, sacroiliac joints, or lumbar neuraxial structures (Fig. 12.3a). Although this transducer produces images with lower resolution, the lower frequency wavelengths (typically 2–5 MHz) compromise with less attenuation and can therefore penetrate into deeper structures. The crystals are aligned along the convex surface of the transducer. The propagating

waves, therefore, fan out from the beam producing an image that is 2D convex or curved similar in shape to a pie slice with a bite taken out of its top (Fig. 12.3b) [5].

Ultrasound Artifacts

Whichever probe is chosen, the examiner must be aware of the possibility of image interference due to artifacts. An ultrasound artifact is a feature that is visualized but may not accurately represent the tissue being scanned. To understand artifacts, it is necessary to understand ultrasound-tissue interaction properties. As the ultrasound wave propagates through tissue, it is partially reflected back to the transducer. The amount of echo reflected is determined by an intrinsic property of the tissue, the acoustic impedance, which is based on the tissue density [3]. The intensity of the reflected echo, or echogenicity, is proportional to the difference in acoustic impedances between adjacent mediums. It is this heterogeneity of a tissue that provides the ability to visualize differing structures in the tissue [6].

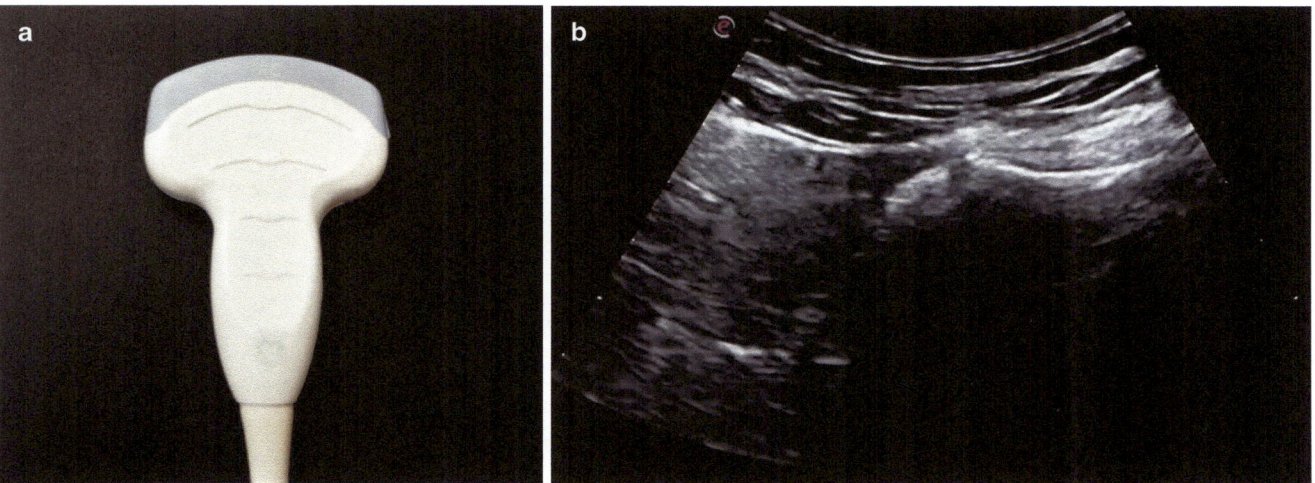

Fig. 12.3 (a) Curvilinear array probe. (b) Sacroiliac joint image. Note the convex-shaped image

Anisotropy

Anisotropy is one of the most common artifacts encountered in musculoskeletal ultrasound. Anisotropy refers to the property of being directionally dependent, as opposed to isotropy, which is uniformity in all directions. Anisotropy occurs when the echogenicity of a material changes depending on the angle of the transducer. This is a well-known phenomenon that occurs especially when viewing tendons and ligaments (and to a lesser extent nerves and muscle) under ultrasound. The tendon fibers appear bright (hyperechoic) when the transducer is perpendicular to the tendon and dark (hypoechoic) when the transducer is angled obliquely at more than 5 degree off perpendicular [7].

Anisotropy can mimic pathology and can be a source of interpretation error. It is especially common when viewing structures around a curved surface, for example, the supraspinatus fibrous attachment on the curved superior facet of the humerus. If the emitted pulse reaches a structure perpendicularly, almost all of the generated echo will travel back towards the transducer. This can be exemplified with a needle entering the field of view parallel to the face of the transducer—the needle appears hyperechoic. If, however, the transmitted pulse encounters the structure at an angle, the redirected pulse will be reflected at an angle equal to the angle of incidence [3]. In this case, only a fraction of the emitted pulses will return to the transducer, resulting in a hypoechoic region. For this reason, it may be difficult to visualize a needle that is inserted into the field of view at a very steep angle, especially when using a linear transducer. Anisotropy can be eliminated by toggling (or angulating) the transducer head so that the beam emitted is perpendicular to the area of interest. Anisotropy, however, can also be used to distinguish a structure from the surrounding tissue, for example, locating the median nerve among the flexor tendons in the carpal tunnel.

Attenuation Artifacts

Two other common types of artifacts encountered in musculoskeletal ultrasound are acoustic enhancement and acoustic shadowing, which are both produced from attenuation error. Recall, attenuation is the absorption and scattering of energy as it propagates through a tissue. Although higher frequency waves result in greater attenuation, at any given frequency the attenuation is also affected by an intrinsic property of the tissue called the acoustic impedance. The acoustic impedance is defined as the density multiplied by the velocity of propagation of the ultrasound wave in the tissue. Similar to acoustic impedance, bone has the highest degree of attenuation, followed by muscle and solid organs, and the lowest attenuation occurs in blood, air, or fluid-filled cavities [3, 7, 8].

Acoustic enhancement occurs when a structure is visualized deep to low attenuating structures. Because sound waves do not attenuate as much in the fluid, for example, the amplitude of the wave is greater in the tissue deep to the fluid compared to adjacent structures viewed at the same depth. Tissues deep to the low-attenuating structure appear hyperechoic. (Fig. 12.4) Some examples of structures that produce increased through transmission are large arteries, fluid-filled cysts, ganglion cysts, nerve sheath tumors, or giant cell tumors [3, 7, 8]. An important artifact that is visualized specifically in the presence of partial or complete supraspinatus tear is called the cartilage-interface sign. Because fiber disruption in the supraspinatus tendon causes it to appear abnormally hypoechoic, a hyperechoic line appears between the tendon and the hyaline cartilage of the humeral head. The

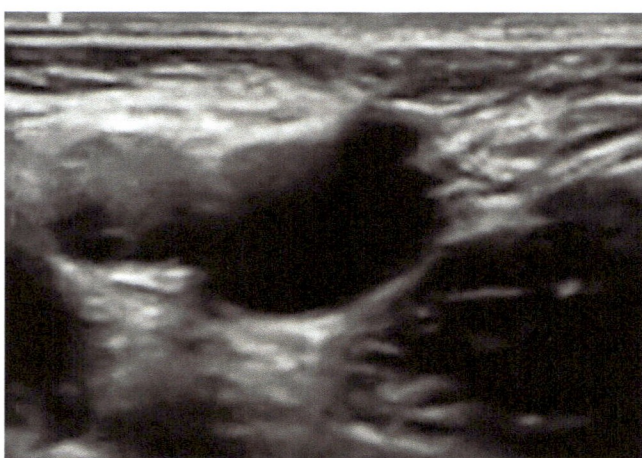

Fig. 12.4 Image of the femoral vessels. Note the acoustic enhancement of the tissues deep to the common femoral artery, on the left, and the larger common femoral vein, on the right

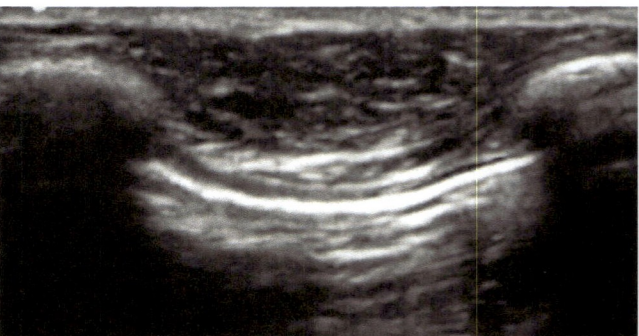

Fig. 12.5 Image of the intercostal neurovascular bundle. Note the acoustic shadows created by the adjacent ribs viewed on either side of the pleural surface

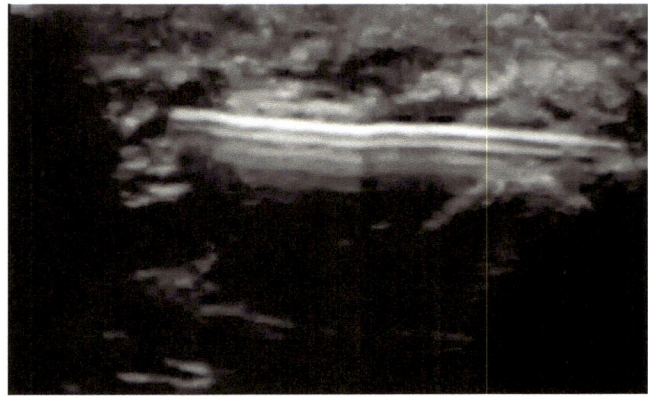

Fig. 12.6 An injection needle can create a reverberation artifact when it is parallel to the transducer. Note the tapering and repeating hyperechoic lines distal to the horizontal needle

cartilage-interface sign has been shown to be 100% specific for a partial or complete tear of the supraspinatus tendon; however, because it is not always seen, the sensitivity is as low as 30% [9].

Acoustic shadowing is the effect opposite of enhancement. It occurs when there is interference in visualization of tissue deep to a structure that strongly attenuates sound waves. The reduced sound waves transmitted beyond that structure result in a shadow of hypoechogenicity (Fig. 12.5). The most common tissue in musculoskeletal ultrasound that produces acoustic shadowing is bone; however, it can also be seen with calcifications, prosthesis, metallic foreign objects, or other dense fibrous tissues.

Artifacts due to Multiple ECHOS

Reverberation and "comet tail" artifacts occur when sound waves repeatedly bounce between any two strong reflectors parallel to each other. These additional echoes are recorded on the image, giving the appearance of a series of bright bands equidistant from each other; however, only the first reflection is spatially correct. Reverberation artifact is most commonly seen when the transmitted ultrasound beam encounters a strongly reflected surface parallel to the transducer face or parallel to another strongly reflected surface, such as an injection or biopsy needle, or metal prosthesis. As a result, the echoes are reflected back and forth, creating a tapering and repeating hyperechoic line distal to the hyperechoic structure (Fig. 12.6) [7, 8]. Each subsequent reflection beam is weaker than the previous one due to beam attenuation. "Comet tail" artifacts are a type of reverberation artifact

created by multiple reflections produced by a small but highly reflective structure, such as small calcifications, dense colloid aggregates, surgical clips, sutures, or staples [7]. It is often seen posterior to air-fluid interfaces and is usually a single long hyperechoic shadow which trails off distally (Fig. 12.7).

A mirror image artifact is produced when a target is located just superficial to a highly reflective smooth surface. The transmitted ultrasound beam is reflected off the hyperechoic structure towards a target. From the target, the echoed beam is first directed back to the highly reflective surface and is ultimately redirected back towards to transducer. True and false images are created on opposite sides of and equidistant from the strong reflecting surface. When scanning the musculoskeletal system, a mirror image artifact can be seen, for example, as an extraosseous structure that is mirrored inside the bony cortex. The false, mirrored image is inverted, often more distorted compared to the true image, and will disappear with subtle changes in transducer positioning [7, 8] (Fig. 12.8).

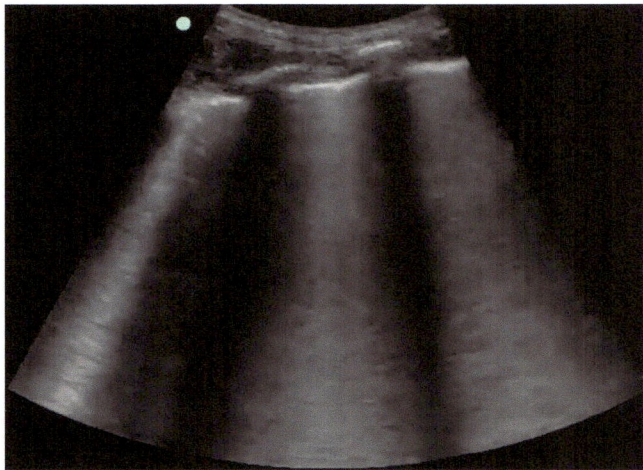

Fig. 12.7 The comet tail artifact can be seen when scanning lung tissue. It is caused by reverberation of the sound beam hitting the strongly reflective pleura and then the air below. Comet tail artifacts appear as echogenic white lines radiating down from the pleura

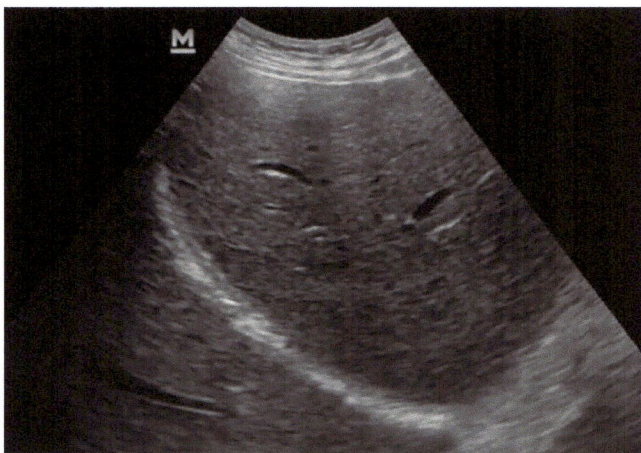

Fig. 12.8 In highly reflective interfaces such as the diaphragm, true and false images are created on opposite sides of and equidistant from the strong reflecting surface. Here the lung parenchyma can be seen on both sides of the diaphragm

Ultrasound Orientation and Handling

It is important to be considerate of positioning for the patient and the examiner. The patient should be comfortable on the exam table to reduce movement during the evaluation. In our experience, it is best to place the patient between the examiner and the display screen to ensure stability and comfort in the most ergonomic position.

There are a few indicator-to-screen general guidelines. First, the top of the screen corresponds to tissue closer to the probe and the bottom of the screen shows structures further from the probe. Second, the indicator on the transducer corresponds to a marking on the display in the upper left or right

corner. The transducer indicator is typically a bump or a groove on the lateral edge of the probe while the marking on the display may be a green dot or white hash mark, for example. Classically, the side with the indicator corresponds to the left side of the screen.

Historically, for sagittal or coronal views, the ultrasound transducer should be oriented such that the indicator on the probe points towards the patient's head. The image produced on the display is therefore oriented such that the left side of the screen is rostral and the right screen is caudal. However, when scanning musculoskeletal structures in the axial or transverse planes, there is no uniform orientation guide. Therefore, it is important the examiner verifies orientation prior to the start of the exam. This can be done by placing a small amount of gel on the lateral face of the probe close to the indicator and confirming that the side with the gel corresponds to the left side of the display screen.

Pathologic findings can be small and are often over an irregular surface on the skin (humeral head, anterior knee, cubital tunnel), thus, probe stability is an important skill for the musculoskeletal examination for high-quality imaging. We recommend holding ultrasound probe between the index finger and thumb with the medial fingers and hypothenar eminence resting directly on the patient's body. This technique helps to anchor the transducer with adequate but not excessive pressure on the skin, allowing for easy toggling of the probe while keeping the area of interest in view.

There are several common terms to describe specific movements of the transducer head when scanning tissues. Tilt or toggle refers to tipping the transducer towards one of its long sides. This technique can be used when viewing anisotropy of tendons as it changes the direction of the ultrasound wave from perpendicular to oblique to the fibers. Translation or slide refers to movement of the entire transducer head horizontally along the skin surface. Transducer rotation is rotation of the head around its central axis. This technique can be used, for example, when scanning the biceps tendon in the bicipital groove. If the biceps tendon is first viewed in short axis in the bicipital groove, the transducer can then be rotated while keeping the biceps tendon in the center of the screen to then scan the tendon in the long-axis. Heel-toeing or rocking raises or lowers one end of the transducer head. This technique can be used when trying to keep the ultrasound beam perpendicular to the needle for a steep angle injection, or when using the step off method with gel for viewing bony landmarks.

References

1. Lawrence JP. Physics and instrumentation of ultrasound. Crit Care Med. 2007;35:S314–22.
2. Ahmed R, Nazarian LN. Overview of musculoskeletal sonography. Ultrasound Q. 2010;26:27–35.

3. Narouze SN. Atlas of ultrasound-guided procedures in interventional pain management. New York: Springer; 2011. p. 14–29.

4. Bianchi S, Martinoli C. Ultrasound of the musculoskeletal system. New York: Springer; 2007. p. 3–7.

5. Szabo TL, Lewin PA. Ultrasound transducer selection in clinical imaging practice. J Ultrasound Med. 2013;2:573–82.

6. Kossoff G. Basic physics and imaging characteristics of ultrasound. World J Surg. 2000;24:134–42.

7. Griffith JF. Diagnostic ultrasound musculoskeletal. Philadelphia: Elsevier; 2015: Part II, Sec 1. p. 6–13.

8. Carmody KA, Moore CL, Feller-Kopman D. Handbook of critical care and emergency ultrasound. San Francisco: McGraw Hill; 2011. p. 11–4.

9. Jacobson JA, Lancaster S, Prasad A, et al. Full-thickness and partial-thickness supraspinatus tendon tears: value of US signs in diagnosis. Radiology. 2004;230(1):234–42.

Diagnostic Ultrasound: Recognizing Musculoskeletal Pathology

Allan Zhang and George C. Chang Chien

Introduction

Ultrasonography has long been considered a valuable method of imaging which offers several distinct advantages over other modalities. This is especially true with certain aspects of musculoskeletal imaging. One of the most important benefits of sonography is its lack of ionizing radiation—which has been shown to increase cancer risks in susceptible individuals with repeated radiation exposure. Prenatal exposure to ionizing radiation is also a known teratogen within the obstetric population. A second advantage of sonography is the real-time dynamic nature of the examination. This ability allows the practitioner to evaluate the origin of the disease process at its precise location, corresponding to patient's reported symptoms. In addition, the dynamic nature of ultrasound permits the evaluation of certain pathology that can only be reproduced with particular positioning or movement. Lastly, ultrasonography offers excellent resolution of superficial structures that can rival and/or exceed cross-sectional CT or MRI in evaluation of anatomy such as the rotator cuff tendons at a fraction of the cost.

This is not to say ultrasonography is without its disadvantages. One of the principal downsides of ultrasonography is its inherent nature of user-dependent variability. The diagnostic quality and ability of ultrasonography is contingent on the proficiency of its operator. In the hands of a skilled master, ultrasonography can be an effective diagnostic asset, while acting as a hindrance and liability in the hands of a novice`. Ultrasound is also ineffective in evaluating certain structures, which may be obstructed by ultrasound-specific artifacts. The most commonly encountered artifact as it relates to musculoskeletal ultrasound is acoustic impedance. Simply put, acoustic impedance describes the differences in tissue density. As this difference increases, more sound waves are reflected back and images are degraded or not seen. This phenomenon is best exemplified with ultrasound evaluation of bones, which will be discussed in detail in the subsequent section.

This chapter will serve as an introduction to diagnostic musculoskeletal ultrasound imaging. Discussion will concentrate on musculoskeletal ultrasound imaging of tendons, ligaments, muscles, cartilages, and bones with focus of normal appearances and different pathological states.

Tendons

Normal Anatomy and Appearance

Tendons are defined as bundles of fibrous collagen tissue attaching muscles to bone. Majority of tendons are surrounded by a synovial membrane termed the tendon sheath. On ultrasound, tendons demonstrate fibrillar architecture and consist of closely spaced, parallel, hyperechoic linear reflections in the longitudinal plane; in the transverse plane, tendons appear as multiple echogenic dots (Fig. 13.1). It is important to note, however, ultrasound appearance of tendons is susceptible to anisotropy, a phenomenon in which tissues demonstrate variable echogenicity depending on the angle at which they are viewed. For example, when tendons are imaged at less than 90 degrees perpendicular angle, they become hypoechoic and lose their fibrillar appearance (Fig. 13.2). Under most circumstances, tendons should be imaged so that its fibrillar pattern is visualized. However, in particular instances when tendons are coalesced with surrounding hyperechoic tissue, it may be helpful to purposely angle the transducer less than 90 degrees to produce anisotropy, and view the targeted tendon as a hypoechoic structure.

A. Zhang (✉)
University of Connecticut, Department of Radiology, Farmington, CT, USA
e-mail: azhang@uchc.edu

G. C. Chang Chien
GCC Institute, Department of Musculoskeletal Medicine and Medical Aesthetics, Newport Beach, CA, USA

© Springer Nature Switzerland AG 2023
C. W Hunter et al. (eds.), *Regenerative Medicine*, https://doi.org/10.1007/978-3-030-75517-1_13

Fig. 13.1 Ultrasound appearance of Achilles tendon in short- and long-axis. The tendon is mildly thickened suggestive of underlying tendinopathy

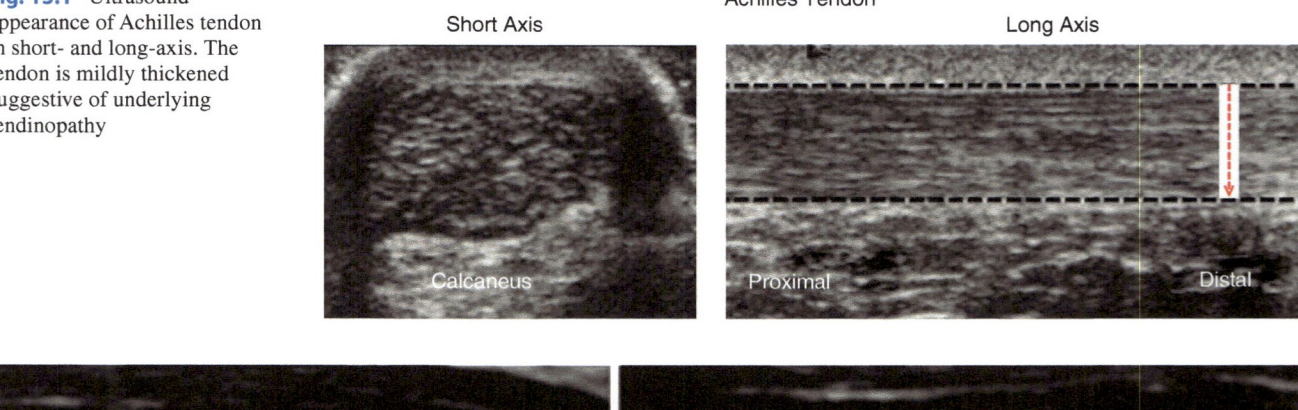

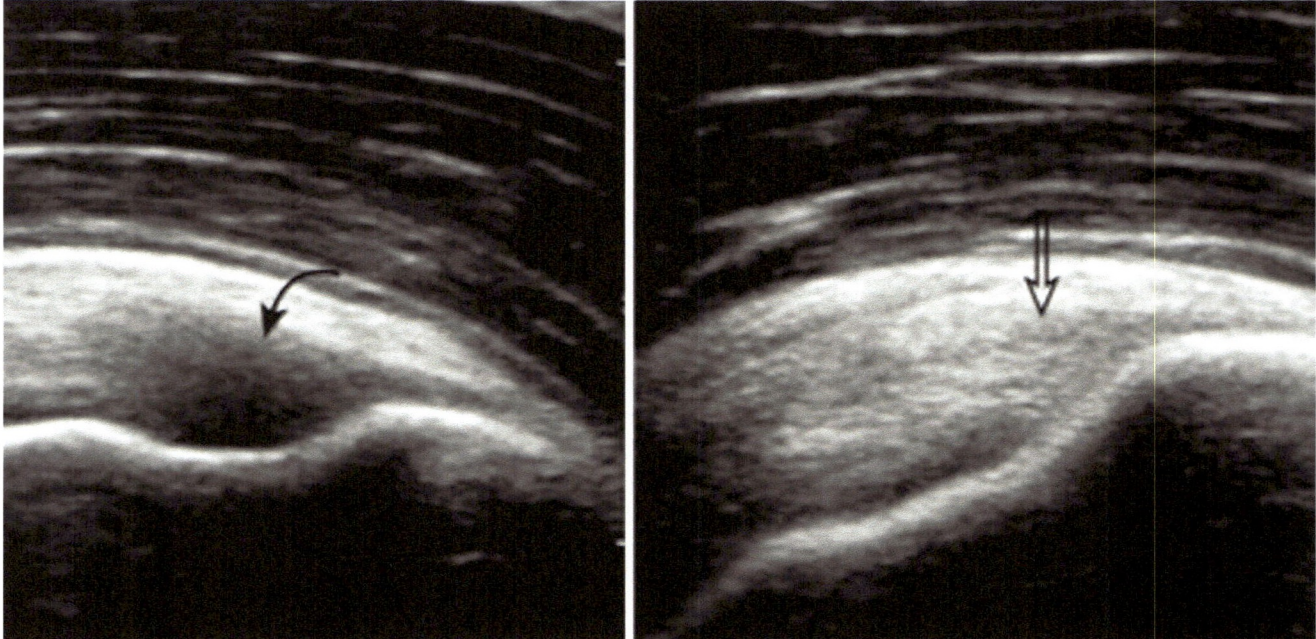

Fig. 13.2 Apparent articular-sided tear of the supraspinatus tendon (Left) imaged at less than 90 degrees demonstrating anisotropy. When imaged at 90 degrees, the tendon is in fact intact and normal in echogenicity (Right)

Tendon Pathology

Ultrasound is highly sensitive and specific for tendon pathology, especially of superficial tendons such that of the rotator cuff tendons. Saraya et al. 2016 demonstrated that ultrasound evaluation of the rotator cuff tendonitis demonstrated sensitivity, negative predictive value, and accuracy of 85%, 86%, and 90%, respectively [1]. For partial thickness tears, the values were 88% sensitivity, 89% specificity, 94% positive predictive value, 80% negative predictive value, and 83% accuracy [1]. In full-thickness tears, however, the sensitivity and specificity were 100% each [1]. These values are comparable to MRI in both sensitivity and specificity, while ultrasound costs a fraction of the expense. Furthermore, Khan et al. 2015 also demonstrated similar sensitivity and specificity values when comparing ultrasound and MRI in evaluation of Achilles tendon pathology [2].

Partial and full-thickness tendon tears will demonstrate partial disruption or complete disruption of the normal fibrillar pattern on ultrasound. In full-thickness tears, the end of the retracted proximal tendon will appear blunted on long-axis view (Fig. 13.3). Dynamic motion of the tendon may be useful in confirming tendon tears. This can be achieved by viewing the target tissue while activating the muscle/tendon. Isometric contraction against resistance can also demonstrate pathologic tears that are not readily visible under direct visualization. Other helpful signs include nonvisualization of the tendon at its expected anatomic location or adjacent fluid collections at the site of tendon tear indicative of hematoma formation or tendon sheath effusion. In partial-thickness tears, the disruption will be focal and regional and will not cause tendon retraction. Additional signs include the cartilage interface sign which describes partial articular-sided or complete tear of the supraspinatus tendon at its insertion on

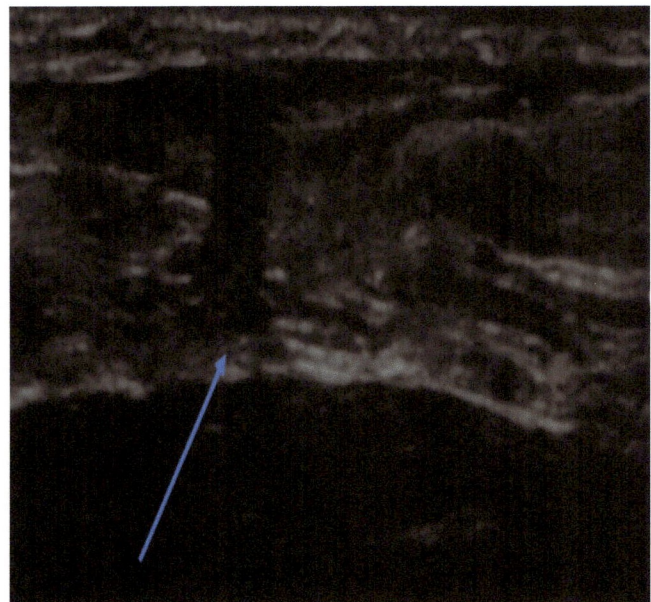

Fig. 13.3 Partially torn (blue arrow) image of the Achilles tendon in long-axis

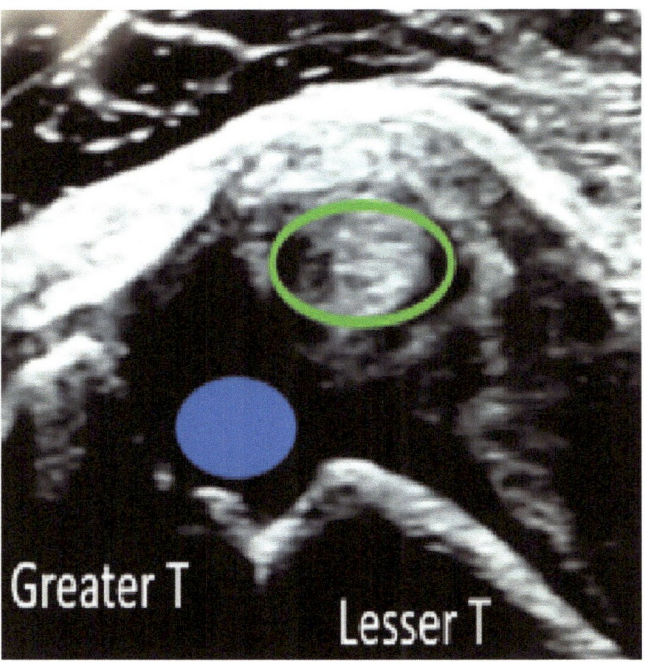

Fig. 13.4 US images demonstrating medial subluxation of the long head bicep tendon. Normally, the long head of the bicep tendon (green circle) should be situated within the bicipital groove (blue circle) within the greater and lesser tuberosities (Greater T, Lesser T)

the greater tuberosity. In this case, a hypoechoic region is seen within the supraspinatus tendon. Deep to this region is a linear hyperechoic line which is created by the reflection of sound waves at the fluid-cartilage interface. Chronic tears of the supraspinatus tendon are associated with cortical irregularity of the greater tuberosity, termed the cortical irregularity sign.

Tendon dislocations and subluxations occur most often in the setting of trauma and secondary to disruption of the adjacent ligamentous stabilizers. An example of this is the long head bicep brachii tendon. Anatomically, the long head bicep brachii tendon is normally situated within the bicipital groove and held in place by the superior glenohumeral ligament. When this ligament is deficient, the tendon can freely dislocate medially (Fig. 13.4). Often, a snapping sensation can be reproduced when the extremity is moved in certain positions.

Tendonitis and tenosynovitis are inflammation involving either the tendon or the tendon sheath secondary to a wide range of causes including inflammatory arthropathies, crystalline arthropathies, infection, trauma, autoimmune processes, and even foreign bodies [3]. On ultrasound, tenosynovitis appears as thickening of the tendon sheath or fluid distending the tendon sheath (Fig. 13.5). As an aside, the Achilles tendon is the only tendon in the human body that is not encapsulated by a tendon sheath, and therefore, will not demonstrate tenosynovitis. Tendonitis, on the other hand, may appear as hypervascularity indicative of increased inflammation of the tendon seen on color Doppler imaging. Furthermore,

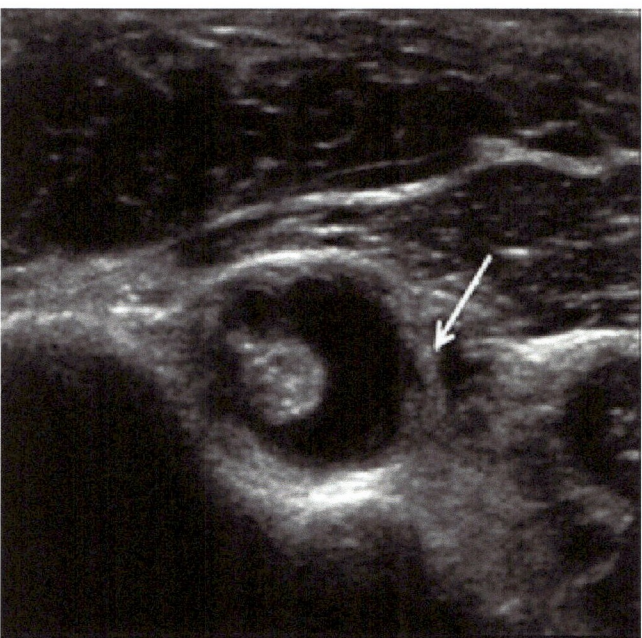

Fig. 13.5 Abnormal volume of fluid surrounding the bicep tendon sheath suggestive of tenosynovitis

tendinosis or tendinopathy may be secondary to chronic degeneration of a tendon, and will appear as tendon thickening with heterogeneous echogenicity on ultrasound (Fig. 13.6).

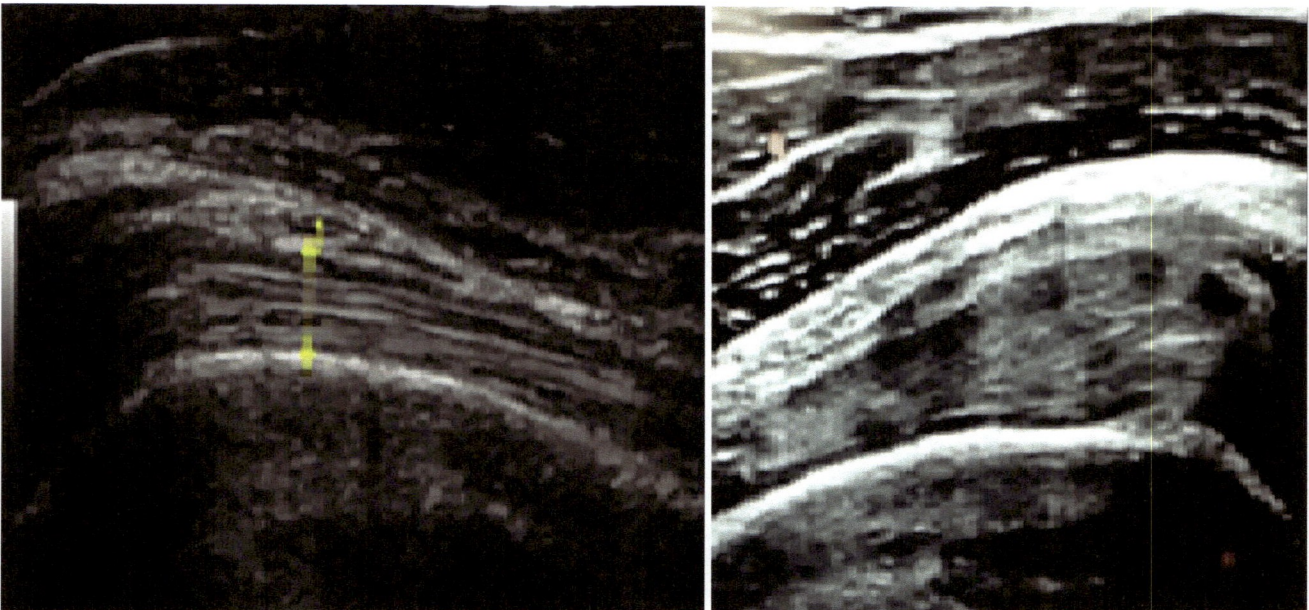

Fig. 13.6 Left image demonstrating normal appearance of the supraspinatus tendon. Right image demonstrating supraspinatus tendon with thickened and heterogenous appearance consistent with tendinopathy

Muscles

Normal Anatomy and Appearance

Skeletal muscles are composed of individual muscle fibers grouped into fascicles that are separated by fibrous connective tissue termed perimysium [3]. Its primary functions are that of movement and maintaining posture. These linear internal fibrous bands of perimysium converge to form a central tendon. On ultrasound, muscles are hypoechoic with intervening linear bands of hyperechoic perimysium (Fig. 13.7). On short-axis view, this appears as diffuse speckles perimysium on a hypoechoic background of muscle (Fig. 13.8), so called "starry sky" appearance.

Muscle Pathology

As with evaluating tendons, ultrasound offers several important potential advantages over MRI such as portability, convenience, and cost. Furthermore, the dynamic capabilities of ultrasound allow for diagnosis of certain muscular pathology not identifiable on static imaging. This is particularly true of muscular herniations which can require patient's active muscle contraction for diagnosis [4–6].

Muscle injuries can be secondary to direct impact trauma (i.e., contusion and laceration) and/or distraction of muscle fibers from forceful muscle contraction (i.e., strains) [7, 8]. Muscle strains are separated into three grades, with each sequential grade corresponding to increase in tears of the muscle fibers. Grade 3 strains are complete disruption of the

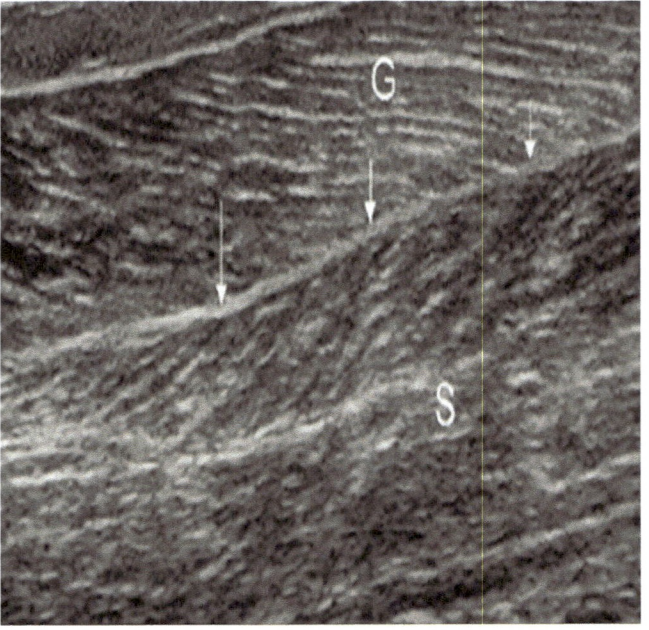

Fig. 13.7 Longitudinal views of the gastrocnemius (G) and soleus(S) muscles

entire muscle often with associated retraction of the proximal muscle stump (Fig. 13.9). Acutely, low-grade muscular strains appear on ultrasound as mild incomplete disruption of the fibers with areas of low echogenicity corresponding to fluid and edema. Over time, intramuscular hematomas may develop with varied appearance on ultrasound based on chronicity. More chronic muscle strains will result in focal areas of scarring that manifest as areas of increased echogenicity

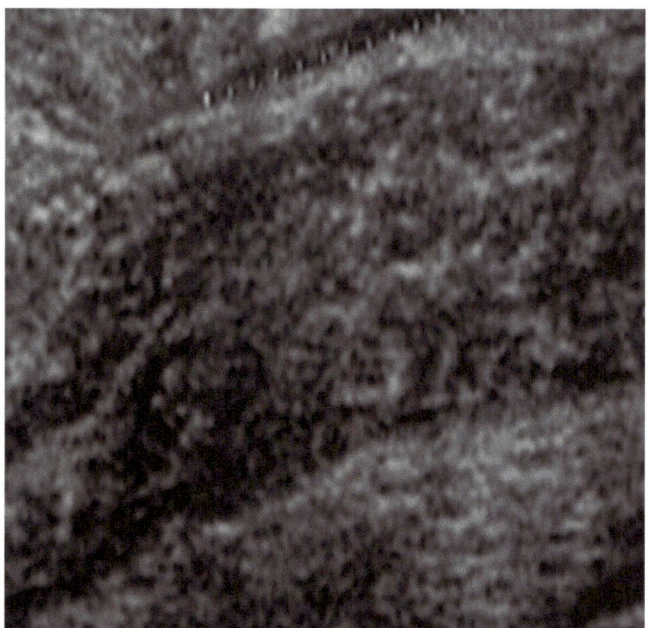

Fig. 13.8 Short-axis view of the same muscle demonstrating the starry sky appearance

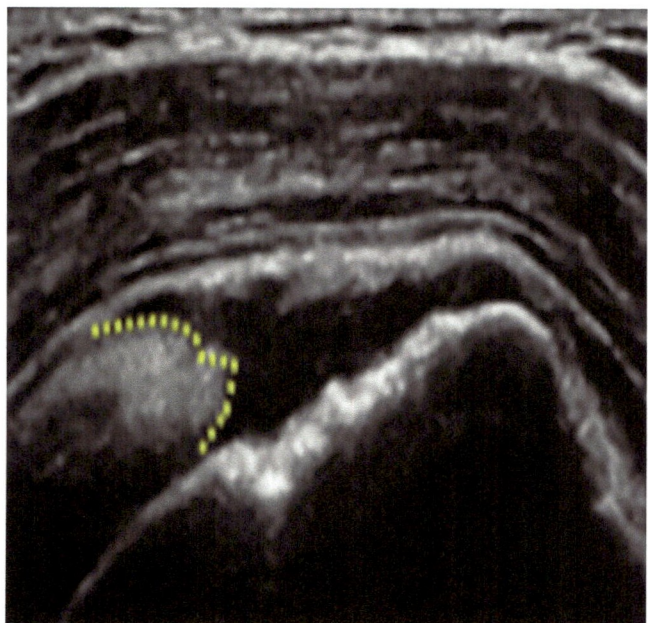

Fig. 13.9 Supraspinatus muscle and tendon retraction

within the muscle [7]. Direct muscular trauma or muscular contusion may have associated intramuscular hematomas. Detailed history is paramount in assessing these patients.

Muscle denervation is the sequelae of injury to its corresponding motor nerve [9]. On ultrasound, this appears as loss of muscle bulk with intramuscular fatty infiltration manifesting as areas of increased echogenicity within the muscle or the entire muscle, as in cases of full thickness tendon rupture

leading to gross fatty atrophy of the muscle. Comparison with the contralateral side can offer diagnostic clues when unilateral peripheral nerve injuries are suspected (Fig. 13.10). Of note, correlative MRI imaging appearance of muscle denervation also appears similar, manifesting as fatty infiltration and muscular atrophy.

Muscle hernias occur most often in the lower extremities secondary to fascial defects with focal muscular protrusions through the defect. They are usually asymptomatic and discovered incidentally. Very rarely, muscle hernias can be the source of pain with potential for entrapment and ischemia [4–6]. On dynamic ultrasound, hernias appear as focal herniating muscle that becomes more conspicuous with contraction (Fig. 13.11).

Ligaments

Normal Anatomy and Appearance

Ligaments are fibroelastic bands that connect bones to bones or cartilage [10, 11]. They function mostly as stabilizers, restricting and guiding movements at joints. Examples include the medial and lateral collateral ligaments of the knee, the anterior and posterior cruciate ligaments of the knee, the deltoid ligament of the foot, and glenohumeral ligaments of the shoulder. On ultrasound, ligaments have very similar appearance to tendons demonstrating a fibrillar hyperechoic appearance (Fig. 13.12). However, imaging of ligament is more susceptible to anisotropy and therefore it is not uncommon for ligaments to appear hypoechoic, rather than hyperechoic, on ultrasound [10]. Ligaments are generally found deeper compared to the adjacent tendons.

Ligament Pathology

Ligamentous abnormalities are almost always secondary to acute trauma or chronic repetitive microtrauma. These can range from low-grade sprains to partial tears to complete ruptures. Acutely, low-grade ligament sprains can manifest on ultrasound as thickening of the ligament with increased hypoechoic areas indicative of edema. A partial-thickness tear may be seen as a hypoechoic area of focal disruption of the ligament (Fig. 13.13) whereas complete ruptures will demonstrate full-thickness ligamentous disruptions. Additionally, there may be secondary-associated injuries including bone contusion, fractures, and/or joint effusions.

Chronic ligamentous injuries such as those from repetitive microtrauma appear as thickened ligaments with irregular fibers and increased intrasubstance echogenicity. They often demonstrate increased laxity and may be associated with joint instability.

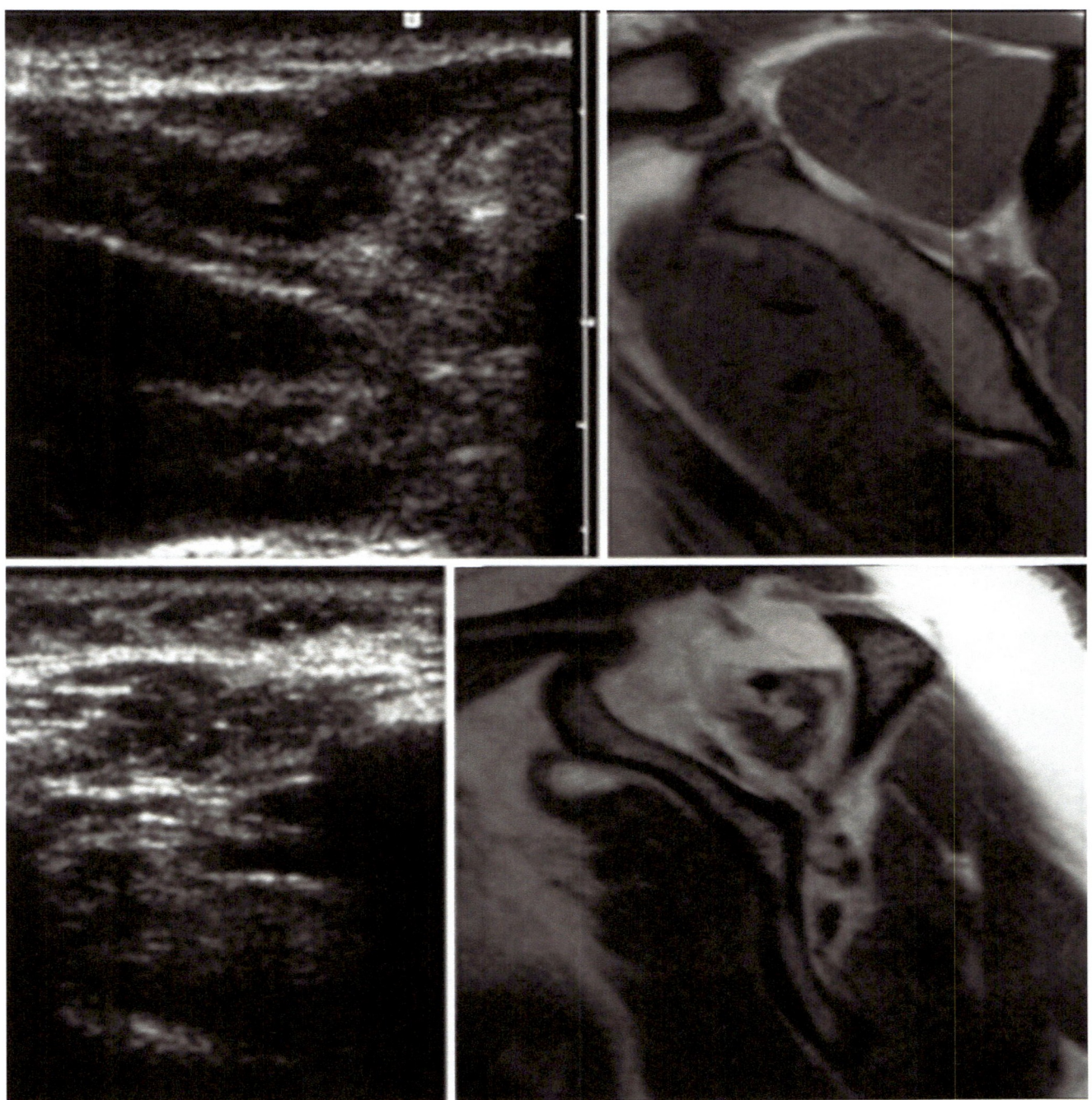

Fig. 13.10 Normal (top) and fatty atrophied (bottom) supraspinatus muscle. Ultrasound and corresponding MRI appearance

Bones and Cartilage

Normal Anatomy and Appearance

Bones are calcified connective tissue consisting of osteocytes embedded in a matrix of ground substance and collagen fibers [9, 10]. They are composed of an outer layer of compact cortical bone and inner layer of intramedullary spongy marrow. The external cortical surface impedes the penetration of sound waves and results a very bright appearance on ultrasound (Fig. 13.14). This limitation renders evaluation of intramedullary pathology more challenging and other modalities such as CT and MRI should be utilized when clinically indicated.

Cartilage is a smooth viscoelastic tissue composed of extracellular matrix and chondrocytes. They are designed to weight bear and act as shock absorbers to distribute loads across joints [12]. Cartilage also functions as lubricants to minimize frictional forces. They are high in fluid content and as such, appear as thin, smooth hypoechoic to anechoic lay-

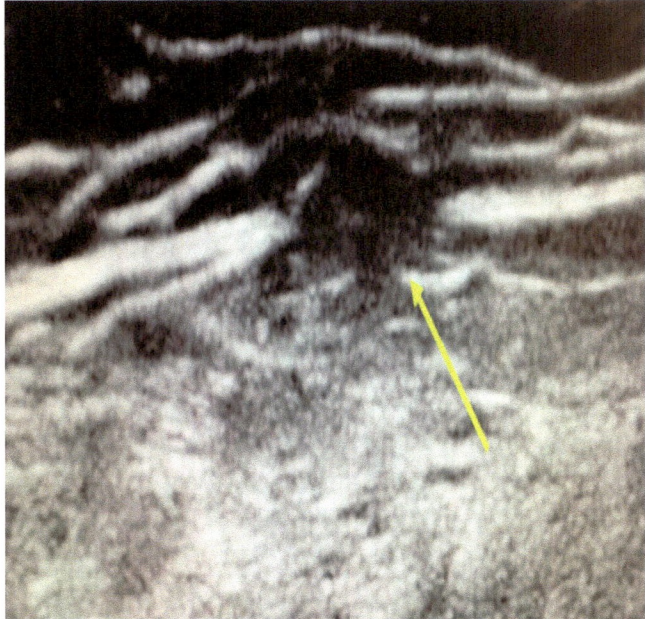

Fig. 13.11 Ultrasound appearance of focal muscle herniation through the fascia of the anterior tibialis in this frequent runner

ers overlying cortical bone (Fig. 13.14). While ultrasonography is not the primary imaging modality for evaluating cartilage, it is important to recognize pathology when encountered during routine examinations.

Bones and Cartilage Pathology

Overlying bones, soft tissues, and vascular structures can make nondisplaced fractures difficult to detect on plain radiographs. Ultrasound offers direct and precise site of examination and any disruption in the cortex of the bone can be readily identified (Fig. 13.15). Furthermore, abnormal calcification of tendons or ligaments can be easily evaluated by ultrasound as areas of increased echogenicity and posterior acoustic shadowing. For example, calcific tendinitis of the rotator cuff tendon, most commonly the supraspinatus tendon, will demonstrate shadowing hyperechoic lesion in the substance of the tendon (Fig. 13.16).

Osteophytes appear as well-corticated bony protuberances at the margins of the involved joints. When osteophytes are identified along with thinning of the articular cartilage, these constellations of findings are highly suggestive of degenerative joint disease. Crystalline arthropathies such as gout and pseudogout can also be reliably evaluated by ultrasound. In gout, the hyperechoic monosodium urate crystals deposit on the surface of the hyaline cartilage, giving rise to the double contour sign. Unlike gout, pseudogout manifests as calcification deposition within the cartilage or meniscus.

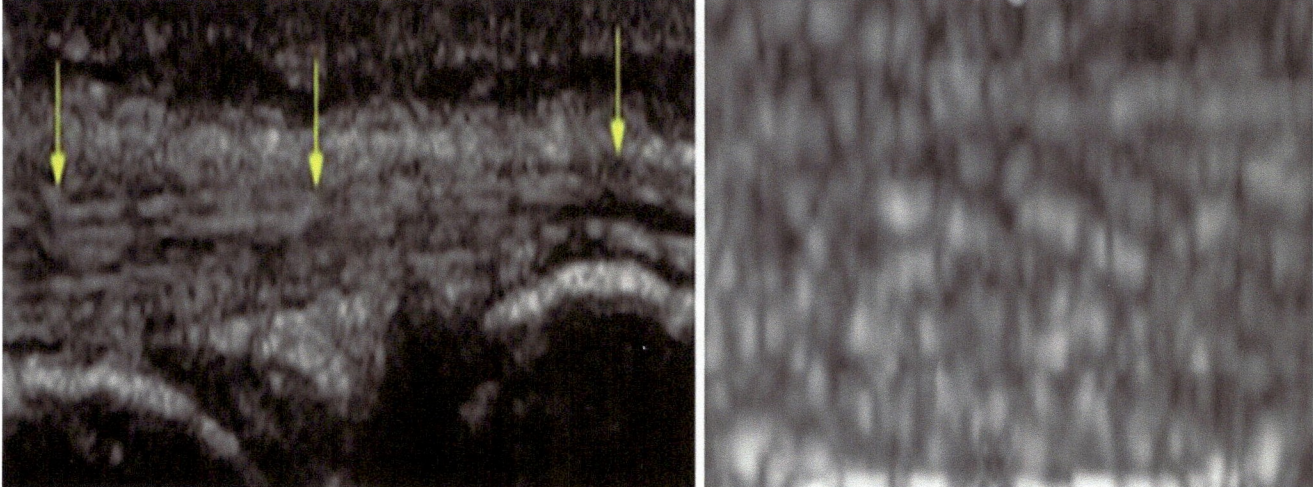

Fig. 13.12 Normal sonographic appearance of the medial collateral ligament in longitudinal (left) and axial (right) planes

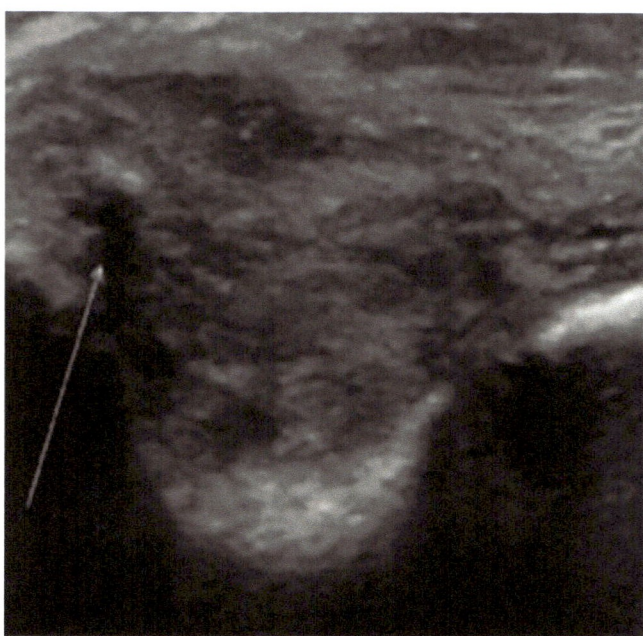

Fig. 13.13 Focal partial thickness tear along the undersurface of the ulnar collateral ligament of the elbow

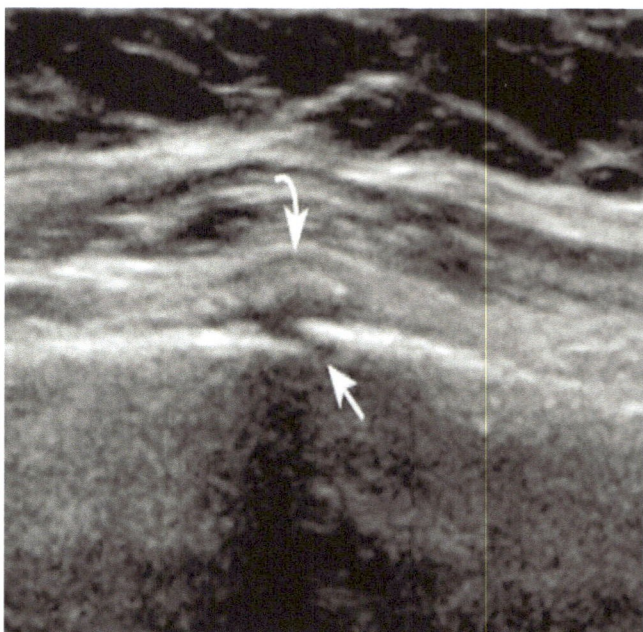

Fig. 13.15 Focal disruption of the echogenic cortex consistent with minimally displaced fracture (arrows)

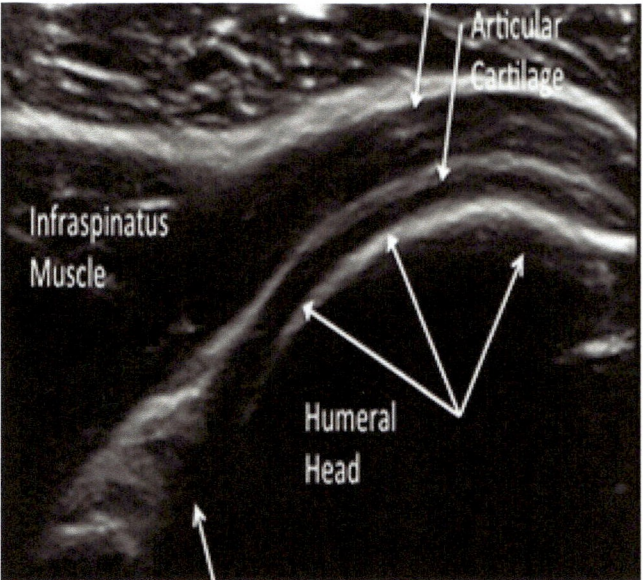

Fig. 13.14 The cortex of the humeral head is echogenic with acoustic shadowing posterior to it

Conclusion

Ultrasound is a valuable asset in diagnosing musculoskeletal-related injuries and pathology. It provides a number of advantages over other imaging modalities such as MRI and CT, with fraction of the cost. Ultrasound does not emit ionizing radiation and can be used in real-time dynamic evaluation of tissue pathology. Furthermore, ultrasound is a mobile imaging tool that has no known contraindications. It provides comparable diagnostic sensitivity and specificity with certain musculoskeletal pathologies including evaluation of rotator cuff tears. Knowing its limitations and drawbacks will allow the clinician to add ultrasound as an invaluable asset to their imaging repertoire.

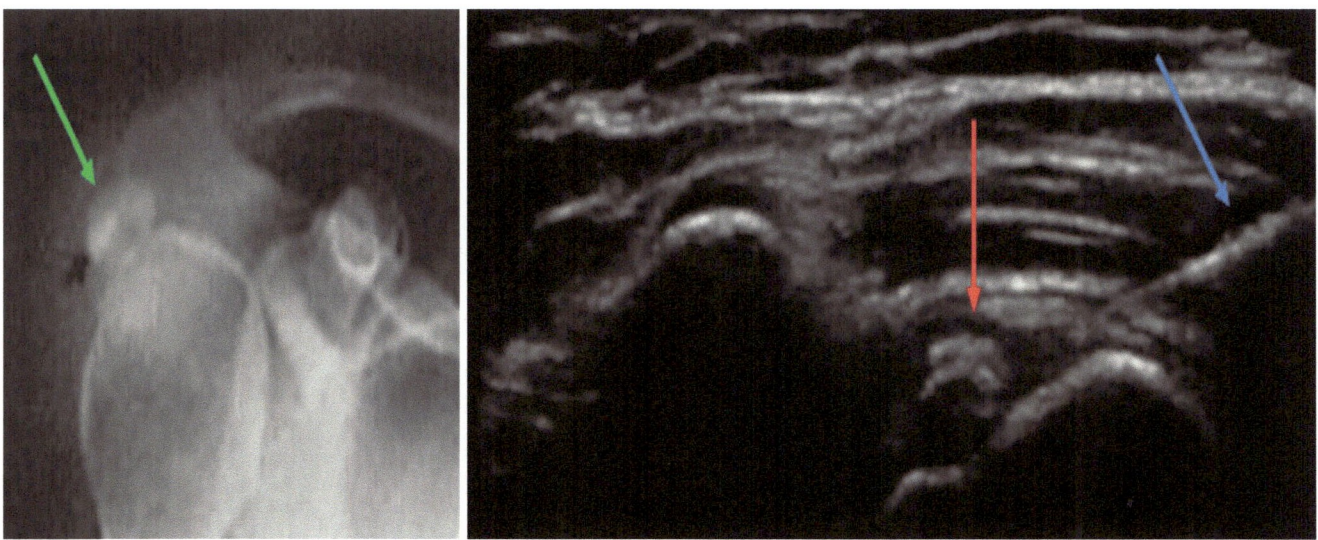

Fig. 13.16 Calcific tendinosis of the supraspinatus tendon. Green arrow points to the calcification seen on radiograph. Red arrow points to the same calcification seen on ultrasound. Blue arrow points to a needle inserted in an attempt to fenestrate the calcification

References

1. Saraya S, Bakry RE. Ultrasound: can it replace MRI in the evaluation of the rotator cuff tears? Egypt J Radiol Nucl Med. 2016;47(1):193–201. https://doi.org/10.1016/j.ejrnm.2015.11.010.
2. Khan K, Forster B, Robinson J, et al. Are ultrasound and magnetic resonance imaging of value in assessment of Achilles tendon disorders? A two-year prospective study. Br J Sports Med. 2003;37(2):149–53. https://doi.org/10.1136/bjsm.37.2.149.
3. Strakowski JA. Introduction to musculoskeletal ultrasound getting started. New York: Demos Medical Publishing; 2016.
4. Petscavage-Thomas J. Clinical applications of dynamic functional musculoskeletal ultrasound. Rep Med Imaging. 2014;7:27–39. https://doi.org/10.2147/rmi.s40194.
5. Artul S, Habib G. The importance of dynamic ultrasound in the diagnosis of tibialis anterior muscle herniation. Crit Ultrasound J. 2014;6(1):14. https://doi.org/10.1186/s13089-014-0014-0.
6. Nguyen JT, Nguyen JL, Wheatley MJ, Nguyen TA. Muscle hernias of the leg: A case report and comprehensive review of the literature. Plast Surg. 2013;21(4):243–7. https://doi.org/10.4172/plastic-surgery.1000834.
7. Lee JC, Mitchell AWM, Healy JC. Imaging of muscle injury in the elite athlete. Br J Radiol. 2012;85(1016):1173–85. https://doi.org/10.1259/bjr/84622172.
8. Zwerver J. Sports medicine and imaging. In: Nuclear medicine and radiologic imaging in sports injuries. Berlin: Heidelberg Springer Berlin Heidelberg; 2015. p. 3–8. https://doi.org/10.1007/978-3-662-46491-5_1.
9. Jacobson JA. Musculoskeletal ultrasound. Philadelphia: Elsevier Saunders; 2007.
10. Hertzberg BS, Middleton WD. Ultrasound: the requisites. Philadelphia: Elsevier; 2016.
11. Hodgson RJ, O'Connor PJ, Grainger AJ. Tendon and ligament imaging. Br J Radiol. 2012;85(1016):1157–72. https://doi.org/10.1259/bjr/34786470.
12. Tiku ML, Sabaawy HE. Cartilage regeneration for treatment of osteoarthritis: a paradigm for nonsurgical intervention. Ther Adv Musculoskeletal Dis. 2015;7(3):76–87. https://doi.org/10.1177/1759720x15576866.

Fluoroscopic Safety

14

Kenneth D. Candido and Tennison Malcolm

Introduction

Fluoroscopy use is a practical necessity for many procedures performed by interventional clinicians; however, when used inappropriately or indiscriminately, it can prove hazardous for both patients and providers. Following a series of rare but serious radiation-induced skin burns, the United States Food and Drug Administration (FDA) issued an advisory statement in 1994 describing the need for sufficient training among caregivers utilizing fluoroscopy [1, 2]. Cancer is the second leading cause of death in the United States and is among the top five causes for middle- and high-income nations worldwide [3, 4]. Prior to significant reductions in occupational risk associated with a greater appreciation of prophylactic measures used to reduce radiation exposure, medical providers with high exposure to ionizing radiation were among the most affected by breast cancer, leukemia, and skin cancer [5–8].

Terminology

An understanding of fluoroscopic safety demands a familiarity with basic radiation principles and nomenclature.

Radiation is the process by which energy is emitted or transmitted either as a wave (e.g., electromagnetic, acoustic, and gravitational radiation) or particle (e.g., alpha, beta, and neutron radiation). Electromagnetic radiation is the traveling wave motion produced by changes in electric and magnetic fields. In order of increasing wavelength, the electromagnetic spectrum ranges from gamma rays and x-rays, to ultra-violet visible light and infrared to microwaves and radio waves (Fig. 14.1). X-rays, gamma rays, alpha particles, and beta particles are forms of ionizing radiation. They characteristically result in electron displacement, free radical formation, and ionization of atoms and molecules following propagation through matter such as air, water, and living tissue.

Radiation exposure is universally expressed in roentgens (R) and in SI units coulomb/kilogram (C/Kg). Radiation absorbed by a person or object is conventionally measured in radiation absorbed dose (rad) and in SI units, gray (Gy). One Gy is equal to 1 joule of energy deposited per kilogram of tissue. As different sources of radiation can have dissimilar medical effects, absorbed radiation is also expressed in dose-equivalents. For x-rays, dose equivalent and absorbed dose are equal. In contrast, the dose equivalent is 20-fold larger than the absorbed dose for alpha radiation as this type of radiation is much more damaging to the human body. Dose equivalents are expressed conventionally as roentgen-equivalent-man (rem) and in SI units as Sievert (Sv). Roentgen-equivalent-man is equal to the radiation absorbed

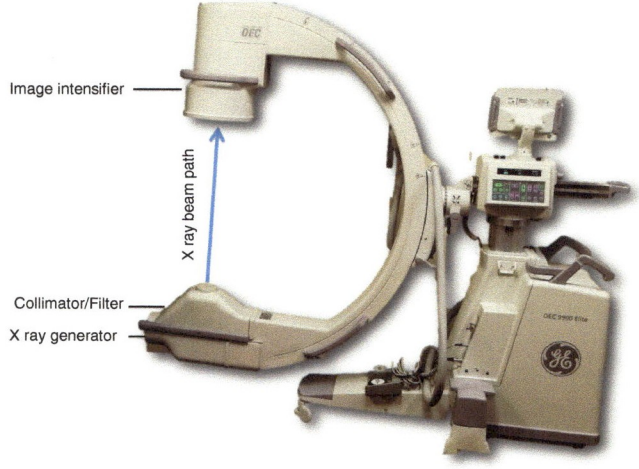

Fig. 14.1 Spectrum of Radiation Energy

K. D. Candido (✉)
Department of Anesthesiology, Department of Anesthesia and Pain Medicine, Advocate Illinois Masonic Medical Center, University of Illinois College of Medicine-Chicago, Chicago, IL, USA
e-mail: kenneth.candido@advocatehealth.com

T. Malcolm
Brigham and Women's Hospital, Department of Anesthesiology, Perioperative and Pain Medicine, Boston, MA, USA

© Springer Nature Switzerland AG 2023
C. W Hunter et al. (eds.), *Regenerative Medicine*, https://doi.org/10.1007/978-3-030-75517-1_14

dose multiplied by a quality factor (QF) specific to the type of radiation used (rem = rad × QF). Radiation exposure is believed to have stochastic and deterministic (also known as nonstochastic) health effects. Stochastic effects (e.g., heritable effects, cancer) occur by chance without a threshold dose and at a rate proportional to the dose received. Deterministic effects (e.g., radiation-induced cataracts) are believed to occur after a threshold amount of radiation is reached and vary in severity proportional to exposure dose.

Biologic Effects

In living tissue, chromosomal DNA is believed to be the principal target mediating cellular effects of ionizing radiation. Error-prone repair of chemically complex double-strand DNA lesions gives rise to chromosomal aberrations, gene mutations, and apoptosis (defined as the death of cells which occurs as a normal and controlled part of an organism's growth or development). DNA damage response and repair processes are major determinants of postinjury effects within the cell. Extensive chromosomal damage and exhausted DNA repair mechanisms favor apoptosis. Oncogenesis (the development of a tumor or tumors) is often a result of perturbations in response, repair, and apoptotic mechanisms.

In tissue, deterministic effects typically involve loss of cellular reproductive capacity, fibrosis, and overall loss of function. Deterministic effects are most likely to be clinically apparent in cells and tissues most sensitive to ionizing radiation; namely highly proliferative cells and tissues such as hematopoietic cells, the gastrointestinal tract, the basal cell layer of skin, and male germ cells. Organs present in pairs or with functional subunits arranged in parallel (e.g., liver, kidney), rather than in series (e.g., gastrointestinal tract) are more resilient and least likely to demonstrate clinical signs of dysfunction. Organ-specific doses of radiation believed to result in a 1% risk of deterministic effects are shown in Table 14.1. After 3 Gy and 6 Gy, 1/100 women and 1/100 men, respectively, may experience permanent sterility [9]. Absorbed radiation doses of 1 Gy are associated with a 1/100 risk of death resulting from sequelae of bone marrow syndrome [9]. Bone marrow contains stem cells. Stem cells are sensitive to radiation exposure and excessive exposure may result in the formation of malignancies. In leukemia, a cancer of the blood, the bone marrow makes abnormal white blood cells. In aplastic anemia, the bone marrow does not make red blood cells. In myeloproliferative disorders, the bone marrow makes too many white blood cells. Each of these conditions can potentially occur in the face of radiation exposure in susceptible hosts.

Oncogenesis is a complex multifactorial process heavily affected by factors intrinsic and extrinsic to the cell. For the purposes of this discussion, it is oversimplified into four

Table 14.1 Estimated exposure to produce 1% risk of morbidity and mortality

Effect	Organ/tissue	Latency period	Exposure (Gy)
Morbidity			1% incidence
Male sterility	Testes	3 weeks	~6[a,b]
Female sterility	Ovaries	<1 week	3[a,b]
Erythema	Skin	1—4 weeks	<3—6[b]
Alopecia	Skin	2—3 weeks	~4[b]
Cataract	Eye	Several years	~1.5[a]
Mortality			
BMS	Bone marrow	30—60 days	~1[b]
GIS	Small intestine	6—9 days	~6[b]
Pneumonitis	Lung	1—7 months	6[b]

Adapted from PMC [14]
BMS, bone marrow syndrome; *GIS,* gastrointestinal syndrome
[a]ICRP (1984) [9]
[b]UNSCEAR (1988) [31]

stages: (1) tumor initiation–irreversible genetic alterations lead to atypical cellular signaling; (2) tumor promotion–changes in the expression of the genome result in enhanced growth and development; (3) malignant conversion; and (4) tumor progression–the final stages are marked by genomic instability and invasive growth [10]. Animal models suggest the role of radiation is primarily limited to tumor initiation [11, 12]. Later, tumor stages are believed less dependent on the mutagenic properties of radiation due to inherent genomic instability characteristic of these advanced stages [11, 12].

X-Ray Generation and Propagation

With fluoroscopy, x-ray generation begins with passing a current (measured in milliamperes [mA]) through a heated, negatively charged filament (cathode). Cathode electrons are accelerated through an x-ray tube towards a positively charged anode. The electric potential energy (measured in kilovolt peak [kVp]) of accelerated electrons is transformed into kinetic energy prior to collision with anode orbital electrons. At the anode, tightly bound inner-shell orbital electrons are ejected after colliding with electrons accelerated through the x-ray tube. The filling of the newly created inner-shell orbital vacancies by outer-shell electrons results in the emission of photons forming the x-ray radiation that is ultimately projected through the patient to an image intensifier responsible for generation of a visual light image. Increasing kVp by 15% has the same effect on image brightness as doubling mAs. During fluoroscopy, high kVp (75 kVp—125 kVp) and low mA (2 mA—6 mA) are typically preferred during fluoroscopy in order to minimize patient exposure without drastic compromises in image quality.

After exiting the x-ray tube, the beam must first pass through a collimator and a filter before reaching the patient

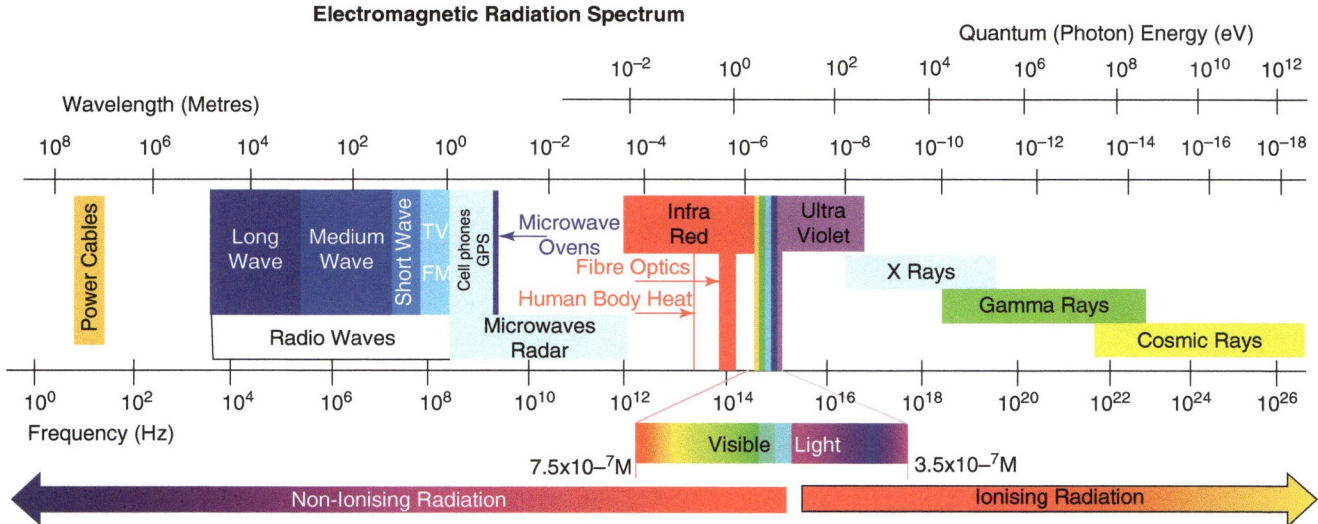

Fig. 14.2 Modern Fluroscope

(Fig. 14.2). The collimator typically contains both round and rectangular radiopaque shutter blades purposed to geometrically restrict the x-ray beam to the targeted anatomic area required for efficient visualization of the structures of interest. The round shutters are commonly known as variable aperture collimators (Iris collimators). Variable aperture collimators are smaller, produce a circular field, and automatically restrict the fluoroscopic beam to the useful field of view despite changes in magnification or source-image-distance. The rectangular shutters are larger, and can be manually adjusted to further limit beam size producing a rectangular field. Filtration helps remove low-energy x-rays that contribute to radiation exposure but do not to image quality. Aluminum and copper are the most popular x-ray filter materials.

In traversing the patient, x-ray radiation can have two important interactions involving either a complete or partial transfer of photon energy. The photoelectric effect entails the atomic process whereby a tightly bound inner-shell orbital electron completely absorbs the energy of an incident photon. The electron is ejected from the orbital and is now termed a photoelectron. In the filling of the newly created inner-shell vacancy by an outer-shell electron, a photon is emitted (secondary radiation) [13]. Compton scattering involves the collision and partial energy transfer between an incident photon and a loosely bound outer-shell orbital electron. The loosely bound electron is ejected and the incident photon deflected. Ejected electrons are responsible for radiobiologic effects associated with x-ray radiation [13]. Secondary radiation from the photoelectric effect and scattered radiation from the Compton effect do not lend to diagnostic value, but instead add to radiation exposure of nearby personnel.

Principles of Radiation Safety

There is no "safe" dose of ionizing radiation. The objectives of radiologic protection are complete prevention of deterministic effects and ensuring the risks of stochastic effects are as maximally diminished "as low as reasonably achievable, societal and economic factors being taken into account"—the ALARA principle [14]. According to the "linear-non-threshold" or LNT model, at exposure doses less than 100 mSv per year, the risks of stochastic effects are believed proportional to dose [14]. Even low doses are believed to carry an attributable risk of hereditary and oncogenic effects.

As per the 2007 International Commission on Radiation Protection (ICRP) recommendations, radiation protection can be subdivided into three core principles: justification, optimization, and application of dose limits. Justification is a principle common to the entire practice of medicine [14]. The benefit to an individual and society from an activity should outweigh the associated potential harm. In the context of fluoroscopic safety, the benefits of utilizing ionizing radiation should also outweigh the occupational risks imposed on the provider.

Dose limit recommendations are made for the United States by the National Council for Radiation Protection (NCRP) and internationally by the ICRP [14, 15]. Occupational dose limit recommendations made by the NCRP are shown in Table 14.2. The NCRP and ICRP share similar recommendations. The maximum permissible dose (MPD) of radiation to a provider is 20 mSv averaged over 5 years (i.e., 5 year MPD = 100 mSv) with no year exceeding 50 mSv [14, 15]. Maximum permissible doses should be considered extreme values with most interventionalists

Table 14.2 NCRP Recommendations for Maximum Permissible Doses during Occupational Exposure

Dose Quantity	Maximum Permissible Dose
Effective dose	
Annual	20 mSv/yr averaged over 5 years with no single year exceeding 50 mSv
Cumulative	10 mSv × age (yr)
Equivalent dose	
Lens of the eye[a]	150 mSv/yr
Skin[b]	500 mSv/yr
Hands and feet	500 mSv/yr

mSv, milliSievert; *yr*, year

[a]Likely to be changed to 50 mSv/yr

[b]Average dose over 1 cm^2 of the most highly irradiated area of the skin

experiencing less than 10% of maximum doses (i.e., between 2 and 4 mSv per year) [16]. Within the United States, x-ray regulations are governed by the Occupational Safety and Health Administration (OSHA) and all sources of ionizing radiation by the Nuclear Regulatory Commission (NRC). NRC requirements take precedence and are therefore most often implemented by hospital radiation safety officers. As per NRC regulations, all personnel likely to experience greater than 5 mSv are required to use an individual monitoring device (e.g., film badge, thermoluminescence dosimeter, etc.)

Optimization of radiation protection entails maintaining exposure remains "as low as reasonably achievable". Patient exposure can be minimized without undue concessions in image quality via optimization of equipment, x-ray beam filtration, and collimation, maximizing the source-object distance (at least 30 cm; optimum >182 cm [approximately 6 feet]), minimizing the object-image distance, limiting the field of direct radiation to only that of clinical interest, and reducing overall fluoroscopy time. The concepts driving these principles are discussed above.

Radiation exposure experienced by the provider is essentially all scatter from the patient. Maximizing the provider's distance from the irradiated field, use of all appropriate shielding devices, and limiting fluoroscopy time and images are the mainstays of optimizing a minimization of radiation exposure to the provider. Maintenance of appropriate distance from the source to the provider is simple yet effective. As exposure follows the inverse square law, doubling distance from the source quarters exposure. Standing a distance of 1 meter (100 cm) from the source yields an exposure dose approximately 0.1% of the entrance skin exposure. When shooting films in the lateral position, scatter doses up to 4 times higher occur on the side of the source compared to the image intensifier [17].

Appropriate shielding involves the use of personal protective shielding (i.e., aprons, leaded eyewear, thyroid shields, leaded gloves), patient-mounted shields, and movable room shields (ceiling-suspended shields, floor-mounted shields, and table-mounted shields). Due to decreasing limits for eye exposure (i.e., ICRP guidelines recommend an average 20 mSv over 5 years), eye shielding is likely to be of increasing importance in the future [14, 18]. A cumulative dose of 0.5 Gy is estimated to be the threshold dose for radiation cataracts [19–21]. Multiple studies have shown this limit may be easily reached without the use of proper protective equipment [22–24]. Within the ORAMED project, exposure levels of interventional radiologists and cardiologists at different hospitals throughout Europe were evaluated. Approximately half of interventional radiologists performing endoscopic retrograde cholangiopancreatography were exposed to eye radiation doses surpassing new ICRP recommendations [24]. Protective eyeglasses and ceiling-suspended shields have been shown to be an effective method of reducing exposure to the lens of the eye [25]. Koukorava et al. demonstrated a 90% decrease in eye exposure with the use of 0.5 mm lead glasses and a 93% decrease with the use of ceiling-suspended screens [19]. Protective eyewear is currently recommended for those expected to experience ocular exposure greater than 4 mSv per month. This threshold will likely be lowered with expected future decreases in MPD [20, 26, 27].

The MPD for the hands is 500 mSv per year [14, 15]. Wearing a ring badge is the current recommended method of measuring hand exposure. Minimizing exposure to the hands is best achieved with distance and shielding. Protective gloves with minimum 0.25 mm lead equivalent are useful but should not lull the wearer into a false sense of security. With automated brightness control, the lead gloves may be detected, resulting in automatically increased radiation output, at least partially negating the protection afforded by the gloves. Alternative measures such as the use of forceps or other holding devices are encouraged.

The recommended MPD for the torso and legs is 500 mSv [14, 15]. Exposure in these areas is significantly reduced with the use of single and two-piece lead apron. The apron is lead-impregnated vinyl or rubber with a shielding equivalent of at least 0.25 mm. Annual inspection of the lead apron is required by the Joint Commission on the Accreditation of Healthcare Organizations, with recommended disposal for defects greater than 15 mm^2 [28]. Thyroid shields with minimum 0.5 mm lead equivalent have been shown to significantly decrease thyroid exposure dose [29, 30]. The protection conferred with the use of a lead apron or thyroid shield is offset by the increased weight and decreased maneuverability associated with these devices.

Summary

Justification, optimization, and application of dose limits are the basic principles underlying radiation safety. Radiation exposure should be given judiciously for the sake of the patient, provider, and society. In with safe practices, radiation exposure should be maintained at doses "as low as reasonably achievable, societal and economic factors being taken into account"—the ALARA principle [14]. Provider exposure is best optimized by maximizing distance from the source of radiation, use of all appropriate shielding devices, limiting fluoroscopy time, and images. With appropriate safety practices, individual providers should rarely, if ever, exceed 10% of established maximum permissible doses of radiation.

References

1. Shope TB. Radiation-induced skin injuries from fluoroscopy. Radiogr Rev Publ Radiol Soc N Am Inc. 1996;16:1195–9.
2. United States Food and Drug Administration. Public Health Advisory: Avoidance of Serious X-Ray Induced Skin Injuries to Patients During Fluoroscopically Guided Procedures. Rockville: Center for Devices and Radiological Health, United States Food and Drug Administration; 1994.
3. FastStats. 2018. Available at: https://www.cdc.gov/nchs/fastats/leading-causes-of-death.htm. Accessed 23 Sept 2018.
4. The top 10 causes of death. World Health Organization. Available at: http://www.who.int/news-room/fact-sheets/detail/the-top-10-causes-of-death. Accessed 23 Sept 2018.
5. Fishman SM, Smith H, Meleger A, Seibert JA. Radiation safety in pain medicine. Reg Anesth Pain Med. 2002;27:296–305.
6. Lewis EB. Leukemia and ionizing radiation. Science. 1957;125:965–72.
7. Ulrich H. The incidence of leukemia in radiologists. N Engl J Med. 1946;234:45.
8. Yoshinaga S, Mabuchi K, Sigurdson AJ, Doody MM, Ron E. Cancer risks among radiologists and radiologic technologists: review of epidemiologic studies. Radiology. 2004;233:313–21.
9. Nonstochastic effects of ionizing radiation: a report of a task group of Committee 1 of the ICRP. Pergamon Press; 1984.
10. Pitot HC. The molecular biology of carcinogenesis. Cancer. 1993;72:962–70.
11. Ellender M, Harrison JD, Edwards AA, Bouffler SD, Cox R. Direct single gene mutational events account for radiation-induced intestinal adenoma yields in Apc(Min/+) mice. Radiat Res. 2005;163:552–6.
12. Sources and effects of ionizing radiation: United Nations Scientific Committee on the Effects of Atomic Radiation: UNSCEAR 2000 report to the General Assembly, with scientific annexes. United Nations; 2000.
13. Bushong SC. BOPOD—radiologic science for technologists: physics, biology, and protection. Elsevier Health Sciences; 2013.
14. PMC E. The 2007 recommendations of the international commission on radiological protection. ICRP publication 103. Ann ICRP. 2007;37:1–332.
15. Report No. 116—Limitation of exposure to ionizing radiation (Supersedes NCRP Report No. 91) (1993) | NCRP | Bethesda. Available at: https://ncrponline.org/shop/reports/report-no-116-limitation-of-exposure-to-ionizing-radiation-supersedes-ncrp-report-no-91–1993/. Accessed 30 Oct 2018.
16. Miller DL. Make radiation protection a habit. Tech Vasc Interv Radiol. 2018;21:37–42.
17. Valentin J. Avoidance of radiation injuries from medical interventional procedures. Ann ICRP. 2000;30:7–67.
18. Dauer LT, et al. Guidance on radiation dose limits for the lens of the eye: overview of the recommendations in NCRP Commentary No. 26. Int J Radiat Biol. 2017;93:1015–23.
19. Koukorava C, Carinou E, Ferrari P, Krim S, Struelens L. Study of the parameters affecting operator doses in interventional radiology using Monte Carlo simulations. Radiat Meas. 2011;46:1216–22.
20. Authors on behalf of ICRP et al. ICRP publication 118: ICRP statement on tissue reactions and early and late effects of radiation in normal tissues and organs—threshold doses for tissue reactions in a radiation protection context. Ann ICRP. 2012;41:1–322.
21. Neriishi K, et al. Radiation dose and cataract surgery incidence in atomic bomb survivors, 1986–2005. Radiology. 2012;265:167–74.
22. Vano E, Gonzalez L, Fernández JM, Haskal ZJ. Eye lens exposure to radiation in interventional suites: caution is warranted. Radiology. 2008;248:945–53.
23. Dauer LT, Thornton RH, Solomon SB, St Germain J. Unprotected operator eye lens doses in oncologic interventional radiology are clinically significant: estimation from patient kerma-area-product data. J Vasc Interv Radiol JVIR. 2010;21:1859–61.
24. Vanhavere F, et al. Measurements of eye lens doses in interventional radiology and cardiology: final results of the ORAMED project. Radiat Meas. 2011;46:1243–7.
25. Richman AH, Chan B, Katz M. Effectiveness of lead lenses in reducing radiation exposure. Radiology. 1976;121:357–9.
26. Center for History and New Media. Zotero Quick Start Guide. Available at: http://zotero.org/support/quick_start_guide.
27. Status of NCRP Scientific Committee 1–23 Commentary on Guidance on Radiation Dose Limits for the Lens of the Eye. Available at: https://www-ncbi-nlm-nih-gov.proxy.library.cornell.edu/pmc/articles/PMC4697269/. Accessed 29 Oct 2018.
28. Lambert K, McKeon T. Inspection of lead aprons: criteria for rejection. Health Phys. 2001;80:S67–9.
29. Müller LP, Suffner J, Mohr W, Degreif J, Rommens PM. Effectiveness of lead thyroid shield for reducing roentgen ray exposure in trauma surgery interventions of the lower leg. Unfallchirurgie. 1997;23:246–51.
30. Yang MS, et al. Evaluation of usability of the shielding effect for thyroid shield for peripheral dose during whole brain radiation therapy. J Korean Soc Radiother Technol. 2014;26:265–72.
31. United Nations Scientific Committee on the Effects of Atomic Radiation. UNSCEAR, 1988. Sources, effects and risks of ionizing radiation. United Nations, New York; 1988.

Pre-procedural Imaging

15

Alexander Ghatan, Ian D. Dworkin, and George C. Chang Chien

Introduction

Regenerative medicine presents exciting new opportunities in the treatment of a variety of musculoskeletal (MSK) disorders; however, proper pre-procedural workup cannot be overlooked and must be completed prior to the initiation of such treatments. Pre-procedural imaging is crucial in both identifying the pathology that can be targeted by various regenerative techniques and ruling out pathology that will not benefit from treatment options. Additionally, pre-procedural imaging will help identify contraindications to regenerative treatments and evaluate for any "red flag" pathology. Conventional radiography has traditionally been helpful at identifying pathology; however, there are many MSK disorders that cannot be properly evaluated early enough with these modalities when regenerative therapies can provide the greatest benefit [1].

Patient Factors and Selection

Though there are no commonly accepted guidelines specific to regenerative medicine injections, there are such factors that are commonly evaluated before conducting conventional interventional procedures such as epidurals. For these spinal procedures, the ideal time to discontinue anticoagulation agents such as Coumadin, clopidogrel, and aspirin is unique to the pharmacokinetics of each individual medication; however, the North American Spine Society (NASS) recom-

mends that an interval of approximately 1 week prior to surgery is prudent [2]. The risks involved in holding anticoagulation are also unique to each patient and must be weighed against the potential benefits of the treatment being provided. Contraindications for steroid injections have been well described in the literature, but there is little evidence for any particular contraindications for regenerative techniques. Table 15.1 lists several common contraindications and patient pre-procedure recommendations that many clinicians use to guide injection candidacy.

While regenerative medicine has an enormous capacity for healing various MSK disorders, it is important to recognize regenerative medicine's limitations and select patients and pathology that will best respond to these various techniques. Regenerative medicine is generally most effective for mild-to-moderate disease, including osteoarthritis Kellgren–Lawrence grade 1 or 2 or grade 1 or 2 ligamentous sprains (discussed below). Surgical management may be more appropriate for complete tears or end-stage, grade 4 osteoarthritis, and thus, pre-procedural imaging can assist in this patient selection. Additional patient characteristics that would impair the body's ability to heal or degrade its regenerative capacity include smoking cigarettes, uncontrolled blood glucose, immunosuppressed states, or active infections. Areas that lack adequate blood supply, such as eschars, or dysvascular or necrotic limbs, are also unlikely to respond to regenerative medicine techniques given their poor capacity to receive and utilize the necessary nutrients for repair.

Common Soft Tissue Injuries (Sprains/Strains)

Many common soft tissue injuries can be treated using regenerative techniques, and therefore, accurately localizing, identifying, and quantifying various injuries are valuable in any clinical setting. Imaging will help localize pathology, but a thorough history and physical examination are necessary in deciding what imaging to obtain. Every patient presents differently,

A. Ghatan (✉)
UCLA/VA of Greater Los Angeles, Department of Physical Medicine and Rehabilitation, Los Angeles, CA, USA

I. D. Dworkin
Board Certified in Physical Medicine and Rehabilitation and Pain Medicine, Department of Interventional Pain Physiatry, Newport Care Medical Group, Newport Beach, CA, USA

G. C. Chang Chien
GCC Institute, Department of Musculoskeletal Medicine and Medical Aesthetics, Newport Beach, CA, USA

© Springer Nature Switzerland AG 2023
C. W Hunter et al. (eds.), *Regenerative Medicine*, https://doi.org/10.1007/978-3-030-75517-1_15

Table 15.1 Pre-procedure patient preparation [2]

Strong Recommendations

Avoid NSAIDs and other anti-inflammatory agents for at least 7 days prior to Platelet Rich Plasma

No eating or drinking 6 hours before the procedure.

Patients are encouraged to hydrate well the day before the procedure.

Patients are encouraged to shower in the morning prior to their procedure.

Patients are advised to avoid using any products on their skin (lotion, makeup, sprays, anything topical to the area) the day of the procedure.

Relative Contraindications

Fever

Cancer

Rash over injection site

Elevated INR or actively taking anticoagulants

Poorly Controlled Type II Diabetes Mellitus or Elevated hemoglobin A1C

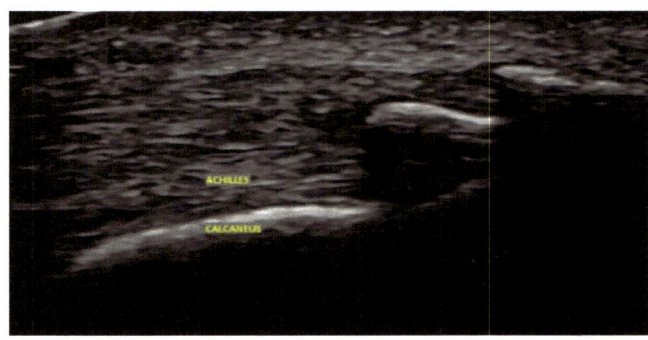

Fig. 15.1 Long axis ultrasound view of chronic Achilles tendonitis with enthesophyte irregularity and calcifications at the Achilles tendon insertion. (Reproduced from Benjamin et al. [61])

and thus, it is important for a clinician to be able to properly evaluate and describe various injuries in a standardized fashion. Below, we briefly discuss common terminology used in describing sprains, strains, and other common soft tissue injuries.

Tendinopathy

Tendinopathies are the various conditions associated with tendon pain caused by overuse. Tendinopathy is associated with histopathologic changes such as minimal inflammation, degeneration and disorganization of collagen fibers, and increased cellularity [3, 4]. Macroscopic changes include pain, tendon thickening, and the loss of mechanic [4]. Some suggest that tendon overuse leads to an imbalance between the protective/regenerative changes of the tissue, and pathologic responses from overuse, which results in pain, tearing, weakness, and degeneration [5].

Tendon and ligament abnormalities are widely assessed by MRI and ultrasound. The high levels of type I collagen in healthy tendons and ligaments, arranged in a cross-linked triple-helix structure, coupled with a structured orientation, provide their characteristic imaging appearances as well as cause particular imaging artifacts on various imaging modalities [6]. Tendons that pass through tight tunnels or around corners are typically covered in a tendon sheath, which is comprised of 2 layers of synovium. Otherwise, tendons are covered by a thin layer of loose fatty connective tissue called the paratenon [6]. The orientation of a tendon's fibers depends on the tension to which the tendon is subjected [7]. For tendons in which the force is directed along the tendon, the collagen is typically aligned along the tendon's long axis. Some tendons have a more complex structure with fibers running in discrete bundles. This is the case for tendons with origins from more than one muscle, such as the quadriceps tendon and the Achilles tendon (Fig. 15.1) [6].

With age, changes in collagen structure such as a loss of water content predispose them to damage [8]. Vascularity also decreases with age, and tendon disease often occurs at these hypovascular areas. Instability or impingement leads to abnormal and excessive loading of the tendon which predisposes to injury [9, 10]. Collagen fibrils can rupture, and these regions may together form intrasubstance tears. These intrasubstance tears may extend to the surface, eventually progressing to full-thickness tears [9, 10]. Though ingrowth of vessels into the tendon is common, there is no evidence of inflammatory mediators [11–14]. Generally, degenerative changes occur before macroscopic tendon tears develop, and as such, it is unusual for a tear to occur in a nondegenerated tendon [6].

Ligament Sprains

Though ligaments are functionally different from tendons as they connect bone to bone, they are structurally similar [6]. The main differences are that ligaments have higher proteoglycan content, higher water content, lower in collagen content, and are less uniform [15]. An additional feature of ligamentous injuries is that because ligaments guide movement at joints, injury is typically associated with joint derangement.

Acute trauma typically causes ligament abnormalities and is often marked by fluid surrounding the ligament, although chronic repetitive microtrauma may be a factor as with tendon injuries [16, 17]. Potential damage includes interstitial tearing of collagen fibers and partial tears that extend to the surface and full-thickness ligament ruptures. Over time, the ligament can become elongated and lax. Other evidence of injuries includes bone contusions, fractures, or joint effusion. After healing, the ligament may appear thickened, weakened, and prone to further damage [6].

Table 15.2 describes the American Academy of Orthopedic Surgeons classification of ligamentous sprains [18]. Each

Table 15.2 AAOS classification of ligamentous sprains [18]

Grade	Description
Grade 1: Mild sprain	Typically described as stretching of the fibrils which may include microscopic damage and swelling, but the gross integrity of the ligament is usually not compromised.
Grade 2: Moderate sprain	Involves partial tearing of the ligament, which can result in laxity
Grade 3: Severe sprain	Complete tear of the ligament usually resulting in instability and interferes with joint function.

Table 15.3 Classification of muscle strains based on functional loss

Grade	Description
Grade 1: Mild	Stretch injury which results in less than 5% functional loss
Grade 2: Moderate	Partial muscle tear with 5–50% loss of function
Grade 3: Severe	Near-complete to complete rupture where there is greater than 50% loss of function. Typically seen at musculotendinous junction with a hematoma filling the space between the two ends.

grade is based on the extent to which the ligament fibrils are interrupted and damaged. Of note, grade 3 injuries also include avulsion injuries, where a piece of the bone is pulled off along with the ligament.

Muscle Injuries

A strain is defined as an injury to the muscle and/or tendon, commonly at the musculotendinous junction [18]. Similar to sprains, strains are graded on a continuum. There can be a mild stretch injury with microscopic damage to the muscle fibers, or the injury can be more severe with partial or complete tear of the muscle–tendon complex. Chronic sprains and strains are common sources of pain. Patients may present with chronic pain, weakness, pain-limited range of motion (ROM), muscle spasms, muscle weakness, edema, or cramping. Repetitive strains and sprains can lead to further functional loss and can be a major pain generator that can be targeted with regenerative medicine.

When an indirect muscle injury occurs, there is a sudden onset muscle pain. It is usually localized to a single muscle and often occurs during an eccentric muscle contraction. The most commonly strained muscles in athletes are the biceps femoris, rectus femoris, and medial gastrocnemius [19]. Muscle strain grading systems can be based on function or imaging which will be discussed in later sections of this chapter and in the ultrasound chapter. Strains can be classified based on the amount functional loss from the patient's baseline (Table 15.3). Of note, grade 3 injuries are the rarest type of muscle injuries and often require surgical intervention. Avulsion injuries are occasionally described as Grade 3b muscle strain injuries [19].

Please see the chapters on MRI and ultrasound for additional information regarding muscle strain grading systems based on these modalities. MRI and ultrasound will also be further reviewed below. Unlike bone, muscles have a limited capacity for muscle regeneration and the majority of healing is by scar formation [19]. Thus, old or chronic muscle injuries may appear like an area of scar tissue within the normal-appearing muscle.

Pre-Procedural Imaging and Common Imaging Findings

X-Ray and Computed Tomography (Ct)

For most musculoskeletal conditions, X-ray is often the first imaging used, but when it comes to regenerative treatments, the utility of X-ray is limited. Plain radiographs are useful at identifying gross deformity, fracture, dislocation, severe osteoarthritis, and ruling out osteoarthritis vs. adhesive capsulitis [20]. It is also useful in assessing joint space narrowing seen in osteoarthritis, and the severity of disease is commonly described using Kellgren–Lawrence classification.

The Kellgren–Lawrence classification of osteoarthritis, or KL grading, uses 4 grades of classification (Table 15.4) [21]. This classification system was originally described using AP views of knee radiographs but is commonly used to describe osteoarthritis in other joints as well (Fig. 15.2).

There are several limitations in using KL grading. One limitation is that the system assumes a linear progression of disease, which is often not the case. A second limitation is that there are times when patients may have osteophyte formation and/or sclerosis without joint space narrowing. Third, if a patient has joint space narrowing without any osteophytes, the KL grading system cannot be applied. X-ray fluoroscopy is also important in evaluating intervertebral disk integrity during diskography. Please see the following section for more information regarding diskography.

Computed tomography (CT) scans provide detailed visualization of bony structures and may assist in visualizing fractures not visible on X-ray [22]. They are furthermore readily available and quickly obtainable if the patient is unable to have an MRI; however, X-ray and CT are not typically used in imaging soft tissue injuries as they provide little insight into soft tissue pathology vital to pre-regenerative medicine procedures.

Magnetic Resonance Imaging (MRI)

When it comes to regenerative medicine, healing and repairing soft tissue are paramount, and therefore, the best imaging modality of soft tissues is with MRI. In this section, we will discuss the basics of how MRIs work, the different types of MRI, and some common pathological soft tissue findings that may be targeted with regenerative medicine.

The basis of MRI is in the magnetic resonance of hydrogen protons within the tissue being imaged [22]. Hydrogen protons, similar to tiny magnets with north and south poles, are susceptible to external magnetic fields. When hydrogen protons enter a strong external magnetic field, like an MRI scanner, most of the protons will align themselves in parallel to the strong field. An additional magnetic field, called a gradient, can be manually added to the MRI's native magnetic field, which creates an additional subdivision in the total magnetic field. The protons can then be triggered to flip or spin by radio-frequent pulses with a specific frequency. This causes the hydrogen protons to spin simultaneously, shifting/flipping back and forth in different axis, and is termed excitation and relaxation. Eventually, these induced magnetic fields/signal changes are registered by receiver coils and processed into the MRI image on a gray scale based on signal intensity. High signal intensity is seen as white, intermediate signal intensity appears gray, while low signal intensity appears dark gray or black.

Table 15.4 Kellgren–Lawrence classification of osteoarthritis [21]

Grade	Description
Grade 1	Doubtful narrowing of joint space and possible osteophyte formation.
Grade 2	Definite osteophytes and possible narrowing of joint space.
Grade 3	Moderate/multiple osteophyte formation, definite narrowing of joints space, some sclerosis, and possible deformity of bone contour.
Grade 4	Large osteophytes, severe narrowing of joint space, severe sclerosis, and definite deformity of bone contour are apparent.

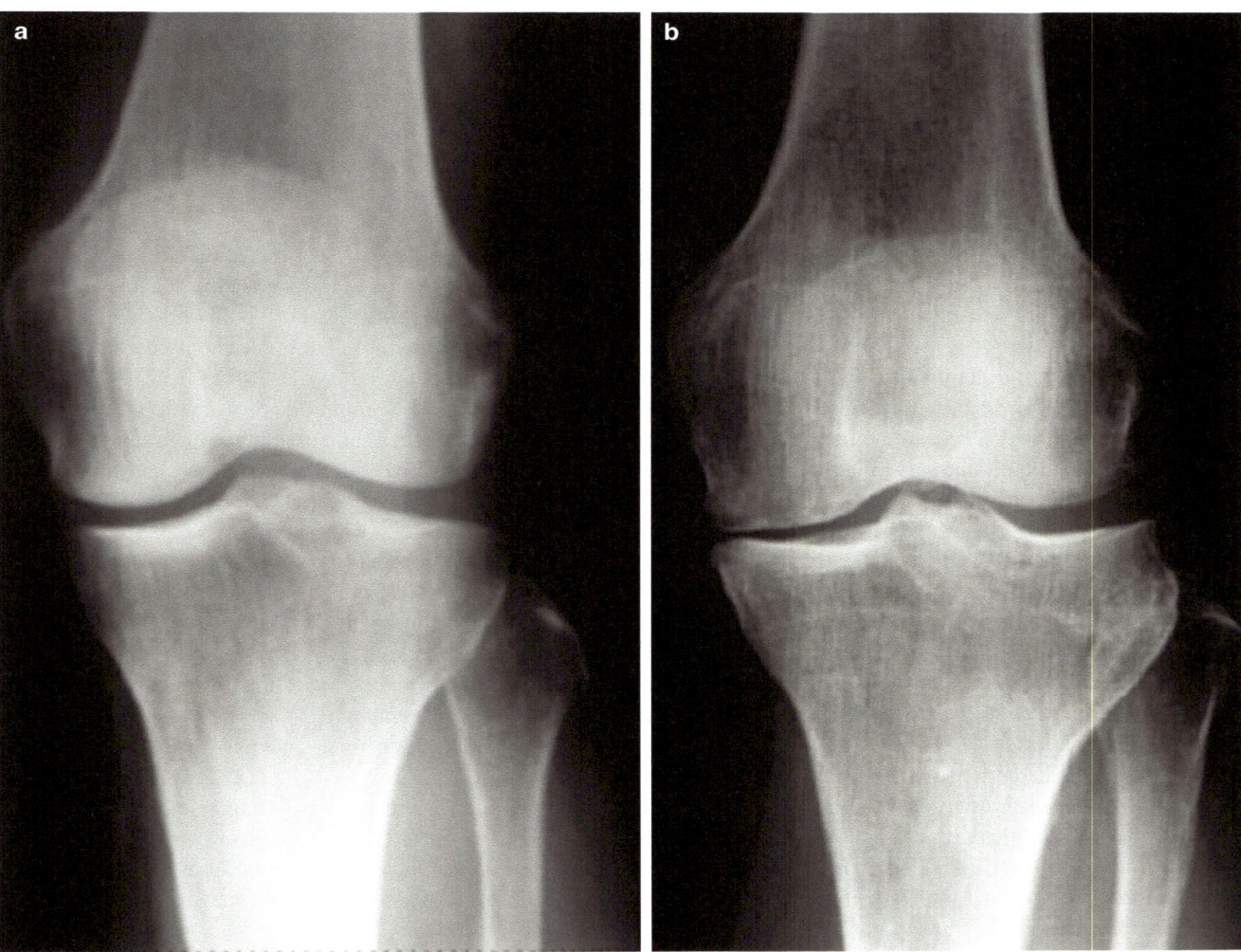

Fig. 15.2 AP plain X-rays of Kellgren–Lawrence grade 1 (**a**) and grade 2 (**b**). (Reproduced from Akira Horikawa et al. [62])

Table 15.5 T1-Weighted MRI sequences [22]

Very low signal intensity (black)	Low signal intensity (dark gray)	Intermediate signal intensity (light gray)	High signal intensity (white)
Calcium Dense Cortical Bone Intravascular/Flowing Blood Air	Water CSF Collagen Cartilage Tendons Ligaments Scars Bone Marrow Edema	Protein Dense Tissue Abscesses/Cysts Normal Synovial Fluid	Fat Normal Bone Marrow Blood (static) Contrast (Gadolinium)

Table 15.6 T2Weighted MRI sequences. [22]

Very low signal intensity (black)	Low signal intensity (dark gray)	Intermediate signal intensity (light gray)	High signal intensity (white)
Calcium Dense Cortical one Intravascular/ Flowing Blood Air	Cartilage Tendons Ligaments	Cartilage Fat Muscles	Fluid Edema CSF

MRI Sequences

Individual MRI sequences are based on the combinations of various radio-frequent pulses and gradients which allow visualization of varying pathology [22].

T1-Weighted Sequences The most common use of T1-weighted imaging is in the visualization of normal musculoskeletal anatomy [22]. In this sequence, the image is determined by the differences in relaxation times between water and fat. Fat has a high signal intensity (white), and water has a low signal intensity (black). This is because in a T1 series, fat has a shorter relaxation time than water. Table 15.5 describes the expected signals for various anatomical structures.

T2-Weighted Sequences On T2-weighted images, water has high signal intensity (white) which makes it useful to highlight the edema and inflammation associated with pathology (Table 15.6). In T2, similar to T1, air and calcifications have very low signal intensity (dark) [22]. Fig. 15.3 demonstrates the differences between T1 and T2 MRI sequences.

Proton Density (PD)-Weighted Imaging Proton density-weighted imaging is a visual representation of protons per volume within tissue [22]. Tissues with lower proton density will have a low signal intensity and will appear dark. Tissues with higher proton density will have a high signal and appear white. Fat, being a proton-dense tissue, has a relatively high signal intensity (light gray) but not as high as in a T1-weighted image (white). Fluid has intermediate signal intensity rather than the high signal intensity seen on T2-weighted images.

A common use for PD-weighted imaging is in the evaluation of meniscal tears of the knee. PD-weighted imaging is also useful in distinguishing between CSF and pathology [22]. On T2-weighted imaging, CSF and many pathologies have a high signal but on PD-weighted imaging, the contrast between CSF (intermediate signal intensity) and most pathologies (high signal intensity) will be better visualized.

Fat Suppression imaging (STIR and SPIR) The suppression of adipose tissues is an option that can be used in various MRI sequences. Fat suppression images are commonly referred to as fat saturation images or "FatSat." This creates a low-signal intensity of fat which helps in contrasting it from vessels and various pathologies [22]. In musculoskeletal imaging, fat suppression can be useful. For example, bone marrow is high in fat which may mask bone barrow edema on a T2-weighted image. Thus, in suppressing the fat, edema from a fracture, tumor, or other pathology will be more easily visualized.

Short-tau inversion recovery (STIR) and spectral presaturation inversion recovery (SPIR) sequences are the most commonly used fat suppression sequences and are both T2-weighted images [22]. STIR sequences are very useful in detecting bone marrow edema.

Diffusion-Weighted Imaging (DWI) Diffusion refers to the random movement of molecules within a substance. The diffusion behavior of hydrogen molecules is determined by different field strengths [22]. DWI is T2-weighted images. This type of MRI is commonly employed in the diagnosis of acute strokes but is not often employed in the evaluation of MSK disorders.

MRI Contrast

When an MRI is performed with contrast, it will typically rely on a T1-weighted image since use with T2-weighted imagines have little value due to the fact that both fluid/edema and contrast will have a high signal intensity and be generally indistinguishable [22]. The most commonly used contrast type for MRI is gadolinium. It reduces the T1 relaxation time of the protons that absorb the contrast, and thus, these protons will have higher (white) signal intensity.

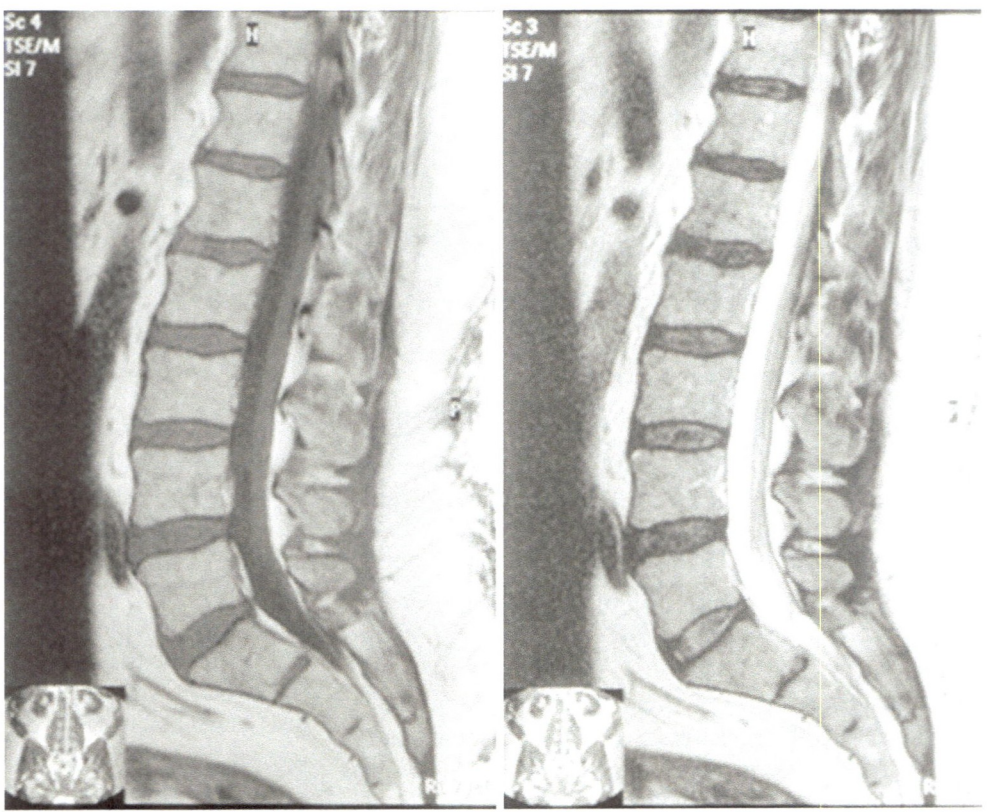

Common indications for MRI contrast include detecting various lesions (tumor, metastases, infection, abscess), characterization of lesions, especially in the viscera, imaging of vessels/vascular pathology, and imagining of intraarticular structures (MR arthrogram) [22].

Tendon and Ligaments on MRI

The structure of tendons determines their appearance on MRI. Due to the abundance and orientation of collagen and water molecules, normal tendons appear as dark (low signal intensity) on most MRI sequences, including T1- and T2-weighted sequences [22]. With injury, the fluid signal within a tendon or ligament tears can be identified with *T2*-weighted images [10]. MRI provides high spatial resolution of tendons and ligaments. There is a direct correlation between image resolution and the strength of the MRI's magnetic field - as the strength of the field increases, so does the resolution of the image. Therefore, an MRI with a stronger the magnetic field is much more likely to detect a partial-thickness tear [6, 23].

Tendinopathies and Ligamentous Sprains on MRI

One of the first signs of a tendon injury on MRI is an increase in signal intensity, which can be seen on T1-weighted images [6]. Additionally, the tendon may appear thickened. The appearance of a tendon tear varies with chronicity. In the more acute setting, T2-weighted or STIR images may show increased fluid signal within tendon tears [24, 25]. In an older tear, scarring within the defect can produce an intermediate signal. Increased signal on T2-weighted images with fat suppression is the best way to diagnose tears on MRI with the best specificity (Fig. 15.4).

Partial-thickness tears often heal with the defects being filled with fluid or granulation tissue [6]. The resulting tissue is weaker than the native tendon and can propagate into full-thickness tears. When the entire tendon is disrupted, the torn ends can retract, altering the normal/expected anatomy, making visualization difficult. When this occurs, the secondary signs of full-thickness tears such as muscle edema, atrophy, tendon contour irregularity, and/or retraction of the musculotendinous junction assist in making the diagnosis.

Ligamentous sprains appear similarly on MRI. In the acute setting, T2-weighted or STIR images may show increased fluid signal around the ligament and may appear thickened with increased signal within the ligament [6]. In an older tear, the ligament may appear irregular, thickened, or possibly thinned.

Muscle Contusions on MRI

The role of imaging in acute muscle injury has changed from merely confirming a clinical diagnosis to defining the precise location and extent of the injury. Being able to measure the size and extent of soft tissue disruption assists in predicting outcome and determining treatment. When assessing muscle injury by MRI, either a STIR, fat sat PD-weighted, or fat sat T2-weighted sequence should be utilized [19]. T1 should be included when assessing for blood products or atrophy. It is always important to compare the T1 and STIR series in suspected areas of muscle injury as a focal area of fatty infiltration (which may be due to atrophy) may be misinterpreted as an intramuscular scar.

Contusions typically occur when there is a blunt force trauma to a muscle without disruption to the skin. On MRI imaging, the appearance of contusions depends upon the blood products and fluid characteristics within the lesion, which changes with time (Table 15.7) [19].

In the hyperacute stage (<24 hours) of the injury, the contusion causes edema and interstitial hemorrhage, which leads to the characteristic feather-like high signal within the muscle on fat-suppressed fluid sensitive sequences (i.e., STIR, fat sat PD-weighted, or fat sat T2-weighted) [19]. The feather-like

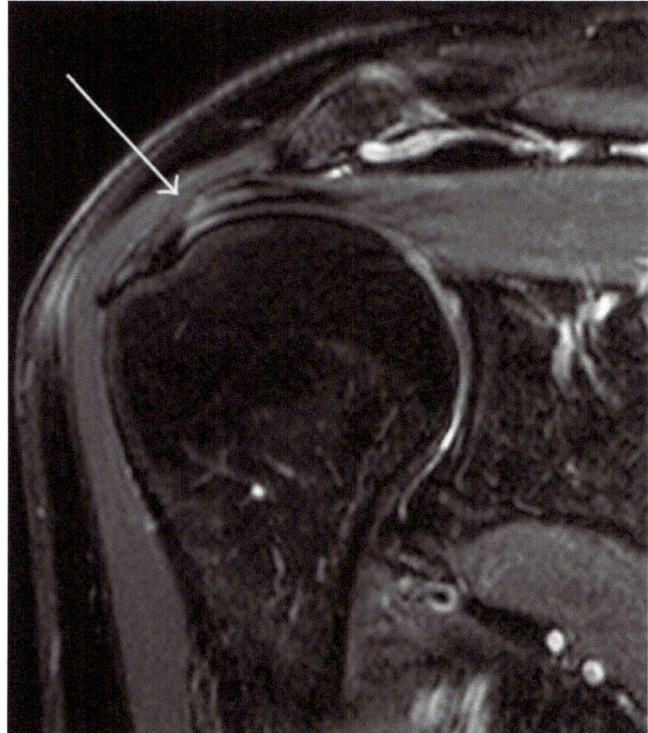

Fig. 15.4 Grade 1 tendinosis on T2-weighted fat-suppressed MRI. (Reproduced from Andrea et al. [64])

appearance occurs due to the high signal of blood and edema spreading between the individual muscle fibers.

In the acute stage (24–48 hours) of the injury, the contusion appears as an irregular muscle laceration [19, 22]. Blood products may result in areas of faint high signal on T1-weighted images; however, the same imaging findings could be seen in a low-grade muscle strain.

In the subacute stage (48–72 hours) of the injury, the contusion becomes a more clearly defined fluid collection within the muscle [19, 22]. The muscle surrounding the site of injury remains diffusely high signal on fluid-sensitive sequences. Characteristics of a hematoma will change with time depending on the nature of the blood product within it based on metabolic breakdown.

As time passes, a hematoma will undergo fibrosis and calcification [19, 22]. Fibrosis of the hematoma margins will contract the lesion over time. Calcification can lead to weakening, making the muscle susceptible to repeat injury.

Muscle Strains on MRI

As previously discussed, a muscle strain is an indirect muscle injury, which often occurs during an eccentric muscle contraction. Muscle strains can be graded via MRI based on the extent of cross-sectional area of disruption of the muscle fascicles as compared to clinical grading which was discussed above based on functional impairment [19]. MRI assists in determining the extent of cross-sectional fiber disruption, which most commonly occurs at the musculotendinous junction.

- *Grade 1 Strain*: There is less than 5% disruption in the cross-sectional area of the muscle. On fluid-sensitive fat-suppressed sequences (i.e., T2-weighted fat sat), there is an increased signal at the site of injury due to the edema and blood products radiating from the injury site which produces the classic feather-type appearance within the muscle on MRI. Perifascial fluid may also be seen.
- *Grade 2 Strain*: There is at least 5% but less than 100% disruption in the cross-sectional area of the muscle causing distortion of the normal muscle architecture. This typically results in hematoma formation at the musculotendinous junction. The feathery-type muscle edema pattern as described in grade 1 injury may also be present. There may also be some laxity of the central tendon within the muscle.
- *Grade 3 Strain*: There is complete disruption of the muscle, typically at the musculotendinous junction with a

Table 15.7 Appearance of muscle contusions on MRI [19]

	Hyperacute (<4 hour)	Acute (4–6 hour)	Early subacute (6–72 hour)	Late subacute (72 hour to 4 weeks)	Chronic (>4 weeks)
T1 Signal Intensity	Intermediate	Intermediate	High	High	Low
T2 Signal Intensity	High	Low	Low	High	Low

hematoma filling the space between the two ends. Grade 3 injuries are the rarest type of muscle injuries and often require surgical intervention. Avulsion injuries are occasionally described as grade 3b muscle strain injuries.

Unlike bone, muscle has a limited capacity for regeneration following injury [19]. The majority of healing is by scar formation. Thus, old or chronic muscle injuries may appear like an area of scar tissue within the normal-appearing muscle. Figure 15.5 demonstrates a T2-weighted MRI of complete rupture of left distal biceps femoris tendon at the musculotendinous junction.

MRI of the Spine

Obtaining MRI images of the spine is crucial for detecting various pathologies as it gives detailed visualization of the soft tissue, and the various aforementioned sequences can help differentiate between different injuries and lesions. Spine degeneration, such as spondyloarthropathies and disk degeneration, can be best visualized using MRI which is why it is the preferred imaging modality in back pain; however, while MRI provides a good visual representation of the spine, it cannot definitively localize patient's pain. Thorough history, clinical exam, and the possible addition of electrodiagnostics in conjunction with the imaging are necessary. There are other provocative exams and invasive tests that can be used to help identify the patient's pain, some of which will be discussed further in this chapter.

Disk degeneration and diskogenic back pain are prime targets for treatment with regenerative techniques. Signal changes of the disk, vertebral endplates, and subchondral bone are seen on MRIs of degenerative spines and are strongly associated with low back pain [26]. These bone marrow and vertebral end place lesions were originally classified in 1988 by Modic et al. and are referred to as "Modic changes." [27, 28] In 1990, Miller further classified these imaging findings into what is now known as "modified Modic changes," and in 2001, Weishupt et al. further classified Modic changes into four degrees based on the percentage of vertebral height involvement in a mid-sagittal image of the spine (Table 15.8) [29, 30].

Relationship between Modic Changes and Lower Back Pain Despite this characterization of spinal changes, only a small proportion of pathology can be diagnosed with certainty based on a pathoanatomical entity alone [31]. There is increasing evidence though that demonstrates the prevalence of Modic changes, especially type 1, increases in people with nonspecific low back pain compared to people without low back pain [32–34]. Modic changes at L5/S1 and, especially

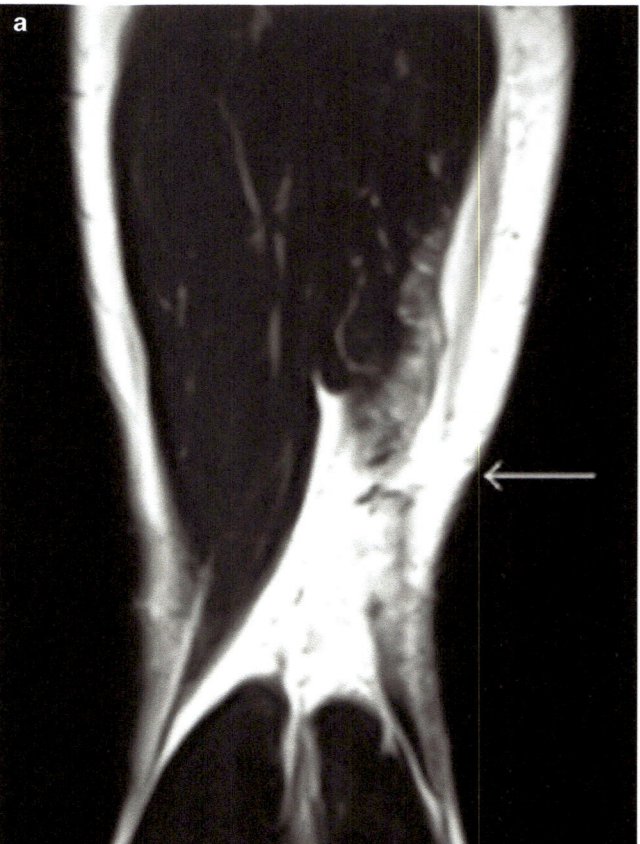

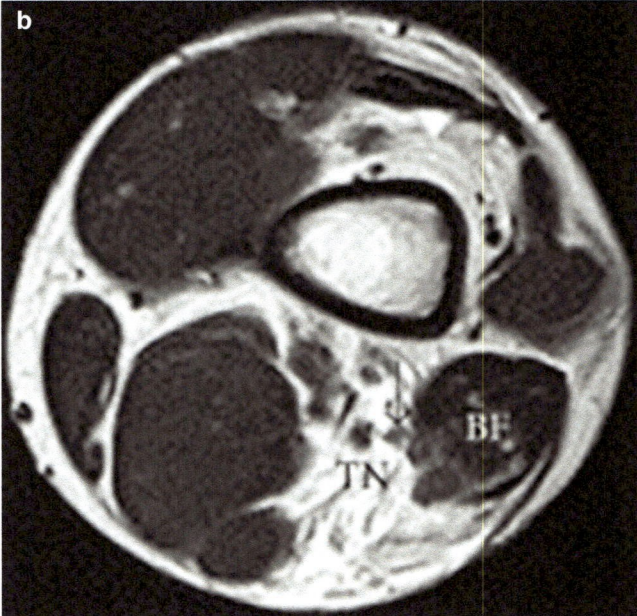

Fig. 15.5 T2-weighted MRI of a complete rupture of the distal biceps femoris tendon at the musculotendinous junction. (Reproduced from Aki Fukuda et al. [65])

Modic Type 1, are more likely related to low back pain than other levels and types of Modic changes (Fig. 15.6) [35]. Additionally, Modic changes are often associated with Schmorl's nodes, which occur when the nucleus pulposus herniates through the vertebral endplate and into the adjacent vertebral body (Fig. 15.7). On MRI, they appear as focal endplate defects (low signal on T1 and high signal on T2). They also have a well-defined herniation pit and a surround-ing wall of high signal on T1 and T2 within the vertebral body [26, 36]. Though there is a lack of consensus regarding Schmorl's nodes clinical significance, Hamanishi et al. stud-ied 400 patients with lower back pain and found that 19% of patients with back pain had Schmorl's nodes compared to only 9% of control patients [37].

Differentiating Modic Changes from Spinal Infections and Tumors Spinal infections and tumors may appear simi-larly to Modic changes on MRI, but there are some important distinguishing characteristics [38]. Spondylodiskitis, an infection of the disk and vertebral body, presents as lesions with high signal on T2 compared to normal or low signal on T2 in disk degeneration. Spondylodiskitis can cause signifi-cant paravertebral soft tissue edema and can even lead to epi-dural mass effect [38, 39]. Erosion of vertebral body and end plates are always seen in intervertebral disk infections, whereas Modic changes may be focal or diffuse along the endplates, but tend to be linear and always parallel to the endplates [26, 40].

The most common type of neoplastic lesion found in the spinal column is secondary to metastasis [26]. Metastatic disk involvement is rare and is therefore easily distinguish-able from Modic changes by the absence of disk space involvement.

Relationship between Modic Changes and Diskography Some authors report that when the signal

Table 15.8 Modified modic changes combining Miller et al. and Weishupt et al. criteria [29, 30]

Modic Type	Description
Type 0 or first-degree changes	Normal; no degeneration. No MRI evidence of bone marrow or vertebral end plate lesions. No T1 or T2 changes
Type 1 or second-degree changes	Vertebral body and bone marrow edema/inflammation and hypervascularity T1: low signal T2: high signal Mild signal intensity changes of less than or equal to 25% of the vertebral height
Type 2 or third-degree changes	Normal haemopoietic bone marrow is replaced by fat infiltration secondary to ischemia. T1: high signal T2: normal-appearing to high signal Moderate changes at 25–50% of the vertebral height
Type 3 or fourth-degree changes	Subchondral bony sclerosis seen T1: low signal T2: low signal Severe changes greater than 50% of the vertebral height

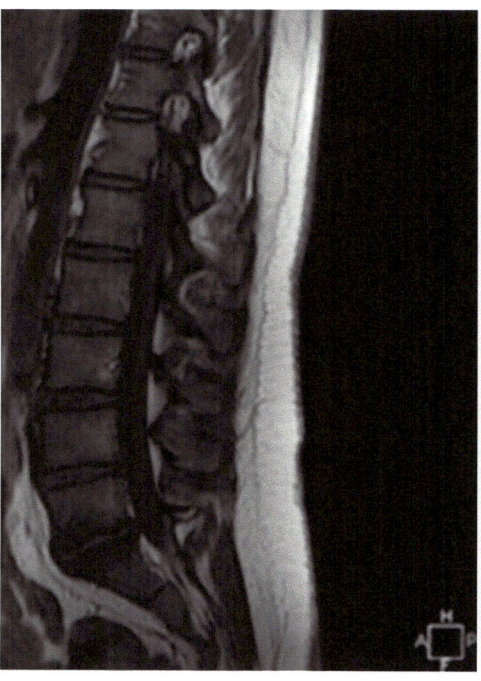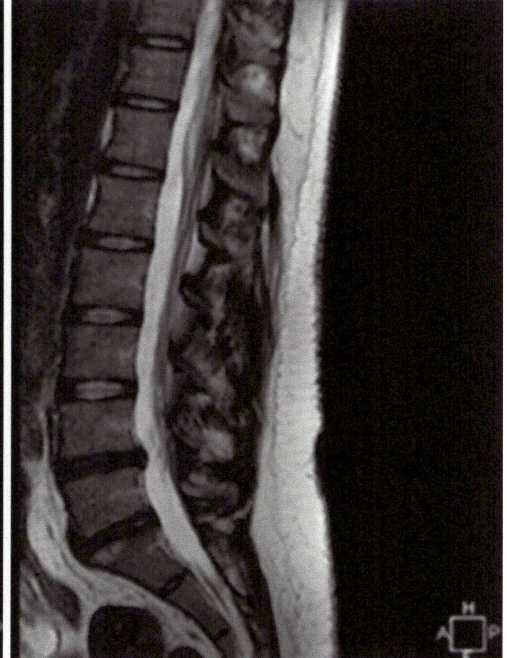

Fig. 15.6 Early reactive endplate changes at L5/S1 (Modic type 1). (Reproduced from Michael [63])

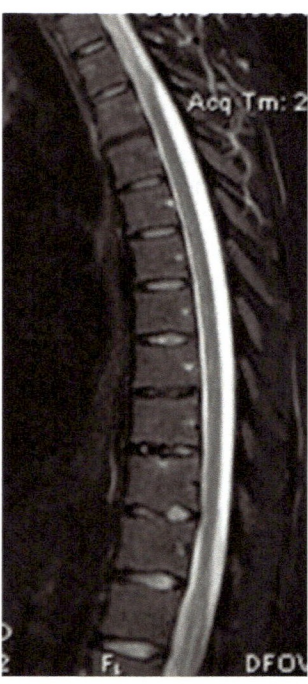

Fig. 15.7 Sagittal T2WI of a 17-year-old male with Scheuermann's disease with multilevel involvement of Schmorl's nodes and endplate irregularities. (Reproduced from Aikaterini et al. [66])

intensity changes in the endplates and decreased signal intensity in degenerative lumbar disks were combined, the specificity of using MRI to diagnose disk pain disease increases significantly [26, 41]. The signal intensity changes in endplates indicate a high degree of specificity, but lack sensitivity in diskogenic low back pain. Therefore, Modic changes are of important value in the diagnosis of diskogenic low back pain, but MRI does not completely replace the diskography due to the lack of the sensitivity. Diskography will be discussed in detail later in this chapter.

Relationship Between High-Intensity Zone on MRI and Diskography in Patients with Low Back Pain The presence of a high-intensity zone on MRI is another imaging finding that may indicate a patient's pain generator. The high-intensity zone (HIZ) was first described in lumbar spine MRI studies [42] and is defined as a focal area of high signal on T2-weighted sequences in the posterior annulus fibrosus. It has a considerably brighter signal intensity than nucleus pulposus from which it is distinctly disassociated [43–45].

The correlation between HIZ on MRI and diskography in patients with low back pain has been examined with varying results (Fig. 15.8).

Some data suggest that the presence of HIZ could be used as an indicator of annular tears and diskogenic low back pain [42, 43, 45–49]. Additionally, some authors posit that the HIZ is caused by the inflammation of annulus fibrosus and that there is a correlation between the presence of HIZ within the poste-

rior annulus of a lumbar disk on MRI and the pain response following diskography in patients with low back pain. There is also evidence that HIZ is indicative of a Grade 3 to 4 annular tear and that the signal change is due to the accumulation of mucoid fluid within the fissure of the annulus. Others counter this, speculating that the value of HIZ is limited to the diagnosis of diskogenic low back pain [50–53]. Regardless, the finding of a HIZ should be investigated by diskography and potentially treated as a patient's pain generator.

MRI Limitations

The MRI is useful for lesion detection and localization; however, it is expensive, time-consuming, and can be uncomfortable, particularly for patients with claustrophobia [19]. It also only acquires static images. Additionally, MRI is contraindicated in patients with certain pacemakers and surgical brain clips.

Ultrasound

Ultrasound imaging has several advantages over MRI including superior spatial resolution, lower cost, patient and practitioner convenience, portability, and is essentially the only imaging modality that can provide dynamic imaging of musculoskeletal injuries and is a crucial tool in needle guidance of various joint injections (Fig. 15.9) [19]. One significant limitation of ultrasounds is operator dependency and the need for an acoustic window which can be difficult to obtain. Images can vary depending on the skill of technique, knowledge of anatomy, and experience. Ultrasound also has limited fields of vision and cannot penetrate bone. Additionally, injuries under ultrasound are less prominent/obvious than in MRI, which can also image both ligamentous injuries and associated intraarticular damage. Ultrasound basics will be reviewed here, but please refer to this text's chapter on ultrasound for additional, more comprehensive information.

Ultrasound Basics

Echogenicity is the ability of a tissue to reflect or transmit ultrasound waves in the context of surrounding tissue [54]. Hyperechoic tissue appears white, hypoechoic tissue appears gray, and anechoic tissue appears black. The following are the appearances of commonly evaluated structures under ultrasound:

- Bone appears anechoic (black) with a hyperechoic rim (bright) because the beam cannot penetrate bone; thus, it casts in acoustic shadow behind it.
- Cartilage is hypoechoic (gray) and is more penetrable than bone.
- Blood vessels appear anechoic (black) and can differentiate between veins and arteries as veins are easily collaps-

Fig. 15.8 Serial T2-weighted MRI findings of a degenerated disk with a slight protrusion is visible; however, originally, no high signal intensity zone (HIZ) is obvious. Subsequent imaging reveals obvious HIZ. (Reproduced from Kosuke et al. [67])

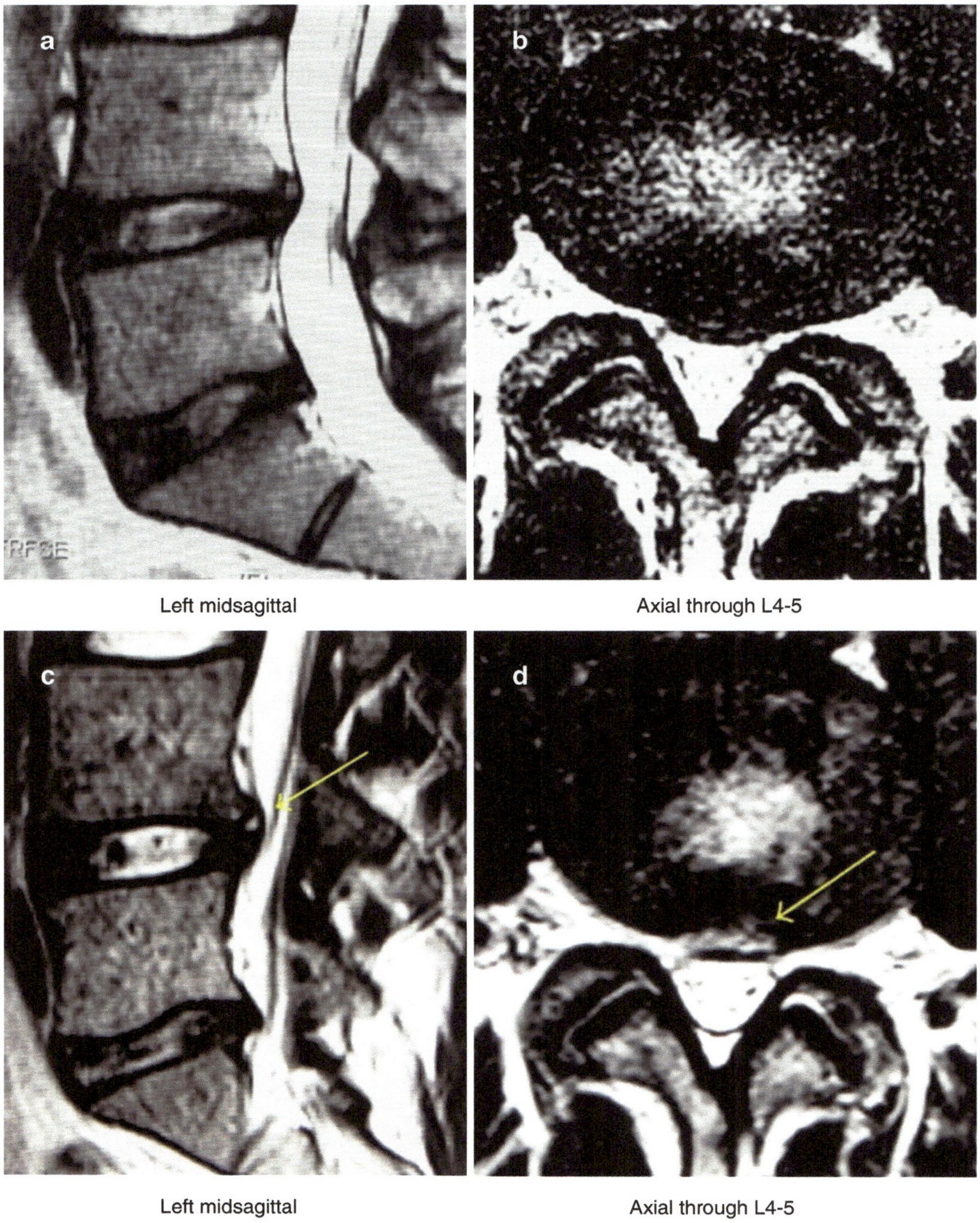

Left midsagittal

Axial through L4-5

Left midsagittal

Axial through L4-5

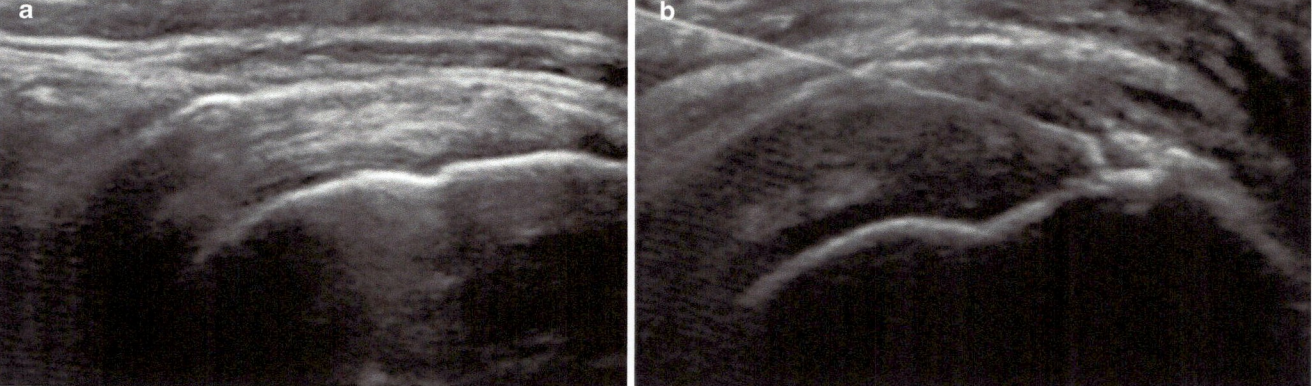

Fig. 15.9 US-guided Injection into the subacromial bursa (**a**) and supraspinatus tendon tear (**b**)

ible when pressure is applied by the transducer, while arteries are pulsatile and are not easily collapsible.

- Muscles are hypoechoic (gray) with striate structure.
- Fat is almost anechoic (black).
- Fascia/connective tissue strands/fascicles appear as hyperechoic (white) lines and bands.

Contusions on Ultrasound

Contusions on ultrasound are Ill-defined areas of hyperechogenicity within a muscle that crosses fascial boundaries [19]. They can be hyperacute, acute, or subacute.

If the contusion is *Hyperacute (<24 hours)*, the injured muscle appears swollen and may be isoechoic with adjacent normal-appearing muscle [19]. If the injury was from a forceful trauma, there may be significant rupture of muscle fibers and bleeding into the potential space resulting in a hematoma.

If the contusion is *acute (24–48 hours)*, hematoma will appear as an irregularly outlined muscle laceration with hypoechoic fluid inside [19]. During this period, the hematoma may solidify and become hyperechoic compared to the surrounding muscle.

Finally, if the contusion is *subacute (48–72 hours)*, it becomes a clearly defined hypoechoic fluid collection with an echogenic margin [19]. Over time, this echogenic margin gradually enlarges and fills in the hematoma in a centripetal fashion.

If the hematoma is causing significant pain, exerting mass effect on neurovascular structures, or is placing the tissue at risk for compartment syndrome, clot evacuation may be considered via ultrasound guidance at 10–14 days after the initial injury [19].

Muscle Strain on Ultrasound

Muscle strains on ultrasound are rated on a three-point grading system as shown in Table 15.9 [55].

Tendons and Ligaments on Ultrasound

The tendon's fascicular structure is seen on ultrasound as closely spaced echogenic lines on longitudinal scanning. In the transverse plane, echogenic dots or lines are seen. While ligaments also appear as echogenic fibrillar structures [56],

they are less echogenic than tendons [57] due to their less regular structure. The reflective fascicles within the tendons and ligaments can be seen best when the ultrasound beam is perpendicular to the fascicles' orientation and a different group of fibers can be seen by changing the probe orientation along the axis. Both tendons and ligaments exhibit anisotropy [6]. There is no echogenic appearance if the beam is not perpendicular which may simulate disease. This must be considered when examining tendons where the fibers change direction or are not parallel to the skin.

Tendinopathy and Ligamentous Sprains on Ultrasound

Under ultrasound, tendinopathy appears as areas of less organized fibrillar structure with increased spacing between the hyperechoic fibrillar lines and overall reduced echogenicity, which are associated with tendon thickening [6]. The appearance of tendon tears depends on the chronicity of the injury. In the acute phase, there may be anechoic fluid within the tear, but with time the echogenicity can increase and the tendon may appear normal. Dynamic visualization can particularly aide in identifying tendon and ligamentous pathology that may otherwise be missed. Also, Doppler imaging is useful in helping distinguish between small intrasubstance tears and vessels that have developed within a tendinopathic tendon.

Under ultrasound, acute ligamentous sprains may appear as thickened areas of the ligament with diffuse hypoechogenicity and surrounding edema [58]. Ligamentous tears may appear as areas with reduced echogenicity that interrupt the ligament fibers [59]. An interruption that extends across the entire thickness of the tendon is considered a complete or full-thickness tear [6]. As healing progresses, the fluid surrounding the injury site dissipates but the thickening and the laxity on dynamic imaging may remain.

On ultrasound, tendinosis appears as heterogeneous areas with reduced echogenicity [60]. In more chronic tendinosis, there may be calcifications within the tendon with varying appearances.

Conclusion

Pre-procedural imaging is vital in the evaluation and diagnosing of various MSK diseases, as well as imperative to rule out other pathology that cannot be treated with regenerative techniques (cancer, abscesses, etc.). By understanding the different uses of X-ray, CT, MRI, and ultrasound, clinicians can choose the most appropriate imaging modality leading to more effective care. X-rays are often the first images obtained but have limited use outside of evaluating fractures and osteoarthritis. CT can provide more detailed visualization of bony structures, fluid collections, and can be used if MRI is

Table 15.9 Ultrasound grading of muscle strains

Grade	Characteristics
Grade 1	May appear normal or show areas of increased echogenicity at the injury site taking up less than 5% of the muscle substance in cross section. Long cavities within the muscle measuring 10 mm or less are also considered Grade 1.
Grade 2	>5% but <100% disruption of the cross-sectional area of the muscle typically visualized at the musculotendinous or myofascial junction.
Grade 3	Ultrasound shows complete disruption of the muscle at the musculotendinous junction. Surrounding muscle is hyperechoic, and intermuscular perifacial and subcutaneous fluid collections are commonly visualized.

contraindicated but is also not typically employed to evaluate soft tissue injuries. MRI is the most important modality in pre-procedural imaging, but proper sequence selection and knowledge of their differences are crucial to their interpretation of underlying pathology. Diskography is an important modality to use for diskogenic pain if intradiskal stem cells are being considered because it is the gold standard in correlating imaging deficiencies with the patient's symptoms. Finally, ultrasound has quickly become a lynchpin of regenerative medicine, providing dynamic visualization of pathology and direct needle visualization to ensure the regenerative techniques reach their desired location. Most importantly though in pre-procedural preparation is the continued use of a thorough and well-documented history and physical examination which no imaging modality can supplant.

References

1. Tan AL, Wakefield RJ, Conaghan PG, Emery P, McGonagle D. Imaging of the musculoskeletal system: magnetic resonance imaging, ultrasonography and computed tomography. Best Pract Res Clin Rheumatol. 2003;17(3):513–28. https://doi.org/10.1016/S1521-6942(03)00021-4.
2. North American Spine Society. Evidence-based clinical guidelines for multidisciplinary spine care antithrombotic therapies in spine surgery. 2009. [Internet] 2018 Sep 1. Available from: https://www.spine.org/Documents/ResearchClinicalCare/Guidelines/AntithromboticTherapies.pdf.
3. Khan KM, Cook JL, Bonar F, Harcourt P, Astrom M. Histopathology of common tendinopathies. Update and implications for clinical management. Sports Med. 1999;27:393–408.
4. Soslowsky LJ, Thomopoulos S, Tun S, Flanagan CL, Keefer CC, Mastaw J, Carpenter JE. Neer award 1999. Overuse activity injures the supraspinatus tendon in an animal model: a histologic and biomechanical study. J Shoulder Elb Surg. 2000;9:79–84.
5. Andres BM, Murrell GAC. Treatment of tendinopathy: what works, what does not, and what is on the horizon. Clin Orthop Relat Res. 2008;466(7):1539–54. https://doi.org/10.1007/s11999-008-0260-1.
6. Hodgson RJ, O'Connor PJ, Grainger AJ. Tendon and ligament imaging. Br J Radiol. 2012;85(1016):1157–72. https://doi.org/10.1259/bjr/34786470.
7. O'Brien M. Anatomy of tendons. In: Maffulli N, Renstrom P, Leadbetter WB, editors. Tendon injuries. London: Springer-Verlag; 2005.
8. Tuite DJ, Renstrom PA, O'Brien M. The aging tendon. Scand J Med Sci Sports. 1997;7:72–7.
9. Pierre-Jerome C, Moncayo V, Terk MR. MRI of the Achilles tendon: a comprehensive review of the anatomy, biomechanics, and imaging of overuse tendinopathies. Acta Radiol. 2010;51:438–54.
10. Calleja M, Connell DA. The Achilles tendon. Semin Musculoskelet Radiol. 2010;14:307–22.
11. Kannus P, Jozsa L. Histopathological changes preceding spontaneous rupture of a tendon. A controlled study of 891 patients. J Bone Joint Surg Am. 1991;73:1507–25.
12. Aström M, Rausing A. Chronic Achilles tendinopathy. A survey of surgical and histopathologic findings. Clin Orthop Relat Res. 1995;316:151–64. PMID: 7634699.
13. Alfredson H, Thorsen K, Lorentzon R. In situ microdialysis in tendon tissue: high levels of glutamate, but not prostaglandin E2 in chronic Achilles tendon pain. Knee Surg Sports Traumatol Arthrosc. 1999;7:378–81.
14. Regan W, Wold LE, Coonrad R, Morrey BF. Microscopic histopathology of chronic refractory lateral epicondylitis. Am J Sports Med. 1992;20:746–9.
15. Simon SR. Orthopaedic basic science. Iowa: Am Acad Orthopaed Surg; 2000.
16. Bredella MA, Tirman PF, Fritz RC, Feller JF, Wischer TK, Genant HK. MR imaging findings of lateral ulnar collateral ligament abnormalities in patients with lateral epicondylitis. AJR Am J Roentgenol. 1999;173:1379–82.
17. Finlay K, Ferri M, Friedman L. Ultrasound of the elbow. Skelet Radiol. 2004;33:63–79.
18. American Academy of Orthopedic Surgeons. Sprains, strains and other soft tissue injuries [Internet] 2018 Sep 1. Available from: https://orthoinfo.aaos.org/en/diseases—conditions/sprains-strains-and-other-soft-tissue-injuries/.
19. Lee JC, Mitchell AWM, Healy JC. Imaging of muscle injury in the elite athlete. Br J Radiol. 2012;85(1016):1173–85. *PMC*. Web. 22 Sept. 2018.
20. O'Kane JW, Toresdahl BG. The evidenced-based shoulder evaluation. Curr Sports Med Rep. 2014;13:307–13. https://doi.org/10.1249/JSR.0000000000000090.
21. Kohn MD, Sassoon AA. Classifications in brief kellgren-lawrence classification of osteoarthritis. Clin Orthop Relat Res. 2016;74(8):1886–93.
22. Westbrook C, Roth C, Talbot J. MRI in practice. Chichester: Wiley-Blackwell; 2011.
23. Magee T, Williams D. 3.0-T MRI of the supraspinatus tendon. AJR Am J Roentgenol. 2006;187:881–6.
24. Stoller DW, Wolf EM, Li AE, Nottage WM, Tirman PFJ. The shoulder. In: Stoller DW, editor. Magnetic resonance imaging in orthopaedics and sports medicine. Baltimore: Lippincott Williams & Wilkins; 2007. p. 1131–462.
25. Reinus WR, Shady KL, Mirowitz SA, Totty WG. MR diagnosis of rotator cuff tears of the shoulder: value of using T_2-weighted fat-saturated images. AJR Am J Roentgenol. 1995;164:1451–5.
26. Zhang Y-H, et al. Modic Changes: A Systematic Review of the Literature. Eur Spine J. 2008;17(10):1289–99. *PMC*. Web. 22 Sept. 2018.
27. Modic MT, Masaryk TJ, Ross JS, Carter JR. Imaging of degenerative disk disease. Radiology. 1988;168:177–86.
28. Modic MT, Steinberg PM, Ross JS, Masaryk TJ, Carter JR. Degenerative disk disease: assessment of changes in vertebral body marrow with MR imaging. Radiology. 1988;166:193–9.
29. Miller G. The spine. In: Berquist T, editor. MRI of the musculoskeletal system. 2nd ed. New York: Raven; 1990.
30. Weishaupt D, Zanetti M, Hodler J, Min K, Fuchs B, Pfirrmann CW, et al. Painful lumbar disk derangement: relevance of endplate abnormalities at MR imaging. Radiology. 2001;218:420–7.
31. Waddell G. 1987 Volvo award in clinical sciences: a new clinical model for the treatment of low-back pain. Spine. 1987;12:632–44. https://doi.org/10.1097/00007632-198709000-00002.
32. Kjaer P, Leboeuf-Yde C, Korsholm L, Sorensen JS, Bendix T. Magnetic resonance imaging and low back pain in adults: a diagnostic imaging study of 40-year-old men and women. Spine. 2005;30:1173–80. https://doi.org/10.1097/01.brs.0000162396.97739.76.
33. Kjaer P, Korsholm L, Bendix T, Sorensen JS, Leboeuf-Yde C. Modic changes and their associations with clinical findings. Eur Spine J. 2006;15:1312–9. https://doi.org/10.1007/s00586-006-0185-x.
34. Albert HB, Manniche C. Modic changes following lumbar disc herniation. Eur Spine J. 2007;16:977–82. https://doi.org/10.1007/s00586-007-0336-8.
35. Kuisma M, Karppinen J, Niinimaki J, Ojala R, Haapea M, Heliovaara M, et al. Modic changes in endplates of lumbar vertebral bodies: prevalence and association with low back and sciatic

pain among middle-aged male workers. Spine. 2007;32:1116–22. https://doi.org/10.1097/01.brs.0000261561.12944.ff.

36. Williams FM, Manek NJ, Sambrook PN, Spector TD, Macgregor AJ. Schmorl's nodes: common, highly heritable, and related to lumbar disc disease. Arthritis Rheum. 2007;57:855–60. https://doi.org/10.1002/art.22789.

37. Hamanishi C, Kawabata T, Yosii T, Tanaka S. Schmorl's nodes on magnetic resonance imaging. Their incidence and clinical relevance. Spine (Phila Pa 1976). 1994;19:450–3.

38. Boden SD, Davis DO, Dina TS, Sunner JL, Wiesel SW. Postoperative diskitis: distinguishing early MR imaging findings from normal postoperative disk space changes. Radiology. 1992;184:765–71.

39. Modic MT, Feiglin DH, Piraino DW, Boumphrey F, Weinstein MA, Duchesneau PM, et al. Vertebral osteomyelitis: assessment using MR. Radiology. 1985;157:157–66.

40. Kuisma M, Karppinen J, Niinimaki J, Kurunlahti M, Haapea M, Vanharanta H, et al. A three-year follow-up of lumbar spine endplate (Modic) changes. Spine. 2006;31:1714–8. https://doi.org/10.1097/01.brs.0000224167.18483.14.

41. Toyone T, Takahashi K, Kitahara H, Yamagata M, Murakami M, Moriya H. Vertebral bone-marrow changes in degenerative lumbar disc disease: an MRI study of 74 patients with low back pain. J Bone Joint Surg Br. 1994;76:757–64.

42. Aprill C, Bogduk N. High-intensity zone: a diagnostic sign of painful lumbar disc on magnetic resonance imaging. Br J Radiol. 1992;65:361–9.

43. Peng B, Hou S, Wu W, et al. The pathogenesis and clinical significance of a high-intensity zone (HIZ) of lumbar intervertebral disc on MR imaging in the patient with discogenic low back pain. Eur Spine J. 2006;15:583–7.

44. Khan I, Hargunani R, Saifuddin A. The lumbar high-intensity zone: 20 years on. Clin Radiol. 2014;69:551–8.

45. Wang H, et al. Correlation between High-Intensity Zone on MRI and Discography in Patients with Low Back Pain. Ed. Weisheng Zhang. Medicine. 2017;96(30):e7222. *PMC*. Web. 22 Sept. 2018.

46. Schellhas KP, Pollei SR, Gundry CR, et al. Lumbar disc high-intensity zone. Correlation of magnetic resonance imaging and discography. Spine (Phila Pa 1976). 1996;21:79–86.

47. Saifuddin A, Braithwaite I, White J, et al. The value of lumbar spine magnetic resonance imaging in the demonstration of anular tears. Spine (Phila Pa 1976). 1998;23:453–7.

48. Walsh TR, Weinstein JN, Spratt KF, et al. Lumbar discography in normal subjects. A controlled, prospective study. J Bone Joint Surg Am. 1990;72:1081–8.

49. Wang ZX, Hu YG. High-intensity zone (HIZ) of lumbar intervertebral disc on T2-weighted magnetic resonance images: spatial distribution, and correlation of distribution with low back pain (LBP). Eur Spine J. 2012;21:1311–5.

50. Carragee EJ, Paragioudakis SJ, Khurana S. 2000 Volvo Award winner in clinical studies: lumbar high-intensity zone and discography in subjects without low back problems. Spine (Phila Pa 1976). 2000;25:2987–92.

51. Lee KS, Doh JW, Bae HG, et al. Diagnostic criteria for the clinical syndrome of internal disc disruption: are they reliable? Br J Neurosurg. 2003;17:19–23.

52. Ito M, Incorvaia KM, Yu SF, et al. Predictive signs of discogenic lumbar pain on magnetic resonance imaging with discography correlation. Spine (Phila Pa 1976). 1998;23:1252–8, discussion 1259–1260.

53. Teraguchi M, Yoshimura N, Hashizume H, et al. The association of combination of disc degeneration, end plate signal change, and Schmorl node with low back pain in a large population study: the Wakayama spine study. Spine J. 2015;15:622–8.

54. Ihnatsenka B, Boezaart AP. Ultrasound: basic understanding and learning the language. Int J Shoulder Surg. 2010;4(3):55–62. *PMC*. Web. 22 Sept. 2018.

55. Peetrons P. Ultrasound of muscles. Eur Radiol. 2002;12:35–43, Jan.

56. Lee JC, Healy JC. Normal sonographic anatomy of the wrist and hand. Radiographics. 2005;25:1577–90.

57. Allison SJ, Nazarian LN. Musculoskeletal ultrasound: evaluation of ankle tendons and ligaments. AJR Am J Roentgenol. 2010;194:W514.

58. Morvan G, Busson J, Wybier M, Mathieu P. Ultrasound of the ankle. Eur J Ultrasound. 2001;14:73–82.

59. Peetrons P, Creteur V, Bacq C. Sonography of ankle ligaments. J Clin Ultrasound. 2004;32:491–9.

60. Papatheodorou A, Ellinas P, Takis F, Tsanis A, Maris I, Batakis N. US of the shoulder: rotator cuff and non-rotator cuff disorders. Radiographics. 2006;26:e23.

61. Buchanan BK, DeLuca JP, Lammlein KP. Technical innovation case report: ultrasound-guided prolotherapy injection for insertional achilles calcific tendinosis. Case Rep Orthop. 2016;2016:1560161, 4 pages.

62. Horikawa A, Miyakoshi N, Shimada Y, Kodama H. The relationship between osteoporosis and osteoarthritis of the knee: a report of 2 cases with suspected osteonecrosis. Case Rep Orthoped. 2014;2014:514058, 6 pages.

63. Hasz MW. Diagnostic testing for degenerative disc disease. Adv Orthoped. 2012;2012:413913, 7 pages.

64. Donovan A, Schweitzer M, Bencardino J, Petchprapa C, Cohen J, Ciavarra G. Correlation between rotator cuff tears and systemic atherosclerotic disease. Radiol Res Pract. 2011;2011:128353, 7 pages.

65. Fukuda A, Nishimura A, Nakazora S, Kato K, Sudo A. Entrapment of common peroneal nerve by surgical suture following distal biceps femoris tendon repair. Case Rep Orthoped. 2016;2016:7909805. , 3 pages.

66. Solomou A, Kraniotis P, Rigopoulou A, Petsas T. Frequent benign, nontraumatic, noninflammatory causes of low back pain in adolescents: mri findings. Radiol Res Pract. 2018;2018:7638505, 5 pages.

67. Sugiura K, Tonogai I, Matsuura T, et al. Discoscopic findings of high signal intensity zones on magnetic resonance imaging of lumbar intervertebral discs. Case Rep Orthoped. 2014;2014:245952, 5 pages.

Discography

16

Aaron Calodney and Andrew T. Vest

Introduction

Approximately 80% of the U.S. population suffers at least one episode of back pain at some time in their lives, while 5–10% of patients develop chronic back pain [1]. In 2011, the Center for Disease Control and Prevention's (CDC) National Center for Health Statistics reported that 28.4% of adults over age 18 experienced lower back pain during the previous 3 months [2].

As reported in the 2016 National Health Interview Survey, back pain significantly limits work and daily activity for 28.4% of Americans [3]. A commonly repeated figure suggests that, cumulatively, Americans lose 149 million workdays each year due to back-related disability [4, 5]. Despite the availability of multiple imaging modalities and clinical examination, ascertaining the source of any given patient's back pain can be challenging.

Discogenic pain is a mechanical pain that is usually experienced in the axial spine distribution. It is exacerbated by activity and relieved by rest. It accounts for 26–42% of chronic low back pain [6–8].

Discography is a diagnostic procedure used to assess discogenic pain by evaluating the intervertebral disc in the cervical, thoracic, and lumbar spine. Disc morphology, pressure, and volume along with the patient's response to injection are recorded and used to confirm or exclude the disc as the source of pain. This allows correlation of findings from spinal imaging studies with the patient's pain symptoms and pattern [9–11]. In theory, discography identifies a painful disc by stimulating nociceptors in the outer third of the annulus, stimulating annular tears extending into the nucleus that have developed neoinnervation or by stimulation of nociceptors within the vertebral endplate [12].

Discography is the only diagnostic technique that directly correlates a patient's symptoms with disc morphology [13]. In this way, discography is conceptually similar to manual palpation. Pain provocation on discography is analogous to tenderness elicited on palpation [14, 15]. Consistent and reproducible pain portends greater diagnostic certainty. Numerous formal investigations have demonstrated that discography performed by experienced interventionalists can improve both surgical and nonsurgical treatment outcomes [16–21].

Advances in MR imaging detect increasingly minute degenerative changes, which often require clinical correlation [14, 22–24]. Any of these findings are asymptomatic. Discography is unique in allowing a link between radiographic findings and clinical presentation. MRI findings including degenerative changes in disc morphology do not correlate with symptoms of lower back pain [25–27].

To better understand the role discography can play in the diagnosis of spinal pain, this chapter will review:

- Historical use of the procedure
- Procedure validation
- Disc anatomy and physiology
- Disc pathophysiology
- Indications and contraindications
- Patient selection criteria
- Pre-and peri-procedure considerations
- Discography procedure
- Post-procedure care
- Complications
- Results: interpretation, documentation, and follow-up
- Correlation of discography with other imaging studies
- Evidence supporting and controversies regarding the procedures
- Use of discography in treatment planning—regenerative/biologic treatment

A. Calodney (✉)
Precision Spine Care, Texas Spine and Joint Hospital, Tyler, TX, USA

A. T. Vest
University of North Texas Health Science Center, Dallas, TX, USA

© Springer Nature Switzerland AG 2023
C. W Hunter et al. (eds.), *Regenerative Medicine*, https://doi.org/10.1007/978-3-030-75517-1_16

History of Discography

The identification of the intervertebral disc as a source of back pain and radiculopathy was advanced by the early work of Schmorl and Junghanns reported in 1932. Their imaging and dissection of 10,000 cadaveric spines demonstrated that the intervertebral disc could be a source of pain and introduced discography as an anatomic study of the internal structure of the cadaveric disc [11]. In 1934, Mixter and Barr further confirmed the intervertebral disc as a pain generator by surgically removing a prolapsed posterior disc, leading to pain relief [28]. Prior to discography, clinicians relied on Myelography with iophendylate (Pantopaque) to visualize spinal pathology; however, the disc and the epidural space remained opaque to this form of imaging inspection. Multiple authors published clinical reports throughout the 1940s [29] and 1950s [30] that advanced understanding of disc physiology, pathophysiology, the use of injected dyes to illuminate disc innervation and degeneration, and the evolving use of discography to diagnose disc pathology [11, 28, 31].

Lindblom has been credited with identifying discogenic pain as a primary source of back pain in 1941. He injected contrast into a cadaveric disc and concluded that injection of an opaque medium into the disc reveals disc ruptures and protrusions and identifies whether the patient's pain emanates from the punctured disc [31, 32]. Wise and Weiford performed the first discography in the USA in 1951 [33].

In 1964, Holt published a study of 50 patients with no history of neck or arm pain that challenged the validity of cervical discography as a diagnostic tool. In an examination of 148 discs, only 10 could be characterized as retaining injectate within the central confines of the annulus typically described as normal by other authors. He concluded that cervical discography is without diagnostic value [34]. Schellhas et al. compared MRI and cervical discography in 10 subjects without painful neck symptoms and 10 subjects with neck pain. The authors found that normal discs were not painful in either symptomatic or nonsymptomatic subjects. When pressurized, painful discs corresponded to the pain reported by the patient [35]. Holt in 1968 published a study questioning the credibility of lumbar discography. Discography was carried out on 30 volunteers from a penitentiary inmate population. He reported a 37% false-positive rate [36]. Simmons and Aprill reassessed Holt's paper finding four major issues. First, Hypaque contrast is irritating and likely irritated surrounding structures. Secondly, needle placement was likely improper as neither CT nor fluoroscopy was utilized. Thirdly, the study population and motivation for participation in this penitentiary study population are problematic, and lastly, errors in accounting of data are noted [37].

Walsh and Aprill replicated Holt's study in 1990 in 10 asymptomatic volunteers and 7 patients with lower back pain. Six of seven low back pain patients had positive discograms, while none of the ten asymptomatic volunteers had positive studies. The false-positive rate was 0%, and specificity was 100% [38].

Validation

There exists a gold standard dilemma. There is no histopathologic correlate of a painful disc against which to measure an imaging or diagnostic test such as discography. This lack of a criterion standard for lumbar discogenic pain—other than discography itself—implies that the validity of discography cannot be directly determined. The false-positive rate can be determined by studying the prevalence of positive responses in a group of asymptomatic volunteers [12].

Methodological variability in study design, clinical techniques, definitions, and interpretations of discography, as well as little consensus about what constitutes a false-positive rate, has made the reliability of systematic review, and thus evidence-based guideline development, challenging [12, 39–46]. Techniques and safeguards to address concerns identified in the clinical literature have developed. Lumbar discography was routinely performed without manometry until Derby demonstrated the importance of pressure measurement and operational criteria [17, 47]. The use of strict criteria including injection pressure and response intensity can decrease false positives and protect against putative risk of damage to the disc [9, 39, 41].

The operational definition of a "positive" vs a "negative" response to disc provocation is important. Derby et al. were able to demonstrate that when discography was applied with appropriate pressure, volume, and response intensity criteria the procedure yielded 0–10% false-positive results [12, 42, 48].

Guidelines by the Spine Intervention Society and the American Society of Interventional Pain Physicians focus on criteria for the use and interpretation of provocative discography [49].

Positive Discography Criteria

In ideal situations, a gold standard or criterion is obtained by tissue confirmation of the presence or absence of a disease. Surgical inspection of a degenerated disc and advanced imaging cannot assess the presence of discogenic pain [1].

Guidelines for provocation discography have been developed by multiple professional medical societies. The technique of lumbar discography has been standardized by the International Association for the Study of Pain as well as the Spine Intervention Society. Comprehensive literature reviews have been provided in the American Society for Interventional Pain Medicine guidelines of 2009, 2013, and 2018 [8, 50, 51].

In 2013, the Spine Intervention Society established the following criteria for definitive diagnosis of discogenic pain using provocation discography: [49].

- Concordant pain response of ≥6/10
- Volume limit of 3 mL
- Pressurization of the disc to no greater than 50 psi above the opening pressure
- Adjacent disc(s) provide controls
  - For one control disc: Painless response or nonconcordant pain that occurs at a pressure greater than 15 psi over opening pressure
  - For two adjacent control discs: Painless response at both levels or one painless disc and one disc with nonconcordant pain that occurs at a pressure greater than 15 psi over opening pressure

Similar criteria have been developed by the American Society of Interventional Pain Physicians (ASIPP) [1]. A discogram can be interpreted as positive only if the target disc:

Produces concordant pain with an intensity of ≥7 on a 10-point numerical pain rating scale or 70% of the highest reported pain (i.e. worst spontaneous pain of 7 = 7 × 70% = 5).

Two adjacent discs do not produce any pain at all with provocation discography or only one disc in the case of L5/S1 with low-volume and low-pressure injection.

"Concordant pain" will be defined here as pain during provocation that closely approximates the patient's usual pain pattern, whereas "nonconcordant pain" is a pain response upon pressurization that does not mirror usual pain pattern.

Disc Anatomy and Physiology

Intervertebral discs function primarily to transmit loads and facilitate movement between vertebral bodies. They are complex structures comprised of a thick fibrous outer ring of cartilage and an annulus fibrosus that surrounds an inner gelatinous centre called the nucleus pulposus. The disc is positioned between the inferior and superior cartilage endplates [31].

The annulus fibrosus is comprised of concentric lamellae of fibrocartilage. Each lamella consists primarily of collagen type I fibres that pass obliquely between vertebral bodies, with the orientation of the fibres reversed in alternating lamellae [52]. Annular fibres provide resistance to vertical, forward, backward, and lateral sliding movements in response to outward expansion of the nucleus pulposa. The annulus fibrosus acts like a ligament to restrain movement and stabilize the vertebral joint [31].

The nucleus pulposus, which is the central core of the disc, is located posteriorly within the disc [28]. It absorbs shock during axial loading by expanding radially and resists spinal compression by spreading axial load evenly across the vertebral body, even when the spine is flexed or extended [31, 52]. The nucleus pulposus consists of a proteoglycan and water gel held together loosely by an irregular network of fine collagen type II and elastin fibres. Aggrecan, the major proteoglycan of the disc, has a high anionic glycosaminoglycan content of chondroitin sulphate and keratan sulphate, which provides the osmotic properties needed to preserve hydration and resist compression [31, 52]. Because the nucleus pulposus does not have its own blood supply, it receives its nourishment via diffusion from the vasculature along the periphery of the annulus fibrosa and vertebral body [28].

The vertebral endplate is a thin (less than 1 mm), horizontal layer of hyaline cartilage that is weakly bonded to the perforated cortical bone of the vertebral body and the collagen fibres of the annulus and nucleus [31, 52]. Only the cartilaginous endplates have blood supply. Biochemically, the important constituents of the disc are collagen fibres, elastin fibres, and aggrecan [53].

Two interconnected nerve plexes innervate the cervical, thoracic, and lumbar discs. Both plexes innervate the annulus fibrosus to a depth of 3.5 mm, with most nerve endings concentrated dorsally and posterior laterally. Branches of two sympathetic trunks, the proximal ends of the lumbar ventral rami and the grey rami communicans, form the plexuses that innervate the anterior part of the disc. The sinovertebral nerve provides the main nerve supply to the posterior intervertebral disc and to every other structure of the spinal canal [31]. The density of receptors within the lumbar endplates and the annulus is similar. Endplate innervation is densest centrally, near the nucleus [53] (Fig. 16.1).

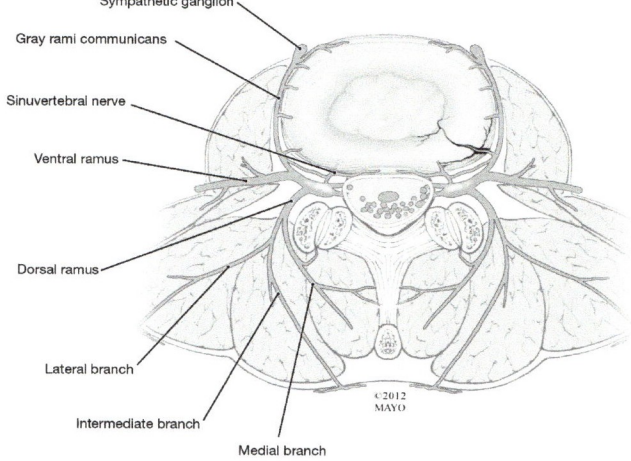

Fig. 16.1 Innervation of the intervertebral disc. (From: Maus and Aprill [106]. Used with permission of Mayo Foundation for Medical Education and Research, all rights reserved)

The intervertebral disc changes degeneratively, morphologically, and biochemically over the course of the human life cycle. With advancing age, proteoglycans and water decrease within the nucleus pulposus, resulting in insufficient hydrodynamic transfer of axial stress to the outer annulus fibrosus [52]. This decreased hydration results in loss of mechanical tension in the annulus fibrosus collagen fibres and results in abnormal spinal axial loading forces and segmental instability. Minor changes in stress forces on the spine can result in the development of neck or back pain and

narrowing of the spinal canal over time. In early stage degeneration, the disc undergoes an imbalance of anabolic and catabolic factors that leads to extracellular matrix degradation [54, 55] (Fig. 16.2).

Abnormal distribution of axial stress results in the tearing of the annulus fibrosus, which reduces the structural integrity of the disc [56]. Excessive mechanical loading, whether through trauma, sustained physical activity such as sports, or activities of daily living, disrupts the disc's structure, precipitating cell-mediated responses that lead to further disrup-

Fig. 16.2 Anatomic relationship of the lumbar disc, endplate, and nerve root. Pathologic changes secondary to neovascularization of the disc. (Source: Netterimages. Role of Inflammation In Lumbar Pain. Image# 7408. Netter illustration used with permission of Elsevier Inc. All rights reserved. www. netterimages.com)

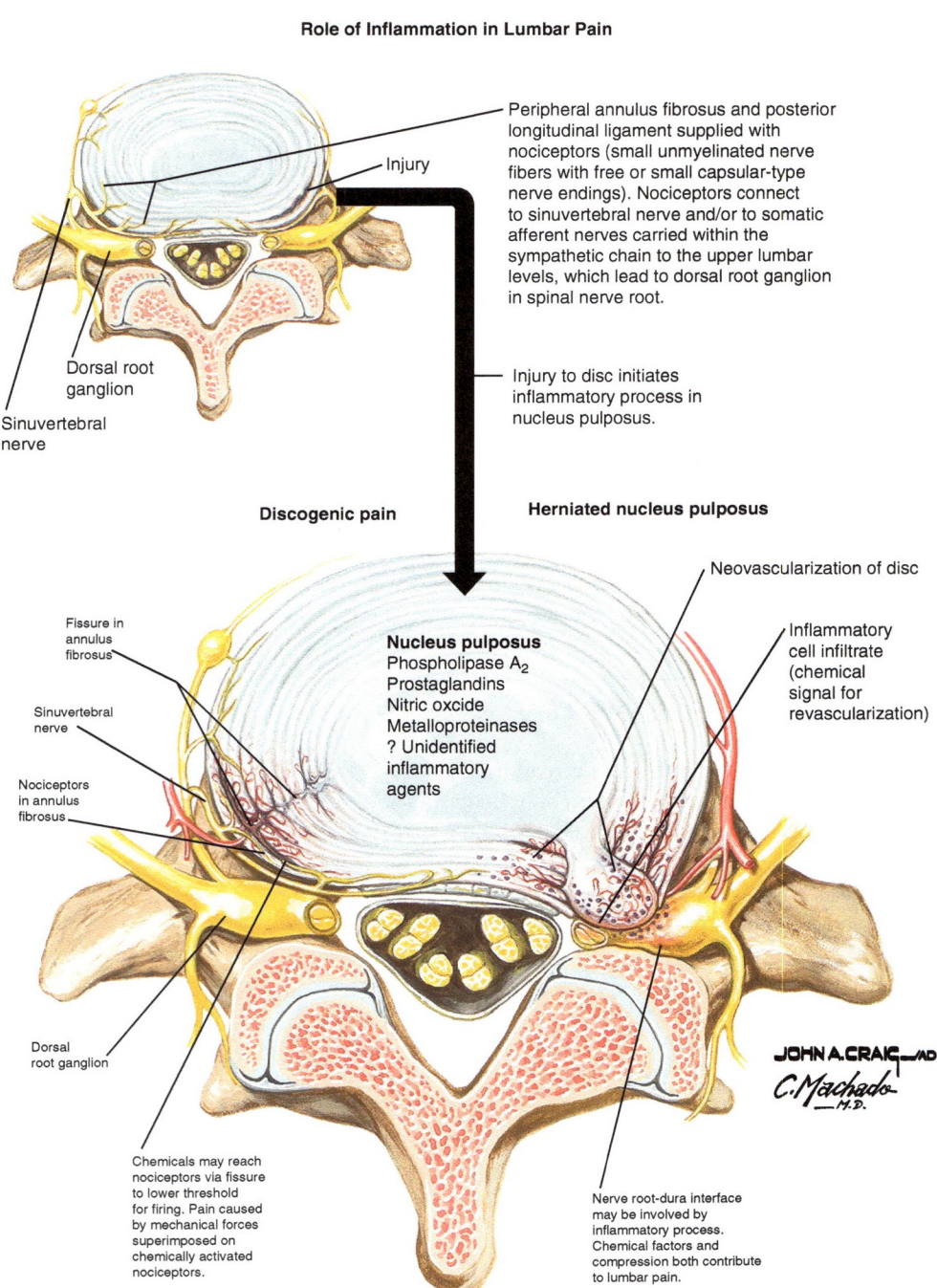

tion. Genetic inheritance likely contributes to degenerative susceptibility [52]. Pathological changes within the disc are distinct from other types of disc degenerative disease such as herniation [6, 57].

Pathophysiology

The pathophysiology of the intervertebral disc is complex, and only a brief summary will be considered here. Degenerated discs exhibit abnormally widespread innervation with sensory nerve fibres penetrating deep into the nucleus pulposus [58, 59]. Most discs with positive pain provocation on discography show radial fissures within the annulus [60].

The process of disc degeneration produces an inflammatory response, generated by cells within the nucleus pulposus, where multiple inflammatory factors are released. Histologic studies reveal ingrowth of vascularized granulation tissue along the annular fissures [59]. Immunohistochemical analyses have demonstrated cytokine-sensitized nociceptors, phagocytic cells, and perivascular neoinnervation (axonogenesis). Small, free nerve fibres may be found in the outer annulus and extend to the inner annulus and nucleus pulposus [59, 61–63].

Patients with discogenic back pain have significantly higher levels of released interleukin-1, interleukin-6, and interleukin-8, compared to patients with disc herniation [61]. Nerve fibres in the disc may contain nociceptive neurotransmitters, such as substance P, calcitonin gene-related peptide, and vasoactive intestinal peptide. These inflammatory factors migrate through fissures into the outer third of the annulus or into the endplate, where stimulation of free nerve endings results in pain [62]. The degenerating discs thus exhibit free nerve endings (pain receptors) and inflammation, which are two of the factors responsible for the pain response [64].

Patient Selection

There is little discussion in the literature about how to identify which patients are suitable candidates for discographic procedures. In general, a high level of suspicion for discogenic pain, where the persistent level of pain is severe enough to consider surgical intervention, is required [31]. Because most patients with low back pain experience improvement and resolution within 3 months, discography is typically reserved for adults who report back pain for an extended period. Earlier discography should be considered rarely and only for specific extraordinary cases, and it should not be used as a routine procedure for patients with nonspecific back pain [31, 64–66]. Prior to discography, the patient

should try multiple more conservative treatment modalities, (e.g. lifestyle and activity modification, medication, physical therapy, fluoroscopically guided injections, and other conservative methods) with insufficient therapeutic success [11, 28, 31]. The patient must also be able to understand the purpose of the procedure, comply with instructions, and provide meaningful feedback during the stimulation [11]. For example, the patient must be able to clearly describe any pain produced during disc stimulation and compare it to their usual pain [10]. The procedure is often performed using conscious sedation. This method supports patient comfort and allows the patient to be responsive during the pressurization phase of procedure.

Indications

Various authors describe the conditions for which discography is a suitable procedure [10, 11, 28, 64, 66]. These recommendations are not entirely uniform and, to some extent, vary based on the clinician's specialty (e.g. anaesthesiologist/interventional pain specialist, pain medicine and rehabilitation specialist, or radiologist).

Updated guidelines by the North American Spine Society enumerate a core set of criteria for which there is widespread agreement and which have withstood the test of time [65]. According to these criteria, indications for discography include, but are not limited to:

- Assessment of demonstrably abnormal discs to help evaluate the extent of abnormality or correlate the abnormality with the clinical symptoms. Such symptoms may include recurrent pain from a previously operated disc and lateral disc herniation.
- Assessment of patients with persistent, severe symptoms in whom other diagnostic tests have failed to reveal clear confirmation of a suspected disc as the source of pain.
- Assessment of patients who have failed to respond to surgical intervention, to determine if there is painful pseudarthrosis or a symptomatic disc in a posteriorly fused segment and to help evaluate possible recurrent disc herniation.
- Assessment of discs before fusion, to determine if the discs within the proposed fusion segment are symptomatic and to determine if discs adjacent to the segment are normal.
- Assessment of candidates for minimally invasive surgical intervention to confirm a contained disc herniation or to investigate dye distribution pattern before chemonucleolysis or percutaneous procedures.

A report by Walker et al. concurs with these criteria and adds that potential candidates for discography should have no contraindications, particularly evidence of psychogenic pain [31].

Contraindications

The main contraindications to discography are similar to those of other interventional procedures:

- Known bleeding disorder and use of anticoagulation/antithrombotic therapy that cannot be with temporary medication discontinuation
- Pregnancy
- Systemic infection or skin infection over the puncture site
- Allergy to radiologic contrast that precludes testing with contrast media, local anaesthetic, or antibiotics (pretreatment with antihistamine and corticosteroids or use of gadolinium may ameliorate these problems in some patients)
- Psychiatric conditions, such as psychogenic pain, posttraumatic stress disorder, or psychotic diagnoses
- Inability or unwillingness to provide informed consent to the procedure
- Inability to assess the patient's response to the procedure, for example, due to sedation or significant analgesic use
- Anatomic features that would preclude a safe and effective procedure, for example, severe spinal stenosis resulting in intraspinal obstruction
- Solid bone fusion that prevents access to the disc
- Severe spinal canal compromise at the disc level to be investigated [11, 28, 31]

Some of the above contraindications, such as infection, can be temporary, and others, such as allergy and anticoagulation/antithrombotic therapy, can be addressed in the pre- and perioperative period.

Preoperative Considerations

Patient preexisting conditions, such as allergies to contrast, latex, iodine, and antibiotics must be addressed. Prophylactic medications, such as diphenhydramine and a steroid agent for allergy management, can be prescribed for those patients whose allergy is not severe. Patient compliance must be ensured [10].

Familiarity with all medications a patient is taking is essential, including herbs and supplements. Instructions regarding the use of approved medications prior to the procedure vary. The potential benefit to the patient receiving discography must outweigh the risk of withholding essential medications. Recommendations for patients receiving antiplatelet/anticoagulant/antithrombotic medications vary and have changed over time. Communication with the clinician managing the patient's medication is essential. Pain, antiinflammatory, sedative, and any other medications or substances that alter the patient's perception of pain should not be used the day of the procedure to ensure test results are not comprised [10].

Perioperative Considerations

Clinicians concur that the usual perioperative protocols and precautions common to other spine interventions also apply to discography. A complete history and physical examination should be performed. The patient's CT and MR imaging should be reviewed to determine the levels to be studied. The patient needs to be informed of the risks and benefits of discography, and it should be made clear to the patient that his/her response to disc stimulation is the basis for the test results. The procedure should be explained to the patient in sufficient detail to convey what to expect, including the likelihood of discomfort or pain during the provocation portion of procedure and soreness for a few days afterwards, so that the patient can provide informed consent. Further, it is critically important for patients to fully understand that they will be required to actively participate in the provocation portion of the procedure by comparing the pain evoked by the procedure with their usual pain. They should also be made aware of the potential for complications, including pneumothorax if thoracic or lower cervical segments are to be tested [10, 11, 28, 31, 64]. Options for patients with severe iodine contrast allergy include gadolinium contrast and saline.

On the day of the discography, the patient can drink fluids, but should not eat for 2 hours before the procedure. Instructions regarding approved medications prior to the procedure vary [10].

Prior to the procedure, the patient is typically positioned in a prone position, prepped, and draped in a sterile manner. To avoid any confusion between needle-induced annular pain and a provocative pain response, the disc can be approached from the asymptomatic side, if the typical pain is predominantly on one side. Patients can also be positioned in a modified lateral decubitus position with the symptomatic side down. This position also facilitates optimal fluoroscopic imaging and keeps the image intensifier out of the way during initial needle placement.

Intravenous sedation can be given to relax the patient, providing that it does not compromise the patient's ability to participate during the procedure. Short-acting analgesics that can be readily reversed are generally preferred. Analgesia should be individually titrated to avoid oversedation. Heart rate, pulse oximetry, blood pressure, and respiration should be monitored throughout the procedure.

Standard infection prophylaxis practice is to administer intravenous and intradiscal antibiotics prior to the procedure, which have been reported to reduce the risk of infection and discitis [28, 49, 67–69]. The skin is also typically prepped with a povidone-iodine solution to further mitigate infection risk as S. aureus/epidermidis are typical constituents of normal skin flora [70–74]. In patients with a known iodine allergy, there are noniodine or alcohol-based solutions that can be used instead. The proceduralist must maintain sterile technique throughout the procedure. At no time should the needle tip be touched with the gloved hand; sterile gauze can be used to manipulate the needle tip. Intradiscal and/or oral antibiotics can be given at the discretion of the physician [40]. The consequences of discitis are so significant that many practitioners consider the use of prophylactic antibiotics to be the standard of care, especially for high-risk patients [28]. In a survey of its members, the Spine Intervention Society reported that 83.81% use preoperative antibiotics and 84.97% use intradiscal antibiotics [75].

The Procedure

Lumbar Discography

The most likely level of the pain generator and the two adjoining levels should be investigated. It is uncommon to study more than four segments. When stimulating the discs, the patient is blinded regarding the onset and level stimulated. If the patient's usual pain is localized to one side, the disc space can be approached from the contralateral, asymptomatic, side. Approaching the asymptomatic side can potentially reduce confusion regarding the source of any provoked pain. The American Society of Interventional Pain Physicians, [28] and the Spine Intervention Society guidelines detail recommendations for this procedure [49].

Patients are typically positioned prone. Foam pillows or pads can be utilized to reduce lumbar lordosis. Both single-needle and two-needle techniques have been described. A two-needle system uses a longer, small-gauge procedure needle passed through a shorter, larger gauge needle into the disc. A single-needle technique uses a single styletted needle passed through the skin and directly into the disc. A retrospective study of 100 thoracic discographies used a single-needle technique, a 24-gauge, 3.5-inch needle inserted directly through the skin into the thoracic disc, using intermittent fluoroscopic guidance and bevel rotation. No patient experienced any serious complications [76]. Both single- and double-needle techniques must utilize styletted needles. The stylet prevents tissue from accumulating in the needle and entering the disc [77]. The fluoroscope is used to obtain an anterior–posterior (AP) image of the target level. The fluoroscope is then angled cephalad or caudad until the

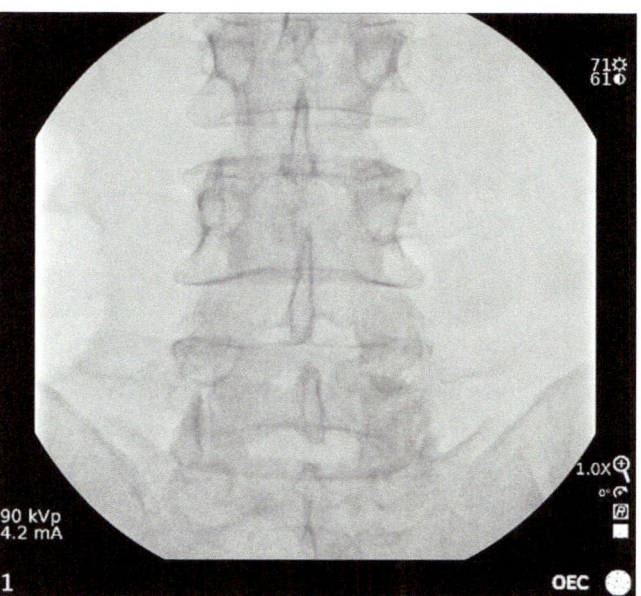

Fig. 16.3 AP View with the L4–5 endplates parallel to fluoroscopy beam

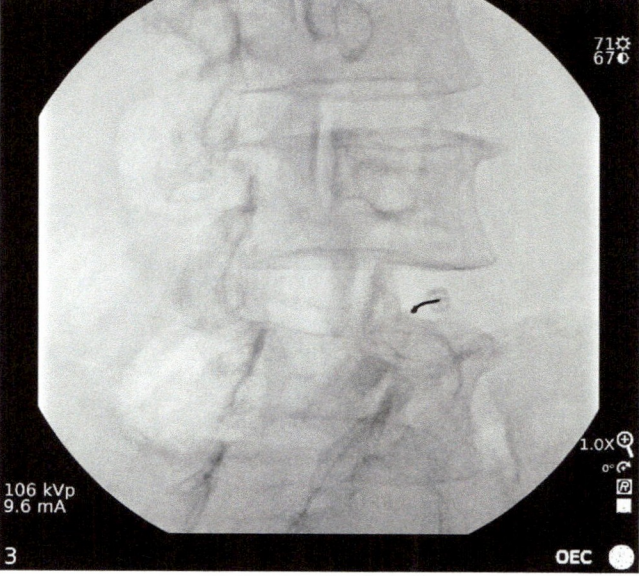

Fig. 16.4 L4–5 Oblique view. The SAP is in the midline of the disc space. The needle is slightly lateral to the SAP

image beam is parallel to the planes of the inferior and superior endplates that surround the target disc. (Fig. 16.3).

The fluoroscope is next rotated obliquely towards the side of needle entry until the facet joint line is in the midline of the target disc. The needle is to be passed just lateral to the lateral aspect of the superior articular process (SAP) at the level of the target disc. (Fig. 16.4) (ring apophysis) [78] of L5 superiorly. (Fig. 16.5) A curved tipped needle can be used to avoid the iliac crest while obtaining disc access. (Fig. 16.6).

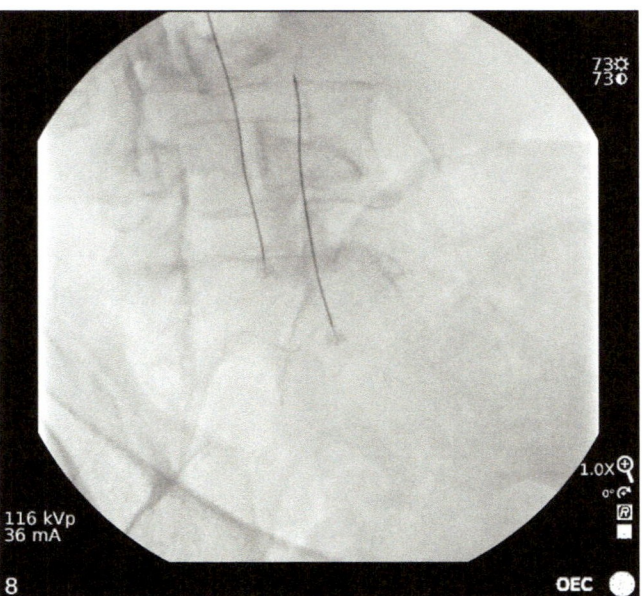

Fig. 16.5 L5-S1 Oblique view. The fluoroscopy beam has been angled cephalad to displace the iliac crest inferior. The existing needles are placed in the L3–4 and L4–5 disc spaces

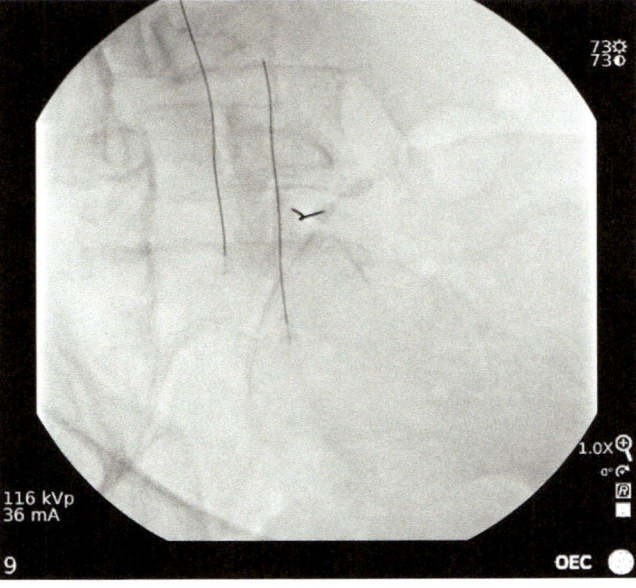

Fig. 16.6 L5-S1 Oblique view with needle inserted. Iliac crest is inferior, SAP medial to the needle. The curved tip needle can be used to avoid the iliac crest

The authors' standard needle length is 7 inches, although shorter needles can be used in slender patients. A longer needle, up to 10 inches, may be needed to access the L5-S1 disc in a large patient. A 22- or 25-gauge needle can be used to obtain disc access; however, 25-gauge needles can be difficult to manipulate due to their compliance but are less traumatic to the annular tissue. The needle is inserted through the

skin parallel to the fluoroscopic beam and advanced just lateral to the SAP. If bone obstructs needle placement, the discographer must determine if the SAP or an endplate has been contacted and make the proper needle correction. Once the needle has been advanced distal to the SAP, the fluoroscopy beam is rotated to obtain a lateral image. (Fig. 16.7)

Care must be taken when crossing the level of the intervertebral foramen not to strike the ventral ramus. If the patient complains of paraesthesia during this portion of the procedure, the needle must be slightly withdrawn and redirected. The needle is then advanced, and, if no paraesthesia is elicited, the next structure that the needle will encounter is the disc annulus. A firm resistance will be felt by the discographer at this point. It is common for the patient to experience a dull ache in the lower back or buttock as the needle passes through the annulus. The needle is then advanced into the centre of the disc, and final needle position is confirmed with both lateral and AP imaging. (Fig. 16.8).

After proper placement of the needle into the target disc, the stylet is removed from the needle. The needle is connected to a syringe that will inject contrast mixed with antibiotic. If the patient has a known allergy to contrast, either saline or gadolinium [79] mixed with antibiotic can be injected. At least one painless disc must be identified as a control level during provocation discography in order to validate the procedure. If all discs studied are painful, the discogram can be considered invalid, and an adjacent level should be tested in order to identify a control. The diagnosis is stronger if the concordant disc displays a grade 3 fissure or greater on a post-discography CT scan. The diagnosis is

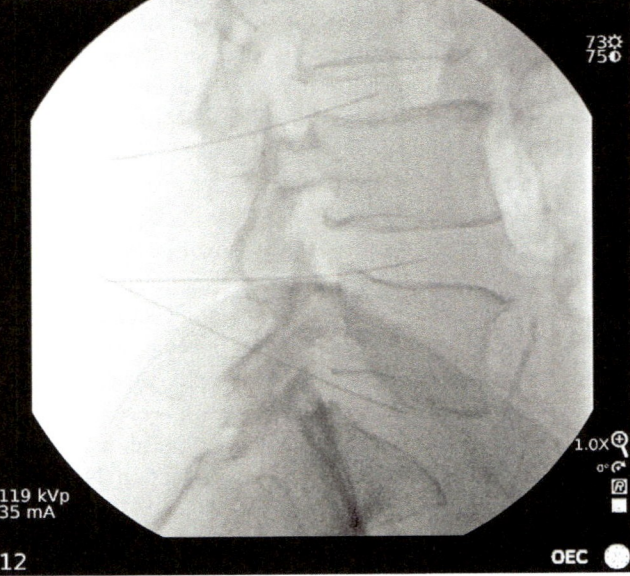

Fig. 16.7 Lateral view with needles inserted in the L3–4, L4–5 and L5-S1 discs. The needle is advanced into the centre of the disc as viewed laterally

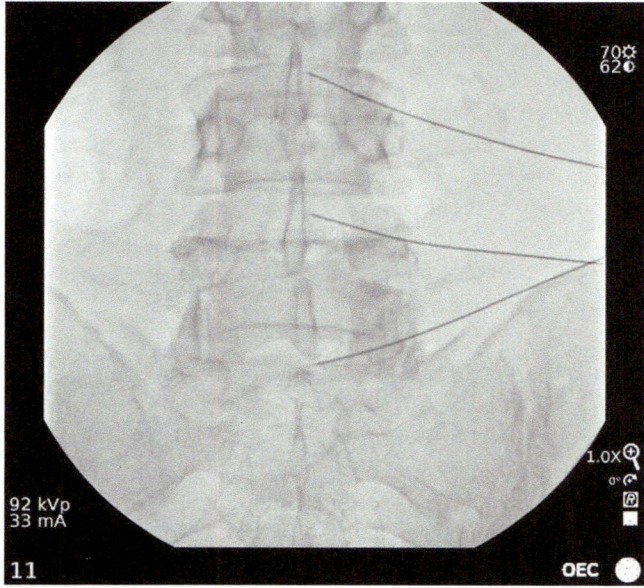

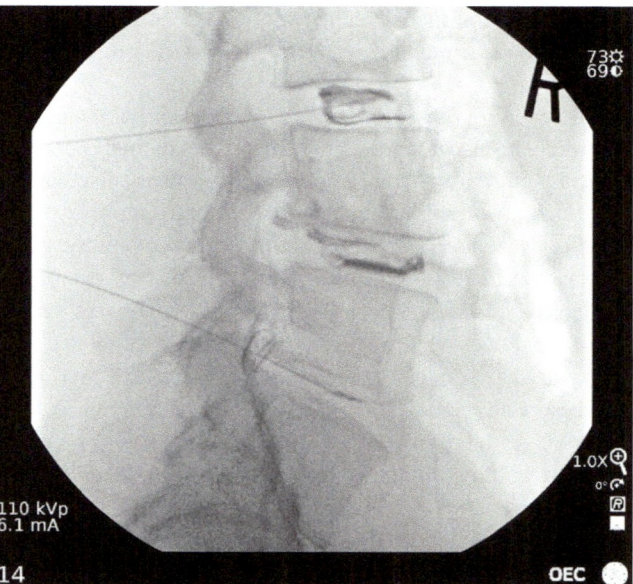

Fig. 16.8 AP view with needle in L3–4, L4–5, and L5-S1 discs. Needle placement should be midline in the AP and the lateral views. Note larger gauge needle used as introducer for FAD catheter

Fig. 16.9 Lateral view of a normal L3–4 disc. The L4–5 disc demonstrates a posterior tear. The L5-S1 disc space is narrowed and shows a posterior tear and posterior disc protrusion

most robust if a single disc demonstrates concordant pain production and the two adjacent discs are nonpainful [12].

Regardless of the technique employed, after the needles are positioned in the disc(s), each disc is evaluated by injecting contrast. Depending on the patient's size, a normal lumbar disc accepts from 0.5 mL to 3 mL with a firm endpoint or high discometric pressure [80]. Lumbar intradiscal pressure can be directly measured with a pressure gauge in psi at the onset of pain or with a firm endpoint. The volume and pressures are recorded while contrast is injected. The patient's response to the injection is noted [81]. In a normal disc, contrast remains in the nucleus and appears as a "cotton ball" (Figs. 16.9 and 16.10).

If the patient experiences pain with injection, the location, severity, and quality are documented. Transient pain can be provoked when fissures are opened. To be truly positive, the pain must be sustained during injection [17]. "Concordant pain" is pain during provocation that replicates the patient's usual pain pattern. "Nonconcordant pain" is pain during provocation that does not replicate the usual pain pattern. Disc morphology, including disc height, tears, and leaks, are also recorded. A confirmatory repressurization of a concordant disc or indeterminant disc is routinely performed to reconfirm the discographer's findings. Another method used to verify the consistency of the patient's response to disc pressurization is the use of a sham injection. The patient is told that the disc is being injected, while the syringe is held in the operator's hands. Any pain response is noted. Patients are expected to survive sham injection without response and to respond consistently to repressurization. A robust result would be one in which the patient

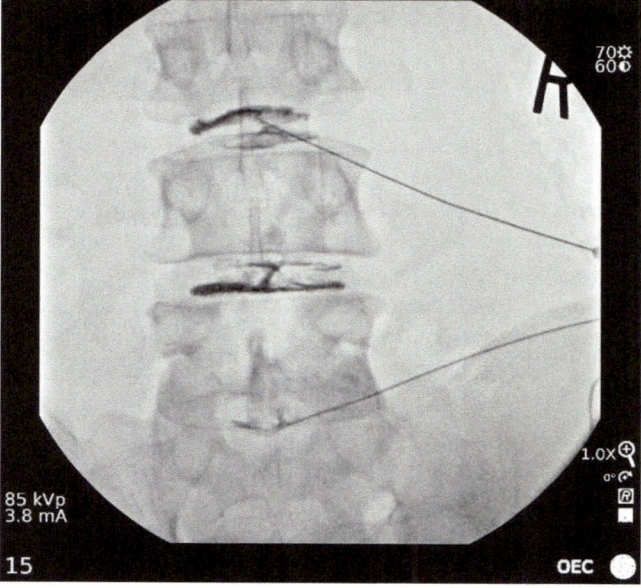

Fig. 16.10 AP view with normal L3–4 disc morphology and degenerative L4–5 and L5-S1 discs

survives sham injection without a painful response but responds consistently with a pain response to pressurization and repressurization to the disc.

Injection is continued until:

1. Pain is reproduced at a level of 6/10 or greater
2. Intradiscal pressure > 50 psi above opening pressure in a disc with a grade 3 annular tear

3. 4.0 ml of volume is reached
4. 80–100 psi is reached in a normal appearing disc [42, 82]

The opening pressure, pressure at onset of pain, and peak pressure are also recorded [79]. The use of manometry to measure intradiscal pressure during lumbar discography generates quantifiable, objective data which improve procedural consistency [17]. Intradiscal pressure monitoring also reduces the incidence of false-positive results by decreasing the likelihood of over-pressurization [83]. Pressures greater than 50 psi over opening pressure have been associated with a very high false-positive rate based on a retrospective study of pressure and pain response by O'Neill and Kurgansky [84].

Upon completion of the injections, x-rays of the lumbar spine are obtained in the posteroanterior and lateral views. A nucleogram of a normal lumbar disc appears as a rounded or bilobular-contained component of the lumbar disc. Annular disruption shows contrast spread beyond the nuclear border, typically in a radial fashion. Nucleogram patterns can range from normal (cotton ball and lobular) to abnormal (irregular, fissured, ruptured, and degenerative).

Post-Lumbar Discography Imaging

Post-lumbar discography CT imaging provides further detailed information about the presence and degree of annular pathology, as well as disc degeneration. The extent of annular pathology on CT-discography correlates with the likelihood of a concordantly painful disc [44]. The modified Dallas discogram classification system assists in assessing patients with lumbar spine pain for annular pathology. Grade 3–5 annular tears demonstrate a high correlation with concordant low back pain.

Graded morphology of internal disc structure:

- Grade 0: Normal disc morphology
- Grade 1: Contrast spreads radially along a fissure to the inner 1/3 of the annulus
- Grade 2: Contrast spreads into middle 1/3 of the annulus
- Grade 3: Contrast spreads to the outer 1/3 of the annulus, involving <30(degrees) of the disc circumference
- Grade 4: Contrast spreads to the outer 1/3 of the annulus, involving >30(degrees) of the disc circumference
- Grade 5: Full-thickness tear with extra-annular leakage into epidural space

Vanharanta and colleagues [60] were able to demonstrate from post-discography CT imaging that increasing disc degeneration was associated with increased likelihood of pain provocation. Discs with severe (grade 3 and above) annular disruption were associated with pain provocation 77% of the time.

Colhoun et al. [20] compared post-discography morphology with improved surgical outcomes. Patients with abnormal disc morphology and consistent response to pain provocation had successful surgical outcomes 89% of the time. Comparatively, they found successful surgical outcomes only 52% of the time in patients with abnormal disc morphology without pain provocation. (Table 16.1).

Cervical Discography

With some alterations, the pre- and peri-operative considerations discussed above also apply to cervical discography. Meticulous sterile technique and wide prep are important. Cervical discography may be performed using the original anterior approach or a modified anterolateral approach. The latter has been associated with less risk and has become the most commonly used approach. (Fig. 16.11).

The original **anterior paratracheal** technique places the patient in the supine position. The C-arm fluoroscope is employed to visualize the cervical spine. The patient's head and neck are placed in extension to widen the anterior disc space for easier access into the disc. Through patient or fluoroscope positioning, the spinous processes are aligned midline with visualization of the vertebral endplates and uncinate processes. (Fig. 16.12).

The oesophagus is typically located left of midline, so using a right-sided approach when performing cervical discography can lessen the risk of puncturing the oesophagus. With the nondominant hand, palpate anterior cervical structures with index and/or middle finger. Move the trachea and oesophagus medially and the carotid artery and internal jugular vein laterally. Direct the needle towards the anterolateral border of the endplate just below the target disc. This safety step is used to prevent overpenetration of the needle, which can travel through disc and directly into the spinal canal. Upon bony contact, the needle is held firmly, and the C-arm rotated into a lateral projection to confirm positioning. The needle is then walked off of the endplate superiorly into the disc annulus. The operator should be able to appreciate the clear tactile difference between hard bone and the more compliant disc. PA and lateral fluoroscopic views are used to

Table 16.1 Discography predicted success of subsequent surgical intervention.

Therapeutic utility			
		Response to treatment	
		Success	Failure
Disc stimulation	Positive	121	16
	Negative	16	15

Colhoun's data [20] showed sensitivity of 0.88 and specificity 0.48 in patients with positive discograms who underwent cervical spine surgery. Note that very few patients with negative discograms underwent surgery which may contribute to a selection bias

Fig. 16.11 Lateral and midline line approach trajectories. Cervical discography is typically done from the right side due to the left-sided location of the oesophagus. (From: Melnik et al. [126]; with permission of Springer Nature)

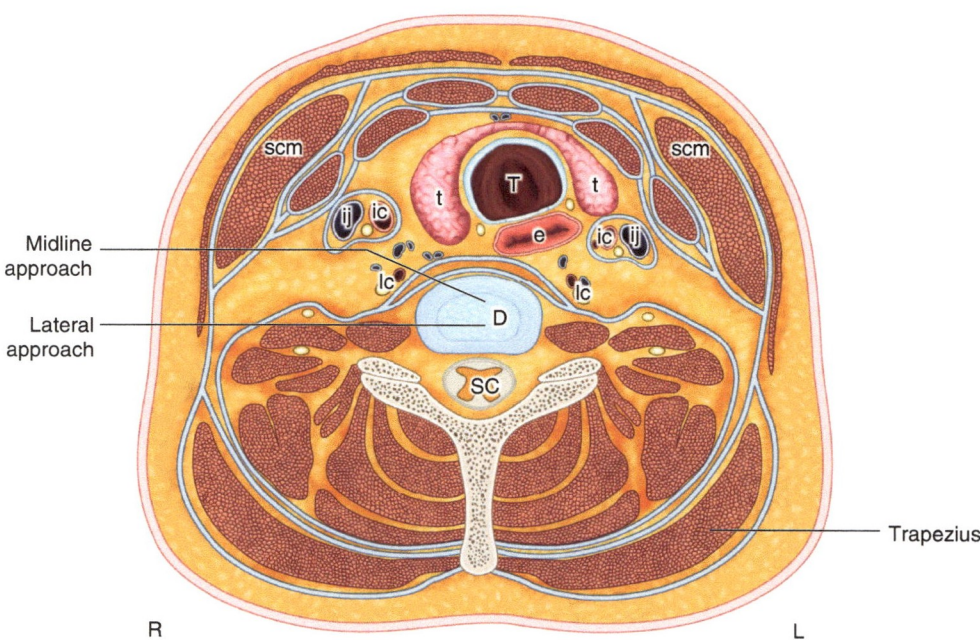

Fig. 16.12 AP C-spine with caudal tilt to square vertebral endplates. (From: Calodney and Griffin [127]; used with permission of Springer Nature)

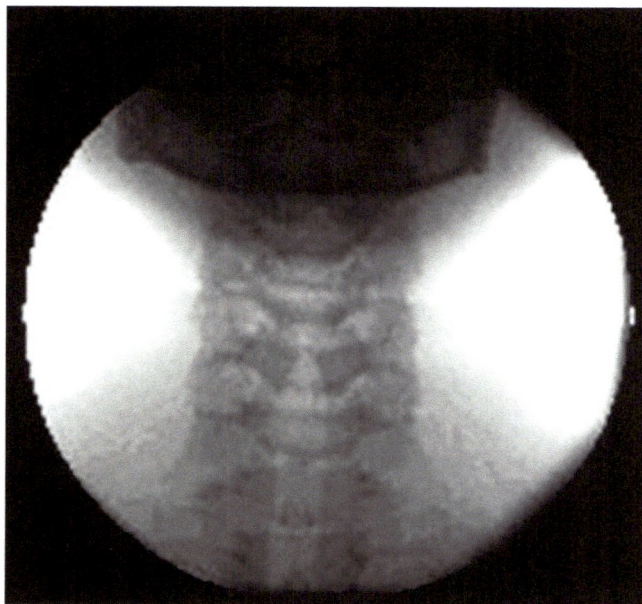

advance tip into the centre of the disc space. A lateral view should be assessed as soon as the firm annulus fibrosis is felt.

Alternatively, the **oblique approach** has the advantage of keeping the operator's hand out of the fluoroscopic beam and is often the only way to enter lower cervical discs in larger patients. The disadvantage is that it puts the carotid artery at greater risk, as the course of the carotid can be anatomically variable. The patient is again positioned supine, and the

C-arm is first positioned for a PA view of the cervical spine and then tilted caudally until the beam is parallel to the target disc space. The C-arm is then rotated towards the side of entry (generally the right side) to obtain open neural foraminal view. Using the focus of the beam as a guide directs the needle towards the medial edge of the uncinate process lateral and inferior to the disc space. Walk the needle off medially and superiorly into the disc annulus. (Figs. 16.13 and 16.14).

Again, once contact with annulus is felt, PA and lateral fluoroscopic views are utilized to advance the needle into the centre and confirm proper positioning. (Figs. 16.15 and 16.16).

After successful placement within the centre of the disc, the stylet is removed and the needle hub is filled with a few drops of contrast. Normal cervical discs can have volumes as small as 0.1 ml. In lieu of these small cervical volumes, even the dead space of the needle hub becomes important to consider. A 3 cc syringe and low volume extension tubing filled with contrast (containing the antibiotic) are attached to the needle. Often significant pressures are needed to reach dye point (the pressure at which contrast is first seen entering the disc) in the cervical spine, particularly in younger, normal discs. Additionally, dye point can be sudden and produce what we term an "opening snap", which can startle the patient and may be uncomfortable. It is important to distinguish this sensation from a positive painful discogenic response. Confirmatory techniques including sham injection and repressurization can help make this distinction and improve validity of patient response. (Fig. 16.17).

Contrast volume used for cervical discography ranges from approximately 0.1–0.5 ml. Occasionally, volumes over

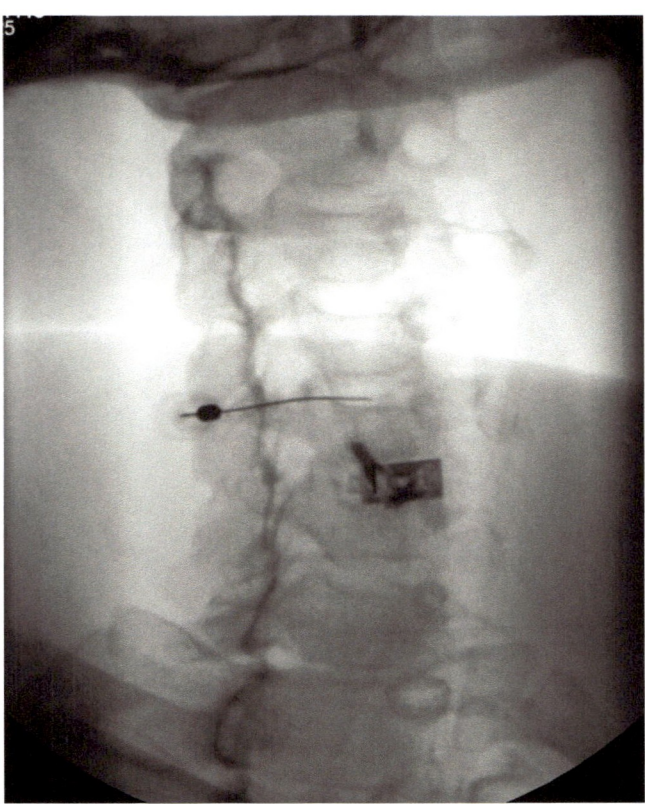

Fig. 16.13 Oblique view with needle tapping the medial edge of the uncinate process

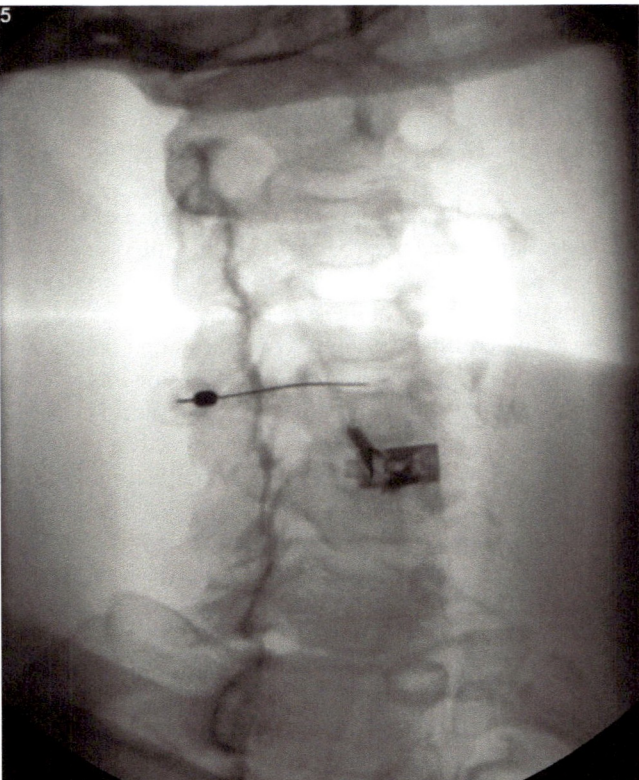

Fig. 16.14 Oblique view, needle walked off uncinate process medially and superiorly into the centre of the disc

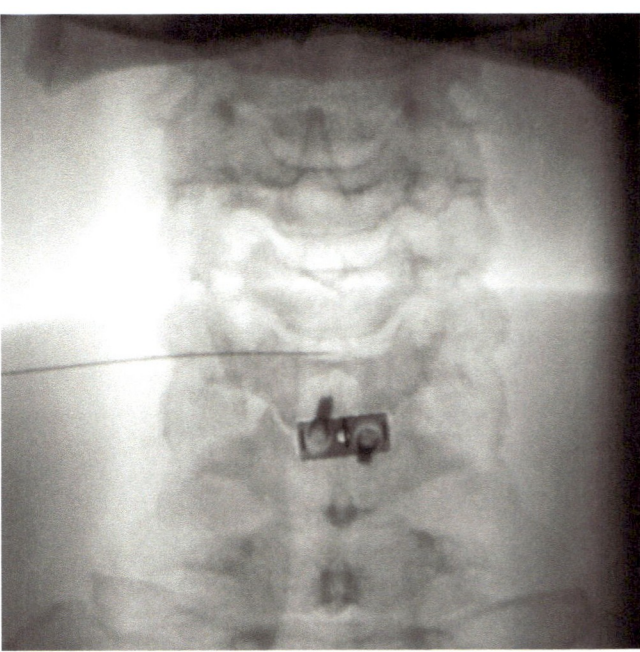

Fig. 16.15 AP view of the needle placed in the centre of the C6–7 disc space

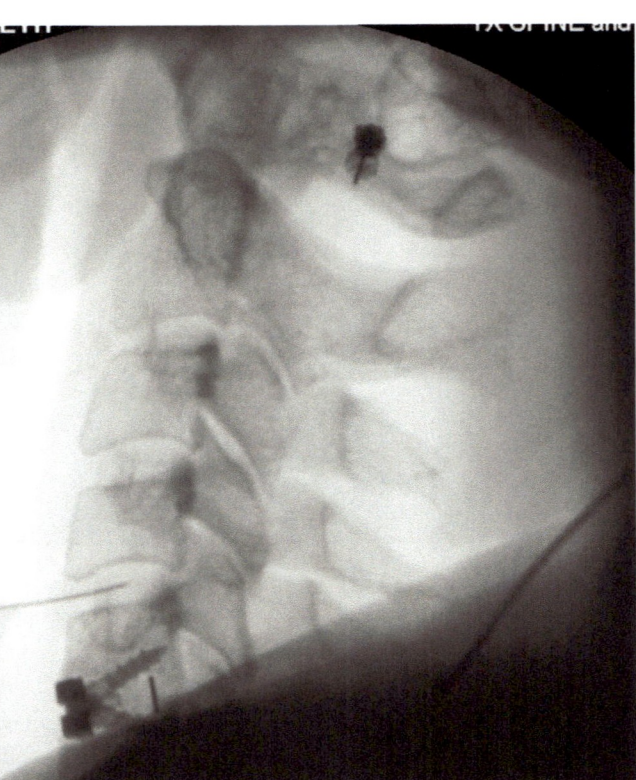

Fig. 16.16 Lateral view with needle in the centre of the disc space

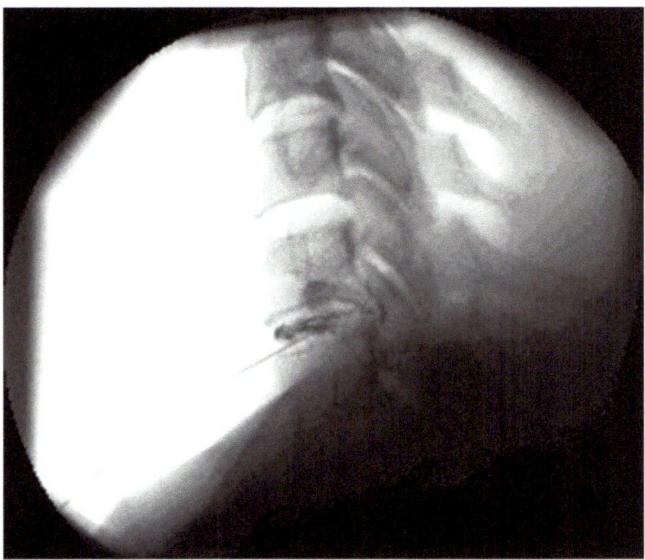

Fig. 16.17 Lateral view with contrast demonstrating small posterior tear and bulge

0.5 ml up to 1 ml may be required to pressurize the disc. These larger volumes can be an indication of incompetent or severely degenerated cervical discs. The average cervical disc volume noted by Ohnmeiss and colleagues [85] and corroborated by the authors' years of clinical experience is about 0.23 ml. Indeed, care should be taken if volumes exceed 0.5 cc. The injection of contrast should be terminated if a large leak is noted, firm resistance develops, or significant pain is experienced by the patient.

When investigating multiple cervical discs, the first needle is typically inserted into the most cephalad disc, followed by needle insertion into the remaining discs in a sequential and caudad direction. A single 25-gauge spinal needle is used to enter each disc level for the study. The needle entry should be more laterally at the C2/3 and C3/4 disc levels to avoid the hypopharynx and more medially at the C7/T1 level to avoid the apex of the lung.

Post-Cervical Discography Imaging

Post-cervical discography CT imaging is more challenging secondary to smaller injected contrast volumes. Generally, the CT should be done within 30–60 minutes of the contrast injection lest the contrast be largely redistributed before obtaining the CT images. Annular tears and protrusions involving the cervical disc that are detected by CT-discography are not always visualized by MRI, because the typical 3–5 mm MRI slice does not provide the necessary detailed information. CT-discography is optimal with 1 mm slices with gantry angles appropriately parallel to each cervical disc.

Thoracic Discography

Thoracic discography is less common and thus less studied than either lumbar or cervical discography. The general pre- and peri-procedural recommendations discussed in the corresponding cervical and lumbar sections also apply here. The technique used is similar to lumbar discography with some alterations to ensure the safety of the spinal cord in the thoracic spine. Another important pre-procedural consideration is that upper thoracic levels may be difficult to enter. Shorter disc heights and the close approximation of the ribs and costovertebral joints make this anatomy difficult to investigate. Additionally, degenerative changes tend to complicate matters further. Lower and midthoracic discs are generally easier and safer to study in most patients.

The patient should be placed in a prone position on the table. Adjust the C-arm to provide a posterior oblique position with the superior articular process one-third to one halfway across the disc space with squared endplates. The endplates at each level are squared off to ensure the parallel orientation of the beam with the sub- and supra-adjacent vertebral body margins. This view creates a "box" configuration formed by the endplates, the superior articular process\lamina, and the rib head. (Fig. 16.18).

The box defines a safe pathway into the annulus while avoiding the spinal cord medially and the lung laterally. Advance the needle within the confines of the box to the outer annulus. The needle must stay medial to the rib head and costovertebral joint in order to avoid the pleura. The needle must stay lateral to the lamina and interpedicular line to avoid entering the spinal canal. (Figs. 16.19, 16.20, 16.21, 16.22, and 16.23).

After encountering, the outer annulus continues into the central third of the disc using fluoroscopic guidance. Subsequent procedural steps are similar to those discussed in detail earlier in the text in regard to cervical contrast injection, as manometry is not generally used in with the thoracic or cervical regions.

Post-Thoracic Discography Imaging

CT imaging provides further detailed post-thoracic discography information on the degree of annular pathology and disc degeneration. The modified Dallas discogram classification system for annular pathology can be used to define the degree of abnormal findings.

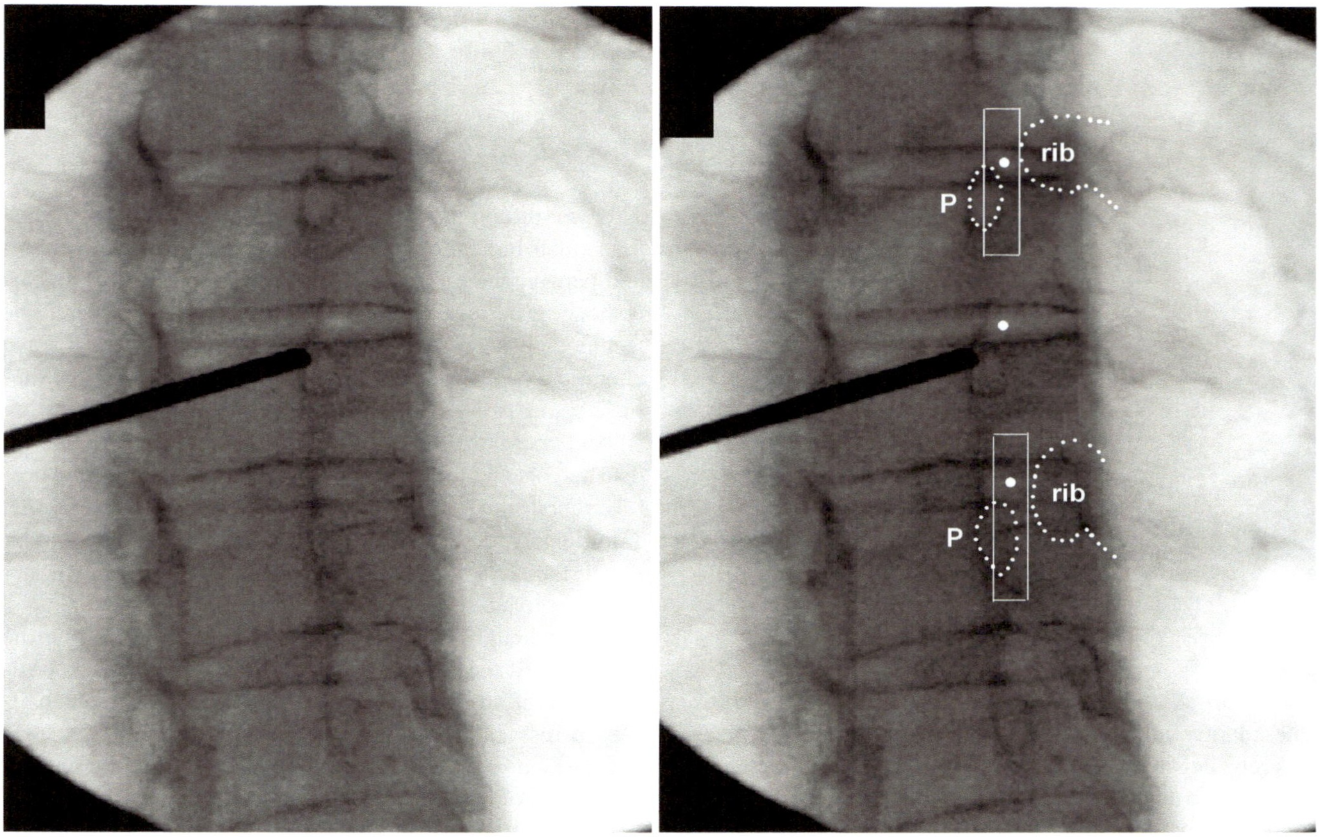

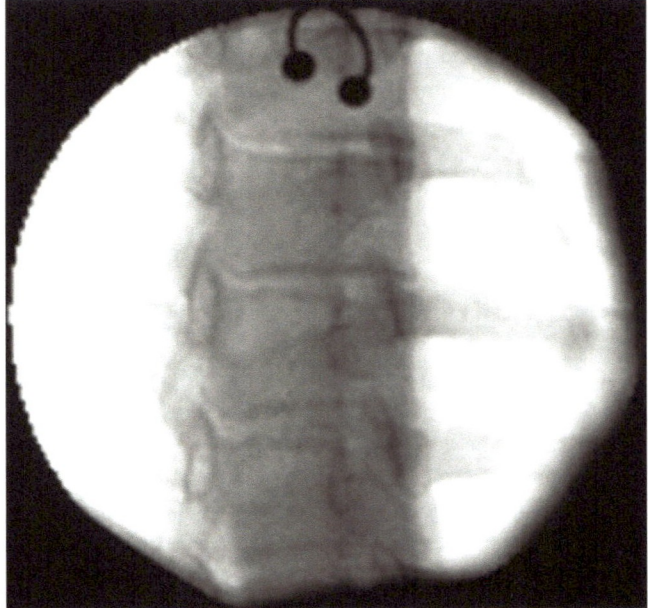

Fig. 16.18 The needle passes through a "box" bounded laterally by rib head, medially by lamina, and superiorly and inferiorly by endplates. (Reproduced with permission from Bogduk [128])

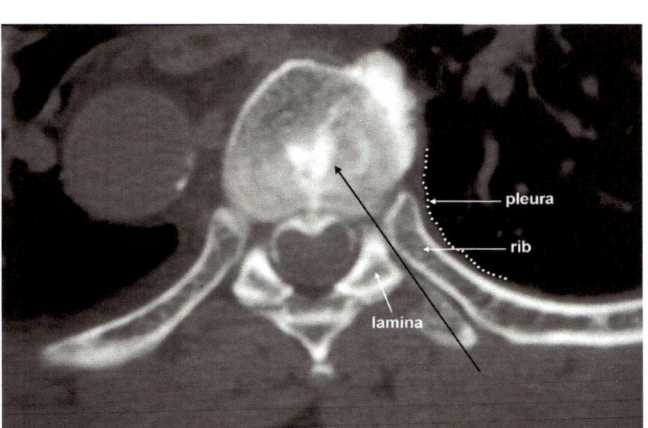

Fig. 16.19 Needle trajectory remains medial to the rib head and lateral to lamina to avoid pleura and spinal canal. (Reproduced with permission from Bogduk [128])

Fig. 16.20 Note that the endplates are squared and the C-arm is rotated towards the side of entry until the "box" is approximately 25% across the disc space

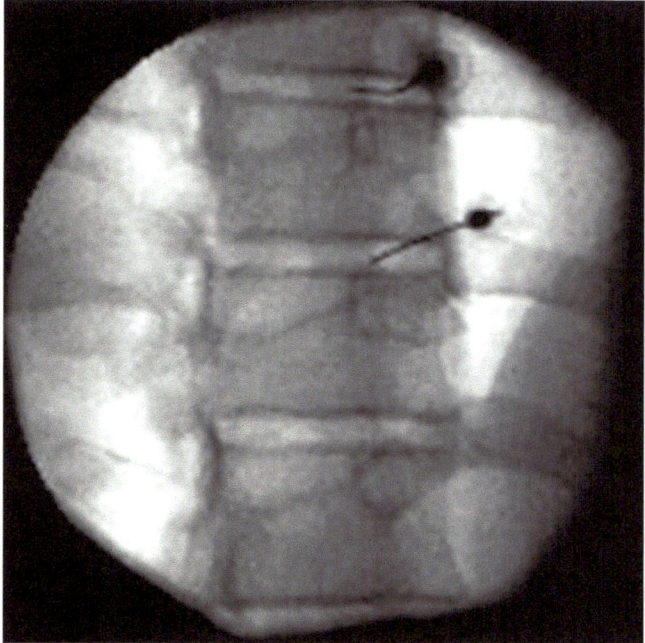

Fig. 16.21 Endplates must be parallel at the level of entry. The needles can be seen passing medial to the rib heads into the thoracic disc

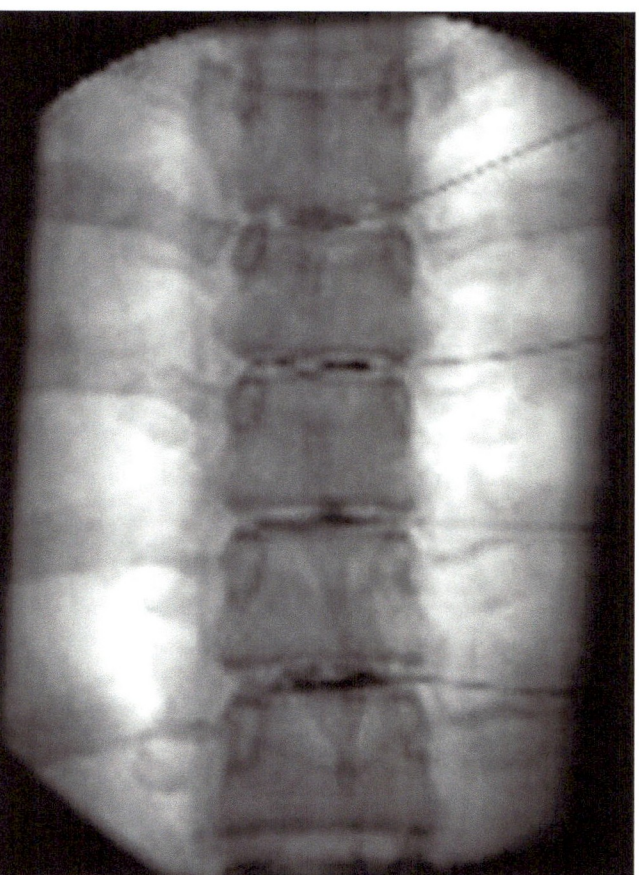

Fig. 16.23 AP view with contrast in the thoracic discs

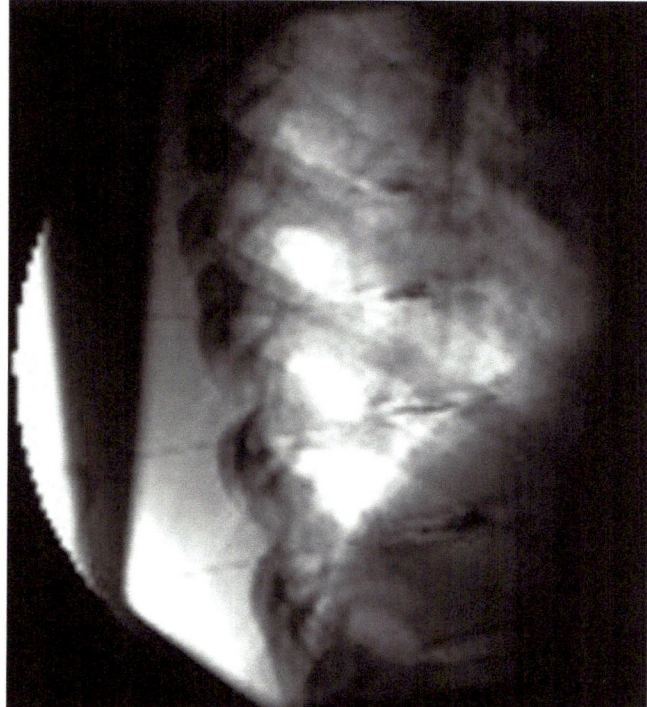

Fig. 16.22 Lateral view with needle in the middle of the disc space. The upper two discs appear normal. The Lower two levels demonstrated posterior annular abnormalities and recreated familiar pain on injection of contrast

Post-Procedure Care

Patients should be observed for at least 30–60 minutes following the procedure and instructed not to drive until the next day following the procedure. If needed, patients should be provided with post-procedure analgesia. Advise patients to call if they experience symptoms such as worsening pain, fever, chills, malaise, and night sweats within 1 week of the procedure, which could indicate a disc infection. Shortness of breath could indicate pneumothorax. Note: Discitis symptoms, which are typically severe back pain, may not appear for weeks to months after discography [10, 28, 76].

Interpretation and Documentation

The disc pressure (discometry or manometry) at the onset of pain during lumbar discography can be measured and recorded with a pressure gauge in pounds per square inch (psi). As mentioned above, manometry provides additional objective measure and its use in lumbar discography is generally considered standard of care. Manometry is infrequently used in cervical and thoracic discography.

Painful lumbar discs can be categorized into one of four categories with the aid of manometry [28]:

1. Normal discs: no pain
2. Chemically sensitive disc: pain <15 psi above opening pressure
3. Mechanically sensitive disc: pain >15 psi and < 50 psi above opening pressure
4. Indeterminate disc: pain >50 psi above opening pressure

Derby and colleagues [17] found that patients with chemically sensitive discs had better outcomes with interbody fusion when compared with intertransverse fusion or nonoperative treatment.

The pain level reaction reflects the pain intensity experienced by the patient during the injection regardless of whether the pain is concordant or discordant. The intensity is graded verbally on a numeric pain rating scale, often using 10 as the greatest degree of pain and 0 as no pain at all.

The pain quality is crucial because it establishes whether pain provoked during the procedure mirrors the pain experienced by the patient. The pain can be vague and discordant, partly concordant, (i.e. merely a component of their typical pain), or an exact reproduction of the patient's concordant pain.

Strict diagnostic criteria are crucial for discography as it is a provocational study; as such, it is inherently prone to the challenges of objectivity. Subjective patient input, the previous lack of standardization, and questionable specificity have fuelled debate among proceduralists [12, 17, 38, 42, 86–88]. The resultant discussions among peers have helped to advance the standards and ultimately the objective measure of discography.

The specificity of discography has historically been a source of controversy. However, data from a recent meta-analysis were able to demonstrate that lumbar discography adhering to updated practice guidelines is associated with a low false-positive rate [51]. A recent Wolfer et al. [42] meta-analysis of all completed data sets involving subjects asymptomatic for lower back pain using ISIS/IASP guidelines found a false-positive rate of 9.3% per patients and 6.0% per disc.

Others have been able to demonstrate a similarly low false-positive rate with cervical and thoracic discography. Schellhas et al., [35] in a study of 40 cervical discs in asymptomatic patients, there were no pain responses. Wood and colleagues [89] found that of asymptomatic volunteers 3 of 40 (7.5%) injections were painful. However, all three of these discs demonstrated prominent Schmorl's nodes, and the provoked response was unfamiliar and nonconcordant.

Strict diagnostic guidelines and procedural modifications have clearly increased the diagnostic accuracy of discogra-

phy. However, controversy remains. Carragee has published multiple papers questioning the validity of discography. His works have suggested that discography may result in misdiagnosis, unnecessary surgery, and potentially accelerate disc degeneration [42].

Carragee demonstrated high false-positive rates of 40% in a sample of 20 post-discectomy or post-laminectomy patients who were asymptomatic at the time of discography [87]. These results are considered by many to be heavily influenced by the study's patient population and Carragee's use of higher-pressure cut-offs [82]. Greater than 75% of his patient sampling had somatization disorder. Discography is ultimately a subjective test as patient participation is a required component. Therefore, caution is needed when interpreting discography response in patients with low pain tolerances or in those with abnormal psychometric profiles. Secondly, this particular observational study was taken from a population who had previously undergone lumbar spine surgery in the form of discectomy or laminectomy. The results demonstrate a high likelihood of having a positive discogram at a previously operated level for both symptomatic and asymptomatic groups alike. False-positive rates may indeed be disproportionately higher in post-discectomy patients. This does not invalidate its efficacy but rather it implies that extra care should be taken when interpreting discography outcomes in these patient populations.

In 2009, prospective longitudinal cohort data were published to investigate the long-term impact of discography by comparing MRI indices in individuals who had undergone discography with matched controls [41]. A cohort of 75 subjects were followed for 7–10 years after baseline workup. The research group showed that individuals who received discography were subsequently found to have higher rates of lumbar disc degeneration, lumbar disc herniation, spine surgery, significant lower back pain episodes, and more medical follow-up compared to the control group [41].

Deeper review found that for the sample sizes used in this study the confidence intervals between study and control groups overlap with regard to higher levels of disc degeneration and Modic changes and are therefore not likely to be statistically significant [40]. Foraminal disc herniations were found to be 2–5 times more common in the post-discography group. However, the rates of foraminal herniation in the general population are nearly equivalent with those found in his treatment group, while his control group was curiously less affected [40]. A similar trend was observed with regard to the prevalence of Modic changes. The control cohort had significantly lower rates of Modic changes (11%) compared to those found in the general population (36%) [9]. Other questions arose with regard to this study's substantial loss to follow-up, lack of adherence to current SIS/IASP procedural guidelines, and the exclusion of appropriate discography candidates [9]. The loss to follow-up rate was substantial.

While a high attrition rate is generally expected in long-term clinical studies, the loss to follow-up was reported as high as 30% in the 2009 data [9]. The magnitude of this attrition rate significantly impairs the ability to comment on true patient outcomes [9]. With regard to procedural technique, inappropriately high disc pressures were produced in a majority of subjects. In fact, 96% of subjects were subjected to pressures of 80 psi or greater. This is an important procedural error, as high disc pressures have been demonstrated to cause annular disruption [90].

Other long-term cohort studies have not demonstrated higher rates of disc degeneration associated with discography. In a small prospective study ($N = 36$), Pfirrmann scores in subjects with symptomatic low back pain who had undergone provocation discography with or without confirmation by intradiscal bupivacaine injection were compared with matched controls [91]. Ohtori et al. found that no significant difference in disc degeneration was observed on MRI between both groups at 3–5-year follow-up intervals. Similarly, a cross-sectional cohort study found no evidence of degenerative disc changes 10–20 years after discography [92]. However, radiography (not MRI) was used to assess degree of change. Without MRI to detect minute changes, it is hard to draw definitive conclusions from this longitudinal study. Data from a 7-year matched cohort study using MRI, likewise, found no relationship between progression of degenerative disc disease and provocative discography. In this study by McCormick et al., 66 discs exposed to provocative discography following SIS/ASIP guidelines were matched to a control cohort of patients with low back pain. There was no difference in proportion of punctured discs that advanced in Pfirrmann scores compared to matched cohort, nor was there a difference between puncture and nonpunctured discs within the provocative discography group. The same study also found no differences in T2-signal-intensity-to-CSF ratio, disc height, new disc herniations, new HIZs, or new Type 1 Modic changes in the group exposed to provocative discography [93].

Published animal data seem to suggest that disc puncture with small-gauge needles does not cause a progressive increase in disc degeneration, [9] These data seem even more relevant when considering needle size to disc height ratio in these smaller animal models [94].

In conclusion, Carragees' studies demonstrate that discography has false positives like any other diagnostic test. Abnormal psychometric testing and patients with previously operated discs have disproportionately higher false rates, and therefore, their results should be interpreted with caution. Likewise, strict adherence to SIS/IASP guidelines is important both to limit number of false positives and to limit risk of over-pressurization injury. The risk of progression of disc degeneration following provocative discography has not been reproduced in similar matched cohort studies. The findings of Carragee may have been influenced with methodologic flaws in study design, lack of adherence to current guidelines, and substantial loss to follow-up.

Complications

The overall complication rate for discography is quite low (i.e. estimated to be less than 1 per cent) [95]. Improved injection techniques, advanced imaging, and better contrast materials have all contributed to a decreased incidence of complications over time. Infection is a potential and well-recognized complication of any interventional procedure. The two most grave complications of discography are discitis and neural injury. Incidence of discitis has significantly declined after widespread use of prophylactic intradiscal and intravenous antibiotics. Likewise, the use of proper technique can avoid neural injury. A paraesthesia is a clear indication to the discographer to withdraw and redirect the needle. Overall reported complications associated with lumbar discography range between 0% and 2.7% of patients.

A retrospective analysis of 4400 cervical disc injections in 1357 patients, to assess the morbidity and mortality of cervical discography, was reported by Zaidman et al. in 1995. The authors found that less than 0.6% of the patients experienced a significant adverse event, and 0.16% of cervical discograms resulted in patient injury [96]. In a systematic review of cervical discography, Kapoor et al. found a discitis rate of 22 in 14,133 disc injections (0.15%) in 21 of 4804 patients (0.44%) [97].

Willems et al. reported on a case series of 200 lumbar discography patients (435 discograms) and also conducted a systematic review of the literature to identify discitis risk and assess the need for prophylactic antibiotics. In nine studies reviewed, the authors found an incidence of 12 cases of discitis in 4891 patients (0.25%) and 12,770 discs (0.094%) where clinicians had not used prophylactic antibiotics. In one study examined, where clinicians used prophylactic antibiotics in 127 patients, no cases of discitis were reported [98].

Thoracic discography is rare, and no recent analysis of complications for thoracic discography was identified. Pobiel et al. conducted a retrospective review of 12,634 discographies performed at all levels in 10,663 patients over a 12-year period. Of these, thoracic discographies were done on 1141 patients and 3083 discs. While 17 cases could not be completed, no instances of thoracic discitis occurred. At all spinal levels and procedures, only two patients experienced discitis, for an overall incidence of 0.016% [69].

Potential, although very rare, complications include:

- allergic contrast allergy
- bleeding

- bowel perforation
- bruising
- discitis
- epidural abscess
- increased pain
- meningitis
- myelopathy
- nerve root injury
- pneumothorax
- retroperitoneal structures, including the kidney and spleen
- subarachnoid puncture
- trauma to the spinal cord
- vagal response [10, 28, 31, 49]

Correlating Imaging with Discography

Disc degeneration is a ubiquitous term often meaning different things to different experts. The process by which a disc becomes painful has not be directly established [40]. The microenvironment within the disc shifts as ageing chondrocytes become less able to maintain the homeostasis of the matrix [99]. Cyclic loading, genetic, epigenetic, and metabolic environment all seem to play a role in disc degeneration [99]. However, age is the strongest correlate of degenerative changes [99]. The epidemiologic evidence demonstrates that these changes are not painful, and the moniker may in itself be a source of distress to patients [99].

Internal disc disruption is not an age-related phenomenon and is associated with axial pain [99]. The aetiology of internal disc disruption is fatigue failure occurring with cyclic loading with the subsequent recruitment of inflammatory cytokines and activation of metalloproteases [99]. Vertebral endplates are susceptible to fatigue failure when subjected to repeated compression loads as small as 50% to 60% of the ultimate tensile strength of the endplate [100]. The endplate can fracture after as few as 100 reps [40]. Stress profilometry can be used to detect and quantify endplate disruption. A pressure transducer is inserted across the diameter of the disc and slowly withdrawn, while intradiscal pressures are monitored. (Fig. 16.24).

Internal disc disruption generally demonstrates a characteristic profilometry profile with posterior endplate fractures [99]. Internal disc disruption is characterized by isolated radial fissures through the annulus fibrosis of lumbar intervertebral discs. These findings can be seen on post-discography CT imaging.

Rapidly advancing imaging technology provides the clinician with vast amounts of digital data; however, in the absence of clear clinical correlates, these data can quickly become a barrier to selecting patient appropriate therapy. Provocation discography remains the reference standard for the diagnosis of discogenic pain; however, it is reasonable to

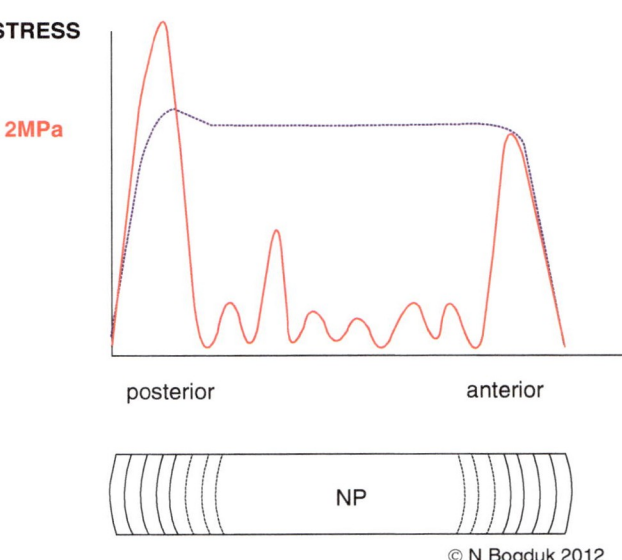

Fig. 16.24 Features of a normal disc and one affected by internal disc disruption (IDD) under stress profilometry. Graph showing the magnitude of stresses within the disc across a diameter as probe passes from the anterior annulus to the posterior annulus. In normal disc, the stresses are uniform. In a disc with IDD, the stresses in the nucleus pulposus are irregular, decreased, and may be zero, but the stress in the posterior annulus is increased substantially more than normal. (From: Bogduk [129]. By permission of Oxford University Press)

consider if the diagnosis can be established based on imaging alone.

Many studies have been able to demonstrate that as disc degeneration advances so too does the potential for discogenic pain [101, 102]. In a population of symptomatic patients with axial pain considered discogenic in nature, severe loss of disc height has a specificity of at least 97% and a PPV of 90%. (Fig. 16.25).

O'Neill and colleagues were able to demonstrate that changes in disc contour, specifically disc bulge, were associated with a + LR of 5.3 [103]. Analysis of available data performed by Maus & April in 2012 found that uniformly dark T2 signal with or without loss of disc height is a likewise a finding of high specificity (88–96%). Discs with severe T2 signal loss are rarely nonpainful. Endplate marrow changes were originally classified by Modic in 1988 [104]:

- Type 1 change represents ingrowth of vascularized granulation tissue into sub-endplate marrow. Type I Modic change exhibits hypointense T1 and hyperintense T2 signals on MR imaging.
- Type II change exhibits elevated T1 and T2 signals and reflects fatty infiltration of the sub-endplate marrow.
- Type III changes are hypointense on T1 and T2 and are likely representative of an area of bony sclerosis. Type III Modic changes are typically not associated with pain (B1).

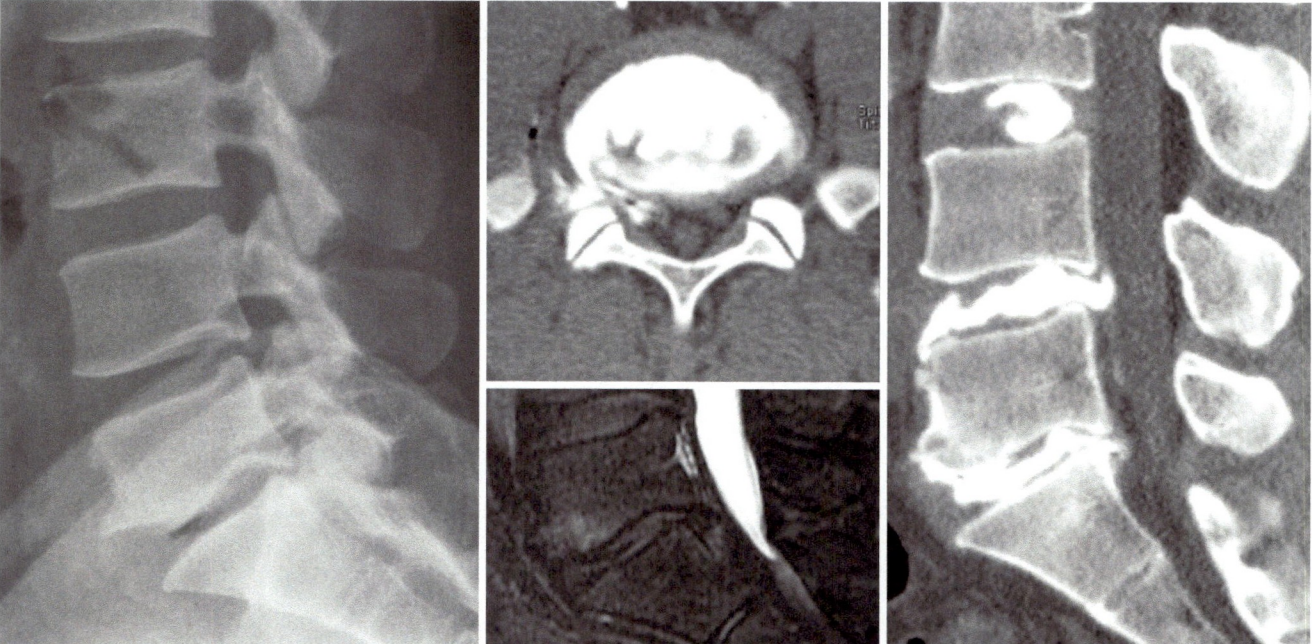

Fig. 16.25 Symptomatic patients who have severe loss of disc height and signal loss on MRI strongly correlate with a positive result on discography. L5 disc space narrowing, L4 nuclear signal loss in a patient with concordant pain at L4 and L5 on discography. (From: Maus and Aprill [106]; with permission from Elsevier)

Type I changes are felt to represent an active inflammatory state compared to type II or type III changes. Toyone and colleagues found that type I or type II Modic changes involving greater than 25% of the vertebral body strongly correlate with a positive result on provocation discography [105]. These findings were associated with high specificity, PPV, and LRs. (Fig. 16.26).

A high-intensity zone (HIZ) is believed to represent a complex grade 4 circumferential tear where nuclear material has been trapped within the annulus fibrosis [106]. The presence of an HIZ was found to have a sensitivity of 82% and a specificity of 89%. Additionally, an HIZ represents a LR of 7.3 [106] (Fig. 16.27).

These five structural changes correlate strongly with a positive result on discography. However, the presence of these features in the symptomatic patient population are rare and are generally felt to represent the advanced stages of internal disc disruption. The absence of these features does not preclude the disc as a potential source of pain. Additionally, some of these features may be seen in asymptomatic patients usually of advanced age [99]. On the opposite end of the spectrum, a normal disc on MRI is associated with high negative likelihood ratios [40, 106]. The normal discs on MRI are rarely painful. Although, a recent publication by Zucherman et al. demonstrated that a normal MRI can still surprise with a positive discography [107]. In syn-thesizing these data, it is reasonable to conclude that MRI is most helpful when characterizing the extremes of internal disc disruption. Intermediate MRI changes do not provide the clinician with definitive evidence for or against the possibility of discogenic sources of pain. Provocative discography continues to be the reference standard for the diagnosis of discogenic pain [106].

Uses of Discography in Regenerative Medicine

Our understanding of the cellular biology of the disc has advanced greatly over the last decade paralleled nicely by the development of new potential regenerative interventions. Despite these advancements, there are still gaps in our understanding of the pathogenesis of disc degeneration. The presumed aetiology of the degenerative process appears to be driven by changes in the behaviour of resident cells, which culminates in the loss of disc hydration, changes in the extracellular matrix, and ultimately changes in gross architecture and load-bearing potential [108]. The majority of this information has largely been extrapolated by examining discs taken at autopsy, removed during surgery, or from large animal studies [108, 109]. Here is a brief summary of the histopathologic changes that have been observed in the degenerative disc [108, 109]:

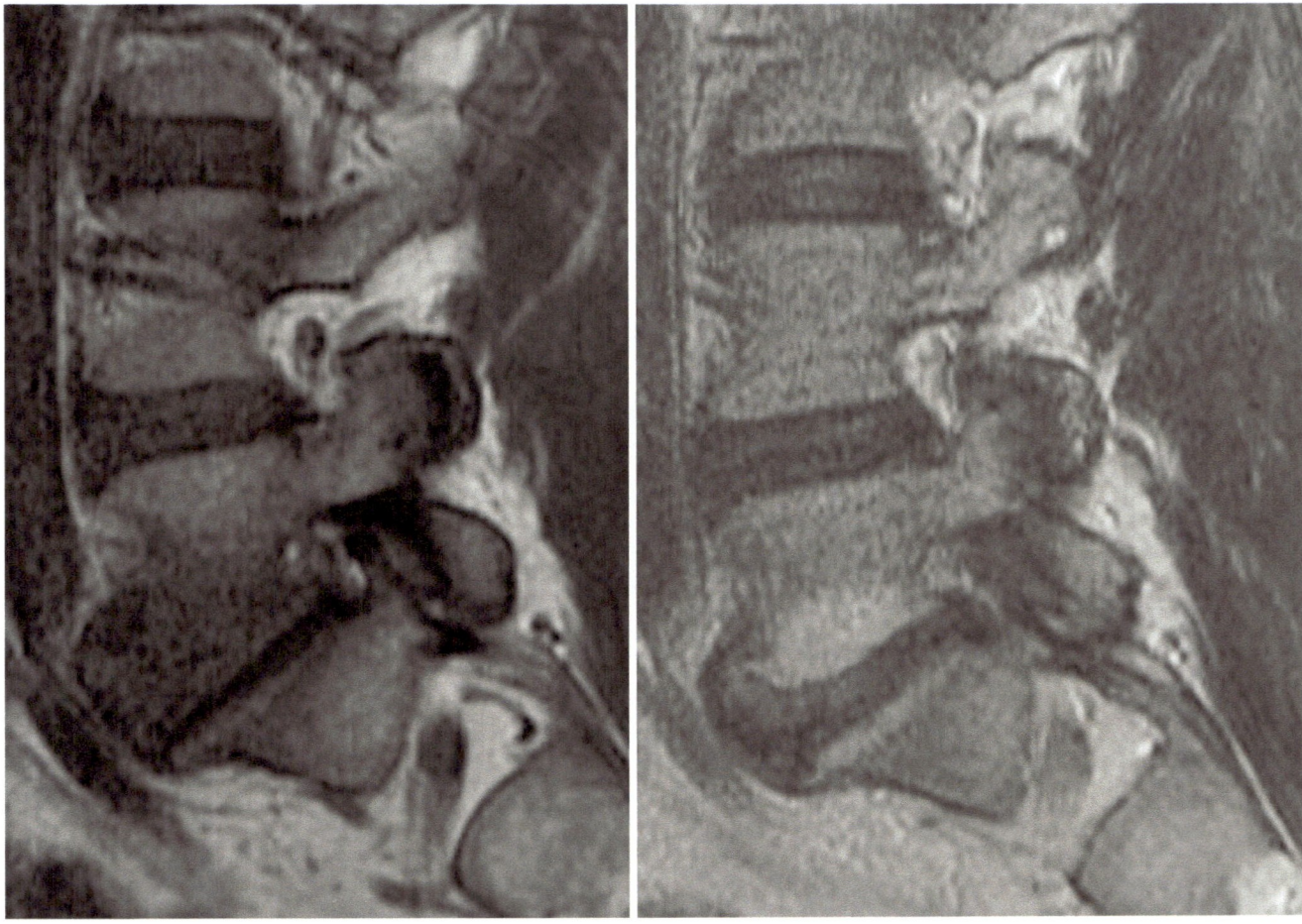

Fig. 16.26 L5 level demonstrating Type I Modic change involving >25% of vertical height of a vertebral body. (From: Maus [130]. Copyright International Spine Intervention Society 2015; used with permission)

- Markedly higher concentrations of proteases (i.e. aggrecanase and metalloprotease)—Macromolecular degradation outpaces the macromolecular synthesis [108]
- Decreased aggrecan (a large polyanionic proteoglycan with a high osmotic pressure [110])—with less of this proteoglycan, there is less osmotic potential. Disc desiccation ensues.
- Lamellar disorganization—the discs lose structural integrity and load-bearing potential [108]
- Cartilaginous endplates calcify—decreasing nutrient transport to cells [111]
- Angiogenesis and neurogenesis occur in response to cellular damage/stress—healthy discs are generally avascular aneural structures [108]
- Recruitment of inflammatory cells (i.e. macrophages) is amplified by angiogenesis—healthy discs are generally avascular and therefore are relatively nutrient poor. Higher concentrations of more metabolically active macrophages deplete native nutrient pools quickly. Glucose and pH decrease [108]

- Inflammatory cytokines at higher concentrations due to larger populations of inflammatory cells in the degenerated disc—this upregulates matrix degradation and exacerbates nutritional stresses [108]
- Decrease in viable and functional cell numbers, with large populations of senescent cells—calcified cartilaginous endplates greatly limit the recruitment of new cells. Specific chemokines, CCL5, and CXCL6 are upregulated and play a role in cellular recruitment [108] (Fig. 16.28)

Cell therapy in the form of mesenchymal stem cells (MSCs), bone marrow aspirate concentrate (BMAC), and platelet-rich plasma (PRP) have shown great potential to slow or even potentially reverse the degenerative process before major structural changes occur [112]. Svanvik et al. found that MSCs co-cultured with native IVD cells have the potential to differentiate towards chondrocyte-like cells that are phenotypically similar to those found within the NP of the disc [113]. These cells are capable of mobilizing endogenous populations of stem/progenitor cells, stimulating ana-

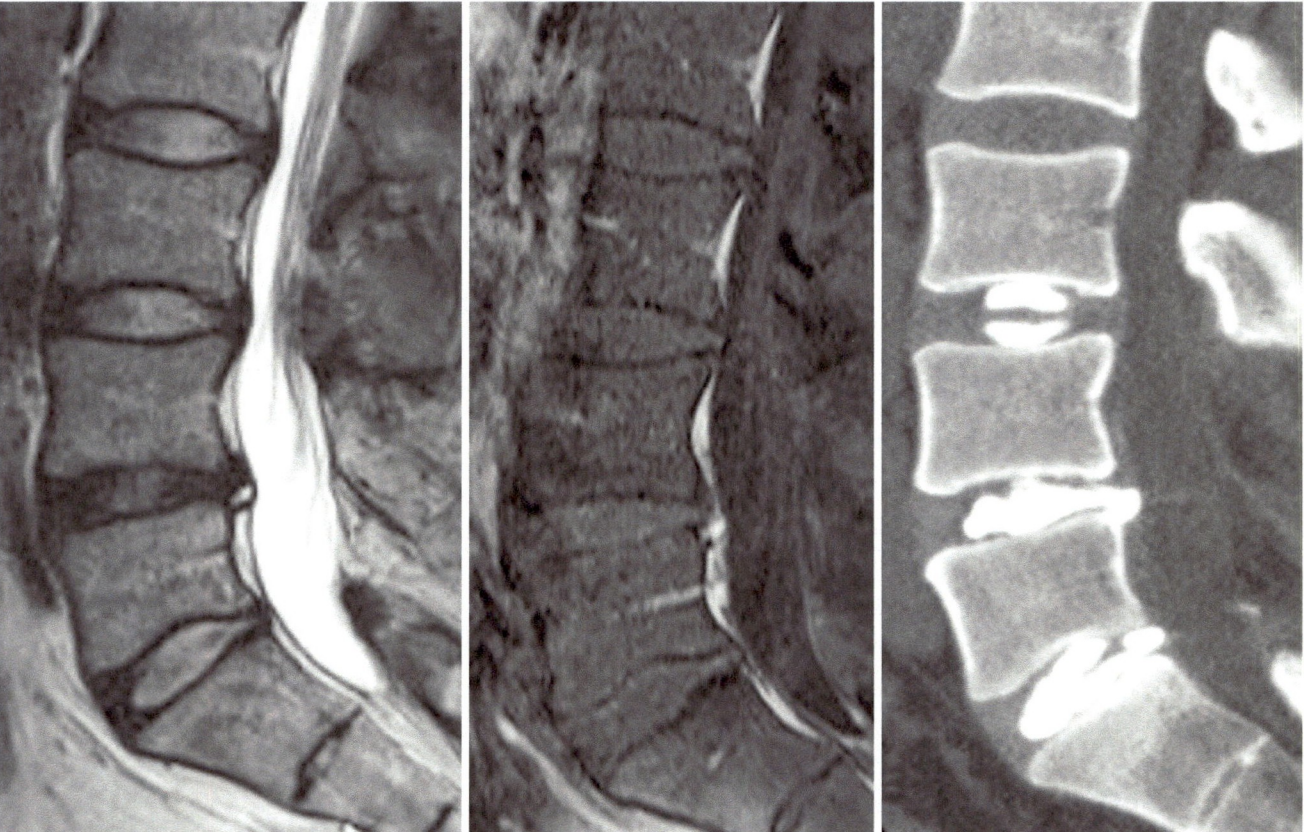

Fig. 16.27 L4 level demonstrating presence of a high-intensity zone (HIZ) and subsequent positive discogram at that level. (From: Maus and Aprill [106]; with permission from Elsevier)

bolic processes, and dampening inflammatory activity [114]. These observations highlight the potential therapeutic benefits of intradiscal MSCs in preventing and possibly even reversing the early steps of the degenerative cascade. Early data for cellular regenerative therapy using various animal models have been promising [115] (Fig. 16.29).

Likewise, positive outcomes have been achieved with human disc cells or mesenchymal stem cell transplantations into porcine models [113]. Mesenchymal stem cells seem to demonstrate some ability to interact with resident cell populations, regulate local homeostasis, and attract additional cells. Stem cells have demonstrated in vitro chondrogenic differentiation potential and may, therefore be, capable of stimulating new ECM; however, the regenerative potential seems to be limited to reversing or slowing earlier degenerative changes before structural remodelling can occur [113]. More research is needed to determine the precise point at which biologic therapy is likely to be of little use. Additional concerns regarding safety and efficacy remain, and much more data are required before definitive statements can be made.

Conclusion

Discography has become an indispensable tool in the evaluation of spinal pain [38, 116–118]. The differential diagnosis when evaluating patients with back pain is broad, and the disc is a common potential culprit. The clinical picture is further complicated by ubiquitous age-related degenerative changes [99]. These changes accumulate and do not necessarily implicate a specific source of pain. There are a handful of radiologic findings which strongly implicate discogenic pain; however, these are rare and may potentially be present in asymptomatic patients. Discography continues to be the reference standard for diagnosing discogenic pain. There are both historical and current controversies surrounding its use [17, 38, 42, 87, 88, 119]. This has generated healthy discussion and advanced the standards of this diagnostic procedure. The use of manometry, sham injection, and strict criteria for identifying positive discs are all intended to limit the likelihood of a false-positive result. It is acknowledged that discography has been interpreted with caution in patients with certain behavioural pathology.

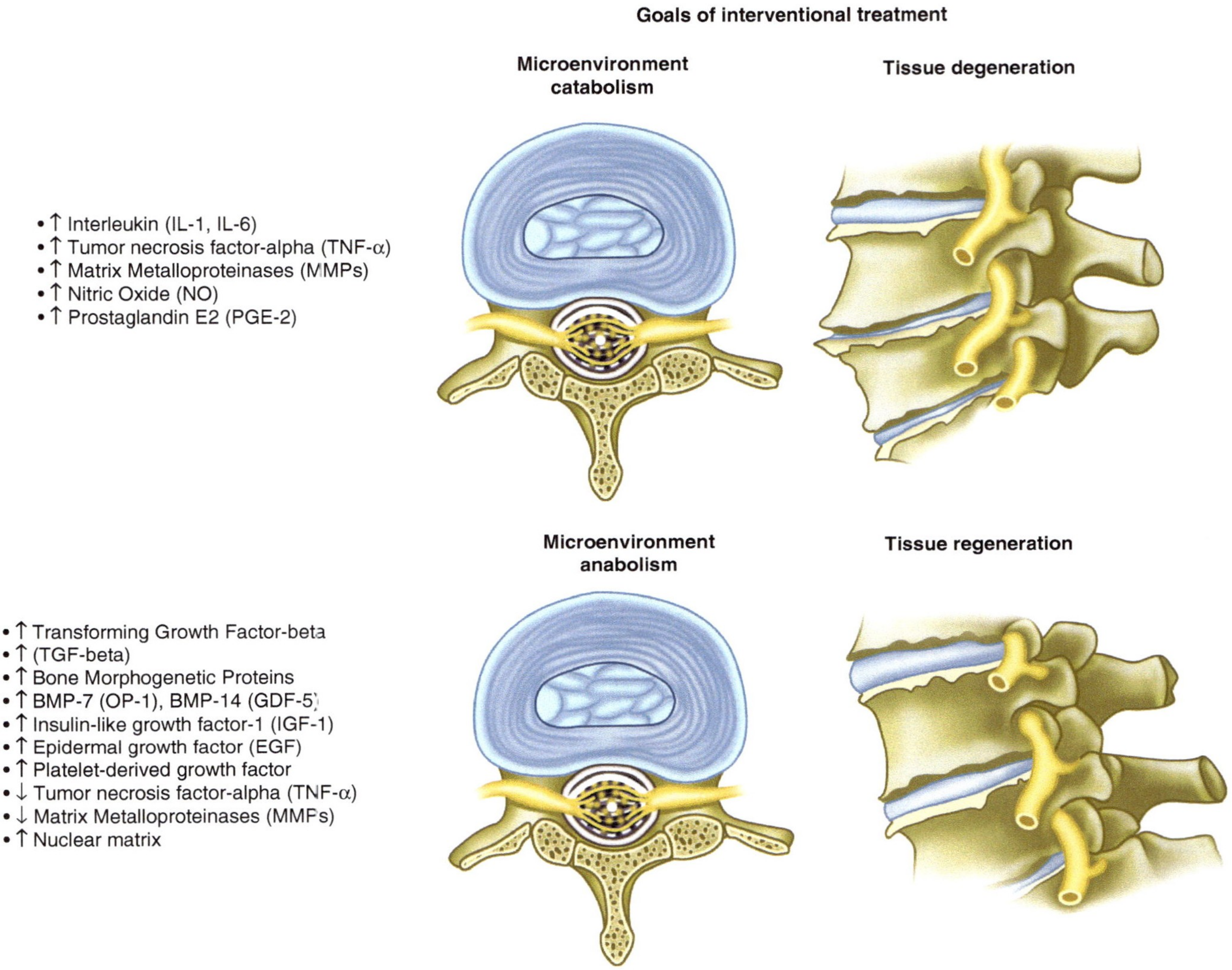

Goals of interventional treatment

**Microenvironment
catabolism**

Tissue degeneration

- ↑ Interleukin (IL-1, IL-6)
- ↑ Tumor necrosis factor-alpha (TNF-α)
- ↑ Matrix Metalloproteinases (MMPs)
- ↑ Nitric Oxide (NO)
- ↑ Prostaglandin E2 (PGE-2)

**Microenvironment
anabolism**

Tissue regeneration

- ↑ Transforming Growth Factor-beta
- ↑ (TGF-beta)
- ↑ Bone Morphogenetic Proteins
- ↑ BMP-7 (OP-1), BMP-14 (GDF-5)
- ↑ Insulin-like growth factor-1 (IGF-1)
- ↑ Epidermal growth factor (EGF)
- ↑ Platelet-derived growth factor
- ↓ Tumor necrosis factor-alpha (TNF-α)
- ↓ Matrix Metalloproteinases (MMFs)
- ↑ Nuclear matrix

Fig. 16.28 Figure 19.28 Goals of interventional treatment are to improve the microenvironment of the disc to allow for tissue regeneration. (**a**) Increasing inflammatory cytokine favours the development of a catabolic microenvironment. Goals of biologic therapy are to down regulate the production of inflammatory cytokines in order to decrease catabolic activity and (**b**) up regulate extracellular matrix proteins that increase anabolic activity

There is an expanding role for new, minimally invasive spinal interventions for the treatment of painful discogenic back pain. Emerging biologic therapies in the form of cellular replacement or cell-rich scaffolding offer potentially therapeutic options where previously there were none. These potential treatments demand a sensitive diagnostic test to select appropriate potential candidates. Discography is the current standard. However, new noninvasive diagnostic modalities are currently being developed. Magnetic resonance spectroscopy (MRS) is a noninvasive study being used to characterize in vivo metabolic features within tissues in several clinical [120]. Keshari and colleagues were able to demonstrate that certain disc chemicals specifically lactate and proteoglycan can provide spectroscopically quantifiable biomarkers for discogenic pain [120]. These biomarkers have well-documented features of the degenerative disc microenvironment. Early data suggest that MRS may be a highly specific screening modality for patients with discogenic sources of pain. MRS as it is used to work up discogenic pain is in the early stages of development, and it may yet be many years before this technology experiences widespread clinical use.

It is the authors' contention that discography when used in conjunction with radiographic imaging is the preferred method to evaluate the lumbar disc as a potential source of axial back pain. Discography can help clarify the clinical picture, identify which patients may benefit from novel regenerative techniques, and guide surgical intervention.

When using a diagnostic test to select patients for treatment, the accuracy of the test is important [121–124]. A rela-

Fig. 16.29 Catabolic environment of the painful degenerative disc is associated with increased levels of pro-inflammatory mediators. Regenerative biologics may help restore a healthy, anabolic phenotype. (From: Richardson et al. [131]; with permission of Elsevier)

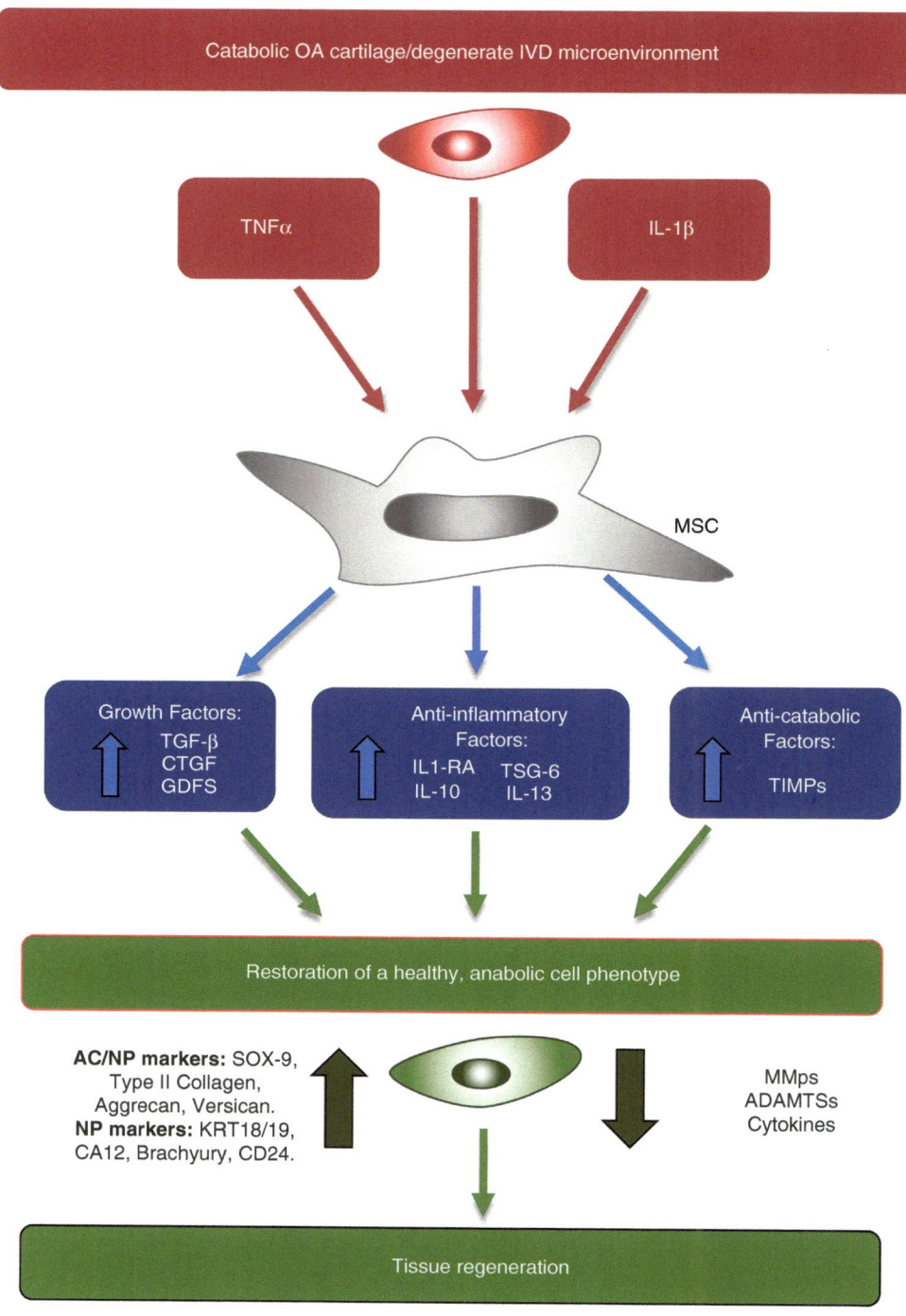

tively inexpensive and low-risk procedure such as an intradiscal biologic favours the use of a highly sensitive screening test. If the diagnosis is missed, the patient may be subject to more costly and invasive treatments such as surgery. A test with high specificity and positive predictive value is preferred for more costly and risky intervention including spine fusion. This would reduce unnecessary exposure to risk for the patient.

There are potential limitations associated with any interventional procedure, discography notwithstanding. A posi-

tive level on discography does not rule out the presence of other sources of pain; it does not prove the clinical significance of the pain, nor guarantee interventional or surgical outcome. A negative discogram effectively rules the disc out as a pain generator. It has diagnostic utility and negative predictive value. It acts as a barrier to excessive surgery and disc-related intervention and provides closure.

The lumbar intervertebral disc is a common cause of chronic lower back pain. Discography can accurately identify appropriate candidates for current and future intradiscal or subchondral therapies [125]. The skill set for performing discography includes disc access, which is needed for any intradiscal injection procedure.

References

1. Manchikanti L, Singh V, Datta S, Cohen SP, Hirsch JA. Comprehensive review of epidemiology, scope, and impact of spinal pain evidence-based medicine. Pain Physician. 2009;12:35–70.
2. National Center for Health Statistics. Health, United States, 2011: with special feature on socioeconomic status and health. Hyattsville; 2012.
3. Dahlhamer J, Lucas J, Zelaya C, Nahin R, Mackey S, DeBar L, et al. Prevalence of chronic pain and high-impact chronic pain among adults — United States, 2016. MMWR Morb Mortal Wkly Rep. 2018;67(36):1001–6.
4. Ma VY, Chan L, Carruthers KJ. Incidence, prevalence, costs, and impact on disability of common conditions requiring rehabilitation in the United States: stroke, spinal cord injury, traumatic brain injury, multiple sclerosis, osteoarthritis, rheumatoid arthritis, limb loss, and back pain. Arch Phys Med Rehabil. 2014;95(5):986–95.
5. Gou HR, Tanaka S, Halperin WE, Cameron LL. Back pain prevalence in US industry and estimates of lost workdays. Am J Public Health. 1999;89(7):1029–35.
6. Verrills P, Nowesenitz G, Barnard A. Prevalence and characteristics of discogenic pain in tertiary practice: 223 consecutive cases utilizing lumbar discography. Pain Med. 2015;16(8):1490–9.
7. Skulpoonkitti B, Day M. Sphenopalatine ganglion blocks. In: Manchikanti L, Kaye AD, FJE F, Hirsch JA, editors. Essentials of interventional techniques in managing chronic pain. 1st ed. Cham: Springer International Publishing; 2018. p. 519–29.
8. Manchikanti L, Benyamin RM, Singh V, Falco FJE, Hameed H, Derby R, et al. An update of the systematic appraisal of the accuracy and utility of lumbar discography in chronic low back pain. Pain Physician. 2013;16(2 Suppl):SE55–95.
9. Mccormick ZL, Defrancesch F, Loomba V, Moradian M, Bathina R, Rappard G. Diagnostic value, prognostic value, and safety of provocation discography. Pain Med. 2018;19:3–8.
10. Calodney A, Griffith D. Discography. In: Mathis JGS, editor. Image-guided spine interventions. New York: Springer; 2010. p. 107–46.
11. Landers MH. Intervertebral disk stimulation. Provocation diskography. In: Pimheiro-Franco JL, Vaccaro AR, Benzel EC, Mayer HM, editors. Advanced concepts in lumbar degenerative disk disease. 1st ed. Philadelphia: Elsevier Inc.; 2011. p. 117–38.
12. Derby R, Lee S-H, Kim B-J, Chen Y, Aprill C, Bogduk N. Pressure-controlled lumbar discography in volunteers without low back symptoms. Pain Med. 2005;6(3):213–21.
13. Lindblom K. Technique and results of diagnostic disc puncture and injection (discography) in the lumbar region. Acta Orthop Scand. 1951;20(4):315–26.
14. Schellhas KP, Pollei S, Gundry C, Heithoff K. Lumbar disc high-intensity zone: correlation of magnetic resonance imaging and discography. Spine (Phila Pa 1976). 1996;21(1):79–86.
15. Bogduk N, Aprill C, Derby R. Discography. In: White A, Schofferman J, editors. Spine care diagnosis and conservative treatment. St. Louis: Mosby; 1995. p. 219–36.
16. KOZAK JA, O'BRIEN JP. Simultaneous combined anterior and posterior fusion, an independent analysis of a treatment for the disabled low-back pain patient. Spine (Phila Pa 1976). 1990;15(4):322–8.
17. Derby R, Howard MW, Grant JM, Lettice JJ, Van Peteghem PK, Ryan DP. The ability of pressure-controlled discography to predict surgical and nonsurgical outcomes. Spine (Phila Pa 1976). 1999;24(4):364–71; discussion 371–2
18. BLUMENTHAL SL, BAKER J, DOSSETT A, SELBY DK. The role of anterior lumbar fusion for internal disc disruption. Spine (Phila Pa 1976). 1988;13(5):566–9.
19. Newman MH, Grinstead GL. Anterior lumbar interbody fusion for internal disc disruption. Spine (Phila Pa 1976). 1992;17(7):831–3.
20. Colhoun E, McCall IW, Williams L, Cassar Pullicino VN. Provocation discography as a guide to planning operations on the spine. J Bone Joint Surg Br. 1988;70(2):267–71.
21. Cohen SP, Hurley RW. The ability of diagnostic spinal injections to predict surgical outcomes. Anesth Analg. 2007;105(6):1756–75, table of contents.
22. Lam KS, Carlin D, Mulholland RC. Lumbar disc high-intensity zone: the value and significance of provocative discography in the determination of the discogenic pain source. Eur Spine J. 2000;9(1):36–41.
23. Saifuddin A, Braithwaite I, White J, Taylor BA, Renton P. The value of lumbar spine magnetic resonance imaging in the demonstration of annular tears. Spine (Phila Pa 1976). 1998;23(4):453–7.
24. Ito M, Incorvaia KM, Yu SF, Fredrickson BE, Yuan HA, Rosenbaum AE. Predictive signs of discogenic lumbar pain on magnetic resonance imaging with discography correlation. Spine (Phila Pa 1976). 1998;23(11):1252–8.
25. Jensen MC, Brant-Zawadzki MN, Obuchowski N, Modic MT, Malkasian D, Ross JS. Magnetic resonance imaging of the lumbar spine in people without back pain. N Engl J Med. 1994;331(2):69–73.
26. Modic MT, Obuchowski NA, Ross JS, Brant-Zawadzki MN, Grooff PN, Mazanec DJ, et al. Acute low back pain and radiculopathy: MR imaging findings and their prognostic role and effect on outcome. Radiology. 2005;237(2):597–604.
27. Borenstein DG, O'Mara JW, Boden SD, Lauerman WC, Jacobson A, Platenberg C, et al. The value of magnetic resonance imaging of the lumbar spine to predict low-back pain in asymptomatic subjects. J Bone Jt Surg-Am Vol. 2001;83(9):1306–11.
28. Manchikanti L, Kaye AD, Falco FJE, Hirsch JA. Discography. In: Manchikanti L, Kaye AD, Falco FJE, Hirsch JA, editors. Essentials of interventional techniques in managing chronic pain. 1st ed. Cham: Springer International Publishing; 2018. p. 273–300.
29. Roofe PG. Innervation of annulus fibrosus and posterior longitudinal ligament. Arch Neurol Psychiatr. 1940;44(1):100.
30. Cloward RB. The anterior approach for removal of ruptured cervical disks. J Neurosurg. 1958;15(6):602–17.
31. Walker J, El Abd O, Isaac Z, Muzin S. Discography in practice: a clinical and historical review. Curr Rev Musculoskelet Med. 2008;1:69–83.
32. Lindblom K. Diagnostic puncture of intervertebral disks in sciatica. Acta Orthop. 1948;17(1–4):231–9.
33. Wise RE, Weiford EC. X-ray visualization of the intervertebral disk; report of a case. Cleve Clin Q. 1951;18(2):127–30.
34. Holt EP. Fallacy of cervical discography. Report of 50 cases in normal subjects. JAMA. 1964;188:799–801.
35. Schellhas KP, Smith MD, Gundry CR, Pollei SR. Cervical discogenic pain: prospective correlation of magnetic resonance imaging and discography in asymptomatic subjects and pain sufferers. Cerv Spine. 1996;21:300–11.
36. HOLT EP. The question of lumbar discography. J Bone Jt Surg. 1968;50(4):720–6.
37. Simmons J, Aprill C, Dwyer A, Brodsky A. A reassessment of Holt's data on: the question of lumbar discography. – PubMed – NCBI. Clin Orthop Relat Res. 1988;237:120–4.

38. Walsh TR, Weinstein JN, Spratt KF, Lehmann TR, Aprill C, Sayre H. Lumbar discography in normal subjects. A controlled, prospective study. J Bone Joint Surg Am. 1990;72(7):1081–8.

39. Cuellar JM, Stauff MP, Herzog RJ, Carrino JA, Baker GA, Carragee EJ. Does provocative discography cause clinically important injury to the lumbar intervertebral disc? A 10-year matched cohort study. Spine J. 2016;16:273–80.

40. Bogduk N, Aprill C, Derby R. Lumbar discogenic pain: state-of-the-art review. Pain Med. 2013;14:813–36.

41. Carragee EJ, Don AS, Hurwitz EL, Cuellar JM, Carrino JA, Carrino J, et al. 2009 ISSLS prize winner: does discography cause accelerated progression of degeneration changes in the lumbar disc: a 10 year matched cohort study. Spine (Phila Pa 1976). 2009;34(21):2338–45.

42. Wolfer L, Wolfer LR, Derby R, Lee J-E, Lee S-H. Systematic review systematic review of lumbar provocation discography in asymptomatic subjects with a meta-analysis of false-positive rates. Pain Physician. 2008;11(4):513–38.

43. Carragee EJ, Lincoln T, Parmar VS, Alamin T. A gold standard evaluation of the discogenic pain diagnosis as determined by provocative discography. Spine (Phila Pa 1976). 2006;31(18):2115–23.

44. Derby R, Kim B-J, Chen Y, Seo K-S, Lee S-H. The relation between annular disruption on computed tomography scan and pressure-controlled diskography. Arch Phys Med Rehabil. 2005;86(8):1534–8.

45. Carragee EJ, Alamin TF. Discography. A review. Spine J. 2001;1(5):364–72.

46. Massie W. A critical evaluation of discography. J Bone Jt Surg. 1967;49:1243–4.

47. Derby R, Kine G, Schwarzer A. Pain provocation in normal nucleograms during discography. Prevalence and relationship of intradiscal pressure. Sci Newsl Int Spine Inject Soc Newsl. 1993;1:8–17.

48. Derby R, Kim B-J, Lee S-H, Chen Y, Seo K-S, Aprill C. Comparison of discographic findings in asymptomatic subject discs and the negative discs of chronic LBP patients: can discography distinguish asymptomatic discs among morphologically abnormal discs? Spine J. 2005;5(4):389–94.

49. Bogduk N. In: Bogduk N, editor. Practice guidelines for spinal diagnostic and treatment procedures. 2nd ed. International Spine Intervention Society: San Francisco; 2013. 685 p.

50. Manchikanti L, Glaser SE, Wolfer L, Derby R, Cohen SP. Systematic review of lumbar discography as a diagnostic test for chronic low back pain. Pain Physician. 2009;12:541–59.

51. Manchikanti L, Soin A, Benyamin RM, Singh V, Falco FJ, Calodney AK, et al. An update of the systematic appraisal of the accuracy and utility of discography in chronic spinal pain. Pain Physician. 2018;21:91–110.

52. Adams MA, Roughley PJ. What is intervertebral disc degeneration, and what causes it? Spine (Phila Pa 1976). 31(18):2151–61.

53. Raj PP. Intervertebral disc: anatomy, physiology, pathophysiology, treatment. Pain Pract. 2008;8(1):18–44.

54. Pennicooke B, Moriguchi Y, Hussain I, Bonssar L, Härtl R. Biological treatment approaches for degenerative disc disease: a review of clinical trials and future directions. Cureus. 2016;8(11):e892.

55. Sivan SS, Hayes AJ, Wachtel E, Caterson B, Merkher Y, Maroudas A, et al. Biochemical composition and turnover of the extracellular matrix of the normal and degenerate intervertebral disc. Eur Spine J. 2014;23(Suppl 3):S344–53.

56. Kim S-M, Lee S-H, Lee B-R, Hwang J-W. Analysis of the correlation among age, disc morphology, positive discography and prognosis in patients with chronic low Back pain. Ann Rehabil Med. 2015;39(3):340–6.

57. Deer TR, Mekhail N, Provenzano D, Pope J, Krames E, Leong M, et al. The appropriate use of neurostimulation of the spinal cord and peripheral nervous system for the treatment of chronic pain and ischemic diseases: the neuromodulation appropriateness consensus committee. Neuromodulation. 2014;17(6):515–50; discussion 550

58. Edgar MA. The nerve supply of the lumbar intervertebral disc. J Bone Jt Surg. 2007;89-B(9):1135–9.

59. Freemont AJ, Peacock TE, Goupille P, Hoyland JA, O'Brien J, Jayson MIV. Nerve ingrowth into diseased intervertebral disc in chronic back pain. Lancet. 1997;350:178–81.

60. Vanharanta H, Sachs BL, Spivey MA, Guyer RD, Hochschuler SH, Rashbaum RF, Johnson RG, Ohnmeiss DMV. The relationship of pain provocation to lumbar disc deterioration as seen by CT discography. Spine (Phila Pa 1976). 1987;12(3):295–8.

61. Burke J, Watson R, McCormack D, Dowling F, Walsh M, Fitzpatrick J. Intervertebral discs which cause low back pain secrete high levels of proinflammatory mediators. J Bone Jt Surg [Br]. 2002;84:196–201.

62. Peng B, Wu W, Hou S, Li P, Zhang C, Yang Y. The pathogenesis of discogenic low back pain. J Bone Joint Surg Br. 2005;87-B(1):62–7.

63. Coppes MH, Marani E, Thomeer RTGG. Innervation of painful lumbar discs. Spine (Phila Pa 1976). 1997;22(20):2342–9.

64. Ivie CS, Gianoli D, Pino CA. Provocative discography as predictor of discogenic pain and therapeutic outcome. Tech Reg Anesth Pain Manag. 2011;15:12–9.

65. Guyer RD, Ohnmeiss DD, Guyer RD, Ohnmeiss DD. Lumbar discography. Spine J. 2003;3:11S–27S.

66. Resnick D, Malone D, Ryken T. Guidelines for the use of discography for the diagnosis of painful degenerative lumbar disc disease. Neurosurg Focus. 2002;13(2):E12.

67. Fraser RD, Osti OL, Vernon-Roberts B. Discitis after discography. J Bone Joint Surg Br. 1987;69(1):26–35.

68. Osti OL, Vernon-Roberts B, Fraser RD. Volvo Award in experimental studies. Annulus tears and intervertebral disc degeneration. An experimental study using an animal model. Spine (Phila Pa 1976). 1990;15(8):762–7.

69. Pobiel RS, Schellhas KP, Pollei SR, Johnson BA, Golden MJ, Eklund JA. Diskography: infectious complications from a series of 12,634 cases. Am J Neuroradiol. 2006;27:1930–2.

70. Arrington JA, Murtagh FR, Silbiger ML, Rechtine GR, Nokes SR. Magnetic resonance imaging of postdiscogram discitis and osteomyelitis in the lumbar spine: case report. J Fla Med Assoc. 1986;73(3):192–4.

71. Guyer RD, Collier R, Stith WJ, Ohnmeiss DD, Hochschuler SH, Rashbaum RF, et al. Discitis after discography. Spine (Phila Pa 1976). 1988;13(12):1352–4.

72. Guyer RD, Ohnmeiss DD. Lumbar discography. Position statement from the North American Spine Society Diagnostic and Therapeutic Committee. Spine (Phila Pa 1976). 1995;20(18):2048–59.

73. Klessig HT, Showsh SA, Sekorski A. The use of Intradiscal antibiotics for discography: an in vitro study of gentamicin, cefazolin, and clindamycin. Spine (Phila Pa 1976). 2003;28(15):1735–8.

74. Grogan J, Hemminghytt S, Williams A, et al. Another strategy recommended by Balderstone et al. J Spinal Disord Tech. 2004;17:248–50.

75. Kim D, Wadley R. Variability in techniques and patient safety protocols in discography. J Spinal Disord Tech. 2010;23(6):431–8.

76. Schellhas K, Pollei S, Dorwart R. Thoracic discography. A safe and reliable technique. Spine (Phila Pa 1976). 1994;19(18):2103–9.

77. Sharma SK, Jones JO, Zeballos PP, Irwin SA, Martin TW. The prevention of discitis during discography. PMID:19643677. https://doi.org/10.1016/j.spinee.2009.06.001.

78. Pauza K. Nomenclature and terminology for spine specialists (appropriate words meant to replace the most commonly misused words of the spine specialists). PASSOR Educational Guidelines Task Force. 2005;

79. Khot A, Bowditch M, Powell J, Sharp D. The use of intradiscal steroid therapy for lumbar spinal discogenic pain. Spine (Phila Pa 1976). 2004;29(8):833–6.

80. Derby R, Lee SHH, Lee JE, Lee SHH. Comparison of pressure-controlled provocation discography using automated versus manual syringe pump manometry in patients with chronic low back pain. Pain Med. 2011;12:18–26.

81. Zhou Y, Abdi S. A review of the literature. Clin J Pain. 2006;22(5):468–81.

82. Bogduk N. Lumbar disc stimulation (provocation discography). In: Bogduk N, editor. Practice guidelines for spinal diagnostic and treatment procedures. ISIS; 2004. p. 20–46.

83. Cavanaugh JM, Kallakuri S, Özaktay AC. Innervation of the rabbit lumbar intervertebral disc and posterior longitudinal ligament. Spine (Phila Pa 1976). 1995;20(19):2080–5.

84. O'Neill C, Kurgansky M. Subgroups of positive discs on discography. Spine (Phila Pa 1976). 2004;29(19):2134–9.

85. Ohnmeiss DD, Guyer RD, Mason SL. The relation between cervical discographic pain responses and radiographic images. Clin J Pain. 2000;16(1):1–5.

86. Carragee EJ, Chen Y, Tanner CM, Hayward C, Rossi M, Hagle C. Can discography cause long-term Back symptoms in previously asymptomatic subjects? Spine (Phila Pa 1976). 2000;25(14):1803–8.

87. Carragee EJ, Alamin TF, Miller J, Grafe M. Provocative discography in volunteer subjects with mild persistent low back pain. Spine J. 2002;2(1):25–34.

88. Carragee EJ, Tanner CM, Yang B, Brito JL, Truong T. False-positive findings on lumbar discography. Reliability of subjective concordance assessment during provocative disc injection. Spine (Phila Pa 1976). 1999;24(23):2542–7.

89. Wood KB, Schellhas KP, Garvey TA, Aeppli D. Thoracic discography in healthy individuals. A controlled prospective study of magnetic resonance imaging and discography in asymptomatic and symptomatic individuals. Spine (Phila Pa 1976). 1999;24(15, 1548):–55.

90. Veres SP, Robertson PA, Broom ND. ISSLS prize winner: how loading rate influences disc failure mechanics: a microstructural assessment of internal disruption. Spine (Phila Pa 1976). 2010;35(21):1897–908.

91. Ohtori S, Inoue G, Orita S, Eguchi Y, Ochiai N, Kishida S, et al. No acceleration of intervertebral disc degeneration after a single injection of bupivacaine in young age group with follow-up of 5 years. Asian Spine J. 2013;7(3):212–7.

92. Flanagan MN, Chung BU. Roentgenographic changes in 188 patients 10–20 years after discography and chemonucleolysis. Spine (Phila Pa 1976). 1986;11(5):444–8.

93. McCormick ZL, Lehman VT, Plastaras CT, Walega DR, Huddleston P, Moussallem C, et al. Low-pressure lumbar provocation discography according to spine intervention society/international association for the study of pain standards does not cause acceleration of disc degeneration in patients with symptomatic low back pain: a 7 year matched cohort study. Spine (Phila Pa 1976). 2019;1

94. Elliott DM, Yerramalli CS, Beckstein JC, Boxberger JI, Johannessen W, Vresilovic EJ. The effect of relative needle diameter in puncture and sham injection animal models of degeneration. Spine (Phila Pa 1976). 2008;33(6):588–96.

95. Peh WCG. Provocative discography: current status. Biomed Imaging Interv J. 2005;1(1):e2.

96. Zeidman S, Thompson K, Ducker T. Complications of cervical discography analysis of 4400 diagnostic disc injections. Neurosurgery. 1995;37(3):414–7.

97. Kapoor SG, Huff J, Cohen SP. Systematic review of the incidence of discitis after cervical discography. Spine J. 2010;10(8):739–45.

98. Willems PC, Jacobs W, Duinkerke ES, De Kleuver M. Lumbar discography: should we use prophylactic antibiotics? A study of 435 consecutive discograms and a systematic review of the literature. J Spinal Disord Tech. 2004;17(3):243–7.

99. Bogduk N. Degenerative joint disease of the spine. Radiol Clin N Am. 2012;50(4):613–28.

100. Hansson TH, Keller TS, Spengler DM. Mechanical behavior of the human lumbar spine. II. Fatigue strength during dynamic compressive loading. J Orthop Res. 1987;5(4):479–87.

101. Videman T, Nummi P, Battié MC, Gill K. Digital assessment of MRI for lumbar disc desiccation\a comparison of digital versus subjective assessments and digital intensity profiles versus discogram and macroanatomic findings. Spine (Phila Pa 1976). 1994;19(Supplement):192–8.

102. Lei D, Rege A, Koti M, Smith FW, Wardlaw D. Painful disc lesion: can modern biplanar magnetic resonance imaging replace discography? J Spinal Disord Tech. 2008;21(6):430–5.

103. O'Neill C, Kurgansky M, Kaiser J, Lau W. Accuracy of MRI for diagnosis of discogenic pain. Pain Phys. 11(3):311–26.

104. Modic MT, Steinberg PM, Ross JS, Masaryk TJ, Carter JR. Degenerative disk disease: assessment of changes in vertebral body marrow with MR imaging. Radiology. 1988;166(1):193–9.

105. Toyone T, Takahashi K, Kitahara H, Yamagata M, Murakami M, Moriya H. Vertebral bone marrow changes in degenerative lumbar disc disease. An MRI study of 74 patients with low back pain. J Bone Joint Surg Br. 1994;76(5):757–64.

106. Maus TP, Aprill CN. Lumbar diskogenic pain, provocation diskography, and imaging correlates. Radiol Clin North Am. 2012:681–704.

107. Zucherman J, Derby R, Hsu K, Picetti G, Kaiser J, Schofferman J, et al. Normal magnetic resonance imaging with abnormal discography. Spine (Phila Pa 1976). 1988;13(12):1355–9.

108. Bendtsen M, Bunger C, Colombier P, Le Visage C, Roberts S, Sakai D, et al. Biological challenges for regeneration of the degenerated disc using cellular therapies. Acta Orthop. 2016.

109. Masuda K, Lotz JC. New challenges for intervertebral disc treatment using regenerative medicine. Tissue Eng Part B Rev. 2010;16(1):147–58.

110. Sivan S, Merkher Y, Wachtel E, Ehrlich S, Maroudas A. Correlation of swelling pressure and intrafibrillar water in young and aged human intervertebral discs. J Orthop Res. 2006;24(6):1292–8.

111. Kletsas D. Senescent cells in the intervertebral disc: numbers and mechanisms. Spine J. 2009;9(8):677–8.

112. Yoshikawa T, Ueda Y, Miyazaki K, Koizumi M, Takakura Y. Disc regeneration therapy using marrow mesenchymal cell transplantation. Spine (Phila Pa 1976). 2010;35(11):E475–80.

113. Svanvik T, Henriksson HB, Karlsson C, Hagman M, Lindahl A, Brisby H. Human disk cells from degenerated disks and mesenchymal stem cells in co-culture result in increased matrix production. Cells Tissues Organs. 2010;191(1):2–11.

114. Hohaus C, Ganey TM, Minkus Y, Meisel HJ. Cell transplantation in lumbar spine disc degeneration disease. Eur Spine J. 2008;17(S4):492–503.

115. Wang Z, Perez-Terzic CM, Smith J, Mauck WD, Shelerud RA, Maus TP, et al. Efficacy of intervertebral disc regeneration with stem cells – a systematic review and meta-analysis of animal controlled trials. Gene. 2015;564(1):1–8.

116. Buirski G, Silberstein M. The symptomatic Lumbar disc in patients with low-back pain. Spine (Phila Pa 1976). 1993;18(13):1808–11.

117. Milette PC, Fontaine S, Lepanto L, Cardinal É, Breton G. Differentiating lumbar disc protrusions, disc bulges, and discs with normal contour but abnormal signal intensity. Spine (Phila Pa 1976). 1999;24(1):44–53.

118. Manchikanti L, Singh V, Pampati V, Fellows B, Beyer C, Damron K, et al. Provocative discography in low back pain patients with or

without somatization disorder: a randomized prospective evaluation. Pain Physician. 2001;4(3):227–39.

119. Carragee EJ, Tanner CM, Khurana S, Hayward C, Welsh J, Date E, et al. The rates of false-positive lumbar discography in select patients without low back symptoms. Spine (Phila Pa 1976). 2000;25(11):1373–80; discussion 1381

120. Gornet MG, Peacock J, Claude J, Schranck FW, Copay AG, Eastlack RK, et al. Magnetic resonance spectroscopy (MRS) can identify painful lumbar discs and may facilitate improved clinical outcomes of lumbar surgeries for discogenic pain. Eur Spine J. 2019;28(4):674–87.

121. Manchikanti L, Boswell MV, Singh V, Benyamin RM, Fellows B, Abdi S, et al. Comprehensive evidence-based guidelines for interventional techniques in the management of chronic spinal pain. Pain Phys. 12(4):699–802.

122. Falco FJE, Manchikanti L, Datta S, Sehgal N, Geffert S, Onyewu O, et al. An update of the systematic assessment of the diagnostic accuracy of lumbar facet joint nerve blocks. Pain Phys. 15(6):E869–907.

123. Simopoulos TT, Manchikanti L, Singh V, Gupta S, Hameed H, Diwan S, et al. A systematic evaluation of prevalence and diagnostic accuracy of sacroiliac joint interventions. Pain Phys. 15(3):E305–44.

124. Hancock MJ, Maher CG, Latimer J, Spindler MF, McAuley JH, Laslett M, et al. Systematic review of tests to identify the disc, SIJ or facet joint as the source of low back pain. Eur Spine J. 2007;16(10):1539–50.

125. Lorio M, Clerk-Lamalice O, Beall DP, Julien T. International society for the advancement of spine surgery guideline—intraosseous ablation of the basivertebral nerve for the relief of chronic low back pain. Int J Spine Surg. 2020;

126. Melnik I, Derby R, Baker RM. Provocative discography. In: Deer TR, Leong MS, editors. Comprehensive treatment of chronic pain by medical, interventional, and integrative approaches. New York: Springer Nature; 2013. p. 461–77.

127. Calodney A, Griffin D. Discography. In: Mathis JM, Golovac S, editors. Image guided spine interventions. New York: In, Springer Science+Business Media; 2010.

128. Bogduk N. Practice guidelines for spinal diagnostic and treatment procedures. 2nd ed. San Francisco: International Spine Intervention Society; 2013.

129. Bugden N. Degenerative joint disease of the spine. Pain Med. 2013;14(6):813–36.

130. Maus TP. SIS presentation on modic changes. Int Spine Intervent Soc. 2015.

131. Richardson SM, Kalamegam G, Pushparaj PN, et al. Mesenchymal stem cells in regenerative medicine: focus on articular cartilage and intervertebral disc regeneration. Methods. 2016;99:69–80.

Ultrasound-Guided Injections: Preprocedure Planning

17

Steve M. Aydin and George C. Chang Chien

Introduction

Ultrasound has become more popular in musculoskeletal medicine; its use for regenerative techniques is easy to incorporate. Ultrasound is beneficial in both diagnostic purposes and image-guidance for procedures. Its use has the ability to demonstrate different tissue types such as ligament, tendon, bone, and muscle, as well as nerve and vascular tissues [1]. The demonstration of pathology, such as swelling, effusion, tears, or inflammation, can be seen with experienced ultrasound sonography. Identifying the pathologic tissues is often difficult with physical examination and even advanced imaging such as MRI. Millimeter scale resolution combined with dynamic testing is critical advantages and is of special use in the evaluation of pain for regenerative techniques. We will discuss in this chapter the thought process equipment preparation for ultrasound-guided regenerative medicine procedures (USGRMP) [2].

Equipment

When planning on any procedure, the risk and benefit of the treatment should always be considered and discussed with the patient. After informed consent and the risk and benefits have been discussed with the patient, one will need to prepare to do the injection. A thorough understanding of the functions of your ultrasound machine and how to optimize the image will allow for ease of use.

Preparing the equipment for the proper image and identification of the pathology of which is being addressed is paramount. What is most important is maximizing and visualizing the area of interest. Having a formal understanding of different frequencies, the proper depths, different probes are important. Based on the type of tissue, location, depth, and type of injury, different settings on the machine can be utilized.

Proper ergonomics will simplify procedure process. This will include proximity of the machine, screen placed within appropriate line of sight, and easy access to controls for maximizing proper imaging. This should also be done in a fashion in which the patient can be protected from cross-contamination and/or infection. A "Patient sandwich" wherein the patient is positioned in between the proceduralist and the ultrasound machine is a convenient way to optimize ergonomics and line of sight for the physician performing the procedure, but will likely require an additional set of hands in the room to operate the machine. If additional help is unavailable, an alternative to the "patient sandwich" is the "triangle," wherein the machine is within reach of the physician, and the physician, patient, and ultrasound machine make up the 3 points of the triangle. To maintain sterility, a foot pedal, voice-command, or control buttons on the ultrasound probe can be utilized to adjust the machine once sterile gloves are donned (Figs. 17.1 and 17.2).

Set up for the proper imaging will require an initial evaluation and screening with the ultrasound. The proper probe should be utilized, and image should be demonstrated before the technique or injection therapy is conducted. For example, an ultrasound evaluation of the joint, tendons, and ligaments should be done prior to the injection, to identify the area of pathology should be marked. Once this is demonstrated with the proper probe placement and location, injection therapy should be prepared so that limited needle trauma will be done and proper access with appropriate length and location of the needle or injection therapy can be done.

Preparing in this fashion will allow for ease of injection therapy and limit discomfort to the individual undergoing the treatment options. This will also limit the amount of

S. M. Aydin (✉)
Zucker School of Medicine at Hofstra University/Northwell Health, Department of Medicine and Rehabilitation, Manhasset, NY, USA

G. C. Chang Chien
GCC Institute, Department of Musculoskeletal Medicine and Medical Aesthetics, Newport Beach, CA, USA

C. W Hunter et al. (eds.), *Regenerative Medicine*, https://doi.org/10.1007/978-3-030-75517-1_17

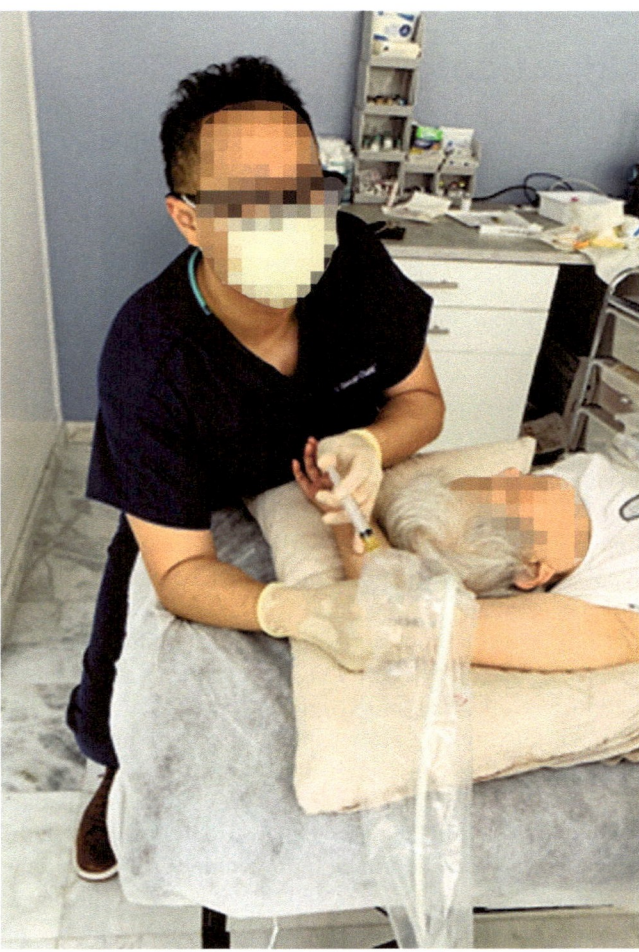

Fig. 17.1 Patient between physician and the ultrasound machine which allows for direct line of site to the image. Note the sterile probe cover. This ultrasound device is equipped with a probe with programmable buttons for basic commonly used functions

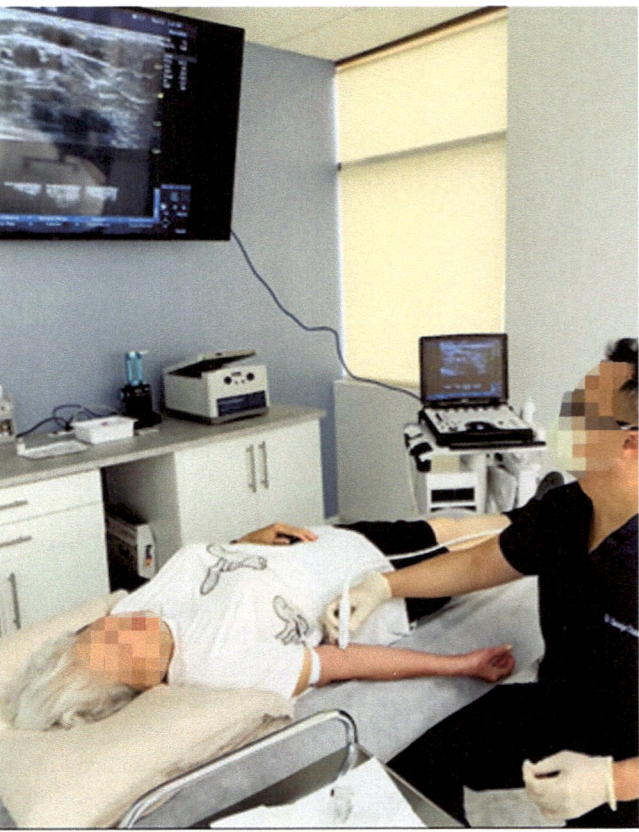

Fig. 17.2 The physician, the patient, and the ultrasound machine make a "triangle." The ultrasound device is within arms' length for the physician to make further adjustments if necessary. In this set up, a larger monitor also allows for improved visualization and re-demonstration of the "patient sandwich"

struggle the practitioner may have performing the therapy and allow for efficient injection and therapy treatment. The goal of these treatments is to induce regenerative healing option with the product that is being injected and limit the risk of increased pain and trauma to the peripheral regions. With less preparation, difficulty can develop during the injection therapy, which in turn can result in inefficient treatment and or higher risk of complication and/or infection.

Typically, the target of interest is placed in the center of the screen. Utilizing the depth control of the machine, the area surrounding the target of interest should be viewed to identify vasculature, nerves, bones, cysts, etc. The target does not need to be in the center of the screen for the procedure, as they may result in excessive unnecessary distance traversed with the needle. An ultrasound gel bridge may be useful in instances to create a trajectory that minimizes needle length utilized.

Ultrasound-Guided Injection Supplies

Aside from syringes, needles, and gloves, a sterile probe cover is often useful to maintain sterility and reduce risk for infection. If a sterile gel bridge is utilized, it would best be performed utilizing a sterile probe cover to reduce contaminating the injection site. Further tools to assist in the USGRMP include needle guides which maintain the needle in the middle of the probe.

Consent, Risk, and Benefits

In any type of medical procedure, informed consent is obtained from the patient. This is the presentation of the risks, benefits, and outcomes of the procedure that the patient will be undergoing. As a treating individual, you will have the responsibility of presenting the benefits, the potential negative outcomes, and the potential complications that can occur with undergoing this treatment. This explanation is not limited to risk and benefits but will also include the procedural process and expectations following the procedure in the short and long term.

The patient will be given a consent form which will outline the risks and benefits, along with an opportunity to read and review and then sign providing informed consent.

A typical consent form will include diagnosis, nature, and purpose of the treatment, as well as the risk and expected benefits from the procedure. The site of treatment will also be placed on the consent. Furthermore, it will include other options such as conservative treatments which may have been foregone or already done. Many times, it will have a description of the procedure and the type of procedure that is going to be done. In some cases, there will be sections of initialing certain areas along with a signature on the consent form. This shows a specific understanding of certain risks that can be present during the procedure and outcome. Finally, it should include a section where signatures by the patient and the provider will be noted and signed with the date.

Once the informed consent has been completed, with both parties having signed it, then the procedure may then be started.

Depending on your state medical board recommendations, your consent should acknowledge that these procedures may be considered investigational in your state and your patients should be made aware.

Clean, Aseptic, and Surgical Aseptic Techniques

The type of procedure being done should familiar to the clinician, and they should be aware of the type of technique to be used to limit infection and cross-contamination. In most cases, "clean aseptic" and "surgical aseptic" techniques are used in the medical field to limit cross-contamination and infection. By definition, aseptic surgical technique is a process of which strict precautions are taken to limit microorganisms contamination in an operating room. The goal of this is to maintain aseptic environment and minimize the introduction of infectious processes and pathogens. Aseptic technique aims to prevent microorganisms on hands and surfaces from being introduced into a specific area. While clean technique aims to reduce contamination of microorganisms that are pathogenic [2–5].

With "surgical aseptic technique," sterile gloves and gowns are used, cleaning solutions for surfaces, and work spaces are all maintained in sterile environments. "Aseptic technique" utilizes sterile gloves for key parts and sites and limits cross-contamination of other sterile products that are in the field [6].

When utilizing "clean technique," this involves limiting and minimizing the transmission of microorganisms from environment or healthcare provider to person receiving treatment. This involves appropriate hand hygiene as well as clean gloves. Efforts are made to prevent contamination of supplies and materials while the procedure is done.

Procedure Set Up

In preparing for the procedure, setup is crucial for efficiency and limiting risk of cross-contamination and infection. Preparation of the blood products such as platelet-rich plasma is done, and it should be done with aseptic technique given that the preparation will eventually result in the product that is going to be injected. Sterile technique will maintain sterile product from initial draw to time of injection [7].

Based on the location and the type of injection and the comfort level of the individual providing the treatment, clean or sterile technique may be utilized. In many case of peripheral injections, clean technique may be utilized. This would allow for proper preparation of the ultrasound equipment as well as the blood product going to be injected. Most commonly, clean technique will be used for peripheral joint, ligament, and tendon injection therapies. However, sterile technique may be utilized in specific cases and may be based on practitioner preference.

Positioning of the individual should be done to allow for ease of procedure for the practitioner. This would include positioning of the patient, proper placement of the ultrasound machine, as well as clean or sterile fields for medication and injection products. In many cases, local anesthetic, gloves, syringes, needles, and extension tubing are present and ready on a sterile field. When ultrasound guidance is used, the ultrasound probe may be prepared with sterile gel and a sterile cover to prevent cross-contamination while using imaging for placement and needle guidance. Dependent on the image needs for the injection therapy, a larger or smaller sterile field for the injection site may be needed. For example, if the region of pathology is identified and the ultrasound probe was placed and maintained, a small sterile region may be prepared for needle access, and then under needle guidance with ultrasound, the needle can be brought into the field under sterile conditions to the proper visualized area of pathology. This would allow for minimal movement of the ultrasound probe and potential cross-contamination. On the other hand, in cases where a larger field is needed or prepared, and the ultrasound probe needs to be removed to identify regions of pathology in multiple areas, this may not be possible and a larger sterile field on the skin will need to be prepped [8, 9].

In general, proper setup will result in efficient injection therapy as well as limit cross-contamination and potential bad outcomes. There are different ways and options to conduct these interventions, and a level of comfort by the practitioner will determine what is most effective and efficient.

Conclusion

Pre-procedural planning by scanning the affected area prior to the procedure, and setting up your tray, and patient positioning in a routine standardized format will allow for efficient use of your time and reduced discomfort for your patients. Following your state medical board's regulations outlining regenerative medicine is a necessary part to maintaining compliance of your practice with state and federal regulating bodies.

References

1. Chen Z, Deng Z, Ma Y, Liao J, Li Q, Li M, Liu H, Chen G, Zeng C, Zheng Q. Preparation, procedures and evaluation of platelet-rich plasma injection in the treatment of knee osteoarthritis. J Vis Exp. 2019;143
2. Food and Drug Adminstration (FDA). U.S. Department of Human Services. Informed Consent Web Page. https://www.fda.gov/forpatients/clinicaltrials/informedconsent/default.htm. 2019.
3. CDC guidelines could cut bloodstream infections from dialysis. 2013. cdc.gov/media/releases/2013/p0513-dialysis-infections.html.
4. Zimlichman E, Henderson D, Tamir O, Franz C, Song P, Yamin C, Bates D. Health care-associated infections: a meta-analysis of costs and financial impact on the US Health Care System. JAMA Intern Med. 2013;173(22):2039–46.
5. Joint Commission. Aseptic vs. clean technique. 2013. jointcommission.org/assets/1/6/CLABSI_Toolkit_Tool_3-8_Aseptic_versus_Clean_Technique.pdf.
6. O'Grady NP, et al. Healthcare Infection Control Practices Advisory Committee [HICPAC]. Guidelines for the prevention of intravascular catheter-related infections. Clin Infect Dis. 2011;52(9):e162–93. Epub 2011 Apr 1
7. Narouze S. Atlas of ultrasound guided procedures in interventional pain management. New York: Springer; 2011.
8. Peng PW, Narouze S. Ultrasound-guided interventional procedures in pain medicine: a review of anatomy, sonoanatomy, and procedures: part I: nonaxial structures. Reg Anesth Pain Med. 2009;34(5):458–74.
9. Sites BD, Gallagher JD, Cravero J, Lundberg J, Blike G. The learning curve associated with a simulated ultrasound-guided interventional task by inexperienced anesthesia residents. Reg Anesth Pain Med. 2004;29(6):544–8.

Effects of Local Anesthetics and Contrast Agents on Regenerative Medicine Procedures

18

Allan Zhang and George C. Chang Chien

Introduction

Regenerative medicine (RM) is an emerging area of medical practice with the goal of restoring or establishing normal function by replacing, engineering, or regenerating human cells, tissues, or organs that have been lost or injured due to age, disease, or congenital defects. RM injections include the use of platelet-rich plasma (PRP), hypertonic dextrose, and mesenchymal stem cells. When used in the treatment of musculoskeletal conditions, RM therapies focus on promoting the body's innate healing capacity. Level 1 evidence supports the use of RM in the treatment of osteoarthritis and tendinopathy [1–8]. Emerging evidence indicates that there may be a positive response in the treatment of intervertebral disk degeneration (IDD) and ligamentous pathology. Studies on RM injectates such as PRP have demonstrated upregulation of anabolic genes, transcription of new proteins, and down regulation of markers of catabolism and apoptosis [1, 6]. Importantly, RM therapies reflect a shift away from the use of conventional destructive techniques, such as corticosteroid injections, local anesthetic injections, and neurolysis which are currently the mainstay treatment for many painful nonsurgical musculoskeletal conditions.

Local anesthetics, corticosteroids, and contrast agents are routinely used during interventional orthopedic and pain management procedures for both diagnostic and therapeutic purposes. A growing body of literature suggests that these routinely used injectates promote catabolic processes including apoptosis which are thought to accelerate the disease process. It is therefore paramount to understand the effect of these agents on RM injectates and on target tissues including tenocytes, chondrocytes, nucleus pulposus, and ligamentous

tissue. Numerous studies have shown time- and dose-dependent chondrotoxicity of both local anesthetics and contrast agents on human and animal soft tissues.

This chapter evaluates the current literature on the effects of local anesthetics contrast agents with an attempt to establish a cohesive recommendation regarding their usage in regenerative medicine procedures.

Local Anesthetics

Degenerative musculoskeletal conditions such are osteoarthritis, tendinopathy, and degenerative disk disease are widely prevalent and associated with debilitating symptoms that affect all age groups. The general stepwise treatment of these conditions involves conservative medical management and may eventually progress to surgical interventions such as total joint replacement. Intra-articular and peri-tendinous injections of local anesthetics and corticosteroids are popular procedures but have demonstrated toxic effects on tissues exposed to them. Current data suggest that these accelerate the disease processes of osteoarthritis and tendinopathy.

Piper et al. described the chondrotoxic effects of local anesthetics in 2008 by exposing human femoral head articular cartilage explants and cultured chondrocytes to 0.5% ropivacaine, 0.5% bupivacaine, and normal saline for 30 minutes [16]. Chondrocyte viability was measured after 24 hrs. The results demonstrated statistically significant reduction in cell viability of both femoral head articular cartilage explants and cultured chondrocytes when exposed to both ropivacaine and bupivacaine. Subsequent studies by other investigators further expanded on this subject to include additional anesthetic agents such as lidocaine and mepivacaine. The results of these studies further corroborated with the findings that local anesthetics are chondrotoxic in dose-, time-, and type-dependent fashion, with bupivacaine being the most chondrotoxic. The effects of local anesthetics have also been investigated with other cell types including tenocytes and collagen fibers from intervertebral disks. Scherb

A. Zhang
University of Connecticut, Department of Radiology, Farmington, CT, USA

G. C. Chang Chien (✉)
GCC Institute, Department of Musculoskeletal Medicine and Medical Aesthetics, Newport Beach, CA, USA

© Springer Nature Switzerland AG 2023
C. W Hunter et al. (eds.), *Regenerative Medicine*, https://doi.org/10.1007/978-3-030-75517-1_18

187

et al. in 2009 treated harvested human tendon tissues with increasing concentrations of bupivacaine [9]. Tenocyte proliferation and extracellular production were significantly lower when compared with the saline control group. The study by Zhang el al in 2016 demonstrated similar findings as incubation with bupivacaine resulted in substantial reduction in cell viability of both fibroblasts and tenocytes [10]. Furthermore, the study by Iwasaki et al. showed significant human nucleus pulposus cell apoptosis when treated with increasing concentrations of bupivacaine and lidocaine [11].

Interestingly, ropivacaine exhibited less cytotoxicity when compared with its counterparts. Ropivacaine, like lidocaine, bupivacaine, and mepivacaine, is an amid-type local anesthetic that exerts its anesthetic effect by inhibiting the opening of voltage-gated sodium channels on cell membranes and blocking the propagation of action potentials generated by pain neurons [12, 13]. Unlike lidocaine and bupivacaine, however, ropivacaine demonstrated no significant decrease in cell viability regardless of dose or duration in the study by Dregella et al. [14]. In the study by Zhang el al in 2016, ropivacaine again demonstrated no statistically significant reduction in all three experimented cell lines of dermal fibroblasts, human mesenchymal stem cells, and tenocytes [10]. This was also seen in the studies by Grishko et al. and Piper et al. [15, 16]. At higher concentrations of ropivacaine, such as 0.5% and greater, however, Breu et al. and Rao et al. did show significant chondrocyte death albeit to a lesser extent than both lidocaine and bupivacaine [17, 18] .

Contrast Agents

Chondrotoxic effects are also present with the use of radio-contrast media. By incubating human mesenchymal stem cells (hMSCs) to varying concentrations of ionic and nonionic contrast media, Kim et al. demonstrated that contrast media exerted chondrotoxic effects in a dose- and type-dependent manner with ionic contrasts being most detrimental [19, 20]. The nonionic agents Iopromide (Ultravist™), Iodixanol (Visipaque™), and Iopamidol (Isovue™) demonstrated mild dose-dependent chondrotoxicity to hMSCs and nucleus pulposus cells as well as bovine intervertebral disk cells. The nonionic agents Iotrolan (Isovist™) and Iohexol (Omnipaque™), however, showed no chondrotoxic effects as evidenced by the studies of Iwasaki et al. and Chee et al. [11, 21]. The difference in chondrotoxic effect between these two groups of nonionic contrasts is unclear. While osmolarity has been implicated as a potential culprit in toxicity among ionic agents, among the nonionic agents, iohexol is the highest in osmolarity and yet demonstrated no toxicity in the study by Chee et al. [21]. Conversely, all ionic contrast media demonstrated chondrotoxicity in studies on both human and animal cells. The ionic agents are hyperosmolar

when compared with their nonionic counterparts. This difference in osmolarity has been shown to induce DNA fragmentation in canine renal cells [20]. Furthermore, the chemotoxic effect of carboxyl groups, which are found only in ionic contrast media, could be a factor affecting chondrocyte viability. Nonionic contrasts have no carboxyl groups and have a higher number of hydroxyl groups than ionic contrasts, resulting in less protein binding [20]. Additionally, all gadolinium-based agents demonstrated dose-dependent chondrotoxicity.

Discussion

Local anesthetics have demonstrated detrimental effects on regenerative medicine injectates including plasma-rich platelets (PRP) and its constituents as well as mesenchymal stem cells (MSCs) utilized in cell therapy. PRP is an autologous preparation of patient's whole blood which has been filtered, allowing for extraction of supra-physiological concentration of patient's platelets. Proteins released by platelets function as catalysts for cell proliferation, growth factor release, tissue repair, and regeneration. MSCs, similarly, can also be utilized in a multitude of different ways in regenerative medicine therapy. They have been shown to effectively treat degenerative joint diseases as well as enhance recovery of ligamentous and tendinous injury. The beneficial effects of these injectates depend on the viability of the cells. As such, any agents that can reduce the cell viability can decrease the effectiveness of these injectates. Based on the collection of studies reviewed here, MSCs experienced cellular apoptosis and/or necrosis after exposure to local anesthetics in dose-, time-, and type-dependent model. Likewise, local anesthetics inhibit platelet aggregation and functionality and have the potential to compromise the therapeutic effect of PRP.

Contrast agents are considered necessary for some procedures to confirm appropriate and accurate needle placement. The use of contrast agents can be avoided by basing accurate needle placement on radiographic imaging or utilizing alternative image guidance such as ultrasound. Based on our review, contrast agents exerted chondrotoxic effects in a dose- and type-dependent manner with ionic contrasts being most detrimental. Nonionic contrast media mild dose-dependent chondrotoxicity and were the least harmful of those studied.

Conclusion

All investigated local anesthetics demonstrated chondrotoxicity to some extent. For clinical considerations, local anesthetics should be avoided when possible or utilized with the lowest effective concentration and dose. Ropivacaine and mepivacaine were found to be the least chondrotoxic local anesthetics in multiple studies. These agents also have the potential to reduce the effectiveness of regenerative medicine injectates including stem cell therapy and PRP. In regards to radiocontrast, all ionic agents demonstrated cytotoxicity to nucleus pulposus cells. This has clinical relevance as diskograms for localizing lower back pain often require multiples injections of radiocontrast. Among the nonionic agents, Iotrolan (Isovist™) and Iohexol (Omnipaque™) were shown to be the least toxic. The intra-articular injection of gadolinium-based agents for MRI arthrography should be avoided for the agents described above. While this is true, newer gadolinium agents need to be further delineated before more recommendations can be suggested.

References

1. Cook CS, Smith PA. Clinical update: why PRP should be your first choice for injection therapy in treating osteoarthritis of the knee. Curr Rev Musculoskelet Med. 2018;11(4):583–92. https://doi.org/10.1007/s12178-018-9524-x.
2. Zhao L, Kaye AD, Abd-Elsayed A. Stem cells for the treatment of knee osteoarthritis: a comprehensive review. Pain Physician. 2018;21:229–41.
3. Lai LP, Stitik TP, Foye PM, Georgy JS, Varun P, Boqing C. Use of platelet-rich plasma in intra-articular knee injections for osteoarthritis: a systematic review. PM&R. 2015;7(6):637–48. https://doi.org/10.1016/j.pmrj.2015.02.003.
4. Laver L, Marom N, Dnyanesh L, Mei-Dan O, Espregueira-Mendes J, Gobbi A. PRP for degenerative cartilage disease: a systematic review of clinical studies. Cartilage. 2016;8:341–64. https://doi.org/10.1177/1947603516670709.
5. Jang S, Kim J, Cha S. Platelet-rich plasma (PRP) injections as an effective treatment for early osteoarthritis. Eur J Orthop Surg Traumatol. 2012;23:573–80. https://doi.org/10.1007/s00590-012-1037-5.
6. Moussa M, Lajeunesse D, Hilal G, El Atat O, Haykal G, Serhal R, et al. Platelet rich plasma (PRP) induces chondroprotection via increasing autophagy, anti-inflammatory markers, and decreasing apoptosis in human osteoarthritic cartilage. Exp Cell Res. 2017;352:146–56. https://doi.org/10.1016/j.yexcr.2017.02.012.
7. Wichan K, Alisara A, Kornkit C, Niti P, Manusak B, Peerapong P, Jatupon K. Short-term outcomes of platelet-rich plasma injection for treatment of osteoarthritis of the knee. Knee Surg Sports Traumatol Arthrosc. 2015;24(5):1665–77. https://doi.org/10.1007/s00167-015-3784-4.
8. Levy D, Petersen K, Scalley Vaught M, Christian D, Cole B. Injections for knee osteoarthritis: corticosteroids, Viscosupplementation, platelet-rich plasma, and autologous stem cells. Arthroscopy. 2018;34:1730–43. https://doi.org/10.1016/j.arthro.2018.02.022.
9. Michael BS, Seung-Hwan H, Jean-Paul C, Gregory PG, Lew S. Effect of bupivacaine on cultured tenocytes. Orthopedics. 2009;32:26. https://doi.org/10.3928/01477447-20090101-19.
10. Zhang A, Ficklscherer A, Pietschmann M, Jansson V, Müller P. Apoptosis and necrosis-inducing cell toxicity of ropivacaine, bupivacaine and triamcinolone in fibroblasts, tenocytes and human mesenchymal stem cells. Sports Orthop Traumatol Sport-

Orthopädie—Sport-Traumatologie. 2016;32(2):201. https://doi.org/10.1016/j.orthtr.2016.03.019.
11. Iwasaki K, Sudo H, Yamada K, Ito M, Iwasaki N. Cytotoxic effects of the radiocontrast agent iotrolan and anesthetic agents bupivacaine and lidocaine in three-dimensional cultures of human intervertebral disc nucleus pulposus cells: identification of the apoptotic pathways. PLoS ONE. 2014;9(3) https://doi.org/10.1371/journal.pone.0092442.
12. Aguirre JA, Votta-Velis G, Borgeat A. Practical pharmacology in regional anesthesia. Essent Reg Anesth. 2011:121–56. https://doi.org/10.1007/978-1-4614-1013-3_5.
13. Patel N, Sadoughi A. Pharmacology of local anesthetics. Essent Pharmacol Anesth Pain Med Crit Care. 2014:179–94. https://doi.org/10.1007/978-1-4614-8948-1_11.
14. Dregalla RC, Lyons NF, Reischling PD, Centeno CJ. Amide-type local anesthetics and human mesenchymal stem cells: clinical implications for stem cell therapy. Stem Cells Transl Med. 2014;3(3):365–74. https://doi.org/10.5966/sctm.2013-0058.
15. Grishko V, Xu M, Wilson G, Pearsall AW 4th. Apoptosis and mitochondrial dysfunction in human chondrocytes following exposure to lidocaine, bupivacaine, and ropivacaine. J Bone Joint Surg Am. 2010;92(3):609–18. Spine J. 2010;10(7):653. doi:https://doi.org/10.1016/j.spinee.2010.05.024.
16. Piper Samantha L, Kim Hubert T. Comparison of ropivacaine and bupivacaine toxicity in human articular chondrocytes. J Bone Joint Surg. 2008;90(5):986–91. https://doi.org/10.2106/JBJS.G.01033.
17. Breu A, et al. Cytotoxicity of local anesthetics on human mesenchymal stem cells in vitro arthroscopy. 29(10):1676–84.
18. Rao AJ, Johnston TR, Harris AH, Smith RL, Costouros JG. Inhibition of chondrocyte and synovial cell death after exposure to commonly used anesthetics. Am J Sports Med. 2013;42(1):50–8. https://doi.org/10.1177/0363546513507426.
19. Kim K-H, Kim Y-S, Kuh S-U, et al. Time- and dose-dependent cytotoxicities of ioxitalamate and indigocarmine in human nucleus pulposus cells. Spine J. 2013;13(5):564–71. https://doi.org/10.1016/j.spinee.2013.01.019.
20. Kim K-H, Park J-Y, Park H-S, et al. Which iodinated contrast media is the least cytotoxic to human disc cells? Spine J. 2015;15(5):1021–7. https://doi.org/10.1016/j.spinee.2015.01.015.
21. Chee AV, Ren J, Lenart BA, Chen E-Y, Zhang Y, An HS. Cytotoxicity of local anesthetics and nonionic contrast agents on bovine intervertebral disc cells cultured in a three-dimensional culture system. Spine J. 2014;14(3):491–8. https://doi.org/10.1016/j.spinee.2013.06.095.

Suggested Reading

Augereau O, Rossignol R, Degiorgi F, Mazat J, Letellier T, Dachary-Prigent J. Apoptotic-like mitochondrial events associated to phosphatidylserine exposure in blood platelets induced by local anaesthetics. Thromb Haemost. 2004;92(07):104–13. https://doi.org/10.1160/th03-10-0631.
Az-Ma T, Hardian M, Yuge O. Inhibitory effect of lidocaine on cultured porcine aortic endothelial cell-dependent antiaggregation of platelets. Anesthesiology. 1995;83(2):374–81. https://doi.org/10.1097/00000542-199508000-00018.
Baker JF, Walsh PM, Byrne DP, Mulhall KJ. In vitro assessment of human chondrocyte viability after treatment with local anaesthetic, magnesium sulphate or normal saline. Knee Surg Sports Traumatol Arthrosc. 2011;19(6):1043–6. https://doi.org/10.1007/s00167-011-1437-9.

Bausset O, Magalon J, Giraudo L, et al. Impact of local anaesthetics and needle calibres used for painless PRP injections on platelet functionality. Muscles Ligaments Tendons J. 2014;4(1):18–23. Published 2014 May 8

Carofino B, Chowaniec DM, Mccarthy MB, et al. Corticosteroids and local anesthetics decrease positive effects of platelet-rich plasma: an in vitro study on human tendon cells. Arthroscopy: J Arthroscop Relat Surg. 2012;28(5):711–9. https://doi.org/10.1016/j.arthro.2011.09.013.

Dragoo JL, Braun HJ, Kim HJ, Phan HD, Golish SR. The in vitro chondrotoxicity of single-dose local anesthetics. Am J Sports Med. 2012;40(4):794–9. https://doi.org/10.1177/0363546511434571.

Girard A, Atlan M, Bencharif K, Gunasekaran MK, Delarue P, Hulard O, et al. New insights into lidocaine and adrenaline effects on human adipose stem cells. Aesthetic Plast Surg. 2012;37(1):144–52. https://doi.org/10.1007/s00266-012-9988-9.

Gray A, Marrero-Berrios I, Ghodbane M, et al. Effect of local anesthetics on human mesenchymal stromal cell secretion. Nano LIFE. 2015;05(02):1550001. https://doi.org/10.1142/s1793984415500014.

Greisberg JK, Wolf JM, Wyman J, Zou L, Terek RM. Gadolinium inhibits thymidine incorporation and induces apoptosis in chondrocytes. J Orthop Res. 2001;19(5):797–801. https://doi.org/10.1016/s0736-0266(01)00025-0.

Gugerell A, Kober J, Schmid M, Nickl S, Kamolz L, Keck M. Botulinum toxin A and lidocaine have an impact on adipose-derived stem cells, fibroblasts, and mature adipocytes in vitro. J Plast Reconstr Aesthet Surg. 2014;67(9):1276–81. https://doi.org/10.1016/j.bjps.2014.05.029.

Haasters F, Polzer H, Prall WC, et al. Bupivacaine, ropivacaine, and morphine: comparison of toxicity on human hamstring-derived stem/progenitor cells. Knee Surg Sports Traumatol Arthrosc. 2011;19(12):2138–44. https://doi.org/10.1007/s00167-011-1564-3.

Jacobs TF, Vansintjan PS, Roels N, et al. The effect of lidocaine on the viability of cultivated mature human cartilage cells: an in vitro study. Knee Surg Sports Traumatol Arthrosc. 2011;19(7):1206–13. https://doi.org/10.1007/s00167-011-1420-5.

Keck M, Zeyda M, Gollinger K, Burjak S, Kamolz L, Frey M, Stulnig TM. Local anesthetics have a major impact on viability of preadipocytes and their differentiation into adipocytes. Plast Reconstr Surg. 2010;126(5):1500–5. https://doi.org/10.1097/prs.0b013e3181ef8beb.

Liou J, Mao C, Liu F, Lin H, Hung L, Liao C, Day Y. Levobupivacaine differentially suppresses platelet aggregation by modulating calcium release in a dose-dependent manner. Acta Anaesthesiol Taiwanica. 2012;50(3):112–21. https://doi.org/10.1016/j.aat.2012.07.001.

Lo B, Hönemann CW, Kohrs R, Hollmann MW, Polanowska-Grabowska RK, Gear AR, Durieux ME. Local anesthetic actions on thromboxane-induced platelet aggregation. Anesth Analg. 2001;93(5):1240–5. https://doi.org/10.1097/00000539-200111000-00040.

Midura S, Schneider E, Sakamoto F, Rosen G, Winalski C, Midura R. In vitro toxicity in long-term cell culture of MR contrast agents targeted to cartilage evaluation. Osteoarthr Cartil. 2014;22(9):1337–45. https://doi.org/10.1016/j.joca.2014.07.010.

Oznam K, Sirin DY, Yilmaz I, et al. Iopromide- and gadopentetic acid-derived preparates used in MR arthrography may be harmful to chondrocytes. J Orthop Surg Res. 2017;12(1) https://doi.org/10.1186/s13018-017-0600-5.

Pinto LM, Pereira R, Paula ED, Nucci GD, Santana MH, Donato JL. Influence of liposomal local anesthetics on platelet aggregation in vitro. J Liposome Res. 2004;14(1–2):51–9. https://doi.org/10.1081/lpr-120039697.

Porter JM, Crowe B, Cahill M, Shorten GD. The effects of ropivacaine hydrochloride on platelet function: an assessment using the platelet function analyser (PFA-100). Anaesthesia. 2001;56(1):15–8. https://doi.org/10.1046/j.1365-2044.2001.01760.x.

Rahnama R, Wang M, Dang AC, Kim HT, Kuo AC. Cytotoxicity of local anesthetics on human mesenchymal stem cells. J Bone Joint Surg Am Vol. 2013;95(2):132–7. https://doi.org/10.2106/jbjs.k.01291.

Syed HM, Green L, Bianski B, Jobe CM, Wongworawat MD. Bupivacaine and triamcinolone may be toxic to human chondrocytes: a pilot study. Clin Orthop Relat Res. 2011;469(10):2941–7. https://doi.org/10.1007/s11999-011-1834-x.

Tayton ER, Smith JO, Aarvold A, Kalra S, Dunlop DG, Oreffo RO. Translational hurdles for tissue engineering. J Bone Joint Surg. Br Vol. 2012;94-B(6):848–55. https://doi.org/10.1302/0301-620x.94b6.28479.

Medicolegal Aspects of Regenerative Medicine

19

Matthew B. Murphy and Theodore T. Sand

Introduction

The focus of this chapter is on the regulatory and ethical aspects of physicians in the USA acquiring, manipulating, and using materials as regenerative therapeutic agents in treating musculoskeletal conditions. In this respect, materials or processes that are considered by the Food and Drug Administration (FDA) as drugs or biological drugs will not be included in this review. However, the framework for assessing whether or not a particular patient- or donor-derived material is or is not a drug or biological drug will be covered. Challenges facing the physician as they encounter materials and/or processes for providing regenerative therapies will be discussed, along with mechanisms to report clinicians or corporations that may be marketing noncompliant biologic products.

Regulatory Framework for Regenerative Medicine

The FDA has a mandate from Congress to protect the health and welfare of Americans, with several key laws that have been in place for decades. In particular, two rules govern much of the field of regenerative medicine: Public Health Service Act and 21 CFR 1271.

Public Health Service Act (PHSA), Sections 361 and 351

The PHSA was enacted in 1944 in Title 42 U.S.C. Public Health and Social Welfare. This is the key statue that grants the FDA the authority to protect citizens from the introduction, transmission, or spread of communicable diseases in matters related to healthcare. In particular, Section 361 of the PHSA [1] has been used by the FDA to ensure the safety of biological materials used as regenerative therapeutic treatments. Section 351 of the PHSA sets out what constitutes "biological products" [1], which are regulated as drugs.

21 CFR 1271 Human Cells, Tissues, and Cellular and Tissue-Based Products (HCT/Ps)

The FDA established a comprehensive framework for dealing with human biological tissues in 21 CFR 1271, which was published in the Federal Register in January 19, 2001 [2]. The final rule became effective in 2005.

1271.3 "How Does FDA Define Important Terms in This Part?"

Critical definitions of terminology used in 1271 were published in 1271.3 [3], including what materials are considered to be HCT/Ps (human cells, tissues, and cellular and tissue-based products) and what materials are excluded from being considered HCT/Ps. For example, the definition of an HCT/P is:

(d) *Human cells, tissues, or cellular or tissue-based products (HCT/Ps)* means articles containing or consisting of human cells or tissues that are intended for implantation, transplantation, infusion, or transfer into a human recipient. [4]

However, of particular interest to the autologous regenerative medical community is the exclusion of whole blood and bone marrow from being governed by 1271:

The following articles are not considered HCT/Ps:

(1) Vascularized human organs for transplantation;
(2) *Whole blood or blood components or blood derivative products subject to listing under parts 607 and 207 of this chapter, respectively*;

M. B. Murphy (✉)
Murphy Technology Consulting, Austin, TX, USA
e-mail: mbmurphy@utexas.edu

T. T. Sand
Sand Consulting, Poway, CA, USA

(3) Secreted or extracted human products, such as milk, collagen, and cell factors; except that semen is considered an HCT/P;

(4) *Minimally manipulated bone marrow for homologous use and not combined with another article (except for water, crystalloids, or a sterilizing, preserving, or storage agent, if the addition of the agent does not raise new clinical safety concerns with respect to the bone marrow); ....* [4]

The implications of what is and is not considered by the FDA to be an HCT/P will be reviewed below.

1271.10 (a and b) "Are My HCT/Ps Regulated Solely Under Section 361 of the PHS Act and the Regulations in This Part, and If So What Must I Do?"

Section 1271.10(a) establishes the criteria that need to be met in order that an HCT/P is considered to be a Section 361 material. There are four criteria that need to be met:

1. The HCT/P is minimally manipulated.
2. The HCT/P is used in a homologous manner.
3. The HCT/P is not adulterated by materials other than fluids like buffers or water, and
4. The HCT/P is used either in the donor (autologous therapy) or in the donor's first- or second-degree relatives, or the HCT/P does not depend on the metabolic activity of living cells. [5]

Section 1271.10(b) indicates that if an establishment meets the requirements of 1271.10(a), their HCT/P is a Section 361 HCT/P and the establishment needs to follow the rules outlined in the relevant other sections of 1271 [6]. If the HCT/P does not meet the requirements in 1271.10(a), the material is a Section 351 (PHSA) biological drug and will require registration of both the manufacturer and the material with the FDA, submitting an application for and receiving approval of a Biologics License Application and completion of relevant premarket studies (e.g., Investigational New Drug; IND).

1271.15 "Are There Exceptions from the Requirements of This Part?"

When the FDA began to establish the framework for dealing with human tissues that became 21 CFR 1271, it had to sort out how to address the issue of physicians who had for decades routinely used a patient's tissue during the course of a surgical procedure. This was a problem, since the FDA is not allowed to regulate a physician's practice of medicine. However, this exception was not included in 1271 as a mechanism to administer Section 351 HCT/Ps without proper FDA approval.

In order to allow for a physician's practice of medicine, the FDA indicated in 1271.15(b) [7] that a physician would be able to invoke an exception from the requirements of 1271 (including registering with the FDA as an HCT/P manufacturer, following all requirements of Current Good Tissue Practices,

product labeling, reporting, and establishment inspections) if the physician removed an HCT/P from the patient and returned it to the same patient during the same surgical procedure. For example, a vascular surgeon removes a segment of a patient's saphenous vein, trims it to alter its size and shape, and inserts it as a vascular graft elsewhere in the patient's body. In this example, the FDA would consider the saphenous vein to be an HCT/P, but acknowledges through 1271.15(b) that the surgeon does not need to register with the FDA.

Minimal Manipulation and Homologous Use—Practical Considerations

Qualification for sole regulation as a Section 361 product has always required minimal manipulation and homologous use of the HCT/P among other requirements. As indicated in 1271.3 [3], the following are the FDA's definitions of minimal manipulation and homologous use:

(f) *Minimal manipulation* means: (1) For structural tissue, processing that does not alter the original relevant characteristics of the tissue relating to the tissue's utility for reconstruction, repair, or replacement; and (2) For cells or nonstructural tissues, processing that does not alter the relevant biological characteristics of cells or tissues.

(c) *Homologous use* means the repair, reconstruction, replacement, or supplementation of a recipient's cells or tissues with an HCT/P that performs the same basic function or functions in the recipient as in the donor.

Within their 2017 Final Guidance document entitled "Regulatory Considerations of Human Cells, Tissues and Cellular and Tissue-Based Products: Minimal Manipulation and Homologous Use" (MM/HU) , FDA provided explicit direction and examples of both requirements to offer clarity to clinicians and industry on the current definition of each [8]. Minimal manipulation of the tissue applies to extraction, handling, ex vivo processing, and delivery of the HCT/P to the patient. As part of its explanation of HCT/Ps and the regulation thereof, FDA has classified human-derived biological materials as either structural or nonstructural (cellular) tissues. Structural HCT/Ps provide a physical function in their native environment, including connecting, covering, supporting, cushioning, or separating other tissues. Nonstructural or cellular tissues have no physical function per se, but may have a paracrine, hormonal, or metabolic purpose in the body.

Examples of nonstructural tissues are blood, bone marrow, umbilical cord blood, pancreatic, parathyroid, and lymph node tissues. Examples of structural tissues are adipose, tenon, cartilage, umbilical cord, placental membranes, skin, etc. However, as indicated in section "Regulatory Framework for Regenerative Medicine" above, peripheral blood and bone marrow are not considered to be HCT/Ps by definition, as long as bone marrow is minimally manipulated and used homologously.

If any "manufacturing" processes change the original relevant characteristics of the tissue (e.g., strength, consistency, flexibility, compressibility, cushioning, or response to shear and friction of structural tissues), the process would be considered more than minimal manipulation and trigger Section 351 status for the HCT/P. Changes to the size or shape of the tissue by cutting, milling, or grinding, so long as the original relevant characteristics remain unchanged, constitute minimal manipulation, and are allowable under 1271.10(a).

FDA has provided examples of minimal and above minimal manipulation with regard to structural tissues in the MM/HU Guidance. For example, crushed bone or washed and centrifuged adipose is considered minimally manipulated, because their physical and structural characteristics are essentially the same, while micronizing cartilage or decomposing amniotic tissues into a slurry are more than minimal manipulation because their physical characteristics were substantially altered. According to the final guidance, processing techniques of nonstructural tissues, such as hematopoietic mobilization, apheresis of the enriched blood (including centrifugation), and the readministration of cells in order to repopulate bone marrow do not exceed minimal manipulation. Centrifugation and filtration processes are explained to be only minimal manipulation so long as the biological characteristics of the cells are not affected or altered. An interesting example of overlap is the centrifugation or washing of a structural tissue (e.g., adipose) yielding a structural connective tissue and a fluid containing cells. In this situation, the tissue maintains its relevant characteristics and is minimally manipulated, while the cells and fluid separated from the tissue are not considered to be structural in nature and are therefore more than minimally manipulated. The isolated cells also do not meet the homologous use of "cushioning and supporting" that the FDA has associated with adipose tissue.

In order to qualify for regulation solely under Section 361, cell and tissue products also must be used in a manner homologous to their original role at the harvest site. Homology applies to both form and function according to 1271.10 [5]. Of particular importance is the phrase from the definition for homologous use shown above that HCT/Ps should "…perform the same basic function or functions in the recipient as in the donor." The Agency further indicates HCT/Ps in the donor might have a variety of "…biological/physiological…" functions, but just one of which needs to be active in the recipient for it to be considered homologous use. However, a basic function of the HCT/P present in the recipient that is not found in the donor site is a nonhomologus use. For example, using mechanically disrupted adipose to treat musculoskeletal pathologies associated with pain is considered nonhomologous, since reducing pain is not recognized by the FDA to be a basic function of adipose tissue in the donor.

FDA has offered several examples of homologous functions including autologous bone, tendon, or blood vessel transplants, allograft heart valve transplants, and allograft neonatal membranes (i.e., placental) used as protective dressings. FDA also provided examples of nonhomologous function, such as using any form of adipose tissue to treat intraarticular osteoarthritis or amniotic membranes implanted for the purpose of growing bone. If the use of the cell or tissue product "as reflected by the labeling, advertising, and other indications of the manufacturer's objective intent" is not considered homologous relative to its original function, the product is regulated under Section 351 of the PHSA and would require approval as a drug or device prior to marketing and clinical use. From a regulatory and legal perspective, it is imperative to note that a physician may be considered the "manufacturer" if the tissue is processed in a manner that changes the original relevant characteristics of "such HCT/P" or the HCT/P is utilized in a nonhomologous manner. As such, the physician must comply with all the rules and regulations applied to other tissue product manufacturers, except where the Same Surgical Procedure Exception applies.

Same Surgical Procedure Exception—The Gotcha Section of 1271?

There are two facets to Section 1271.15(b) that are important in understanding how the Same Surgical Procedure Exception is applied. The first point is that the FDA created the exception to cover physicians who had been using a patient's tissue during the same surgical procedure as a routine practice of medicine, since the FDA cannot regulate a physician's practice of medicine. The second point is that the basis for granting an exception when a physician removes a patient's HCT/P and returns it during the same surgical procedure with minimal processing of the HCT/P comes from the FDA's belief that this type of activity presents no greater risk of infection than the surgery itself [9].

While stepping back from regulating a physician's practice of medicine, the FDA has explained that there are limits to what a physician can do with the HCT/P after it has been removed from the patient in order to manage the risk of infection. Consider the following statement from the Same Surgical Procedure Exception Guidance that the FDA issued on November 16, 2017 [9]:

In sum, FDA's view is that autologous cells or tissues that are removed from an individual and implanted into the same individual without intervening processing steps beyond rinsing, cleansing, sizing, or shaping raise no additional risks of contamination and communicable disease transmission beyond that typically associated with surgery. FDA considers the same surgical procedure exception to be a narrow exception to regulation under Part 1271.

What is especially problematic for physicians working with HCT/Ps is the part of that statement that confines what the physician can do physically to or with the HCT/P, namely limiting their actions to "…rinsing, cleansing, sizing, or shaping…". Referring to the previous example of the vascular surgeon taking a saphenous vein for a vascular graft, the surgeon will cut out the saphenous vein segment from the donor's leg, trim it up to fit in the new site, and stitch it in place, all of which should be acceptable to the FDA based on their list of appropriate processing actions, so the vascular surgeon can invoke the exception and does not need to register with the FDA.

In order to further emphasize that 1271.15(b) is not an open invitation for physicians to do whatever they want and think they are compliant with 1271.15(b) requirements, the FDA stated the following in the Guidance [9] to clarify how 1271.15(b) can be satisfied:

(a). Remove and implant the HCT/Ps into the same individual from whom they were removed (autologous use).
(b). Implant the HCT/Ps within the same surgical procedure.
(c). The HCT/Ps remain "such HCT/Ps"; they are in their original form.

While it is not a problem for a physician to meet the requirements indicated in Items (a) and (b), the requirement that the HCT/P remain "such HCT/P" so that it is "in [its] original form…" is a serious limitation.

Let us review a couple of examples to illustrate just how restrictive 1271.15(b) is. Consider a process in which a patient's adipose tissue is removed, rinsed, and cleansed with an enzyme preparation, which digests the adipose tissue and releases single cells (aka SVF, stromal vascular fraction). In addition to other concerns the FDA would have with this type of "cellular" HCT/P, a physician using the SVF preparation would not be able to invoke the exception outlined in 1271.15(b), because the single-cell preparation produced in this process is not at all in its "original form", which was the adipose tissue removed during a lipoaspiration procedure from the patient. The other example also involves lipoaspiration, but in this case mechanically released cells present in the fluid of the lipoaspiration are collected by centrifugation and used to treat the patient. A physician might argue that nothing has been done to the cells after being separated from the adipose tissue in the lipoaspirate, so this must be an acceptable (i.e., 361) HCT/P. However, the fact that the cells were obtained in a lipoaspiration, in which the original form is adipose tissue, means that the cell-based HCT/P does not meet the "such HCT/P" standard, and the physician would not be able to invoke the exception.

There is one other aspect to the Same Surgical Procedure Guidance [9] [that might create confusion among healthcare professionals. The FDA states that a physician should consider whether the HCT/P and the processing of it meet the criteria laid out in the Same Surgical Procedure Guidance, before considering the elements outlined in 1271.10(a).

Furthermore, the FDA indicates that if the physician meets the requirements of "such HCT/P" and the other two elements listed above, there is no need for the physician to consider the other sections of 1271, since they do not apply to a physician who can invoke the exception in 1271.15(b).

Thus, the FDA has appeared to simplify the process for a physician to determine if they can invoke the same surgical procedure exception when working with autologous HCT/Ps: just determine that the HCT/P returned to the patient is "such HCT/P", meaning that it is in its original form. However, it is important not to lose sight of the fact that very little processing of the HCT/P can be performed and still have it satisfy the FDA's standard of it being in its original form. The rules for considering if a physician can invoke the Same Surgical Procedure Exception are summarized in Table 19.1.

On the other hand, consider a physician who is working with a donor-derived HCT/P, implanted during a surgical procedure. Does the physician have to register? Clearly, the donor-derived material is not autologous in origin, so 1271.15(b) would not apply. Assuming that the physician does not make the donor-derived HCT/P in the clinic, the physician would just be using a product that a manufacturer is selling, which becomes the practice of medicine. However, it is important that the physician knows or is assured that the donor-derived HCT/P meets all of the criteria outlined in 1271.10(a). If the product does not meet all of those criteria, the physician is working with a drug, which probably has not been cleared through a premarket process like an IND. In which case, the physician is treating patients with unapproved drugs or biological drugs.

In summary, 1271.15(b) provides a physician with a way to avoid registering with the FDA when working with autologous tissue preparations for regenerative purposes. However,

Table 19.1 Summary of conditions for invoking the same surgical procedure exception and what happens if a physician cannot

What the physician should know about the Same Surgical Procedure Exception:
1. The exception applies only to patient-derived HCT/Ps (i.e., adipose tissue) that are returned to the patient.
2. Implanting of autologous HCT/Ps should occur within the same surgical procedure.
3. The processing of the HCT/P material should be limited to rinsing, cleansing, sizing, and shaping, so that it remains "such HCT/P"—in its original form.
What happens if the physician cannot invoke the exception:
1. The physician (or their facility) will need to register with the FDA as a manufacturer of HCT/Ps.
2. The facility will need to register their HCT/Ps and to update the list as needed.
3. The facility will need to implement a quality control system in order to demonstrate that the manufacturing of the HCT/P meets requirements for minimizing exposure to infectious agents.
4. The facility will need to develop, maintain, and use documents that meet the current standard for Good Tissue Practices (CGTP).

Key: *HCT/P* human cells, tissues and cellular and tissue-based products, *FDA* United States Food and Drug Administration, *CGTP* current good tissue practices

it is important that any processing of the recovered HCT/P prior to implantation is limited to rinsing, cleansing, sizing, and shaping procedures. Consequently, meeting the "such HCT/P" requirement in 1271.15(b) may prove to be daunting when working with materials like adipose tissue that are HCT/Ps.

Handicapping Regenerative Medical Materials and Processes

Table 19.2 provides an assessment of various types of biological tissues/fluids that are commercially available in the USA with respect to their 351/361 status and if they are con-

Table 19.2 Assessment of materials frequently used as regenerative therapeutics

Tissue source	Product forms	Clinical application	HCT/P; 351/361	Notes
Bone marrow aspirate (BMA)	Uncentrifuged BMA or concentrated bone marrow (BMC)	Orthopedic indications	No	Excluded by definition from 1271
Whole blood	PRP, PPP	Orthopedic indications	No	Excluded by definition from 1271
Adipose tissue	Autologous microfractured fat	Lipofilling/body contouring	Yes; 361	Limited to cushioning and supporting in subcutaneous tissues
Adipose tissue	Autologous lipoaspirate	Lipofilling/body contouring	Yes; 361	Limited to cushioning and supporting in subcutaneous tissues like the face and hand
Adipose tissue	Enzymatic digestion (SVF)	Depends on the instructions for use provided by the manufacturer	Yes; 351	Digestion of fat tissue to obtain cells is above minimal manipulation and nonhomologous use
Amniotic fluid (no cells)	Amniotic fluid	Depends on the instructions for use provided by the manufacturer	No	Amniotic fluid without cells is considered to be a secretion and is excluded from 1271 by definition
Amniotic fluid (with viable cells)	Cryopreserved vials of fluid	Depends on the indications for use of the IND	Yes; 351	Materials containing donor-derived viable cells do not meet the criterion in 1271.10(a) (4)
Placental tissue (dried/lyophilized)	Sheets	Wound healing, covering, or barrier	Yes; 361	The placental-derived tissues must not contain viable cells
Placental tissue (dried/lyophilized)	Micronized particles that are flowable or injectable	Will depend on the IDE indications for use	Yes; 351	Micronization of placental tissue alters the physical state of the tissue and rises above minimal manipulation
Cord blood	Whole-cell enriched	Recapitulation of a patient's ablated bone marrow	Yes; 351	Whole cord blood and its components are 351 biological drugs
Cord blood	Plasma only	Depends on the IND indications for use	Yes; 351	Any component of cord blood is regulated as a 351 biological drug
Umbilical cord tissue	Micronized	Depends on the IND indications for use	Yes; 351	Umbilical cord is considered a structural tissue and micronization alters the structural nature of the tissue
Demineralized bone matrix	Granular	Depends on the indications for use in the 510(k)	Yes; 361	DBM is handled separately from 1271
Tendon/ligament (autologous)	Intact segments	Tendon or ligament replacement or repair	Yes; 361	Considered the practice of medicine; meets the 1271.15(b) exception requirements with limited processing
Tendon/ligament (donor)	Intact segments	Tendon or ligament replacement or repair	Yes; 361	Considered the practice of medicine when working with commercial sources
Exosomes (obtained from cultured cells)	Fluid	No data to support use and no IND-cleared product	Yes; 351	Exosomes are collected from culture fluid of various cell types like MSCs—which makes them a 351 material
Cultured cells (autologous or donor)	Cryopreserved fluid	Will be defined by the IND indications for use	Yes; 351	Cultured cells, whether autologous or allo-derived, rise above minimal manipulation—351 material
Bone marrow aspirate, adipose, or neonatal	Any	Intravenous or intrathecal injection	Yes; 351	Nonhomologous applications
Allogeneic products containing viable cells with metabolic activity	Any	Any	Yes; 351	HCT/Ps with viable cells with a systemic or metabolic effect must be used on the donor or first- or second-degree blood relative for 361 status

Key: *351* Section 351 of the Public Health Service Act, *361* Section 361 of the Public Health Service Act, *510(k)* An FDA category of clearance for devices, *1271* 21 CFR (Code of Federal Regulations) 1271, *BMA* bone marrow aspirate, *BMC* bone marrow concentrate, *DBM* demineralized bone matrix, *HCT/P* human cells, tissues and cellular and tissue-based products, *IDE* investigational device exemption, *IND* investigational new drug, *MSC* mesenchymal stem (or stromal) cell, *PPP* platelet-poor plasma, *PRP* platelet-rich plasma, *SVF* stromal vascular fraction

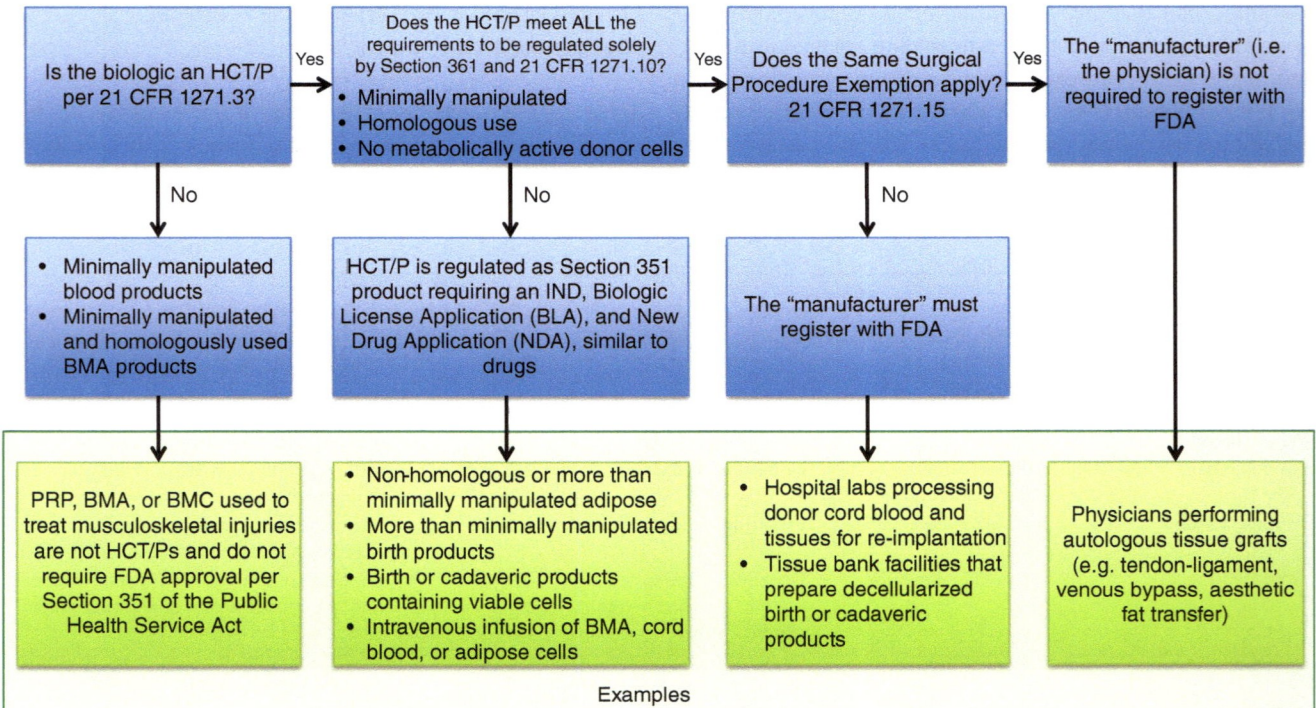

Fig. 19.1 Flowchart of FDA regulation of human biologics under 21 CFR 1271 and Sections 351 and 361 of the Public Health Service Act

sidered by the FDA to be an HCT/P. Some materials, like adipose tissue, are both 351- and 361-category HCT/Ps depending on the handling and intended treatment. For example, enzymatic digestion of autologous adipose tissue rises above minimal manipulation, and the product is considered to be a biological drug [8]. However, adipose tissue processed so that it remains "such HCT/P" can be used for lipofilling in a wide variety of treatment sites (e.g., subcutaneous tissues on the hand and face) [8].

Some of the materials listed in Table 19.2 are obtained from donors. The manufacturers of those materials must meet the requirements laid out in 1271.10(a) for the materials to be FDA compliant. One important element in the classification of donor-derived biological tissues depends on the manufacturer's advertising, which will have an impact on how physicians might use the material to treat patients. For example, placental tissue-derived sheet products when used as a covering in wound care are an acceptable 361-category product. However, if the manufacturer advertises that the identical product can reduce pain, the FDA considers this indication for use to be nonhomologous and the product for reducing pain is a 351-category product. Physicians need to be aware of these distinctions, since sales representatives might not be the best source of information for questions about compliance with FDA regulations. A flowchart of FDA regulation of human biologic products is provided in Fig. 19.1.

Emerging Options in the New Regulatory Environment

Therapeutic materials available to physicians are considered by the FDA to be the practice of medicine (e.g., PRP and BMC), are offered for sale by manufacturers who have met the requirements set out in 21 CFR 1271, or are physician-made at point of care while still satisfying the "such HCT/P" requirement of the same surgical procedure exception (see section "Same Surgical Procedure Exception—The Gotcha Section of 1271?" above). Thus, a physician should not be working with donor-derived materials that have not been established by the manufacturer to fall within the Section 361 product category. Examples of products that do not meet the 361 requirements include materials in which viable cells are present, contain components of cord blood, or incorporate micronized tissues (e.g., injectable tissue).

However, there has been a recent effort at the FDA to make a wider variety of regenerative materials available on an expedited basis for treating patients. The initiative involves a program that was introduced in 2017 in a draft guidance entitled "Expedited Programs for Regenerative Medicine Therapies for Serious Conditions" [10]. The draft guidance describes a new category of therapy known as "regenerative medicine advanced therapy" or RMAT. RMATs are therapies that meet the criteria spelled out in the draft guidance and are eligible for expedited review and other special considerations. An

RMAT designation for a regenerative product will be based in part on preliminary clinical data in which the specific material/product is used to treat serious medical conditions. The FDA acknowledges that this preliminary data might not come from Level 1 clinical studies (i.e., double-blind, placebo-control formatted clinical studies). The FDA also has indicated in a recent publication [11] that RMAT-designated products would be eligible for participation by multisite participants under one investigational umbrella from which the outcomes data would be aggregated for review by the FDA as a part of the Investigational New Drug application for the RMAT.

Make no mistake, however, that a material given RMAT designation still will need to meet the standards established for INDs in terms of efficacy and safety. This means that an RMAT will be treated like any drug in an IND, which implies that the regulatory process will take years and will be associated with a significant cost. Of course, the FDA has indicated that they will interact with manufacturers/physician groups in order to assist in making the process as efficient as possible [10, 11].

What kinds of products might end up as RMATs? Most of the products the FDA has cited with a Warning or Untitled Letter probably could be considered as RMAT candidates. For example, MiMedx has an injectable (e.g., micronized) placental membrane-derived product already designated as an RMAT [12], after receiving an Untitled Letter on their product back in 2013 [13]. Cultured allo-cells or autologous cultured cells also would seem to be good candidates for the RMAT designation. Enzymatic digestion of adipose tissue probably is one of the best candidates for the RMAT process, given its wide-spread, but non-FDA compliant use, while mechanically microfractured adipose tissue would be a close second when used to treat musculoskeletal conditions, since this indication for use currently is not cleared for microfractured adipose tissue.

The fact that the FDA has provided options for expedited review of products or materials it currently considers to be biological drugs (e.g., single-cell suspensions obtained from adipose tissue by enzymatic digestion), but which still are widely used in the regenerative medical community, suggests that the Agency would not be changing its mind on the regulatory status of materials identified in Table 19.2 as "351" category biologics. On the contrary, the FDA has begun to deal with the "bad actors" that continue to use materials that are not appropriately cleared [14], by seeking to shut down clinics or stop the distribution of unapproved products. Their actions should be received by physicians as caveat emptor—be cautious about the materials they use for practicing regenerative medicine.

Ethical Considerations

Physicians contemplating the world of regenerative medicine are confronted by a wide variety of donor-derived therapeutic agents and in-clinic processing options. Frequently, a

physician will turn to their colleagues, sales representatives, or company personnel for guidance as to what is compliant with the FDA. Unfortunately, even their own colleagues could be misinformed, and clearly company personnel will have a bias for selling the physician whatever they get a commission on. What are a physician's options for identifying what is and what is not compliant?

The FDA has weighed in on what they consider to be unethical practitioners of regenerative medicine. For example, in a press release issued in conjunction with the final Guidances on Same Surgical Procedure Exception and MM/HU November 16, 2017, the FDA commissioner made the following statement [14]:

> But the rapid growth and promise of this field have increasingly sowed the ground for the entry of some unscrupulous actors, who have opportunistically seized on the clinical potential of regenerative medicine to make deceptive claims to patients about unproven and, in some cases, dangerous products. By exploiting the lack of consumer understanding of this area, as well as the fear and uncertainties posed by the diseases these bad actors claim to treat, they are jeopardizing the legitimacy and advancement of the entire field.

The commissioner went on to state that those who are exploiting patients will be held accountable. Thus, the FDA intends to do what it can to address the "bad actors", since they are behaving in an unethical manner. The following statement appeared in the Minimal Manipulation and Homologous Use Guidance [8] concerning the danger to the public of physicians willing to skirt ethical obligations and appropriate medical practices to generate revenue:

> FDA intends to focus enforcement actions on products with higher risk, including based on the route and site of administration. For example, actions related to products with routes of administration associated with a higher risk (e.g., those administered by intravenous injection or infusion, aerosol inhalation, intraocular injection, or injection or infusion into the central nervous system) will be prioritized over those associated with a lower risk (e.g., those administered by intradermal, subcutaneous, or intraarticular injection). HCT/Ps that are intended for nonhomologous use, particularly those intended to be used for the prevention or treatment of serious and/or life-threatening diseases and conditions, are also more likely to raise significant safety concerns than HCT/Ps intended for homologous use because there is less basis on which to predict the product's behavior in the recipient, and use of these unapproved products may cause users to delay or discontinue medical treatments that have been found safe and effective through the New Drug Application or BLA approval processes.

Clearly, the FDA has provided detailed information on what types and routes of treatment they believe reflect a lower risk to the patient. Physicians adhering to these general guidelines are unlikely to run afoul of the FDA's efforts to curtail the "bad actors."

Options for Physicians to Assess FDA Compliance and Utility of Therapeutic Agents and Processes

One of the most important steps a physician can take in trying to maintain an ethical position on providing regenerative medical therapies is to understand the basic mechanism of the therapeutic agent. This type of information might be obtained from sales representatives or company officials, but the value of that information needs to be handicapped. For example, there are dozens of amniotic fluid-derived preparations available in the marketplace, but the authors are not aware of published clinical evidence supporting the use of those products for treating any orthopedic condition. Nonetheless, a starting point is to obtain research publications on the product or process.

There are several forums on the Internet that a physician can join and follow in order to receive information on the regulatory status of commercial products (e.g., Biologic Orthopedic Society on LinkedIn). These forums also provide insight into how other physicians practice medicine with the wide variety of products and processes available in the regenerative medical field.

Steps to Take to Ensure FDA Compliance with Regenerative Products and Processes

Assuming a physician has found a donor-derived product that has some clinical support for a specific indication for use, the physician is confronted by the need to determine if the product is compliant with the standards explained in this chapter. The FDA is a willing source of information on the regulatory status of products and processes, so a physician should request that the company provide a letter attesting to the regulatory status of the product, kit, or process. The letter should contain a statement that the product and its indications for use have been reviewed by the FDA and that the product is a 361-category product not requiring an IND before marketing.

More likely, in response to a physician's request for documentation, companies will point to the registration of their company and their product as "evidence" of FDA compliance. They might even point out that their company had been audited. This is just a smoke screen. Any company offering for sale in the US products derived from human donors has to be registered and follow minimal standards for screening donors, handling the material, packaging and storage of the product. Furthermore, when the FDA audits a company, their focus is on the documentation associated with the product and the manufacturing conditions, but not its regulatory status. However, during an audit, the FDA might become aware

of certain facts about the composition or advertised use of the product that raises issues of compliance with 1271. In this situation, the FDA might issue a Warning Letter in which deficiencies in the acquisition, processing, storage, and sale of the product were observed, and if appropriate, stating that the material is not a 361-category material. If there is not a problem with the product itself, but the FDA believes the product does not meet the requirements of a 361-category material, the FDA will issue an Untitled Letter.

Physicians might not find too many companies willing to provide a letter clearly stating the regulatory status of their product or process. However, if a physician is interested enough in the material, but is uncomfortable proceeding without a more detailed assessment of the product, the Tissue Reference Group (TRG) can be contacted in order to receive an informal opinion from the FDA on a product or process for a specific indication for use. The TRG can be contacted by the following link:

https://www.fda.gov/BiologicsBloodVaccines/TissueTissueProducts/RegulationofTissues/ucm152857.htm.

Resources for Physicians in Dealing with the Challenging Regenerative Medicine Marketplace

While physicians might wish to limit their practice of regenerative medicine to those products/processes that are considered to be 1271- or FDA compliant, the field is filled with "bad actors," as indicated by Commissioner Gottlieb [14]. So, what is an ethical physician to do when competing with unethical physicians and clinics? For physicians who believe clinicians are misleading the public by providing outrageous treatment claims or for companies making false claims about the regulatory compliance of their products, the following link provides information for submitting a complaint to the FDA:

https://www.fda.gov/BiologicsBloodVaccines/DevelopmentApprovalProcess/AdvertisingLabelingPromotionalMaterials/ucm118859.htm

On the other hand, if a physician or clinic thinks it might want to treat patients with an HCT/P for which there is uncertainty about the appropriate regulatory path, the following link will provide information on how to obtain preliminary information on the HCT/P prior to submitting a "Request for Designation" (the mechanism by which a manufacturer asks the FDA to provide a formal statement as to how the HCT/P will be regulated by the Agency):

https://www.fda.gov/BiologicsBloodVaccines/TissueTissueProducts/RegulationofTissues/ucm152857.htm.

Conclusion

While the field of regenerative medicine is being embraced by a wide range of physicians and surgeons, there is a need for these healthcare providers to understand not just the mechanics of practicing regenerative medicine, but to understand the regulatory elements and framework that govern the use of regenerative therapeutic agents in the USA. The physician has a variety of materials to consider, ranging from a patient's own tissues to donor-derived cells, fluids, and tissues. The regulation of each of these materials will depend on how the materials are obtained, processed, and used to treat the patient and for a specific indication for use.

21 CFR 1271 was promulgated to legally establish the FDA's authority to regulate human tissue-derived materials, and its role is vital in protecting the public from noncompliant regenerative products. It is important that physicians know what they are using, the rationale for the material's use, and the regulatory requirements for the material. In particular, physicians should challenge company representatives to provide definitive statements as to the regulatory status of the products they wish to sell to the physician. If such documentation is not provided, the physician runs the risk of using an unapproved drug or biological drug, thereby exposing their patients to potential safety issues. Furthermore, physicians should not be shy about complaining to the FDA when they encounter physicians or clinical facilities that satisfy the emerging concept of "bad actors" who seek to exploit a naïve public to pay for noncompliant therapies that lack even minimal clinical efficacy data.

References

1. U.S. Code, 42 U.S.C. Section 264 Public Health Service Act.
2. U.S. Federal Register, 66FR: 5447, 5478 (Jan. 19, 2001).
3. U.S. 21 CFR Section 1271.3 "How does FDA define important terms in this part". https://www.accessdata.fda.gov/scripts/cdrh/cfdocs/cfcfr/CFRSearch.cfm?fr=1271.3.
4. U.S. 21 CFR Section 1271.3(d) "How does FDA define important terms in this part"; https://www.accessdata.fda.gov/scripts/cdrh/cfdocs/cfcfr/CFRSearch.cfm?fr=1271.3.
5. U.S. 21 CFR Section 1271.10(a) "Are my HCT/Ps regulated solely under Section 361 of the PHS Act and the regulations in this part, and if so what must I do?". https://www.accessdata.fda.gov/scripts/cdrh/cfdocs/cfcfr/CFRSearch.cfm?fr=1271.10.
6. U.S. 21 CFR Section 1271.10(b) "Are my HCT/Ps regulated solely under Section 361 of the PHS Act and the regulations in this part, and if so what must I do?". https://www.accessdata.fda.gov/scripts/cdrh/cfdocs/cfcfr/CFRSearch.cfm?fr=1271.10.
7. U.S. 21 CFR Section 1271.15(b) "Are there exceptions from the regulations in this part?" https://www.accessdata.fda.gov/scripts/cdrh/cfdocs/cfcfr/CFRSearch.cfm?fr=1271.15.
8. U.S. FDA. Regulatory considerations for human cells, tissues and cellular and tissue-based products: minimal manipulation and homologous use. Guidance for Industry and Regulatory Staff (Corrected December 2017) https://www.fda.gov/downloads/biologicsbloodvaccines/guidancecomplianceregulatoryinformation/guidances/cellularandgenetherapy/ucm585403.pdf.
9. U.S. FDA. Same Surgical Procedure Exception under 21 CFR 1271.15(b): Questions and Answers Regarding the Scope of the Exception, Guidance for Industry (November 2017) https://www.fda.gov/downloads/biologicsbloodvaccines/guidancecomplianceregulatoryinformation/guidances/tissue/ucm419926.pdf.
10. U.S. FDA. Expedited programs for regenerative medicine therapies for serious conditions. Draft Guidance for Industry (November 2017). https://www.fda.gov/downloads/biologicsbloodvaccines/guidancecomplianceregulatryinformation/guidances/cellularandgenetherapy/ucm585414.pdf.
11. Marks P, Gottlieb S. Balancing safety and innovation for cell-based regenerative medicine. N Engl J Med. 2018;378:954–9. https://doi.org/10.1056/NEJMsr1715626.
12. Hildreth C. What is an RMAT designation and who has one? 2018. https://bioinformant.com/rmat/#list. Accessed 7 Aug 2018.
13. U.S. FDA. Untitled Letter, Surgical Biologics, a MiMedx Group Company, dated August 28, 2013 https://www.fda.gov/biologicsbloodvaccines/guidancecomplianceregulatoryinformation/complianceactivities/enforcement/untitledletters/ucm367184.htm.
14. U.S. FDA. Statement from FDA Commissioner Scott Gottlieb, M.D. on FDA's comprehensive new policy approach to facilitating the development of innovative regenerative medicine products to improve human health. 16 Nov 2017. https://www.fda.gov/NewsEvents/Newsroom/PressAnnouncements/ucm585342.htm.

Anti-Platelet and Anticoagulation Medications

20

George C. Chang Chien and Raj Panchal

Introduction

Regenerative medicine encompasses a branch of medicine that deals with replacing or restoring injured human cells, tissues, and organs in an attempt to improve function, pain, and healing in patients. It relies on the body's own reparative mechanisms to heal previously damaged tissues. The regenerative milieu incorporates progenitor cells, signaling molecules, and structural "scaffolding." Progenitor cells (mesenchymal, hematopoietic) are capable of differentiating into a variety of cells (osteoblasts, chondrocytes, myocytes) in the presence of appropriate signaling molecules. These cells also respond to signaling molecules during injury to coordinate the healing cascade. Signaling molecules include various growth factors (e.g., beta transforming growth factor, insulin-like growth factor, and platelet-derived growth factor) that will promote specific cell differentiation or tissue remodeling. The scaffolding refers to the carrier environments that provide the framework for cell proliferation and growth (e.g., collagen, extracellular matrix, fat).

Currently, regenerative therapies include a wide array of available treatments including bone marrow and adipose-derived stem cells and platelet-rich plasma (PRP). The foundation for PRP involves isolating the plasma portion of blood with higher concentrations of platelets, in an attempt to bring forth the effects of various growth factors stored in platelets. These growth factors play formative roles in hemostasis and the healing cascade. When considering anticoagulation and its relation to regenerative medicine, PRP is of particular importance due to the role platelets play within these pathways.

The steps of hemostasis mentioned below involve an intricate array of enzymes, proteins, and activation pathways, many of which serve as targets of action for different anticoagulants and inhibitors of platelet function. Similarly, the healing cascade allows various points of entry for medications to impair this normal series of events. These may include nonsteroidal anti-inflammatory drugs (NSAIDs), clopidogrel, heparin, warfarin, enoxaparin, selective serotonin reuptake inhibitors (SSRIs), and other medications that can affect serotonin function (e.g., tramadol, tricyclic antidepressants, selective serotonin norepinephrine reuptake inhibitors). For this reason, it is widespread belief that such medications will likely affect the efficacy of autologous PRP. As a result, practitioners currently encourage the discontinuation of NSAIDs 1 week prior to PRP treatment and anywhere from 2 to 6 weeks posttreatment. Thorough clinical studies on the effects of SSRI use before and after PRP injections have not yet been performed. Formalized recommendations on the aforementioned medications are still inconclusive, but will be discussed in greater detail later in the chapter.

Before delving into various medications, specifically anticoagulants, and how they may impact the efficacy of PRP and other regenerative therapies, we must first understand the steps and pathways of hemostasis, the healing cascade, and the role platelets play within the process.

Coagulation and Hemostasis

Normal hemostasis consists of three primary steps: vasoconstriction, platelet plug formation, and coagulation. Vasoconstriction takes place via two primary mechanisms. A localized sympathetic response from nociceptors will initially cause reflex vasospasm in vascular smooth muscle cells in order to minimize blood loss at the site of injury. Secondly, vasoconstriction can take place when damage to the endothelial walls disrupts the balance of release between endothelin (vasoconstrictor) and nitric oxide, prostacyclin,

G. C. Chang Chien
GCC Institute, Department of Musculoskeletal Medicine and Medical Aesthetics, Newport Beach, CA, USA

R. Panchal (✉)
NYU Langone Medical Center, Rusk Rehabilitation, Department of Physical Medicine and Rehabilitation, New York, NY, USA
e-mail: raj.panchal@nyumc.org

© Springer Nature Switzerland AG 2023
C. W Hunter et al. (eds.), *Regenerative Medicine*, https://doi.org/10.1007/978-3-030-75517-1_20

and CD39 (vasodilators). This imbalance will result in release of unopposed endothelin and thus further vasoconstriction [1].

The next step in hemostasis involves the formation of a platelet plug. This can be further broken down into platelet adhesion, activation, and degranulation. Pertinent platelet anatomy in this step includes surface glycoproteins GP1B/5/9 and GP2B3A as well as alpha and dense granules carried within the platelets. During endothelial cell damage, subendothelial collagen becomes exposed and binds von Willebrand factor (vWF). vWF then subsequently binds to GP1B/5/9 from circulating platelets, initiating platelet adhesion. This leads to platelet activation and degranulation of the alpha and dense granules. The alpha granules contain various growth factors that serve as the foundation of PRP as well as cytokines, coagulation factors, fibrinolytic factors, and antibacterial proteins. These cytokines and growth factors act in an autocrine or paracrine fashion to modulate cell signaling and stimulate the healing cascade; bound and activated platelets degranulate to release platelet chemotactic agents to attract more platelets to the site of injury.

Alpha granules are the most abundant granules and there are about 50–80 alpha granules per platelet. Dense granules are also released and contain serotonin (vasoconstrictor), platelet activator and aggregator adenosine diphosphate (ADP), and calcium ions (needed for secondary hemostasis). During this step, thromboxane A2 (TXA2) is also released which serves as another vasoconstrictor and increases expression of GP2B3A. GP2B3A, a platelet surface glycoprotein, also becomes activated and contributes to platelet aggregation by binding fibrinogen, which is then able to also bind GP2B3A from other platelets and form a platelet plug. Formation of the platelet plug also marks the completion of events collectively identified as *primary hemostasis* [1].

Secondary hemostasis involves clot formation via the coagulation cascade. Calcium released from the previous step activates various coagulation factors and initiates the coagulation cascade. This results in the formation of fibrin from fibrinogen in the presence of thrombin and eventual creation of a fibrin mesh and clot formation [1].

The Healing Cascade

The healing cascade lays the framework for regenerative medicine and occurs over the course of weeks, with final tissue remodeling taking potentially many months before restoration of full tissue strength and integrity. Healing involves many of the same activating signals and growth factors released by platelets during degranulation. The process of wound healing can be subdivided into inflammation, proliferation, and maturation stages [2].

Differing models of the cascade may separate out the process of hemostasis (coagulation) as occurring to prior to the inflammation stage, while others include it. Despite the taxonomical variance, the entire process occurs as part of the spectrum of healing. After the formation of a fibrin mesh and a clot, cytokines and growth factors previously released during platelet degranulation stimulate the complement cascade and recruit leukocytes (primarily neutrophils), macrophages, and fibroblasts to the injured area. Local histamine release leads to increased capillary permeability via vasodilation and leakage, allowing migration of mesenchymal stem cells (MSCs) to the site. Neutrophils then lead the process of decontamination through bacterial lysis and scavenging of cellular debris. Monocytes previously activated by platelet growth factors also migrate to the area and may differentiate into macrophages. These macrophages play various important roles: bacterial phagocytosis, cytokine and collagenase secretion for tissue remodeling, and secretion of growth factors that contribute toward angiogenesis and formation of granulation tissue [2]. Among the factors secreted by macrophages are many that are associated with bone repair, such as interleukins (ILs), tumor necrosis factor alpha (TNF-α[alpha]), transforming growth factor beta (TGF-β[beta]), platelet-derived growth factor (PDGF), endothelial growth factor (EGF), and vascular endothelial growth factor (VEGF). [3]

The proliferation stage then begins with epithelialization by migratory epithelial progenitor cells as well as epithelial cells from the wound periphery. Angiogenesis takes place under the signals from previously released platelet growth factors. Fibroblasts drive the production of granulation tissue and collagen deposition around 4 days after an injury. Mesenchymal stem cells are integral in coordinating the healing response, but can also be activated to begin differentiation down chondrogenic, osteogenic, or angiogenic pathways.

During the final maturation or remodeling stage, a wound contracts as collagen continues to be deposited by fibroblasts, granulation tissue compresses into smaller and newly formed scar tissue, and the strength of this new wound increases. This process is also driven by various growth factors that were present for the previous steps. Overall, while it is easier to comprehend all these steps linearly, in reality many of them overlap and occur simultaneously providing an onslaught of regeneration and remodeling [2].

Platelets: Function, Preparation, Activation

The power of PRP stems from the ability to harness bioactive factors stored in platelet alpha granules once they are activated. These include growth factors such as insulin-like growth factor (IGF), platelet-derived growth factor (PDGF), beta transforming growth factor (TGF beta), vascular endothelial growth factor (VEGF), and numerous interleukins. The effects of these include enhancing DNA synthesis, promoting chemotaxis, angiogenesis, and mitogenesis for fibroblasts, chondroblasts and osteoblasts. Platelets also contain dense granules which contain adenosine diphosphate (ADP), adenosine triphosphate (ATP), ionized calcium, and serotonin which are which is necessary for several steps of the coagulation cascade.

The preparation of PRP involves isolating the plasma component of whole blood, with a concentration of platelets 3–6 times above baseline. However, the concentration will vary depending on the method of procuring PRP. During the process of plasma isolation, white blood cells and other blood components will also be obtained and exact concentration of these components can greatly influence outcomes and success of PRP administrations [4].

Obtaining patient's blood will be the first step of harvesting PRP. Generally, about 25–50 cc of venous blood will be obtained in order to yield about 3–6 cc of PRP depending on the platelet count of the individual, device used, and technique employed. Retrieval of blood will generally be followed by one or two rounds of centrifugation. The goal of centrifugation is to separate the whole blood into a layer of plasma, buffy coat, and red blood cells. The most superficial layer, the platelet-poor plasma (PPP), is typically removed and may be saved to obtain other bioactive proteins such as alpha-2 macroglobulin. The buffy coat contains the platelet-rich plasma, as well as leukocytes. Further processing through a second centrifugation process may then be possible to help remove leukocytes from the serum.

An anticoagulant is required in order to prevent early coagulation. The type of anticoagulant used can influence time to activation, yield of platelets, and the volume of growth factors released. Commonly used anticoagulants include sodium citrate and acid citrate dextrose (ACD-A). EDTA has been demonstrated to have potential to damage the platelet membrane. ACD-A generally has a more acidic composition than sodium citrate and has a lower citrate ion concentration, yielding impaired platelet aggregation in vitro. PRP obtained in sodium citrate has been demonstrated in one in vitro study to produce higher platelet recovery after centrifugation step and minimal change in MSC gene expression [5].

As aforementioned, the step of platelet activation involves degranulation of platelets and the subsequent release of growth factors from alpha granules. Platelets are generally activated by one of many ways. Thrombin, collagen, calcium chloride, or even light have been shown to serve as potent activators of platelets. There is no empirical data supporting the use of one mechanism of activation or the other, and practitioners today differ greatly in their mechanism of choice. A caveat to exogenous, preinjection platelet activation is the need for injection prior to clotting of the platelets. Some practitioners may intentionally wait for the clotting to begin to utilize the platelets as part of a fibrin clot.

A 2016 study performed by Cavallo et al. [6] revealed how the release of different growth factors in leukocyte-rich PRP is influenced by the activating agent used. They compared CaCl alone, thrombin alone, their combination, and autologous collagen type 1. The tests were performed in vitro and across the board demonstrated lower release of growth factors and inflammatory mediators when collagen was used compared to the other agents [6]. Another finding was that thrombin, collagen, and CaCl/Thrombin had a rapid release of growth factors that remained stable for up to 24 hours, whereas the release of growth factors with CaCl was more gradual initially, however reaching a similar level of growth factor release at 24 hours. In addition, the activators in this study all demonstrated different levels of clot formation. The collagen group did not demonstrate platelet aggregation over a 24 hour period, while the CaCl and thrombin groups showed clot formation within 15–30 mins [6].

Another less common and novel form of activation is low-level light irradiation (photoactivation). Minimal literature currently exists for this method, but practitioners are increasingly more aware of its benefits. A 2016 double-blind, randomized controlled pilot study by Paterson et al. [7] revealed preliminary evidence of the feasibility and safety of photoactivated platelet-rich plasma in knee osteoarthritis, warranting larger studies.

Medications that May Interfere with Platelet Success

Many medications have been shown to interfere with platelet function, whether directly or indirectly. Formalized recommendations on the continuation or discontinuation of medications when administering PRP are currently based on limited literature. Research is scant on this topic considering the difficulty of executing a well-designed study that accounts for drug-drug interactions, individual platelet levels, comorbidities, age, etc. In spite of this, it is important to understand the existing literature as well as mechanisms of how various medications may influence and impair normal platelet function. In the following section, we will discuss many medications, review their mechanisms of action, and examine how they may influence PRP integrity and efficacy.

NSAIDS

Nonsteroidal anti-inflammatory drugs (NSAIDs) are generally the first class of medications that come to mind when considering medications that may influence PRP efficacy. NSAIDs act by impairing the function of the enzymes cyclooxygenase 1 and 2 (COX1 and COX2), two enzymes involved in converting arachidonic acid into prostaglandin H2 and eventually prostacyclin and TXA2. TXA2 has been shown to play a role in platelet aggregation by promoting vasoconstriction and increasing expression of GP2b/3a. Classic NSAIDs such as aspirin, ibuprofen, and naproxen are all inhibitors of both COX-1 and COX-2, whereas newer NSAIDS such as celecoxib are selective inhibitors of COX-2. NSAIDs have thus been shown to impair platelet aggregation and reduce the PRP quality of bioactive factors delivered to an injured area [8].

Nonselective NSAIDs also affect prostaglandin synthesis through their inhibition of COX1 and COX2. Their adverse effects are well known and well documented, including erosion of the gastric and colonic mucosa, elevated blood pressure, and increased risk for bleeding. During bony fractures, it is believed that injured bone becomes hypoxic, and this drives the production of prostaglandin E2 from osteoblasts. Studies have even shown increased local COX enzyme expression during these hypoxic events, suggesting increased yield of prostaglandins. Prostaglandins promote bone regeneration by means of angiogenesis, chemotaxis, and vasodilation, among other roles. Specifically, PGE2 has been found to have the highest expression in bones. They contribute toward limiting bony resorption as well as the development of osteoblasts. NSAIDs may disrupt this balance of enzymes and impair bone regeneration, potentially shifting the equilibrium toward bony resorption [9]. A systematic review of the existing literature on this topic performed by Pountos et al. [10] found no definitive evidence to withhold NSAIDs in patients with bone fractures; however, they caution that most of the available research has been performed on animals and not human subjects, and thus, NSAIDs should still be considered a risk factor for impairment of bone healing. There is also evidence that shows that NSAIDs negatively affect the structural healing of injured tendons via prostaglandin E2 (PGE2), which is essential for early tendon healing through control of vascular flow [11].

Clopidogrel

Clopidogrel is a prodrug that is biotransformed in vivo and works by acting as an ADP receptor irreversible inhibitor. As you may recall, ADP is an important metabolite in normal platelet function that binds to one of three different ADP receptors (P2Y1, P2Y12, P2X1) on the platelet surface and contributes toward platelet activation and aggregation [12]. Clopidogrel specifically binds the P2Y12 ADP receptor, preventing activation of GP2B/3A and eventual cross-linking by fibrin. Other medications in this class with similar mechanisms include prasugrel and ticagrelor.

By altering platelet activation and aggregation, clopidogrel affects primary hemostasis and normal platelet function.

Medications that Affect Serotonin

There are a number of commonly prescribed medications that will alter serotonin pathways and, in turn, potentially interfere with platelet function. Medications in this category include SSRIs, SNRIs, tricyclic antidepressants, tramadol, migraine abortive medications such as ergotamine, and triptans.

In recent years, SSRIs have been increasingly discussed as another class of medications that may influence PRP success. They work by inhibiting the function of presynaptic serotonin transporters (SERTs) which serve as monoamine transporter proteins and reuptake serotonin from synaptic spaces and back into various cells, including platelets. The net effect of SSRIs ultimately allows for more serotonin to bind with postsynaptic receptors. Serotonin is released from dense granules after platelets are activated and promotes platelet aggregation and vasoconstriction. In the presence of SSRIs, it is believed there is a higher plasma concentration of serotonin and decreased intraplatelet concentration of serotonin due to less reuptake, overall leading to decreased serotonin release during platelet degranulation and less amplified platelet aggregation. Retrospective studies on the issue have yielded results indicating an increased abnormal bleeding risk in patients on SSRIs, specifically upper GI bleeding [13].

By altering normal serotonin function, these medications affect primary hemostasis and normal platelet activation and aggregation.

Anticoagulants

Many of the medications that classify as anticoagulants act upon the pathways of secondary hemostasis, specifically the coagulation cascade rather than primary hemostasis or platelet plug formation. Warfarin works by acting as a competitive inhibitor of VKORC-1 an enzyme that aids in the creation of vitamin K, which is necessary for the synthesis of factors 2,7,9,10, Protein C, and Protein S. Rivaroxaban and apixaban are newer anticoagulants that act as direct, selective, and reversible inhibitors of both free and clot-bound factor Xa. Heparin binds by activating antithrombin III, which then inactivates thrombin and factor Xa, effectively diminishing conversion of fibrinogen to fibrin. Enoxaparin works similarly to Heparin; however, it has a stronger anti-factor Xa effect and less inhibition of thrombin, as thrombin inhibition is heparin size-dependent. Argatroban is a selective and reversible direct thrombin inhibitor. Abciximab affects primary hemostasis and is a chimeric human monoclonal antibody that acts as a platelet 2b/3a receptor inhibitor.

Similar to the effect NSAIDs have on fracture healing, it has been widely believed anticoagulants demonstrate a similar effect. A systematic review by Lindner et al. [14] demonstrated that in a small number of in vivo animal model studies heparin and LMWH were found to reduce trabecular bone volume in a dose-dependent manner; however, the effects of heparin are far greater than LMWH. The studies also revealed warfarin to have similar effect on trabecular bone volume. Histologic analysis suggested heparin impaired osteoblastic bone formation and increased osteoblastic bone resorption, and LMWH inhibits osteogenesis. Warfarin's effects were found to be from inhibition of Vitamin K reductase and Vitamin K epoxide reductase, two enzymes found in osteoblasts. The study also revealed LMWH inhibited VEGF directed angiogenesis. Overall, they concluded that the negative effects of warfarin, heparin, and aspirin on bone healing were unequivocal. While there was enough evidence to suggest the same for LMWH, it was to a lesser degree, and thus, they recommended utilizing this for thromboprophylaxis in a patient with a healing fracture [14].

Evidence suggests that anticoagulant medications have direct effects on wound healing, but their effects on the coagulation cascade do not directly impair platelet aggregation and activation.

Discussion

Clinicians currently vary in their standard of practice regarding which medications to hold prior to PRP administration and after, as well as the length of time to hold these medications for. Research into this topic is very limited, and often, patients' more serious comorbidities are rightly given priority prior to withholding medications that could have grave consequences. These risks should not be taken lightly and should always be taken into consideration prior to administering PRP or deciding to hold a medication for PRP treatment.

The information provided in this chapter thus far has laid the groundwork to understanding the following recommendations. A review article in the September 2016 issue of Pain Physician Journal by Ramsook and Danesh [15] has previously set forth guidelines regarding which medications to hold and for how long, both prior to and after PRP administration. The guidelines proposed assume that the antithrombotic medications included in the article affect platelet degranulation and thus PRP success. Their proposed length of time to hold the medications is based on the medication half-lives as well as the lifespan of a platelet. However, the primary dilemma with this method is that there is not enough empirical data to suggest that many of the medications they recommend holding actually affect the outcomes of success when administering PRP.

In regard to NSAIDs, a pilot study by Schippinger et al. [8] revealed inhibition of platelet aggregation with no difference in platelet count when obtaining platelets from patients taking NSAIDs after orthopedic surgeries vs. obtaining platelets from age-matched healthy controls. They deduced platelet aggregation is an important step in the function of platelets by localizing platelets to a site of injury and leading to increased degranulation at the site. If this step is impaired, they concluded that overall platelet quality and release of growth factors would suffer. The study examined patients taking either dexibuprofen or diclofenac, two nonselective reversible inhibitors of COX-1 and COX-2. A 2017 study by Ludwig et al. [16] concluded that COX-2 selective inhibitors do not affect platelet activation or growth factor release in dogs, warranting further studies to examine whether a similar outcome would be seen in humans. The study did, however, cite a few previous abstracts on humans that showed COX-2 selective NSAIDs such as celecoxib did not significantly affect platelet aggregation or hemostasis.

Another study sought to quantify and compare normative catabolic and anabolic factor concentrations in leukocyte-rich platelet-rich plasma (LR-PRP) at various time points, including baseline, 1 week after initiating naproxen use, and after a 1-week washout period. The anabolic factors vascular endothelial growth factor, fibroblast growth factor 2, platelet-derived growth factor AB (PDGF-AB), and platelet-derived growth factor AA (PDGF-AA) and the catabolic factors interleukin (IL) 1β[beta], IL-6, IL-8, and tumor necrosis factor α[alpha] in LR-PRP were measured at 3 time points: baseline, after 1 week of naproxen use, and after a 1-week washout period. These authors found that naproxen diminished PDGF-AA (44% decrease in median) and PDGF-AB (47% decrease). However, a 1-week washout period was sufficient for the recovery of PDGF-AA, PDGF-AB, and IL-6 to return to baseline levels. Tumor necrosis factor α[alpha], IL-1β[beta], IL-8, vascular endothelial growth factor, and fibroblast growth factor 2 did not show differences between the 3 time points of data collection.

A case report by di Matteo et al. [17] described a 53-year-old active male runner with a medical history significant for a rare metabolic disorder status post multiple aortocoronary bypass surgeries who had been taking aspirin 160 mg daily for 9 years. The runner had been experiencing chronic left knee pain forcing him to stop training for 3 months and failed conservative treatments, eventually opting for three rounds of PRP injections, each 1 week apart. In this case, aspirin was not withheld, and 14 days after the last injection, the patient had no swelling, was pain-free, and returned to training [17]. He was eventually able to run a half marathon 35 days after the last injection. While this is only a case report, it forces clinicians to reexamine previously accepted profiles of patients who can be successfully treated with PRP.

Ex vivo studies have been performed on patients taking SSRIs in order to monitor their effects on platelet quality; however, scanty research exists examining in vivo studies. A 2008 study by McCloskey et al. [18] examined differences in platelet aggregation and ATP release in subjects taking either an SSRI or bupropion for at least 6 weeks. In the presence of agonists such as arachidonic acid and collagen, the study found statistically significant decreases in both platelet aggregation and ATP release in the SSRI group. This does not necessarily provide enough information to safely recommend withholding SSRIs prior to PRP administration, especially in patients suffering from depression, but it does legitimize the possibility for a larger in vivo study.

With regards to various antithrombotic agents, it is essential to consider the patient holistically when determining candidacy for a PRP injection. If someone is taking a medication such as clopidogrel, warfarin, heparin, or enoxaparin, the patient is most likely treating a more critical disease process. The risk of withholding any of the previously mentioned antithrombotic agents for a defined time period will far outweigh the benefit of administering an injection of PRP to treat a musculoskeletal condition. In addition, the mechanisms of action of these medications do not directly overlap with the understood mechanism of action of PRP, making it largely inconsequential to continue these medications when injecting PRP to treat a patient. There is minimal data regarding performing studies examining this relationship and the benefits vs. risks of withholding an antithrombotic agent when administering PRP.

While the majority of this chapter focuses on PRP and the medications that may influence its success, another major application of regenerative medicine involves mesenchymal stem cells. As a growing number of physicians utilize the benefits of harvesting and injecting bone marrow MSCs (BMMSCs) in their patients to stimulate cell proliferation and tissue regrowth, the effects of NSAIDs should be taken into consideration before its usage as well. One 2008 study by Chang et al. [19] showed that in vitro testing in mice found that MSCs treated with dexamethasone, nonselective NSAIDs, and COX-2 selective NSAIDs all suppressed cell proliferation and arrested cell cycle kinetics in the G0/G1 phase. Subsequent addition of prostaglandin E1 and E2 did not reverse the effects, hinting at an alternative mechanism

of action than being prostaglandin mediated. More recent animal studies, however, have yielded different results. A 2017 study by Zhang [20] that suggested that MSCs pretreated with aspirin and injected in rats with periodontitis had the effect of reducing inflammation while promoting periodontal bone and tissue regeneration. Another study by Geetasala in 2017 [21] showed that BMMSCs injected into mice wounds in the presence of celecoxib demonstrated a higher percentage of wound closure, decreased inflammation, and increased cellular proliferation and differentiation vs. BMMSCs injected without celecoxib. These studies leave much to be desired. They cannot be fully compared side by side given the various differences among them; however, they yield overall contradictory results. Given the minimal evidence thus far and lack of human studies examining the impact of anti-inflammatory and anticoagulant medications on the effect and function of BMMSCs, further large-scale human studies are necessary before medication withholding decisions can be confidently made.

Conclusion

The effects of various medications on platelet function are well-documented and established. These medications not only impair normal platelet function, but can negatively affect the quantity and quality of the growth factor profile released from the platelets. General pharmacodynamic guidelines suggest that it takes over 5 half-life periods for a compound to be completely eliminated from the system. Additionally, variance in metabolism and excretion can affect medication half-life duration in different individuals. Please refer to Table 20.1 for a summarization of various medications mentioned in this chapter and their understood effects on platelet function.

Current recommendations for holding medication prior to regenerative medicine procedures should focus not only on the medication half-life, but on the medication mechanism of action, and direct effects on platelet function, and growth factor release. Prior to the regenerative medicine intervention, a well-thought-out risk/benefit analysis should be performed to determine the safety of discontinuing any medication.

Table 20.1 Effects of various medications on platelet function

Medication	Half-life	Mechanism	Comments
Aspirin	15 minutes—9 hours. Based on dose.	Irreversible inhibitor of COX-1 and COX-2	Aspirin irreversibly inhibits platelet function and can effectively last the entire lifespan of the platelet (8–9 days) while it stays in circulation.
Ibuprofen	2–4 hours	Nonselective, reversible inhibitor of COX-1 and COX-2	We recommend the discontinuation of NSAIDs at least 1 week prior to PRP administration and anywhere from 2–6 weeks after.
Naproxen	12–17 hours	Nonselective, reversible inhibitor of COX-1 and COX-2	NSAIDs should not be held if the patient is taking them for life-saving indications such as stroke or cardiac prophylaxis.
Celecoxib		Selective reversible inhibitor of COX-2	No evidence that celecoxib has effect on platelet function
Clopidogrel	7–8 hours	Irreversible inhibitor of ADP receptor	Clopidogrel should not be held if the patient is taking them for life-saving indications such as stroke or cardiac prophylaxis.
Warfarin	20–60 hours	Competitive inhibitor of VKORC-1 (enzyme involved in creation of vitamin K, which is necessary for synthesis of factors 2,7,9,10, protein C, S)	Effects are on secondary hemostasis. Unclear effects on PRP intervention.
Enoxaparin	4.5–7 hours	Binds to antithrombin 3 and acts as irreversible factor Xa inactivator	Effects are on secondary hemostasis. Unclear effects on PRP intervention.
Abciximab	10–30 mins	Chimeric human monoclonal antibody; platelet 2b/3a receptor inhibitor	Effects are on secondary hemostasis. Unclear effects on PRP intervention.
Apixaban	8–15 hours	Direct, selective, and reversible inhibitor of free and clot-bound factor Xa	Effects are on secondary hemostasis. Unclear effects on PRP intervention.
Rivaroxaban	5–9 hours	Direct, selective, reversible free and bound factor Xa inhibitor	Effects are on secondary hemostasis. Unclear effects on PRP intervention.
Heparin	1.5 hours	Antithrombin III inhibitor	Effects are on secondary hemostasis. Unclear effects on PRP intervention.
Argatroban	40–50 mins	Direct, selective, reversible thrombin inhibitor	Effects are on secondary hemostasis. Unclear effects on PRP intervention.
Fluoxetine	1–3 days (acute), 4–6 days (chronic)	Selective serotonin reuptake inhibitor	Will affect normal platelet function. Risk/benefit for discontinuation should be clearly discussed with the patient and the prescriber of the antidepressant medication.
Sertraline	23–26 hours	Selective serotonin reuptake inhibitor	
Paroxetine	21 hours	Selective serotonin reuptake inhibitor	
Duloxetine Venlafaxine		Selective serotonin norepinephrine reuptake inhibitor	
Tricyclic antidepressants Amitriptyline Nortriptyline Desipramine		Serotonin norepinephrine reuptake inhibitor	

References

1. Yun S-H, Sim E-H, Goh R-Y, Park J-I, Han J-Y. Platelet activation: the mechanisms and potential biomarkers. Biomed Res Int. 2016;2016:9060143. https://doi.org/10.1155/2016/9060143.

2. Sinno H, Prakash S. Complements and the wound healing cascade: an updated review. Plast Surg Int. 2013;2013:146764. https://doi.org/10.1155/2013/146764.

3. Bhat A, Wooten RM, Jayasuriya AC. Secretion of growth factors from macrophages when cultured with microparticles. J Biomed Mater Res A. 2013;101(11):3170–80.

4. Dhurat R, Sukesh M. Principles and methods of preparation of platelet-rich plasma: a review and author's perspective. J Cutan Aesthet Surg. 2014;7(4):189–97. https://doi.org/10.4103/0974-2077.150734.

5. do Amaral RJ, da Silva NP, Haddad NF, Lopes LS, Ferreira FD, Filho RB, Cappelletti PA, de Mello W, Cordeiro-Spinetti E, Balduino A. Platelet-rich plasma obtained with different anticoagulants and their effect on platelet numbers and mesenchymal stromal cells behavior in vitro. Stem Cells Int. 2016;2016:7414036.

6. Cavallo C, Roffi A, Grigolo B, et al. Platelet-rich plasma: the choice of activation method affects the release of bioactive molecules. Biomed Res Int. 2016;2016:6591717. https://doi.org/10.1155/2016/6591717.

7. Paterson KL, Nicholls M, Bennell KL, Bates D. Intra-articular injection of photo-activated platelet-rich plasma in patients with knee osteoarthritis: a double-blind, randomized controlled pilot study. BMC Musculoskelet Disord. 2016;17:67. https://doi.org/10.1186/s12891-016-0920-3.

8. Schippinger G, Prüller F, Divjak M, et al. Autologous platelet-rich plasma preparations: influence of nonsteroidal anti-inflammatory drugs on platelet function. Orthop J Sports Med. 2015;3(6):2325967115588896. https://doi.org/10.1177/2325967115588896.

9. van Esch RW, Kool MM, Svan A. NSAIDs can have adverse effects on bone healing. Med Hypotheses. 2013;81(2):343–6. https://doi.org/10.1016/j.mehy.2013.03.042.

10. Pountos I, Georgouli T, Calori GM, Giannoudis PV. Do nonsteroidal anti-inflammatory drugs affect bone healing? A critical analysis. Sci World J. 2012;2012:606404. https://doi.org/10.1100/2012/606404.

11. Chan KM, Fu SC. Anti-inflammatory management for tendon injuries—friends or foes? Sports Med Arthrosc Rehabil Ther Technol. 2009;1(1):23. Published 2009 Oct 13. https://doi.org/10.1186/1758-2555-1-23.

12. Murugappa S, Kunapuli SP. The role of ADP receptors in platelet function. Front Biosci. 2006;11:1977–86.

13. Halperin D, Reber G. Influence of antidepressants on hemostasis. Dialogues Clin Neurosci. 2007;9(1):47–59.

14. Lindner T, Cockbain AJ, El Masry MA, Katonis P, Tsiridis E, Schizas C, Tsiridis E. The effect of anticoagulant pharmacotherapy on fracture healing. Expert Opin Pharmacother. 2008;9(7):1169–87. https://doi.org/10.1517/14656566.9.7.1169.

15. Ramsook RR, Danesh H. Timing of platelet rich plasma injections during antithrombotic therapy. Pain Physician. 2016;19:E1055–61.

16. Ludwig HC, Birdwhistell KE, Brainard BM, Franklin SP. Use of a cyclooxygenase-2 inhibitor does not inhibit platelet activation or growth factor release from platelet-rich plasma. Am J Sports Med. 2017;45(14):3351–7. https://doi.org/10.1177/0363546517730578.

17. di Matteo B, Filardo G, Lo Presti M, Kon E, Marcacci M. Chronic anti-platelet therapy: a contraindication for platelet-rich plasma intra-articular injections? Eur Rev Med Pharmacol Sci. 2014;18(1, supplement):55–9.

18. McCloskey DJ, Postolache TT, Vittone BJ, et al. Selective serotonin reuptake inhibitors (SSRIs): measurement of effect on platelet function. Transl Res: J Lab Clin Med. 2008;151(3):168–72. https://doi.org/10.1016/j.trsl.2007.10.004.

19. Chang J, Li C, Wu S, Yeh C, Chen C, Fu Y, Ho M. Effects of anti-inflammatory drugs on proliferation, cytotoxicity and osteogenesis in bone marrow mesenchymal stem cells. Biochem Pharmacol. 2007;74(9):1371–82. https://doi.org/10.1016/j.bcp.2007.06.047.

20. Zhang Y, Xiong Y, Chen X, Chen C, Zhu Z, Li L. Therapeutic effect of bone marrow mesenchymal stem cells pretreated with acetylsalicylic acid on experimental periodontitis in rats. Int Immunopharmacol. 2018;54:320–8. https://doi.org/10.1016/j.intimp.2017.11.028.

21. Geesala R, Dhoke NR, Das A. Cox-2 inhibition potentiates mouse bone marrow stem cell engraftment and differentiation-mediated wound repair. Cytotherapy. 2017;19(6):756–70. https://doi.org/10.1016/j.jcyt.2017.03.072.

Patient Factors Affecting Regenerative Medicine Outcomes

21

Roya S. Moheimani, Jason Kajbaf,
and George C. Chang Chien

Introduction

Lifestyle changes such as exercise, smoking, or heavy alcohol consumption are known to affect quality of life. However, many of these choices and how they affect platelet function and thus platelet-rich plasma (PRP) injections are yet to be elucidated. What is known, on the other hand, is that healthy lifestyles, the details of which will be discussed later in this chapter, promote tissue regeneration and platelet function. Although there may be a genetic predisposition to factors such as depression or anxiety, type 2 diabetes mellitus, and obesity, these factors will often negatively affect platelet function, which is why these morbidities are often avoided in large PRP injection trials. The following chapter will further detail the effects of these and other lifestyle habits and comorbidities on platelet function.

Factors

Exercise

A joint statement from the American College of Sports Medicine and the American Heart Association recommends all adults aged 18 to 65 years of age partake in moderate intensity aerobic exercise of at least 30 minutes for a minimum of 5 days a week [1]. This is a general recommendation for the reduction of chronic disease and disabilities. Additionally, this is recommended as a means to prevent

unhealthy weight gain. It was noted that there is a dose–response relationship between physical activity and overall health. Furthermore, exercise has been found to modulate platelet function [2].

There is evidence that regular exercise suppresses platelet adhesiveness and aggregation; on the other hand, the opposite is seen in deconditioned individuals. Platelet aggregation is more likely in sedentary individuals, which causes an upregulation in monocytes and further differentiation into macrophages [3], which in turn promotes degradation of the extracellular matrix and upregulation of other catabolic inflammatory cytokines. In cases of tendon and ligament strains and sprains, respectively, this upregulation and accumulation of macrophages have been found to hinder tissue repair, which may lead to more chronic injuries [4]. Therefore, when platelets are introduced to these structures, via platelet-rich plasma (PRP) injections, their potential benefits may be affected either positively or negatively by one's level of physical activity. In cases of acute, strenuous exercises, platelets are typically activated; however, in cases of regular moderate physical activity, platelet activation is inhibited. This was then replicated by another study comparing high-intensity interval training (HIIT) to moderate continuous exercise, where participants performing HIIT were found to have a higher likelihood of exercise-induced thrombosis when compared to the moderate continuous exercise group. Furthermore, specifically in cases of platelet-rich plasma injections, Van Ark et al. performed a study that demonstrated with a combination of endurance training and physical therapy there are improved effects of platelet-rich plasma injections for patients with patellar tendinopathy [5].

Type 2 Diabetes Mellitus

Comorbidities, such as type 2 diabetes mellitus and obesity, have been shown to affect overall wound healing and tissue regeneration [5]. As the number of comorbidities increase, this effect is further compounded. Type 2 diabetes mellitus

R. S. Moheimani (✉)
UCLA/GLA VA, Department of Physical Medicine and Rehabilitation, Los Angeles, CA, USA

J. Kajbaf
David Geffen School of Medicine—UCLA, Department of Physical Medicine and Rehabilitation, Los Angeles, CA, USA

G. C. Chang Chien
GCC Institute, Department of Musculoskeletal Medicine and Medical Aesthetics, Newport Beach, CA, USA

© Springer Nature Switzerland AG 2023
C. W Hunter et al. (eds.), *Regenerative Medicine*, https://doi.org/10.1007/978-3-030-75517-1_21

additionally has been shown to affect pro-inflammatory cytokines and cause platelet aggregation and dysfunction [6, 7]. Specifically, thromboxane production has been found to be higher in patients with type 2 diabetes mellitus, which in turn leads to increased platelet aggregation. Furthermore, in those with type 2 diabetes mellitus, there are changes in the number of leukocyte populations and their activation state, leading to increased apoptosis and tissue fibrosis. Furthermore, type 2 diabetes mellitus, similar to the effects of a lack of exercise discussed earlier, is associated with platelet dysfunction in the form of increased aggregation. This dysfunction is mediated by insulin resistance seen in diabetes, which leads to increased insulin levels that ultimately lead to the sensitization and dysfunction of platelets. Interestingly, it has been noted that PRP has growing popularity to help with poorly healing wounds in diabetic patients, such as diabetic foot ulcers. This suggests that perhaps not only is the poor circulation in type 2 diabetes mellitus leading to poor wound healing but also it is compounded by the platelet dysfunction [8]. It should be noted that type 2 diabetes mellitus is often selected as an exclusionary criteria in many PRP studies and named as a comorbidity that would affect results and hinder healing [9].

Obesity

There is a lot of overlap noted in the pathogenesis of increased platelet dysfunction and increased expression of inflammatory cascade for both type 2 diabetes and obesity [10]. Of note, there is dysfunction of platelets in obesity that is attributed to increased insulin sensitivity but also from the overall increase in chronic inflammation from higher amounts of adipose tissue [11]. Overall, obesity leads to increased platelet aggregation but also paradoxically dysfunctional in the inflammatory cascade. This stems from the conclusion that adipose tissue acts as an endocrine organ, thus affecting the overall cellular milieu of cytokines [12]. In obesity, there is a tendency for increased macrophage activity, which as stated earlier hinders tendon and ligament healing.

Depression and Anxiety

There appears to be similarities in terms of platelet dysfunction in patients with depression and anxiety to those with diabetes and obesity. Anxiety and depression follow the same pattern of increased expression of the inflammatory cascade [13, 14]. Patients with depression have long been noted to have higher levels of pro-inflammatory cytokines, acute phase proteins, chemokines, and cellular adhesion molecules. This has also been reproduced in socially defeated rats with increased anxiety. The inflammatory cascade as a result of depression and anxiety, similar to the comorbidities discussed above, thus leads to platelet dysfunction [15]. Furthermore,

given that nearly all of the human body's serotonin is stored in platelets, there have been many studies that have demonstrated that depression leads to alterations in platelet aggregation. This is why those with depression who forgo treatment for their depression with selective serotonin reuptake inhibitors (SSRIs) after a myocardial infarction have increased rates of adverse cardiovascular events [16]. This notion is supported by the finding that platelets from patients with depression have increased serotonin receptors as well as increased sensitivity to serotonin; thus when exposed to serotonin, they may have a more exaggerated response [17]. This is the case of patients on selective serotonin reuptake inhibitors (SSRIs), which will be discussed in the next section.

Medications

Serotonin is a key neurotransmitter that plays a pivotal role in the central and peripheral nervous system, cardiovascular system, gastroenterology system, and genitourinary system [17] and has also been found to be a key mediator of platelet function. As serotonin is released from platelet granules, it induces platelet aggregation and facilitates local vasoconstriction. Selective serotonin reuptake inhibitors (SSRIs), serotonin and norepinephrine reuptake inhibitors (SNRIs), and tricyclic antidepressants (TCAs) are all agents used to treat mood disorders, and all interact with endogenous serotonin levels [18], and as such have been associated with platelet dysfunction, such as decreased platelet aggregability [16]. Nonsteroidal anti-inflammatory drugs (NSAIDs) are consistently held in PRP studies over the concern for their interference with platelet function [19]. NSAIDs work by inhibiting platelet cyclooxygenase, decreasing thromboxane A2, among other pathways [20]. Thromboxane A2 is key to platelet aggregation; thus, NSAIDs depress platelet aggregability, thereby affecting platelet function (Fig. 21.1). Furthermore, a study by Jayaram et al. specifically explored the effects of aspirin on the release of various growth factors in PRP solutions. The PRP concentrations of VEGF, PDGF, and TGF-beta1 obtained from healthy individuals were compared to the levels of these growth factors after a 14-day course of aspirin. They demonstrated that aspirin and likely other COX inhibitors significantly reduced the release of these growth factors [20]. Thus, when it is medically safe to do so, it is recommended that NSAIDs be held when performing a PRP injection, as they reduce the release of platelet growth factors and thus may mitigate the results of PRP injections.

Nutrition

As stated previously, insulin has been associated with sensitization and disorderly platelet function [7]. This dysfunction was not reproduced after a single carbohydrate-rich

Fig. 21.1 Direct effects of aspirin on platelet function are depicted here. Via decreasing thromboxane A2 (TXA2) and acetylating fibrinogen, among pathways, aspirin and other NSAIDs prevent platelet activation and thus prevent the release of platelet contents, such as vascular endothelial growth factor

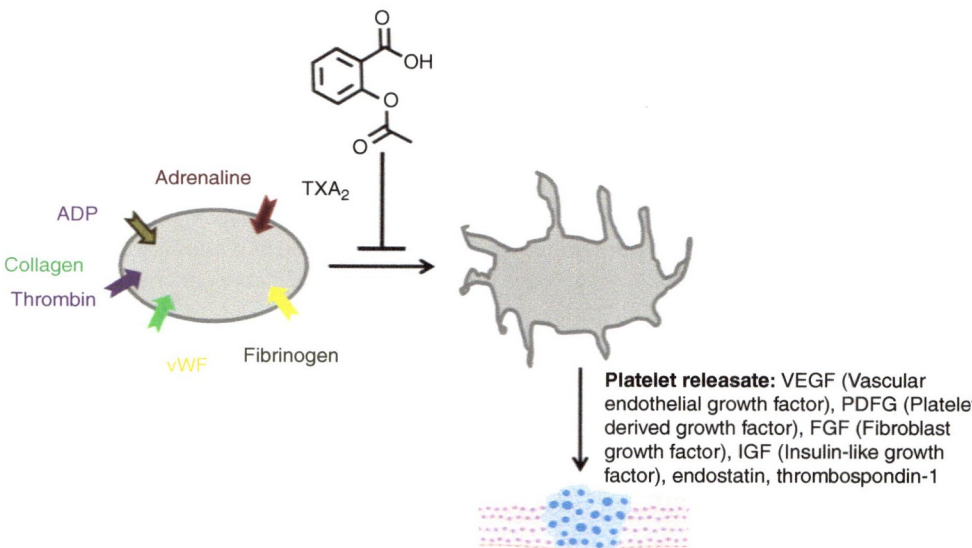

Platelet releasate: VEGF (Vascular endothelial growth factor), PDFG (Platelet derived growth factor), FGF (Fibroblast growth factor), IGF (Insulin-like growth factor), endostatin, thrombospondin-1

meal; however, with long-term consumption of a carbohydrate-rich diet, there is an increase in inflammation and production of cytokines that promote platelet aggregation [21]. Again, this is attributed to prolonged exposure of platelets to elevated insulin levels. The increased propensity of platelet aggregation is hypothesized to explain why diets elevated in refined sugar leads to increased cardiovascular disease, especially when saturated fats are substituted with refined carbohydrates [22]. Refined carbohydrates include fructose containing sugars and high fructose corn syrup, which is found in ultra-processed foods.

Diets that strongly avoid processed carbohydrates include the ketogenic diet, in which the goal is to obtain a state of ketosis, or a metabolic state in which the body's carbohydrate stores are depleted to the point where ketones are produced to power cellular metabolism. One consequence noted, however, is easy bruising, which was seen in up to 30% of participants in a study by Berry-Kravis et al. [23]. It is hypothesized that the diet itself causes platelet depression with probleeding tendency, but also preexisting genetic factors may play a role. This change was attributed to platelet membrane lipid composition.

Furthermore, flavonoids, which are found in many foods such as chocolate, tea, red wine, and beer, have been demonstrated to inhibit platelet activation and aggregation [24]. This has been repeatedly proven with various assays that have shown flavonoids' tendency to bind to thromboxane A2 receptors on platelets, thereby inhibiting its downstream signaling cascade and thus platelet activation.

Caffeine found in tea, coffee, and energy drinks, like the flavonoids discussed above, has been shown to decrease platelet aggregation. Numerous studies have revealed that platelet aggregation that is induced by collagen, ADP, and arachadonic acid in controlled studies is in fact inhibited after caffeine ingestion [24].

Garlic is touted to have significant cardiovascular benefits and thus recommended to introduce in diets to help with coronary artery disease and one mechanism by which garlic has been suggested to help is via its inhibitory effects on platelet aggregation [25]. A study by Rahman et al. revealed that ADP-induced platelet aggregation is significantly inhibited after garlic consumption [26]. Another study demonstrated that garlic also inhibits platelet aggregation in response to adrenaline [25]. Lastly, these results were also evident in another study by Bordia et al., where patients were given garlic three times per day for a one-month period and were subsequently found to have reduced platelet aggregation to both ADP and epinephrine [27].

Omega-3 polyunsaturated fatty acids (Omega-3 PUFA) are also known to have significant cardiovascular benefits by various mechanisms, including reduced platelet aggregation. Omega-3 PUFA are a known component of cell membrane phospholipids and thereby cell membrane function [28]. An interesting study performed on Inuits in 1979 revealed significantly longer bleeding times, which was attributed to a reduction of platelet aggregation as a result of higher consumption of Omega-3 PUFA found in fish [29]. Furthermore, a multitude of recent studies have repeatedly confirmed this early study and found that multiple platelet stimulating factors are inhibited with Omega-3 PUFA consumption [28]. Numerous other nutrients and dietary modifications have also been shown to affect platelet function and in similar mechanisms as described above. Turmeric, onions, tomatoes, and ginger, for instance, have all shown to inhibit platelet aggregation [25].

Smoking

Smoking tobacco cigarettes has largely been attributed as one of the most preventable risk factors to cardiac disease and mortality [30]. Among numerous other pathways, one mechanism by which cigarettes cause their known adverse effects is attributed in part to its affect on platelets and inflammation [31, 32]. Cigarette smoking has been shown to increase a multitude of cytokines, C-reactive protein, and prothrombotic factors. Additionally, smoking both chronically and acutely decreases inflammatory cell adhesion and fibrinolytic factors. This, coupled with decreased levels of nitrous oxide, leads to platelet activation and endothelium rigidity. Additionally, it has been shown that increased sympathetic nerve activity seen with cigarette smoking leads to platelet activation through unopposed increase epinephrine platelet activation. These findings have largely been paralleled in secondhand tobacco cigarette smoke exposure as well [31].

Furthermore, electronic cigarettes (e-cigarettes) have recently grown in popularity and are used in place of or in conjunction with tobacco cigarette smoking [33]. Although they have been touted as a safer alternative, there is growing evidence of the cardiovascular implications of both acute and chronic risk with e-cigarettes [34, 35]. Studies in mice have further suggested that there are prothrombotic changes promoting platelet aggregation in e-cigarette smoke, similar to that seen with regular cigarettes [36]. Additionally, there are studies demonstrating increased inflammation, paralleling studies in chronic tobacco cigarettes, for chronic e-cigarette users. This is an area of active research as e-cigarettes were recently introduced in the past 20 years.

Alcohol

Wine has been implicated as a cardioprotective beverage [37]. The cardioprotective element stems from two major hypotheses: the anti-inflammatory and the antithrombotic effects of alcohol. There is evidence that the consumption of red wine or red grapes decreases platelet aggregation, which is attributed to the high concentration of flavonoids and polyphenols [37]. Polyphenols are thought to act both on platelets directly and through an anti-inflammatory mechanism as well. This is why countries with heavy red wine consumption, such as France, have been found to have lower incidence of ischemic heart disease, despite the consumption of a high carbohydrate and saturated fat diet [24]. Overall, the consumption of moderate alcohol is thought to be anti-inflammatory, whereas consumption of high amounts of alcohol is pro-inflammatory and can lead to alcoholic cirrhosis and thrombocytopenia [37, 38]. Therefore, excessive

alcohol consumption may affect the contents of a PRP injection and thus its results.

Conclusion

Ultimately, the lifestyle habits and various comorbidities discussed above are known to affect platelet function, either in positive or negative ways. Furthermore, although little remains known how much these affect the quality of results following PRP injections, it is recommended to counsel all patients to make healthy lifestyle choices to optimize potential outcomes of the PRP injections.

References

1. Haskell WL, Lee I-M, Pate RR, et al. Physical activity and public health. Med Sci Sports Exerc. 2007;39(8):1423–34. https://doi.org/10.1249/mss.0b013e3180616b27.
2. Wang J-S, Li Y-S, Chen J-C, Chen Y-W. Effects of exercise training and deconditioning on platelet aggregation induced by alternating shear stress in men. Arterioscler Thromb Vasc Biol. 2005;25(2):454–60. https://doi.org/10.1161/01.ATV.0000151987.04607.24.
3. Gawaz M, Langer H, May AE. Platelets in inflammation and atherogenesis. J Clin Invest. 2005;115(12):3378–84. https://doi.org/10.1172/JCI27196.
4. Hays PL, Kawamura S, Deng X-H, et al. The role of macrophages in early healing of a tendon graft in a bone tunnel. J Bone Joint Surg Am. 2008;90(3):565–79. https://doi.org/10.2106/JBJS.F.00531.
5. Khalil H, Cullen M, Chambers H, Carroll M, Walker J. Elements affecting wound healing time: an evidence based analysis. Wound Repair Regen. 2015;23(4):550–6. https://doi.org/10.1111/wrr.12307.
6. Donath MY, Shoelson SE. Type 2 diabetes as an inflammatory disease. Nat Rev Immunol. 2011;11(2):98–107. https://doi.org/10.1038/nri2925.
7. Vinik AI, Erbas T, Park TS, Nolan R, Pittenger GL. Platelet dysfunction in type 2 diabetes. Diabetes Care. 2001;24(8):1476–85. http://www.ncbi.nlm.nih.gov/pubmed/11473089. Accessed 9 Mar 2019.
8. Lacci KM, Dardik A. Platelet-rich plasma: support for its use in wound healing. Yale J Biol Med. 2010;83(1):1–9. http://www.ncbi.nlm.nih.gov/pubmed/20351977. Accessed 20 Apr 2019.
9. Tahririan MA, Moezi M, Motififard M, Nemati M, Nemati A. Ultrasound guided platelet-rich plasma injection for the treatment of rotator cuff tendinopathy. Adv Biomed Res. 2016;5:200. https://doi.org/10.4103/2277-9175.190939.
10. Shoelson SE, Herrero L, Naaz A. Obesity, inflammation, and insulin resistance. Gastroenterology. 2007;132(6):2169–80. https://doi.org/10.1053/j.gastro.2007.03.059.
11. Anfossi G, Russo I, Trovati M. Platelet dysfunction in central obesity. Nutr Metab Cardiovasc Dis. 2009;19(6):440–9. https://doi.org/10.1016/j.numecd.2009.01.006.
12. Fain JN. Release of interleukins and other inflammatory cytokines by human adipose tissue is enhanced in obesity and primarily due to the nonfat cells. Vitam Horm. 2006;74:443–77. https://doi.org/10.1016/S0083-6729(06)74018-3.
13. Patki G, Solanki N, Atrooz F, Allam F, Salim S. Depression, anxiety-like behavior and memory impairment are associated with increased oxidative stress and inflammation in a rat model of social

stress. Brain Res. 2013;1539:73–86. https://doi.org/10.1016/j.brainres.2013.09.033.

14. Raison CL, Capuron L, Miller AH. Cytokines sing the blues: inflammation and the pathogenesis of depression. Trends Immunol. 2006;27(1):24–31. https://doi.org/10.1016/j.it.2005.11.006.

15. Ziegelstein RC, Parakh K, Sakhuja A, Bhat U. Platelet function in patients with major depression. Intern Med J. 2009;39(1):38–43. https://doi.org/10.1111/j.1445-5994.2008.01794.x.

16. Bismuth-Evenzal Y, Gonopolsky Y, Gurwitz D, Iancu I, Weizman A, Rehavi M. Decreased serotonin content and reduced agonist-induced aggregation in platelets of patients chronically medicated with SSRI drugs. J Affect Disord. 2012;136(1–2):99–103. https://doi.org/10.1016/j.jad.2011.08.013.

17. Berger M, Gray JA, Roth BL. The expanded biology of serotonin. Annu Rev Med. 2009;60(1):355–66. https://doi.org/10.1146/annurev.med.60.042307.110802.

18. Mago R, Mahajan R, Thase ME. Medically serious adverse effects of newer antidepressants. Curr Psychiatry Rep. 2008;10(3):249–57. http://www.ncbi.nlm.nih.gov/pubmed/18652794. Accessed 20 Mar 2019.

19. Sampson S, Gerhardt M, Mandelbaum B. Platelet rich plasma injection grafts for musculoskeletal injuries: a review. Curr Rev Musculoskelet Med. 2008;1(3–4):165–74. https://doi.org/10.1007/s12178-008-9032-5.

20. Schafer AI. Effects of nonsteroidal antiinflammatory drugs on platelet function and systemic hemostasis. J Clin Pharmacol. 1995;35(3):209–19. http://www.ncbi.nlm.nih.gov/pubmed/7608308. Accessed 20 Mar 2019.

21. Kopp W. The atherogenic potential of dietary carbohydrate. Prev Med (Baltim). 2006;42(5):336–42. https://doi.org/10.1016/j.ypmed.2006.02.003.

22. DiNicolantonio JJ, Lucan SC, O'Keefe JH. The evidence for saturated fat and for sugar related to coronary heart disease. Prog Cardiovasc Dis. 2016;58(5):464–72. https://doi.org/10.1016/j.pcad.2015.11.006.

23. Berry-Kravis E, Booth G, Taylor A, Valentino LA. Bruising and the ketogenic diet: evidence for diet-induced changes in platelet function. Ann Neurol. 2001;49(1):98–103. http://www.ncbi.nlm.nih.gov/pubmed/11198302. Accessed 21 Mar 2019.

24. Renaud S, de Lorgeril M. Wine, alcohol, platelets, and the French paradox for coronary heart disease. Lancet (London, England). 1992;339(8808):1523–6. http://www.ncbi.nlm.nih.gov/pubmed/1351198. Accessed 21 Mar 2019.

25. Rajaram S. The effect of vegetarian diet, plant foods, and phytochemicals on hemostasis and thrombosis. Am J Clin Nutr. 2003;78(3):552S–8S. https://doi.org/10.1093/ajcn/78.3.552S.

26. Rahman K, Billington D. Dietary supplementation with aged garlic extract inhibits ADP-induced platelet aggregation in humans. J Nutr. 2000;130(11):2662–5. https://doi.org/10.1093/jn/130.11.2662.

27. Bordia A, Verma SK, Srivastava KC. Effect of garlic (Allium sativum) on blood lipids, blood sugar, fibrinogen and fibrinolytic activity in patients with coronary artery disease. Prostaglandins Leukot Essent Fatty Acids. 1998;58(4):257–63. https://doi.org/10.1016/s0952-3278(98)90034-5.

28. Shahidi F, Ambigaipalan P. Omega-3 polyunsaturated fatty acids and their health benefits. Annu Rev Food Sci Technol. 2018;9(1):345–81. https://doi.org/10.1146/annurev-food-111317-095850.

29. von Schacky C, Dyerberg J. ω3 fatty acids. In: Fatty acids and lipids - new findings, vol. 88. Basel: KARGER; i2000. p. 90–9. https://doi.org/10.1159/000059772.

30. Huxley RR, Woodward M. Cigarette smoking as a risk factor for coronary heart disease in women compared with men: a systematic review and meta-analysis of prospective cohort studies. Lancet. 2011;378(9799):1297–305. https://doi.org/10.1016/S0140-6736(11)60781-2.

31. Middlekauff HR, Park J, Moheimani RS. Adverse effects of cigarette and noncigarette smoke exposure on the autonomic nervous system: mechanisms and implications for cardiovascular risk. J Am Coll Cardiol. 2014;64(16):1740–50. https://doi.org/10.1016/j.jacc.2014.06.1201.

32. Ambrose JA, Barua RS. The pathophysiology of cigarette smoking and cardiovascular disease. J Am Coll Cardiol. 2004;43(10):1731–7. https://doi.org/10.1016/j.jacc.2003.12.047.

33. Farsalinos KE, Romagna G, Tsiapras D, Kyrzopoulos S, Voudris V. Evaluation of electronic cigarette use (vaping) topography and estimation of liquid consumption: implications for research protocol standards definition and for public health authorities' regulation. Int J Environ Res Public Health. 2013;10(6):2500–14. https://doi.org/10.3390/ijerph10062500.

34. Moheimani RS, Bhetraratana M, Yin F, et al. Increased cardiac sympathetic activity and oxidative stress in habitual electronic cigarette users. JAMA Cardiol. 2017;2(3):278. https://doi.org/10.1001/jamacardio.2016.5303.

35. Moheimani RS, Bhetraratana M, Peters KM, et al. Sympathomimetic effects of acute E-cigarette use: role of nicotine and non-nicotine constituents. J Am Heart Assoc. 2017;6(9):e006579. https://doi.org/10.1161/JAHA.117.006579.

36. Qasim H, Karim ZA, Silva-Espinoza JC, et al. Short-term E-cigarette exposure increases the risk of thrombogenesis and enhances platelet function in mice. J Am Heart Assoc. 2018;7(15). https://doi.org/10.1161/JAHA.118.009264

37. Renaud SC, Ruf JC. Effects of alcohol on platelet functions. Clin Chim Acta. 1996;246(1–2):77–89. http://www.ncbi.nlm.nih.gov/pubmed/8814972. Accessed 21 Mar 2019.

38. Coltart I, Tranah TH, Shawcross DL. Inflammation and hepatic encephalopathy. Arch Biochem Biophys. 2013;536(2):189–96. https://doi.org/10.1016/j.abb.2013.03.016.

Part III

Treatment Paradigms: Orthobiologics and Additional Utilities

Tendinopathy

22

George C. Chang Chien, Allan Zhang, and Kenneth B. Chapman

Introduction

The tendon serves as the connection between muscle and bone allowing energy transfer for tensile strength and motion. Tendinopathy may be acute, subacute, or chronic, and its clinical presentation is characterized by localized pain and weakened, degenerative tendon tissue. Characteristic findings on pathologic examination include hypervascularity, loss of normal fibrillar architecture, hypercellularity, and degenerative changes in the collagenous matrix. Tendinopathy is often described as tendinitis. The suffix "-itis" suggests the presence of inflammatory process; however, the actual inflammation is minimally present in tendinopathy with chronic overuse as evidenced in histopathological findings in a wide variety of biopsied tendons [1–5].

Overuse, repetitive strain, or mechanical overload to tendons are considered as primary trigger of symptomatic tendinopathy, as implied by the names such as "jumper's knee" and "tennis elbow." Tendons are subject to microtrauma and may result in progressive collagenolytic tendon injury. Abnormal cytokine expression and the development of pain responsive C-fibers contribute to the clinical presentations of chronic tendon pain [6].

Ligaments are bands of durable, flexible fibrous connective tissue connecting bones or cartilage and stabilizing a mobile joint. Extra-capsular and intra-capsular ligaments join together to provide joint stability. Ligaments are visco-elastic. They gradually strain when under tension and return to their original shape when the tension is removed. However, they cannot retain their original shape when extended past a certain point or for a prolonged period of time [7]. If the ligaments lengthen too much, then the joint will be weakened, becoming prone to future dislocations. Tears of ligaments, either partial or complete, can lead to instability of the adjacent joint. Instability of a joint over time can lead to eventual degeneration of cartilage and may progress to osteoarthritis.

Recent advances in regenerative interventional techniques are proving to be efficacious in the treatment of these common musculoskeletal disorders. This chapter will provide a literature review on the most common indications for regenerative therapy and provide an evidence-based review grouped into tissue type and anatomical locations.

Elbow

The most widely studied indication for regenerative medicine and tendinopathy is for the treatment of lateral epicondylitis which afflicts the common extensor tendon of the elbow. Lateral epicondylitis or "tennis elbow" is a painful condition presenting with pain at the lateral side of the elbow joint that increases during gripping, squeezing, supination, and resisted wrist flexion. The common extensor tendon of the elbow is one of the most common tendons treated with ultrasound-guided tendon needle fenestration with positive outcomes in multiple studies [8–21] (Table 22.1) (Fig. 22.1).

Diagnostic ultrasound and/or MRI prior to procedure may be useful to identify the targeted tissue; however, the diagnosis is commonly made on based on clinical presentation. Areas of injury are typically found near the insertion point of the tendons and ligaments. Real-time ultrasound guidance is recommended to target the tissue and assure accuracy of the procedure. Low volume injection (0.1 ml each needle pass, total volume 0.5–2 ml) into the pathologic areas is advised while fenestrating the tendon with a needle. Depending on

G. C. Chang Chien (✉)
GCC Institute, Department of Musculoskeletal Medicine and Medical Aesthetics, Newport Beach, CA, USA

A. Zhang
University of Connecticut, Department of Radiology, Farmington, CT, USA

K. B. Chapman
Pain Medicine, Northwell Health Systems, Department of Pain Medicine, New York, NY, USA

© Springer Nature Switzerland AG 2023
C. W Hunter et al. (eds.), *Regenerative Medicine*, https://doi.org/10.1007/978-3-030-75517-1_22

Table 22.1 Regenerative medicine applications in the elbow

Author	Indication	Study design	Study methods	Outcome measures	Results	Adverse events
Mishra et al. [8]	Lateral epicondylitis	Double-blind, prospective, multicenter RCT	Leukocyte-enriched PRP $N = 112$ Active control $N = 113$ Follow-up periods	VAS pain score, tennis elbow questionnaire Follow-up periods 4, 8, 12, 24 weeks	At 12 weeks, PRP group reported 55.1% improvement in pain scores vs. control 47.4% At 24 weeks, PRP group 71.5% improvement in pain scores vs. Control group 56.1% ($P = 0.019$). At 24 weeks, PRP group reported 29.1% significant elbow tenderness vs. control group 54.0%	No significant complications occurred in either group
Gosens et al. [9]	Lateral epicondylitis	Double-blind, RCT	PRP group $N = 51$ Control group $N = 49$	VAS pain score, DASH score Follow-up periods 4, 8, 12, 26, 52, 104 weeks	PRP group significantly improved during the entire duration of the study ($P < 0.002$) PRP group demonstrated sustained benefits while corticosteroid group returned to baseline levels	No complications were seen concerning the use of PRP, except for the initial worsening of pain which usually lasted for 1–2 weeks
Krogh et al. [10]	Lateral epicondylitis	Double-blind, RCT	PRP group $N = 20$ Glucocorticoid group $N = 20$ Control group $N = 20$	PRTEE questionnaire Follow-up periods 3 months	Pain reduction at 3 months was observed in all three groups, with no statistically significant difference between the groups	No serious adverse events reported
Thanasas et al. [11]	Lateral elbow epicondylitis	RCT	PRP group $N = 14$ autologous blood group $N = 14$	VAS pain score, Liverpool elbow score Follow-up periods 6 weeks, 3 and 6 months	The VAS score improvement was larger in PRP group at every follow-up periods interval, but the difference was statistically significant only at 6 weeks	No adverse events reported
Creaney et al. [12]	Resistant elbow tendinopathy	Double-blind, RCT	PRP group $N = 80$ Autologous blood injection group $N = 70$	PRTEE Follow-up periods 1, 3, 6 months	At 6 months, the authors observed a 66% success rate in the PRP group versus 72% in the ABI group	No adverse events reported
Peerbooms et al. [13]	Lateral epicondylitis	Double-blind, RCT	PRP group $N = 51$ Corticosteroid group $N = 49$	VAS pain score, DASH score	24 of the 49 patients (49%) in the corticosteroid group and 37 of the 51 patients (73%) in the PRP group were successful DASH scores, 25 of the 49 patients (51%) in the corticosteroid group and 37 of the 51 patients (73%) in the PRP group were successful The corticosteroid group was better initially and then declined, whereas the PRP group progressively improved	No adverse complications reported
Chaudhury et al. [14]	Lateral epicondylitis	Prospective pilot study	PRP only $N = 6$	Grayscale images of the injected elbow Follow-up periods 1 and 6 months	Five patients demonstrated improved tendon morphology using ultrasound imaging 6 months after PRP injection	No adverse events reported

Table 22.1 (continued)

Author	Indication	Study design	Study methods	Outcome measures	Results	Adverse events
Mautner et al. [15]	Lateral epicondylitis	Retrospective cross-sectional survey	PRP group $N = 93$	VAS score, assessment of functional pain, overall satisfaction	93% of patients who received an injection at the lateral epicondyle, 100% of patients who received an injection at the Achilles tendon, and 59% of patients who received an injection at the patellar tendon reported moderate to complete resolution of symptoms (50% improvement). More than 80% of patients who received an injection at the rotator cuff, hamstring, gluteus medius, or common flexor tendon at the medial epicondyle reported the same or greater improvement	No adverse events reported
Mautner et al. [15]	Medial epicondylitis	Retrospective cross-sectional survey	PRP group $N = 82$	VAS score, assessment of functional pain, overall satisfaction	Same as above	No adverse events reported
Merolla et al. [16]	Chronic lateral epicondylitis	Prospective comparative study	PRP group $N = 51$ Other group (arthroscopic release $N = 50$)	VAS score, PRTEE, Calibrated hand dynamometer for grip strength	Both patient groups experienced significant improvement in all measures. Between-group comparisons showed a significantly higher value in the PRP group only for grip strength at week 8	No adverse events reported
Dines et al. [17]	Ulnar collateral ligament insufficiency	Retrospective study	PRP group $N = 44$	Mean time to return to play	15 (34%) had an excellent outcome, 17 had a good outcome, 2 had a fair outcome, and 10 had a poor outcome. After injection, 4 (67%) of the 6 professional players returned to professional play	No injection-related complications
Podesta et al. [18]	Ulnar collateral ligament tear	Case series	PRP group $N = 34$	KJOC, DASH score Follow-up periods 11–117 weeks. (average of 70 weeks)	30 of 34 athletes (88%) had returned to the same level of play without any complaints. The average KJOC score improved from 46 to 93. The average DASH score improved from 21 to 1. Sports module of the DASH questionnaire improved from 69 to 3. One player had persistent UCL insufficiency and underwent ligament reconstruction at 31 weeks after injection	No adverse events reported
Varshney et al. [19]	Epicondylar tendinitis	Randomized study	PRP group $N = 33$ Control (local steroid) group $N = 50$	VAS, modified MAYO Follow-up periods 1, 2, 6 months	Six months after treatment with PRP, patients with elbow epicondylitis had a significant improvement in their VAS and MAYO in contrast to steroid, whereas no statistical difference was found between the two groups at 1 and 2 months after intervention	No adverse events reported

(continued)

Table 22.1 (continued)

Author	Indication	Study design	Study methods	Outcome measures	Results	Adverse events
Singh et al. [20]	Tennis elbow	Prospective study	Bone marrow aspirate group $N = 30$	PRTEE score Follow-up periods 2, 6, 12 weeks	Baseline pre-injection mean PRTEE score was 72.8 ± 6.97 which decreased to a mean PRTEE score of 40.93 ± 5.94 after 2 weeks of injection. The mean PRTEE score at 6-week and 12-week follow-up periods was 24.46 ± 4.58 and 14.86 ± 3.48, respectively, showing a highly significant decrease from baseline scores	No adverse events reported
Lee et al. [21]	Lateral epicondylosis	Pilot study	Allogeneic adipose-derived mesenchymal stem cells group $N = 12$	VAS, modified MAYO, longitudinal and transverse ultrasound images of tendon defect areas Follow-up periods 6, 12, 26, and 52 weeks	From baseline through 52 weeks of periodic follow-up periods, VAS scores progressively decreased from 66.8 ± 14.5 mm to 14.8 ± 13.1 mm and elbow performance scores improved from 64.0 ± 13.5 to 90.6 ± 5.8. Tendon defects also significantly decreased through this period	No adverse events reported

VAS visual analogue scale, *PRTEE* patient-rated tennis elbow evaluation, *DASH* disability arm shoulder hand, *MAYO* Mayo Elbow performance score, *KJOC* Kerlan-Jobe Orthopaedic Clinic Shoulder and Elbow, *RCT* randomized controlled trial

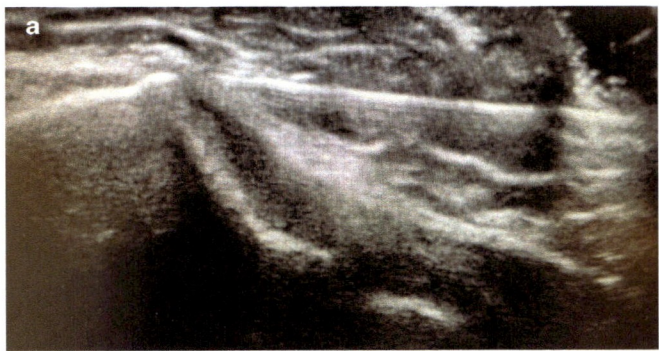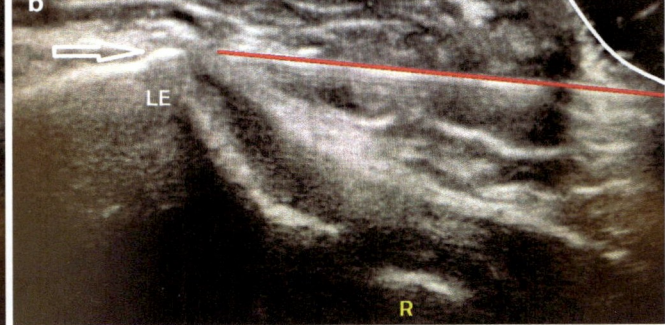

Fig. 22.1 Ultrasound-guided fenestration of the common extensor tendon insertion. Note the needle entering through a gel bridge (dark/anechoic region in the upper right corner), and the needle reaching the enthesopathy (arrow) at the top of the lateral epicondyle (LE). *LE* lateral epicondyle, *R* radial head. www.RegenMedDoctor.org

the baseline pathology, fenestration can be performed with 18–25G needle (Table 22.2).

Alternatively, injectate can also be placed peri-tendon and peri-ligament, a technique in which the injectate is not injected directly into the target tissues, but around them. This will create less trauma to the tissues and should be considered when the target tissues have significant injury (e.g., greater than 50% tendon rupture).

There is evidence to support tendon fenestration alone providing benefits in the treatment of tendinopathy. Thus, current dogma supports injection with low volume into the pathologic tissues. There is no explicit evidence at this time comparing intratendinous versus peri-tendinous injection.

Table 22.2 Tendon fenestration

Indication	Typical pathologic area	Needle gauge	Volume
Lateral epicondylopathy	Common extensor tendon insertion	18-25G	ml per site Total volume 1–2 ml
Medial Epicondylopathy	Common flexor tendon	18-25G	0.1 ml per site Total volume 1–2 ml
Triceps tendinopathy	Triceps tendon insertion	18-25G	0.1 ml per site Total volume 1–3 ml
Radial collateral ligament	Insertion at the lateral epicondyle	25G	ml per site Total volume 0.5–1 ml
Ulnar collateral ligament	Insertion at the medial epicondyle	25G	ml per site. Total volume 0.5–1 ml

Shoulder

Rotator Cuff and Biceps Tendinopathy Rotor cuff consists of four muscles and their tendons: the supraspinatus, infraspinatus, teres minor, and subscapularis. It is an essential structure in stabilizing the shoulder during its wide range of motions. Rotator cuff tendinopathy has been attributed to many factors including hypoperfusion, degeneration, microtrauma, chronic impingement syndrome, and overuse. Left untreated, rotator cuff tendinopathy can progress to partial- or full-thickness rotator cuff tears. Clinical presentation of rotator cuff tendinopathy and/or tear depends on which of the 4 muscles is affected, the most common being the supraspinatus. Biceps tendinopathy may also arise due to overuse via traction, friction, and glenohumeral rotation, which leads to microtrauma. Regenerative medicine has been incorporated into managing both surgical and nonsurgical shoulder pain secondary to rotator cuff and LHB tendinopathy and/or tear [34–40]. Utilizing regenerative medicine principles, platelet-rich plasma (PRP), or other platelet derivatives as well as mesenchymal stem cells (MSCs) have been applied in managing shoulder pain (Table 22.3).

Table 22.3 Clinical Evidence: Sample studies of regenerative medicine therapies for shoulder pathology in humans

Injectate	Author/Year	Indication	Study design	Study methods	Outcome measures	Results
Platelet rich plasma	Verhaegen et al., 2016 [41]	Calcific tendinosis	PRCT	PRP $N = 20$ Control group, no PRP $N = 20$ 6 weeks, 3 and 6 months, and 1 year	Constant score, simple shoulder test, and QuickDASH	All patients improved significantly after surgery ($P < 0.05$). There was no difference in clinical outcome or rotator cuff healing between groups. We observed a high rate of persistent rotator cuff defects after 1 year in both groups
	Rha et al. 2013 [42]	Supraspinatus tendinosis, or tear less than 1.0 cm	Single-center, PCDBRCT	PRP compared to dry needling $N = 19$ $N = 20$ Two dry needling procedures, two PRP injections applied to the affected shoulder at 4-week intervals using ultrasound guidance	Shoulder Pain and Disability Index, passive range of motion of the shoulder, physician global rating scale at the 6-month follow-up periods, adverse effects, and ultrasound measurement	PRP superior to dry needling from 6 weeks to 6 months after injection ($P < 0.05$) At 6 months the mean Shoulder Pain and Disability Index was 17.7 ± 3.7 in the PRP group versus 29.5 ± 3.8 in the dry needling group ($P < 0.05$)
	Tahririan et al., 2016 [43]	<1 cm partial tearing of the bursal side of rotator cuff	Open label case series	17 patients	CSS, before and 3 months after PRP injection	CSS before and after intervention was 37.05 ± 11.03 and 61.76 ± 14.75, respectively ($P < 0.001$)
	Sengodan et al., 2017 [44]	Symptomatic partial rotator cuff tears	Case series	PRP = 20	CSS, UCLA shoulder Baseline, 8, and 12 weeks	Statistically significant improvements in 17 patients in VAS pain score, constant shoulder score and UCLA shoulder score. Healing on radiological evaluation with ultrasonogram 8 weeks
	Shams et al., 2016 [36]	Symptomatic partial rotator cuff tear	PRCT	40 patients randomized to PRP vs. corticosteroid SASD injection	ASES, CMS, SST, VAS. 6, 12, 24 weeks MRI at baseline, and 6 months	Statistically significant difference between PRP group and corticosteroid group 12 weeks after injection, regarding VAS, ASES, CMS, and SST in favor of the PRP group. MRI showed an overall slight nonsignificant improvement in grades of tendinopathy/ tear in both groups

(continued)

Table 22.3 (continued)

Injectate	Author/Year	Indication	Study design	Study methods	Outcome measures	Results
	Lädermann et al., 2016 [45]	Symptomatic partial rotator cuff tear	Prospective case series	25 patients, injection into the supraspinatus tear	Primary outcome was tear size change on MRI arthrogram before and 6 months. Constant score, SANE score, VAS	Tear volume diminution was statistically significant ($P = 0.007$), and a >50% tear volume diminution was observed in 15 patients. A statistically significant improvement of constant score ($P < 0.001$), SANE score ($P = 0.001$), and VAS ($P < 0.001$) was observed. In 21 patients, constant score improvement reached the minimal clinical important difference of 10.4 points
	Kesikburun et al., 2013 [46]	Rotator cuff tear	PCDBRCT	PRP ($n = 20$), saline ($n = 20$). Ultrasound-guided injection into the subacromial space	WORC, SPADI, VAS. 3, 6, 12, and 24, and 52 weeks	No significant difference between the groups in WORC, SPADI, and VAS scores at 1-year follow-up periods ($P = 0.174$, $P = 0.314$, and $P = 0.904$, respectively). Similar results were found at other assessment points
	Nejati et al., 2017 [47]	Rotator cuff impingement syndrome	PRCT	Sixty-two patients were randomly placed into 2 groups, receiving either PRP or exercise therapy	Pain, shoulder (ROM), muscle force, functionality, and MRI findings 1-, 3-, and 6-month follow-up periods	Both treatment options significantly reduced pain and increased shoulder ROM compared with baseline measurements Exercise therapy superior to PRP
	Gumina et al., 2012 [48]	Arthroscopic rotator cuff repair, large full-thickness posterosuperior rotator cuff tears	PRCT	80 patients randomized to treatment either with or without a platelet-leukocyte membrane inserted between the rotator cuff tendon and its footprint	Constant scores and the repair integrity assessed by MRI according to the Sugaya classification. The secondary outcome was the difference between the preoperative and postoperative Simple Shoulder Test (SST) scores	At a mean of 13 months of follow-up periods, rotator cuff re-tears were observed only in the group of patients in whom the membrane had not been used, and a thin but intact tendon was observed more frequently in this group as well. The use of the membrane was associated with significantly better repair integrity ($p = 0.04$)
	Wang et al., 2015 [49]	Arthroscopic supraspinatus repair	PRCT	60 patients underwent arthroscopic supraspinatus tendon repair. Half the patients received 2 ultrasound-guided injections of PRP to the repair site at postoperative days 7 and 14	Structural healing assessed with MRI at 16 weeks, and graded according to the Sugaya classification Oxford Shoulder Score; Quick Disability of the Arm, Shoulder and Hand; VAS; and Short Form-12 quality-of-life score both preoperatively and at postoperative weeks 6, 12, and 16. Isokinetic strength and active range of motion were measured at 16 weeks	PRP treatment did not improve early functional recovery, range of motion, or strength or influence pain scores at any time point after arthroscopic supraspinatus repair. No difference in structural integrity of the supraspinatus repair on MRI at 16 weeks postoperatively ($P = 0.35$)

Table 22.3 (continued)

Injectate	Author/Year	Indication	Study design	Study methods	Outcome measures	Results
	Sanli et al., 2014 [50]	Distal biceps tendinopathy	Prospective case series	12 patients MRI confirmed diagnosis Single US-guided injection of PRP	VAS, rest and activity pain scores, subjective satisfaction scale, elbow functional assessment (EFA) isometric biceps strength Median follow-up periods of 47 months (36–52 months)	All patients showed significant improvement in pain ($p < 0.002$) and functional outcome ($p < 0.004$). Median resting VAS score improved from 6 (3–8) to 0.5 (0–2) and the activity VAS score improved from 8 (6–9) to 2.5 (0–4). EFA improved from 63 to 90. Isometric muscular strength also showed significant improvement. All patients were satisfied with the clinical and functional outcomes at final follow-up periods
	Barker et al., 2015 [51]	Distal biceps tendinopathy	Prospective case series	Six patients with clinical and radiological evidence of distal biceps tendinopathy underwent ultrasound-guided PRP injection	VAS, Mayo elbow performance score	The Mayo elbow performance score improved from 68.3 (range 65 to 85) (fair function) to 95 (range 85 to 100) (excellent function). The VAS at rest improved from a mean of 2.25 (range 2 to 5) pre-injection to 0. The VAS with movement improved from a mean of 7.25 (range 5 to 8) pre-injection to 1.3 (range 0 to 2)
Bone marrow stem cells	Kim et al., 2017 [37]	Symptomatic partial rotator cuff diagnosed with ultrasound >3 months	Prospective single blind	12 patients. 2 ml of BMACs was mixed with 1 ml of PRP	ASES, VAS, manual motor testing. Rotator cuff tear size on ultrasound	The ASES was 39.4 ± 13.0 at baseline to 52.9 ± 22.9 at 3 weeks and 71.8 ± 19.7 at 3 months after the injection ($p < 0.01$) The initial torn area of the rotator cuff tendon was 30.2 ± 24.5 mm², and this area was reduced to 22.5 ± 18.9 mm² at 3 months, but the change was not significant ($p > 0.05$)
Bone marrow stem cells	Hernigiou et al., 2014 [52]	Arthroscopic single row rotator cuff repair	Prospective case series	45 patients compared to 45 match controls	Tendon healing, re-tear rate	45 shoulders (100%) in the BMAC group demonstrated tendon healing by 6 months, compared to only 30 shoulders (67%) in the control group at the same time point. At 10 years, the integrity of the repair was maintained in 39 (87%) shoulders in the BMAC group compared to only 20 (44%) shoulders in the control group
	Centeno et al., 2015 [53]	Osteoarthritis and rotator cuff tears	Retrospective case series	115 shoulders in 102 patients injected with BMC injectate, containing PRP and platelet lysate	DASH, NPS, subjective improvement rating 1 month, 3 months, 6 months, and 12 months and then annually	Average DASH score decreased from 36.1 to 17.1 ($P < 0.001$). Average numeric pain scale value decreased from 4.3 to 2.4 ($P < 0.001$). Average subjective improvement of 48.8%

PDBRCT prospective double-blind randomized controlled trial, *PRCT* prospective randomized controlled trial, *WORC* Western Ontario Rotator Cuff Index, *SPADI* Shoulder Pain and Disability Index, *VAS* visual analog scale, *ASES* American Shoulder and Elbow Surgeons Standardized Shoulder Assessment Form (ASES), *CMS* Constant–Murley Score, *SST* The Simple Shoulder Test, *DASH* disability of the arm, shoulder, and hand score, *NPS* numeric pain scale

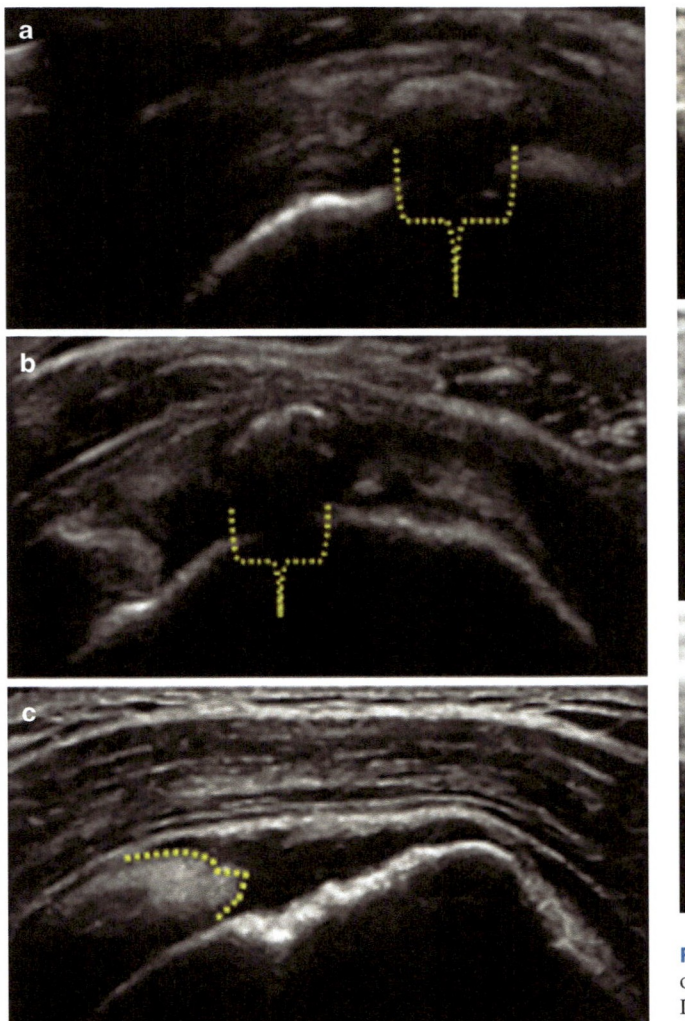

Fig. 22.2 Abnormal supraspinatus. (**a**) Long axis image of the supraspinatus with large deposit of calcium. (**b**) The same calcium deposit when viewed in short axis to the supraspinatus. (**c**) A different shoulder with long axis view of the supraspinatus demonstrating full-thickness rupture of the tendon with retraction of the muscle. Yellow dotted line depicts the edge of the retracted tendon

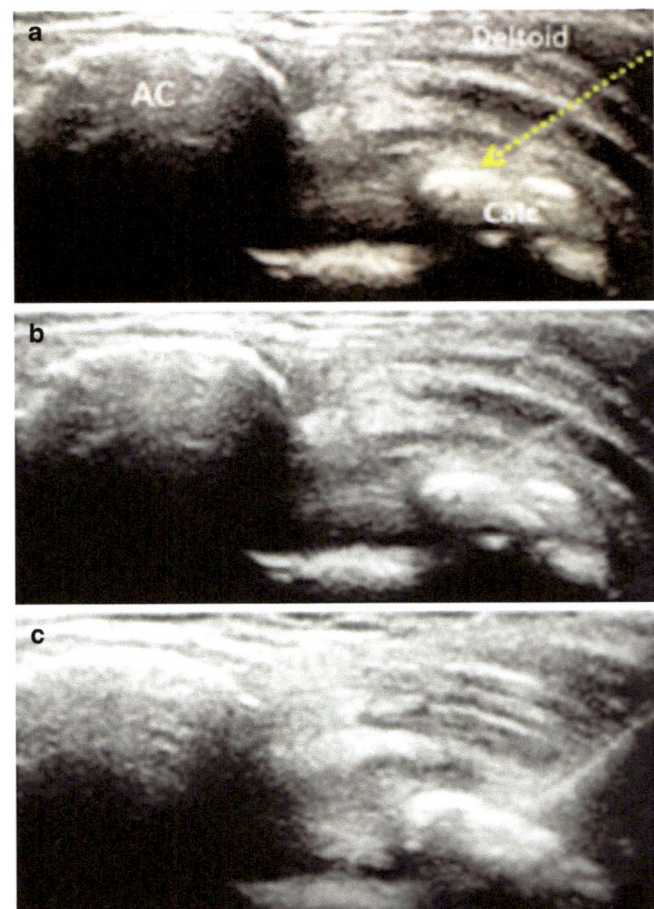

Fig. 22.3 Supraspinatus fenestration in long axis. (**a**) Long axis image of the supraspinatus with large deposit of calcium (Calc). *AC* acromion. Dashed arrow: outline of the needle. (**b**) Fenestration of the calcium deposit. (**c**) Fenestration of the calcium deposit after redirection of the needle

Fenestration of the rotator cuff tendons has also been performed with promising results in the treatment of tendinopathy. Like other fenestration techniques as described in the aforementioned section under elbow, ultrasound guidance should be utilized to improve accuracy. The procedure involves repeated fenestrations of the affected tendon using ultrasound guidance. Fenestrating the injured tendon causes inflammation and localized bleeding, and encourages fibroblastic proliferation. These factors promote ordered collagen formation and ultimately healing of the tendon. Complications include bleeding, infections, and worsening of tendinopathy and/or rupture. An example of supraspinatus fenestration is shown in Figs. 22.2 and 22.3.

Hip

Greater Trochanteric Pain Syndrome

Greater trochanteric pain syndrome (GTPS), also formerly known as trochanter bursitis, is one of the most causes of hip pain. GTPS is an overuse tendinopathy of the gluteus medius and minimus tendons at their connection to the greater trochanter. Patients typically present with lateral hip pain adjacent to the greater trochanter. There are increased levels of pain with ambulation, prolonged standing, and rising from a chair. The pain is reproduced by applying direct pressure on the greater trochanter of the affected side. PRP has shown clinically significant treatment response in patients presenting with GTPS (Table 22.4).

Hamstring Tendinopathy

The main symptom of proximal hamstring tendinopathy is pain in the lower gluteal region, occasionally radiating along the hamstring to the posterior thigh. The pain mainly manifests during activities such as running and jumping or prolonged sitting. Patients can present with palpable tenderness and pain over the ischial tuberosity, and pain may be elicited with resisted knee flexion. Traditional treatment methods include reduction or pause of sports activity, ice for symptomatic relief during the initial phase, nonsteroidal anti-inflammatory drugs (NSAIDs), soft tissue mobilization, physiotherapy, and continuous home exercise program focusing on progressive eccentric hamstring strengthening [22]. Eccentric exercise programs with musculotendinous junction strengthening can promote intratendinous collagen fiber cross-linkage to enable remodeling [22, 23]. Injection with PRP has been demonstrated to facilitate healing and return to play in some athletes, but some studies demonstrate no advantage of PRP to rehabilitation exercises alone [24–33] (Table 22.4).

Table 22.4 Clinical evidence: sample studies of regenerative medicine therapies for hip pathology in humans

Injectate	Author/ year	Indication	Study design	Study methods	Outcome measures	Results/conclusions
Platelet rich plasma	Lee et al./2016 [32]	Gluteus Medius tendinopathy	Prospective case series	Ultrasound-guided intratendinous PRP injections for recalcitrant gluteus medius tendinosis and/or partial tears of the tendon 21 patients were included in the study	The modified Harris hip Score mHHS, HOS-ADL, HOS-sport, iHOT-33. a mean follow-up periods of 19.7 months (12.1–32.3 months)	The mean improvements from pre-injection to post-injection follow-up periods were 56.73 to 74.17 for mHHS, 68.93 to 84.14 for HOS-ADL, 45.54 to 66.72 for HOS-sport, and 34.06 to 66.33 for iHOT-33. All mean outcome measure improvements were clinically and statistically significant ($P < 0.001$)
Platelet rich plasma	Jacobson et al./2016 [33]	Greater trochanteric pain syndrome, Gluteus Medius tendinopathy	Prospective randomized single blinded trial	Ultrasound-guided intratendinous PRP injections ($N = 15$), vs ultrasound-guided tendon fenestration ($N = 15$)	Mean pain score at 1 week, and 2 weeks post procedure	Both ultrasound-guided tendon fenestration and PRP injection are effective for treatment of gluteal tendinosis, with no significant difference between the treatments ($P > 0.99$)
Platelet rich plasma	Davenport et al./2015 [26]	To compare the therapeutic efficacy of autologous PRP or AWB for chronic hamstring tendinopathy	Prospective double-blind randomized controlled trial	Ultrasound-guided injection into the proximal hamstring tendon	mHHS, HOS-ADL, iHOT-33 at 2, 6, and 12 weeks and 6 months after injection Diagnostic ultrasound was used to compare pre-injection and 6-month post-injection	Both PRP and WB groups showed improvements in all outcome measures at 6 months. Significant between-group differences were observed at any time point. Ultrasound imaging showed no significant differences between PRP and WB group tendon appearances
Platelet rich plasma	A Hamid et al./2014 [27]	To evaluate therapeutic efficacy of autologous PRP in grade 2 hamstring muscle injuries	Prospective randomized controlled trial	Twenty-eight patients diagnosed with an acute hamstring injury were randomly allocated to autologous PRP therapy combined with a rehabilitation program or a rehabilitation program only	Return to play	The mean time to return to play was 26.7 ± 7.0 days and 42.5 ± 20.6 days for the PRP and control groups, respectively (t (22) = 2.50, $P = 0.02$) A single autologous PRP injection combined with a rehabilitation program was significantly more effective in treating hamstring injuries than a rehabilitation program alone

(continued)

Table 22.4 (continued)

Injectate	Author/ year	Indication	Study design	Study methods	Outcome measures	Results/conclusions
Platelet rich plasma	Hamilton et al./2015 [28]	To evaluate the efficacy of a single platelet-rich plasma (PRP) injection in reducing the return to sport duration in male athletes, following an acute hamstring injury	Prospective randomized, three-arm (double-blind for the injection arms), parallel-group trial	90 professional athletes with MRI positive hamstring injuries were randomized to injection with PRP-intervention, platelet-poor plasma (PPP-control) or no injection Outcome measures: return to play, re-injury rate after 2 and 6 months	Return to sports	There is no benefit of a single PRP injection over intensive rehabilitation in athletes who have sustained acute, MRI positive hamstring injuries
Platelet rich plasma	Fader et al./2015 [29]	To evaluate the efficacy of ultrasound-guided platelet-rich plasma injections in treating chronic proximal hamstring tendinopathies	Retrospective	18 consecutive patients received a single injection of platelet-rich plasma via ultrasound guidance by a single radiologist	Outcome measures included a questionnaire evaluating previous treatments, visual analog scale (VAS) for pain, subjective improvement, history of injury, and return to activity	The average VAS pre-injection was 4.6 (0–8). Six months after the injection, 10/18 patients had 80% or greater improvement in their VAS. Overall, the average improvement was 63% (5–100)
Platelet rich plasma	Reurink et al./2015 [30]	To evaluate the efficacy of a single platelet-rich plasma (PRP) injection in reducing the return to sport duration in male athletes, following an acute hamstring injury	Multicenter, prospective, randomized, double-blind, placebo-controlled trial		The primary outcome measure was the time needed to return to play during 6 months of follow-up periods. Not previously reported secondary outcome scores included re-injury at 1 year, alteration in clinical and MRI parameters, subjective patient satisfaction and the hamstring outcome score	At 1-year post-injection, we found no benefit of intramuscular PRP compared with placebo injections in patients with acute hamstring injuries in the time to return to play, re-injury rate and alterations of subjective, clinical or MRI measures
Platelet rich plasma	Zanon et al./2016 [31]	To evaluate the efficacy of a single platelet-rich plasma (PRP) injection acute grade 2 hamstring injuries	Twenty-five hamstring injuries sustained by professional football players during a 31-month observation period	Intralesional injection	Sport participation absence (SPA), in days, was considered to correspond to the healing time, re-injury rate, and tissue healing on MRI	The mean SPA for the treated muscle injuries was 36.76 ± 19.02 days. The re-injury rate was 12%. Tissue healing, evaluated on MRI, was characterized by the presence of excellent repair tissue and a small scar. PRP-treated lesions did not heal more quickly than untreated lesions described in the literature, but they showed a smaller scar and excellent repair tissue

VAS visual analogue score, *OA* osteoarthritis, *PRP* platelet-rich plasma, *HA* hyaluronic acid, *WOMAC* Western Ontario and McMaster Universities Osteoarthritis Index, *mHHS* Modified Harris Hip Score, *HOS-ADL* Hip Outcome Score-Activities of Daily Living subscale, *HOS-Sport* Hip Outcome Score-Sport-Specific subscale, *iHOT-33* International HipOutcome Tool-33, *MSCs* mesenchymal stem cells, *AWB* autologous whole blood

Knee and Ankle

Patellar and Achilles tendinopathies are common overuse injuries associated with physical activities such as running and jumping. Presenting symptoms include pain, tenderness, and restricted range of motion. Treatment of patellar and Achilles tendinopathies include conservative, injectional, or potentially surgical intervention. Conservative management includes rest, immobilization, and NSAIDs. The use of ultrasound, low-level laser therapy, and corticosteroid injections has also shown positive responses in patients. More recently, the injection of PRP has become increasingly utilized to treat patellar and Achilles tendinopathies. While many results and trials have demonstrated statistically significant improvement in patellar tendinopathy, the results for Achilles tendinopathy remain somewhat mixed (Table 22.5) [53–66].

Table 22.5 Platelet-rich plasma studies in the treatment of Patellar and Achilles Tendinopathy

Injectate	Authors	Indication	Study design	Study methods	Outcome measures	Results
PRP	Charousset et al. [54]	Patellar tendinopathy	Case series	PRP injections 3x/wk, follow-up periods at 4 wk, 3, 6, 12, 18, and 24 mo	VISA-P; Lysholm score; pain VAS; MRI assessment of tendon	Significantly improved symptoms and function in athletes with chronic patellar tendinopathy
PRP	Dragoo et al. [55]	Patellar tendinopathy	RCT	PRP injections vs dry needling x10; follow-up periods at 3, 6, 9, 12 wk and 6 mo	VISA-P; TAS; Lysholm score; pain VAS	PRP injection with dry needling accelerates the recovery from patellar tendinopathy relative to dry needling alone
PRP	Ventrano et al. [56]	Patellar tendinopathy	RCT	PRP injection ×2 in wks vs. focused ultrasound shock wave; follow-up periods at 2 and 6 mo	VISA-P; pain VAS; modified Blazina scale	Therapeutic injections of PRP lead to better clinical results compared with US shock wave in the treatment of jumper's knee in athletes
PRP	Filardo et al. [57]	Patellar tendinopathy	Case series	PRP injection x3 in two wks span; follow-up periods at 2 and 6 months	VISA-P; TAS; EQVAS; Blazina scale; satisfaction; return to sports; US in 26 patients	Multiple injections of PRP provided a good clinical outcome for the treatment of chronic recalcitrant patellar tendinopathy. Patients affected by bilateral pathology and presenting a long history of pain obtained significantly poorer results.
PRP	Gosens et al. [58]	Patellar tendinopathy	Cohort	PRP injection x5; latest follow-up periods, mean 18.4 mo	VISA-P; VAS [33] for pain during ADL, work, and sport	Patients with patellar tendinopathy showed a statistically significant improvement.
PRP	Ferrero et al. [59]	Patellar tendinopathy	Case series	PRP injections x2, wks apart; follow-up periods at 20 days and 6 mo	VISA-P; US measured tendon thickness	PRP injection in patellar and Achilles tendinopathy results in a significant and lasting improvement of clinical symptoms and leads to recovery of the tendon matrix
PRP	Kon et al. [60]	Patellar tendinopathy	Case series	PRP injection x3, 15 d apart; follow-up periods at 6 mo	TAS; EQ-VAS; patient reported functional recovery and satisfaction	Statistically significant improvements in all scores
PRP	Volpi et al. [61]	Patellar tendinopathy	Case series	PRP injection x1; latest follow-up periods, mean 120 d	VISA-P; MRI appearances	All patients demonstrated improvement on the VISA score; mixed MRI appearance
PRP	De Vos et al. [62]	Achilles tendinopathy	RCT	PRP injection x1 vs. saline injection; follow-up periods at 6 mo	VISA -A	PRP injection did not result in greater improvement in pain and activity
PRP	De Jonge et al. [63]	Achilles tendinopathy	RCT	PRP injection x1 vs. saline injection; follow-up periods at 12 mo	VISA-A	No clinical and ultrasonographic superiority of platelet-rich plasma injection over a placebo injection in chronic Achilles tendinopathy at 1 year combined with an eccentric training program
PRP	Kearney et al. [64]	Achilles tendinopathy	RCT	PRP injection x1 vs. eccentric exercise; follow-up periods at 12 mo	VISA-A, VAS	No statically significant difference

Table 22.5 (continued)

Injectate	Authors	Indication	Study design	Study methods	Outcome measures	Results
PRP	Krogh et al. [65]	Achilles tendinopathy	RCT	PRP injection vs. saline injection; follow-up periods at 6 mo	VISA-A, TT	PRP injection did not result in an improved VISA-A score over a 3-month period in patients with chronic AT compared with placebo
PRP	Boesen et al. [66]	Achilles tendinopathy	RCT	PRP injection x4 vs. saline injection x4; follow-up periods in 6 mo	VISA-A, VAS, TT	Treatment with PRP in combination with eccentric training in chronic AT seems more effective in reducing pain, improving activity level, and reducing tendon thickness and intratendinous vascularity than eccentric training alone

ADL, activities of daily living; EQ-VAS, EuroQol general quality of life Visual Analog Scale; TAS, Tegner Activity Scale; TT, tendon thickness; VAS, visual analog scale; VISA-A, Victoria Institute of Sports Assessment Questionnaire (Achilles); VISA-P, Victoria Institute of Sport Assessment Questionnaire (patella)

Conclusion

Regenerative medicine modalities have demonstrated promising results in the treatment of tendinopathies and should be considered in the treatment algorithm for conservative management. Proficiency in ultrasound and ultrasound-guided interventions should significantly increase the diagnostic and procedural accuracy of the managing physician.

References

1. Khan KM, Cook JL, Bonar F, Harcourt P, Astrom M. Histopathology of common tendinopathies. Update and implications for clinical management. Sports Med. 1999;27:393–408.
2. Kragsnaes MS, Fredberg U, Stribolt K, et al. Stereological quantification of immune-competent cells in baseline biopsy specimens from achilles tendons: results from patients with chronic tendinopathy followed for more than 4 years. Am J Sports Med. 2014;42:2435.
3. Murrell G. Understanding tendinopathies. Br J Sports Med. 2002;36:392–3.
4. Soslowsky LJ, Thomopoulos S, Tun S, Flanagan CL, Keefer CC, Mastaw J, Carpenter JE. Neer Award 1999. Overuse activity injures the supraspinatus tendon in an animal model: a histologic and biomechanical study. J Shoulder Elb Surg. 2000;9:79–84.
5. Andres BM, Murrell GAC. Treatment of tendinopathy: what works, what does not, and what is on the horizon. Clin Orthop Relat Res. 2008;466(7):1539–54. https://doi.org/10.1007/s11999-008-0260-1.
6. Fu SC, Rolf C, Cheuk YC, Lui PP, Chan KM. Deciphering the pathogenesis of tendinopathy: A three-stages process. Sports medicine, arthroscopy, rehabilitation, therapy & technology: *SMARTT*. 2010;2:30. https://doi.org/10.1186/1758-2555-2-30.
7. Hauser RA, Dolan EE, Phillips HJ, Newlin AC, Moore RE, Woldin BA. Ligament injury and healing: a review of current clinical diagnostics and therapeutics (PDF). The Open Rehabilitation Journal. 2013;6(1):5. https://doi.org/10.2174/1874943701306010001.
8. Mishra AK, Skrepnik NV, Edwards SG, et al. Efficacy of platelet-rich plasma for chronic tennis elbow: a double-blind, prospective, multicenter, randomized controlled trial of 230 patients. Am J Sports Med. 2014;42:463–71.
9. Gosens T, Peerbooms JC, van Laar W, den Oudsten BL. Ongoing positive effect of platelet-rich plasma versus corticosteroid injection in lateral epicondylitis: a double-blind randomized controlled trial with 2-year follow-up. Am J Sports Med. 2011;39:1200–8.
10. Krogh TP, Fredberg U, Stengaard-Pedersen K, et al. Treatment of lateral epicondylitis with platelet-rich plasma, glucocorticoid, or saline: a randomized, double-blind, placebo-controlled trial. Am J Sports Med. 2013;41:625–35.
11. Thanasas C, Papadimitriou G, Charalambidis C, et al. Platelet-rich plasma versus autologous whole blood for the treatment of chronic lateral elbow epicondylitis: a randomized controlled clinical trial. Am J Sports Med. 2011;39:2130–4.
12. Creaney L, Wallace A, Curtis M, Connell D. Growth factor-based therapies provide additional benefit beyond physical therapy in resistant elbow tendinopathy: a prospective, single-blind, randomised trial of autologous blood injections versus platelet-rich plasma injections. Br J Sports Med. 2011;45:966–71.
13. Peerbooms JC, Sluimer J, Bruijn DJ, Gosens T. Positive effect of an autologous platelet concentrate in lateral epicondylitis in a double-blind randomized controlled trial: platelet-rich plasma versus corticosteroid injection with a 1-year follow-up. Am J Sports Med. 2010;38:255–62.
14. Chaudhury S, de La Lama M, Adler RS, et al. Platelet-rich plasma for the treatment of lateral epicondylitis: sonographic assessment of tendon morphology and vascularity (pilot study). Skelet Radiol. 2013;42:91–7.
15. Mautner K, Colberg RE, Malanga G, et al. Outcomes after ultrasound-guided platelet-rich plasma injections for chronic tendinopathy: a multicenter, retrospective review. PM R. 2013;5:169–75.
16. Merolla G, Dellabianci F, Ricci A, et al. Arthroscopic debridement versus platelet-rich plasma injection: a prospective, randomized, comparative study of chronic lateral epicondylitis with a nearly 2-year follow-up. Arthroscopy. 2017;33:1320–9.
17. Dines JS, Williams PN, ElAttrache N, et al. Platelet-rich plasma can be used to successfully treat elbow ulnar collateral ligament insufficiency in high-level throwers. Am J Orthop (Belle Mead NJ). 2016;45:296–300.
18. Podesta L, Crow SA, Volkmer D, et al. Treatment of partial ulnar collateral ligament tears in the elbow with platelet-rich plasma. Am J Sports Med. 2013;41:1689–94.
19. Varshney A, Maheshwari R, Juyal A, et al. Autologous platelet-rich plasma versus corticosteroid in the management of elbow epicondylitis: a randomized study. Int J Appl Basic Med Res. 2017;7:125–8.
20. Singh A, Gangwar DS, Singh S. Bone marrow injection: a novel treatment for tennis elbow. J Nat Sci Biol Med. 2014;5:389–91.

21. Lee SY, Kim W, Lim C, Chung SG. Treatment of lateral epicondylosis by using allogeneic adipose-derived mesenchymal stem cells: a pilot study. Stem Cells. 2015;33:2995–3005.

22. Lempainen L, Johansson K, Banke IJ, et al. Expert opinion: diagnosis and treatment of proximal hamstring tendinopathy. Muscles Ligaments Tendons J. 2015;5(1):23–8.

23. Kaux JF, Libertiaux V, Leprince P, Fillet M, Denoel V, Wyss C, Lecut C, Gothot A, Le Goff C, Croisier JL, Crielaard JM, Drion P. Eccentric training for tendon healing AfterAcute lesion: a rat model. Am J Sports Med. 2017;45(6):1440–6.

24. Stasinopoulos D, Manias P. Comparing two eccentric exercise programs for the management of Achilles tendinopathy: a pilot trial. J Bodyw Mov Ther. 2013;17(3):309–15.

25. Guillodo Y, Madouas G, Simon T, Le Dauphin H, Saraux A. Platelet-rich plasma (PRP) treatment of sports-related severe acute hamstring injuries. Muscles Ligaments Tendons J. 2016 Feb 13;5(4):284–8.

26. Davenport KL, Campos JS, Nguyen J, Saboeiro G, Adler RS, Moley PJ. Ultrasound-guided intratendinous injections with platelet-rich plasma or autologous whole blood for treatment of proximal hamstring tendinopathy: a double-blind randomized controlled trial. J Ultrasound Med. 2015;34(8):1455–63.

27. Hamid MSA, Mohamed Ali MR, Yusof A, George J, Lee LP. Platelet-rich plasma injections for the treatment of hamstring injuries: a randomized controlled trial. Am J Sports Med. 2014;42(10):2410–8.

28. Hamilton B, Tol JL, Almusa E, Boukarroum S, Eirale C, Farooq A, Whiteley R, Chalabi H. Platelet-rich plasma does not enhance return to play in hamstring injuries: a randomised controlled trial. Br J Sports Med. 2015;49(14):943–50.

29. Fader RR, Mitchell JJ, Traub S, Nichols R, Roper M, Mei Dan O, McCarty EC. Platelet-rich plasma treatment improves outcomes for chronic proximal hamstring injuries in an athletic population. Muscles Ligaments Tendons J. 2015;4(4):461–6.

30. Reurink G, Goudswaard GJ, Moen MH, Weir A, Verhaar JA, Bierma-Zeinstra SM, Maas M, Tol JL, Dutch HIT-study Investigators. Rationale, secondary outcome scores and 1-year follow-up of a randomised trial of platelet-rich plasma injections in acute hamstring muscle injury: the Dutch Hamstring Injection Therapy study. Br J Sports Med. 2015;49(18):1206–12.

31. Zanon G, Combi F, Combi A, Perticarini L, Sammarchi L, Benazzo F. Platelet-rich plasma in the treatment of acute hamstring injuries in professional football players. Joints. 2016;4(1):17–23.

32. Lee JJ, Harrison JR, Boachie-Adjei K, Vargas E, Moley PJ. Platelet-rich plasma injections with needle tenotomy for gluteus medius tendinopathy: a registry study with prospective follow-up. Orthop J Sports Med. 2016;4(11):2325967116671692.

33. Jacobson JA, Yablon CM, Henning PT, Kazmers IS, Urquhart A, Hallstrom B, Bedi A, Parameswaran A. Greater trochanteric pain syndrome: percutaneous tendon fenestration versus platelet-rich plasma injection for treatment of gluteal tendinosis. J Ultrasound Med. 2016;35(11):2413–20.

34. Cai YZ, Zhang C, Lin XJ. Efficacy of platelet-rich plasma in arthroscopic repair of full-thickness rotator cuff tears: a metaanalysis. J Shoulder Elb Surg. 2015;24:1852–9.

35. Warth RJ, Dornan GJ, James EW, et al. Clinical and structural outcomes after arthroscopic repair of full-thickness rotator cuff tears with and without platelet-rich product supplementation: a meta-analysis and meta-regression. Arthroscopy. 2015;31:306–20.

36. Shams A, El-Sayed M, Gamal O, et al. Subacromial injection of autologous platelet-rich plasma versus corticosteroid for the treatment of symptomatic partial rotator cuff tears. Eur J Orthop Surg Traumatol. 2016;26:837–42.

37. Kim SJ, Song DH, Park JW, et al. Effect of bone marrow aspirate concentrate-platelet-rich plasma on tendon-derived stem cells and rotator cuff tendon tear. Cell Transplant. 2017;26:867–78.

38. Canapp SO Jr, Canapp DA, Ibrahim V, et al. The use of adipose-derived progenitor cells and platelet-rich plasma combination for the treatment of supraspinatus tendinopathy in 55 dogs: a retrospective study. Front Vet Sci. 2016;3:61.

39. Greenspoon JA, Moulton SG, Millett PJ, et al. The role of platelet rich plasma (PRP) and other biologics for rotator cuff repair. Open Orthop J. 2016;10:309–14.

40. Kim SJ, Kim EK, Kim SJ, et al. Effects of bone marrow aspirate concentrate and platelet-rich plasma on patients with partial tear of the rotator cuff tendon. J Orthop Surg Res. 2018;13:1.

41. Verhaegen F, Brys P, Debeer P. Rotator cuff healing after needling of a calcific deposit using platelet-rich plasma augmentation: a randomized, prospective clinical trial. J Shoulder Elb Surg. 2016;25:169–73.

42. Rha DW, Park GY, Kim YK, et al. Comparison of the therapeutic effects of ultrasound-guided platelet-rich plasma injection and dry needling in rotator cuff disease: a randomized controlled trial. Clin Rehabil. 2013;27:113–22.

43. Tahririan MA, Moezi M, Motififard M, et al. Ultrasound guided platelet-rich plasma injection for the treatment of rotator cuff tendinopathy. Adv Biomed Res. 2016;5:200.

44. Sengodan VC, Kurian S, Ramasamy R. Treatment of partial rotator cuff tear with ultrasound-guided platelet-rich plasma. J Clin Imaging Sci. 2017;7:32.

45. Lädermann A, Zumstein MA, Kolo FC, et al. In vivo clinical and radiological effects of platelet-rich plasma on interstitial supraspinatus lesion: case series. Orthop Traumatol Surg Res. 2016;102:977–82.

46. Kesikburun S, Tan AK, Yilmaz B, et al. Platelet-rich plasma injections in the treatment of chronic rotator cuff tendinopathy: a randomized controlled trial with 1-year follow-up. Am J Sports Med. 2013;41:2609–16.

47. Nejati P, Ghahremaninia A, Naderi F, et al. Treatment of subacromial impingement syndrome: platelet-rich plasma or exercise therapy? A Randomized Controlled Trial. Orthop J Sports Med. 2017;5:2325967117702366.

48. Gumina S, Campagna V, Ferrazza G, et al. Use of platelet-leukocyte membrane in arthroscopic repair of large rotator cuff tears: a prospective randomized study. J Bone Joint Surg Am. 2012;94:1345–52.

49. Postoperative platelet-rich plasma injections accelerate early tendon healing and functional recovery after arthroscopic supraspinatus repair? A randomized controlled trial. Am J Sports Med. 2015;43:1430–7. https://pubmed.ncbi.nlm.nih.gov/25790835/.

50. Sanli I, Morgan B, van Tilborg F, et al. Single injection of platelet-rich plasma (PRP) for the treatment of refractory distal biceps tendonitis: long-term results of a prospective multicenter cohort study. Knee Surg Sports Traumatol Arthrosc. 2016;24:2308–12.

51. Barker SL, Connell D, Coghlan JA, et al. Ultrasound-guided platelet-rich plasma injection for distal biceps tendinopathy. Shoulder Elbow. 2015;7:110–4.

52. Hernigou P, Flouzat Lachaniette CH, Delambre J, et al. Biologic augmentation of rotator cuff repair with mesenchymal stem cells during arthroscopy improves healing and prevents further tears: a case-controlled study. Int Orthop. 2014;38:1811–8.

53. Centeno CJ, Al-Sayegh H, Bashir J, et al. A prospective multi-site registry study of a specific protocol of autologous bone marrow concentrate for the treatment of shoulder rotator cuff tears and osteoarthritis. J Pain Res. 2015;8:269–76.

54. Charousset C, Zaoui A, Bellaiche L, Bouyer B. Are multiple plateletrich plasma injections useful for treatment of chronic patellar tendinopathy in athletes?: A prospective study. Am J Sports Med. 2014;42(4):906–11.

55. Dragoo JL, Wasterlain AS, Braun HJ, Nead KT. Platelet-rich plasma as a treatment for patellar tendinopathy: a double-blind, randomized controlled trial. Am J Sports Med. 2014;42(3):610–8.

56. Vetrano M, Castorina A, Vulpiani MC, Baldini R, Pavan A, Ferretti A. Platelet-rich plasma versus focused shock waves in the treatment of jumper's knee in athletes. Am J Sports Med. 2013;41(4):795–803.

57. Filardo G, Kon E, Della Villa S, Vincentelli F, Fornasari PM, Marcacci M. Use of platelet-rich plasma for the treatment of refractory jumper's knee. Int Orthop. 2010;34(6):909–15.

58. Gosens T, Den Oudsten BL, Fievez E, van 't Spijker P, Fievez A. Pain and activity levels before and after platelet-rich plasma injection treatment of patellar tendinopathy: a prospective cohort study and the influence of previous treatments. Int Orthop. 2012;36(9):1941–6.

59. Ferrero G, Fabbro E, Orlandi D, et al. Ultrasound-guided injection of platelet-rich plasma in chronic Achilles and patellar tendinopathy. J Ultrasound. 2012;15(4):260–6.

60. Kon E, Filardo G, Delcogliano M, et al. Platelet-rich plasma: new clinical application: a pilot study for treatment of jumper's knee. Injury. 2009;40(6):598–603.

61. Volpi P, Marinoni L, Bait C, De Girolamo L, Schoenhuber H. Treatment of chronic patellar tendinosis with buffered platelet rich plasma: a preliminary study. Med Sport. 2007;60:595–603.

62. De Vos RJ, Weir A, van Schie HT, et al. Platelet-rich plasma injection for chronic Achilles tendinopathy. A randomized controlled trial. JAMA. 2010;303:144–9.

63. De Jonge S, De Vos RJ, Weir A, et al. One-year follow-up of platelet-rich plasma treatment in chronic Achilles tendinopathy. Am J Sports Med. 2011;39:1623–9.

64. Kearney RS, Parsons N, Costa ML. Achilles tendinopathy management: a pilot randomised controlled trial comparing platelet-rich plasma injection with an eccentric loading programme. Bone Joint Res. 2013;2:227–32.

65. Krogh TP, Ellingsen T, Christensen R, et al. Ultrasound-guided injection therapy of Achilles tendinopathy with platelet-rich plasma or saline. Am J Sports Med. 2016;44:1990–7.

66. Boesen AP, Hansen R, Boesen MI, et al. Effect of high-volume injection, platelet-rich plasma, and sham treatment in chronic midportion Achilles tendinopathy. Am J Sports Med. 2017;45:2034–43.

Joints

23

Naveen S. Khokhar and Michael J. DePalma

Introduction

Osteoarthritis (OA) affects an estimated 10% of males and 18% of women over the age of 60 and is the most common joint disease worldwide [1]. In 2005, 27 million US adults were estimated to have clinical OA affecting quality of life [2]. The annual total direct per patient average cost for treatment of OA in the United States varied from $1442 to $23,335 [3].

OA can be defined as structural or functional dysfunction of a synovial joint including articular joint damage, subchondral bone alteration, synovial inflammatory response, and bone/cartilage overgrowth [4]. Cardinal symptoms suggesting diagnosis of OA include activity-related pain, reduced function, short duration stiffness, joint instability, and reduced range of motion [4]. Risk factors for OA can be divided into person-level (sociodemographic, genetic, obesity, nutrition) and joint-level (joint shape/load/alignment, muscle strength, occupation, injury/trauma) [3, 5, 6] risk factors. Ultimately, OA is a progressive condition with no means to reverse it or replace the lost tissue through conventional medical means. Regenerative medicine, on the other hand, is aimed replacing the cartilaginous lining of these joints which can rejuvenate the joint leading to an alleviation of pain and improved function.

Traditionally, an arthritic joint could only be treated through either palliative means (i.e., corticosteroid injections to suppress the pain) or surgery that typically involves an artificial joint replacement. The introduction of viscosupplementation filled an immediate gap in treatment as it represented the ability to rejuvenate part of the joint by artificially increasing the hyaluronic acid within the joint; however, this treatment does nothing to address any of the other deficiencies within the joint and requires the treatment to be repeated regularly. Using regenerative treatments to restore and regenerate the joint, itself, is the next logical progression in the treatment of arthritic joints.

Indications

The goal of regenerative therapies is to reverse inflammation and transition the joint to an anabolic state to allow for joint remodeling and healing [10]. Three commonly used regenerative injectates or orthobiologic agents are prolotherapy, platelet-rich plasma (PRP), and cell therapies [7, 11].

Prolotherapy
Prolotherapy typically involves injecting hypertonic dextrose to stimulate local irritation, inflammation, and anabolic tissue healing [12]. Dextrose acts as an osmotic agent causing cell dehydration and local trauma leading to connective tissue proliferation [11].

Platelet-Rich Plasma (PRP)
Platelet-rich plasma (PRP) consists of an autologous concentration of platelets isolated with centrifugation from whole blood [11]. The rationale behind use of PRP is that growth factors and bioactive proteins to induce healing, chondrogenesis, and stem cell proliferation as well as reduce proinflammatory mediators such as cytokines? [7]. The efficacy of

---

In reviewing the pathophysiology of osteoarthritis and the basic science for regenerative therapies, treatment options appear promising for intraarticular osteoarthritis of the knee and glenohumeral joint. Evidence for treatment of knee osteoarthritis is more robust than for treatment of glenohumeral osteoarthritis. Therefore, when considering treatment options for the glenohumeral joint, sound judgment will be needed in patient selection and understanding the goals of treatment with regenerative modalities. Further, more long-term outcome studies are needed in the evaluation of these treatments and in understanding the efficacy of regenerative treatment options in the future. Regarding cell therapies, our review focused on BMAC; however, there is a growing body of evidence utilizing stem cell therapies internationally. It is important to consider outcome data regarding these therapies as well as have an understanding of the regulatory processes for these treatments looking forward should these become more accessible in the future.

N. S. Khokhar (✉) · M. J. DePalma
Virginia iSpine Physicians, PC, Interventional Spine Care
Fellowship Program, Richmond, VA, USA

© Springer Nature Switzerland AG 2023
C. W Hunter et al. (eds.), *Regenerative Medicine*, https://doi.org/10.1007/978-3-030-75517-1_23

PRP to treat OA lies in its ability to inhibit catabolic processes such as MMP's inhibition of matrix formation [7].

Cellular Therapy and Allografts

Regarding cell-derived therapies, it is important to consider regulatory guidelines established by the FDA. Therapies that are more than minimally manipulated fall under the regulation of section 351 of the Public Health Service (PHS) act and require a Biologic License Application to demonstrate safety and efficacy of the product [13]. This would include mesenchymal stem cells (MSC) undergoing culture expansion [13]. Partly for this reason, bone marrow aspirate concentrate (BMAC) is an increasingly popular treatment utilizing progenitor cells through a single-stage procedure which is considered minimally manipulated and under the scope of section 361 of the PHS [7]. Section 361 applies to cell/tissue therapies that are minimally manipulated, intended for homologous use, and not combined with other tissues or products except water, crystalloids, or agents for sterilization, preservation, or storage [13]. Therapies under section 361 can be administered without premarket clearance from the FDA [13]. Harvested bone marrow undergoes centrifugation to separate mononucleated cells; however, stem and progenitor cells account for only 0.001–0.01% of the total number of cells in BMAC [7]. BMAC contains MSCs and bone marrow–derived platelets that contain growth factors, chemokines, and cytokines which may contribute to chondrogenesis, collagen synthesis, and suppression of IL-1β [15].

Orthobiologics have been utilized to treat OA throughout the body; however, based on the available evidence, we are only able to make recommendations for use in the shoulder and knee.

Shoulder

Although the basic science supports the rationale for use of prolotherapy, PRP, and BMAC for glenohumeral osteoarthritis, there is limited available evidence at this time to support their effectiveness and safety [7]. The paucity of studies on regenerative therapies for glenohumeral osteoarthritis was demonstrated by Robinson et al. who in a systemic review for non-operative orthobiologic treatments for rotator cuff disorders and glenohumeral OA found that no studies on glenohumeral OA met inclusion criteria [11]. Centeno et al. evaluated autologous BMAC in the treatment of shoulder rotator cuff tears and osteoarthritis in a case series [17]. In total, 34 out of 115 patients were diagnosed with osteoarthritis alone, and within this subgroup there was significant improvement ($p < 0.05$) in numeric pain scale (NPS), disabilities of the arm, shoulder, and hand (DASH), and subjective improvement rating scale [17].

Knee

The bulk of the evidence on intra-articular orthobiologics pertains to the knee. There have been a number of studies on PRP and BMAC for the treatment of knee OA. A single injection of PRP has consistently demonstrated to be at least as good or better than an entire series of hyaluronic acid with respect to pain and patient-reported function. Unfortunately, morphological improvements such as cartilage regeneration within the joint have not been consistently demonstrated with PRP. As a result, stem cells and allografts have gain increased attention due to their potential ability to theoretically "regrow" the joint lining and regenerate the vital portions of the knee.

Prolotherapy

Hauser et al. utilized a modification of Sackett's description of levels of evidence to determine the level of evidence for treatment using dextrose prolotherapy [18]. Level 1 evidence used RCTs with PEDro scores ≥6, whereas level 5 evidence used observational reports, single-subject case reports, or clinical consensus [18]. In reviewing studies for treatment of knee OA, they determined there is level 1 evidence that dextrose prolotherapy significant sustained improvement and level 4 evidence of significant improvement in OA related pain, stiffness, and function [18]. Rabago et al. reported that prolotherapy for knee OA appears safe with no adverse events reported, though the studies were not powered to detect rare events [12].

Platelet-Rich Plasma (PRP)

In a small randomized, double-blind, clinical trial including 42 patients, intra-articular injection of prolotherapy was compared to PRP. Significant improvement was noted within both the prolotherapy and PRP groups in the Western Ontario and McMaster Universities Osteoarthritis Index (WOMAC) scores, functional limitation and pain scores [19]. In a meta-analysis of 26 randomized controlled studies comparing PRP to hyaluronic acid in the treatment of knee OA, WOMAC total, WOMAC physical function, and VAS scores were found to be better in the PRP group at 3, 6, and 12 months, and WOMAC pain, WOMAC stiffness, EQ-VAS, and IKDC scores better in the PRP group at 6 and 12 months [20]. Further, there is level 1 evidence that PRP was more effective in the treatment of knee OA than hyaluronic acid, and there was no statistical difference in adverse events [20]. Kavadar et al. performed a randomized prospective study to evaluate the effectiveness of PRP and effects of different numbers of PRP applications in the treatment of moderate knee osteoarthritis and found PRP to be effective for functional status and pain with a minimum of two injections deemed appropriate [21]. Further, a meta-analysis comparing the effect of leukocyte concentration in PRP for knee OA found no difference in leukocyte-poor or leukocyte-rich in patient reported outcomes [22].

Bone Marrow Aspirate Concentrate (BMAC)

The American Society of Interventional Pain Physicians (ASIPP) published a position policy statement regarding the use of BMAC concluding that there is level 2 evidence for treatment of knee osteoarthritis and level 3 evidence for focal cartilage injuries [23]. Chahla et al. published a systematic review of BMAC for treatment of chondral injuries and osteoarthritis of the knee including 11 studies (3 for osteoarthritis and 8 for chondral injuries), all of which reported good or excellent overall outcomes [24]. In a systematic review and critical analysis of animal and clinical studies regarding BMAC for focal chondral lesions, Cavinatto et al. concluded inconsistent outcomes in animal studies and improved clinical outcomes in the clinical studies, though these had poor methodologic quality [25]. In a systematic review of BMAC for knee osteoarthritis including 8 studies and 299 knees, Keeling et al. found BMAC to be effective in improving pain and patient-reported outcomes; however, it did not demonstrate clinical superiority to other biologic therapies in comparative studies including PRP [26]. In another systematic review including preclinical and clinical studies utilizing 22 studies and 4626 patients, Cavallo et al. noted promising results with regard to effectiveness and safety; however, the studies had significant heterogeneity, few patients, short-term follow-up, and overall poor methodology [27].

Allogeneic Grafts

Allogeneic drugs/grafts have the advantage of being "off-the-shelf" treatments that do not require harvesting and processing like blood or tissue products. At the time of this textbook, there are no allogeneic grafts that are currently FDA-approved for the treatment of knee OA; however, there is one product in a phase III clinical trial called INVOSSA™ (Kolon TissueGene™, South Korea). This product is seeking a disease-modifying osteoarthritis drug (DMOAD) designation and utilizes a combination of allogeneic cell and gene therapy. The graft is composed of injectable genetically engineered chondrocytes virally transduced with TGF-β1 (GEC-TGF-β1), also known as juvenile chondrocytes, to replace lost cartilage, deliver growth factors to damaged cartilage to stimulate growth, and utilize gene therapy to suppress inflammation. In a phase II study, the product demonstrated its ability to slow progression of cartilage damage, and was noted to have statistically significant improvements in both VAS and WOMAC out to 2 years [39–41]. A double-blind, randomized controlled trial is currently underway in the United States.

Ultimately, choice of regenerative therapy requires careful patient selection and understanding the goals of treatment. Further, cost and access to treatment should also be evaluated when selecting a treatment plan. Evidence for treatment of knee osteoarthritis is growing, and consideration of treatment options should be further assessed as more outcome studies are published.

Microanatomy and Biochemistry

Pathophysiology of Osteoarthritis

Osteoarthritis is a complex multifactorial process and pathogenesis involves roles played by the articular cartilage, subchondral bone, and synovium [7].

Cartilage

The formation and breakdown of the cartilaginous matrix are regulated by an equilibrium between anabolic and catabolic influences [8]. Type II collagen is the main structural protein and aggrecans are the most common proteoglycan in articular cartilage [7]. Together, type II collagen and aggrecans contribute to the structural meshwork and hydration of articular cartilage providing tensile strength and compressive resistance [1].

Cartilage destruction has been found to be influenced by various proteases including matrix metalloproteinases (MMPs) which degrade type II cartilage and aggrecanases [7]. Aggrecans diminish in parallel to severity of OA, and although new aggrecans are involved in cartilage repair in early OA, it is lost in synovial fluid in later stages [9]. Chondrocytes regulate cartilage architecture and biochemical composition, and when activated, they produce cytokines, including interleukin (IL) 1β, IL-6, and tumor necrosis factor (TNF) α, as well as MMPs and aggrecan-degrading enzymes [1]. IL-1, TNF-α, and IL-6 have all been implicated in the development of osteoarthritis, and patients with osteoarthritis have been found to exhibit elevated levels of transforming growth factor beta (TGF-β) in synovial fluid which can lead to osteophyte production [7]. It is important to make note of the changing role of TGF-β in the young healthy joint and osteoarthritic joint [14] as TFF-β also has anabolic roles [10]. Further, the formation and accumulation of advanced glycation end products (AGEs) have been suggested for predisposing to the development of cartilage damage [9].

Anti-inflammatory and anabolic molecules within the joint include insulin-like growth factor-1 (IGF-1), TGF-β, fibroblast growth factor 18 (FGF-18), IL-4, IL-10, and platelet-derived growth factor (PDGF) [10].

Subchondral Bone

Subchondral cortical bone forms an interface between cartilage and trabecular bone with features of osteoarthritis including vascular penetration, osteophyte formation, and subchondral cysts [1]. Advanced imaging such as MRI allow for identification of bone marrow lesions which have been shown to localize to areas of severe cartilage damage [1]. Mechanical stimulation of osteoblasts has been shown to lead to expression of inflammatory cytokines and degradative enzymes [1].

Synovium

The role of inflammation in OA associated with low-grade synovitis is of growing interest with symptoms and progression of cartilage degeneration associated with synovitis [9]. Synoviocytes also release inflammatory mediators and degradative enzymes [1].

Basic Concerns and Contraindications

Despite the fact that PRP, BMAC, and allogeneic grafts have shown promise in preclinical and clinical studies, they are all considered off-label and/or investigational in the United States. PRP, in particular, has level 1 evidence to show its efficacy in knee OA – despite this fact, it is not covered by a single insurance carrier in the United States. As such, the authors recommend conventional therapies that are FDA-approved for joint pain and be utilized prior to the offering of these regenerative therapies. PRP can be utilized for knee OA with a Kellgren-Lawrence grade of 3 or less.

Contraindications for using intra-articular orthobiologics are relatively on par with performing any other intra-articular injection – the one major exception is whether or not the patient has cancer (current or in remission) and/or is in treatment. The obvious issues stem from the ability of regenerative therapies to stimulate growth and the potential introduction of stem cells. With respect to the latter, chemotherapy and/or radiation therapy could impact these cells and cause them die off, thus negating the treatment, altogether.

- Systemic infection or local infection
- Coagulopathy or inability to hold anticoagulation
- Current or prior diagnosis of cancer, especially hematologic or lymphatic
- Prolonged NSAID use or contraindications to hold
- Oral steroid or other immunosuppressants
- Patient refusal

Preoperative Considerations

These vary to a large degree depending on the regenerative therapy actually being delivered. For example, NSAIDs are a bigger concern for PRP than for BMAC. Basic preoperative considerations for prolotherapy, PRP, and BMAC are discussed in greater detail in Chaps. 2, 3 and 4, respectively.

Basic preoperative considerations include the following:

- Informed consent and education regarding treatment.
- Holding NSAIDs or oral steroids prior to procedure – time depends on therapy being utilized.
- Time needed for extraction or harvesting of injectate and centrifugation process if relevant.
- Harvest site location (PRP, BMAC, MSCs etc.) and volume to extract in preparation for injectate.
- Concentration or volume of injectate.
- Holding of anticoagulation due to entering a large joint to prevent hemarthrosis.
- Use/concentration of local anesthetic and effects on injectate as may be toxic to MSCs [16] or PRP.
- Aspiration of joint and evaluation of aspirate prior to injection of injectate.
- Type of image guidance and required scheduling.
- Patient positioning and setup.
- Treatment protocol and number of injections depending on injectate.
- Post-procedure monitoring and care.
- Post-procedure instructions including avoiding NSAIDs and oral steroids.

Radiographic Guidance

Needle visualization with image guidance to ensure intra-articular placement for accurate distribution of injectate is recommended. This can be done with ultrasound or fluoroscopic guidance depending on the practitioner's expertise and comfort. Toxicity of contrast medium to regenerative therapies in vivo needs further evaluation when considering options. Cytotoxicity to mesenchymal stem cells has been suggested with iodinated contrast dye [36, 37]. However, this was not suggested at an early time point in vitro with PRP [38].

In a literature review, Berkoff et al. found that all forms of imaging guidance (including ultrasound, air arthrography, fluoroscopy, magnetic resonance arthrography, or magnetic resonance imaging) improved the accuracy of intra-articular injections of the knee (96.7% versus 81.0%, $P < 0.001$) and shoulder (97.3% versus 65.4%, $P < 0.001$). Further, ultrasound guidance of knee injections (95.8% versus 77.8%, $P < 0.001$) and shoulder injections (88.8% versus 61.1%, $P < 0.001$) resulted in better accuracy than did anatomical guidance [28]. Daley et al. also found improved accuracy in glenohumeral joint (95% vs 79% $p < 0.001$) and knee injections (99% vs 79% $p < 0.001$) with all image guidance compared to blind injection [29]. A recent systematic review of level 1 evidence of randomized controlled trials comparing ultrasound-guided and blind knee injections found >95% accuracy for ultrasound guidance vs 77.3–95.74% accuracy for blind approaches [30].

Imaging modality may also be a consideration depending on targeted joint and provider comfort. Amber et al.

published a systematic review with meta-analysis comparing the accuracy of ultrasound-guided and fluoroscopically guided glenohumeral joint injection noting 93% (95% CI, 86–98%) and 80% (95% CI, 63–93%) accuracy with ultrasound and fluoroscopic guidance, respectively; however, differences were not statistically significant [31].

Equipment and Techniques

The overall technique for injecting a biologic into the shoulder or knee is virtually the same as any other conventional injection into them for any other purpose with some minor differences that are outlined below. As such, there are a number of techniques that are well described in textbooks and the literature. The authors advocate the techniques below when injecting biologics into either of these two joints based on the available literature regarding each target (acquisition and processing of injectate have been omitted as they are addressed in other chapters within this text):

Knee
Ultrasound-guided suprapatellar approach [32]:
Equipment

- 25-gauge needle and 2–3 cc local anesthetic
- 18- to 22-gauge, 1.5- to 2-inch needle depending on effusion and injectate
- 2, 10–25 cc syringe for aspirate if needed
- Injectate
- Sterile tray with prep and drape
- High-frequency linear array transducer
- Sterile ultrasound probe cover

Patient Positioning

- Consider the most comfort position for the patient, either sitting upright or in a supine position with the target knee extended.
- Consider scanning landmarks with ultrasound and marking targets with a skin marker prior to sterile preparation.

Sterile Technique

- Patient skin is prepped and draped in standard sterile fashion with chlorhexidine. Povidone-iodine is used in the setting of a chlorhexidine allergy.

Procedure

- Maintain a sterile field and use sterile gel. Apply sterile ultrasound probe cover.

- Place transducer proximal to the patella to visualize the distal quadriceps in short axis.
- Visualize suprapatellar synovial pouch which is the target for this procedure.
- Place skin weal with local anesthetic if indicated proximal to transducer.
- Progress needle with a lateral to medial approach in-plane with transducer until needle is visualized at target.
- When needle is confirmed at target, distribute injectate.
- After withdrawing needle, apply pressure if needed and place bandage/dressing.

Shoulder
Ultrasound-guided posterior approach [35]:
Equipment

- 25-gauge needle and 3–5 cc local anesthetic
- 18- to 22-gauge, 3.5- to 5-inch needle depending on effusion and injectate
- 2, 10–25 cc syringe for aspirate if needed
- Injectate
- Sterile tray with prep and drape
- Medium- to high-frequency curvilinear array transducer
- Sterile ultrasound probe cover

Patient Positioning

- Consider the most comfort position for the patient, either sitting upright or in a lateral recumbent position with the target shoulder accessible.
- Consider scanning landmarks with ultrasound and marking targets with a skin marker prior to sterile preparation.

Sterile Technique

- Patient skin is prepped and draped in standard sterile fashion with chlorhexidine or povidone-iodine.

Procedure

- Maintain a sterile field and use sterile gel. Apply sterile ultrasound probe cover.
- Place transducer over the posterior aspect of the glenohumeral joint parallel to the infraspinatus tendon.
- Visualize the posterior glenohumeral joint with target between humeral head and glenoid labrum which is the target for this procedure.
- Place skin weal with local anesthetic if indicated proximal to transducer.
- Progress needle with a lateral to medial approach in-plane with transducer until needle is visualized at target.
- When needle is confirmed at target, distribute injectate.

- After withdrawing needle, apply pressure if needed and place bandage/dressing.

Fluoroscopic-guided anterior approach [33, 34]:

Equipment

- 25-gauge needle and 3–5 cc local anesthetic
- 18- to 22-gauge, 3.5- to 5-inch needle depending on effusion and injectate
- 2, 10 to 25 cc syringe for aspirate if needed
- Injectate
- Sterile tray with prep and drape.
- C-arm, fluoroscopy suite
- Sterile ultrasound probe cover

Patient Positioning

- Position patient in the prone position.
- Slight external rotation of shoulder with arm at patient's side with palm upward.

Sterile Technique

- Patient skin is prepped and draped in standard sterile fashion with chlorhexidine or povidone-iodine.

Procedure

- Obtain AP view of the shoulder.
- Target is the superomedial humeral head.
- Place skin weal with local anesthetic and anesthetize track if indicated.
- Advance needle to target with fluoroscopic guidance.
- When at target, administer contrast with live fluoroscopy to confirm intra-articular placement.
- When needle is confirmed at target, distribute injectate.
- After withdrawing needle, apply pressure if needed and place bandage/dressing.

Post-procedure Considerations

Regenerative therapies have the propensity to cause inflammation – whether that be inflammation specifically orchestrated to induce healing (i.e., prolotherapy and PRP) or inflammation resulting as a side effect/by-product of the treatment (i.e., BMAC and allogeneic graft). Regardless of the intention, it is an expected consequence when utilizing this particular treatment. Consequently, one should anticipate joint inflammation and potentially swelling in the postoperative period. Moreover, inform the patient that their joint

may swell for a short period of time afterward to alleviate concern.

As mentioned previously, NSAIDs are contraindicated prior to and following the injection of orthobiologics, as NSAIDs are believed to reduce the effectiveness of these agents by limiting the initial inflammatory phase that results immediately after treatment, reducing localized debridement of the injected tissue, decreasing local collagen creation, and the subsequent repair of the target tissue. Baby aspirin (81 mg) for cardiac prevention is generally not contraindicated, as removing the therapy may have unintended consequences that are greater than the potential benefit that may occur from the treatment of the joint. Furthermore, the dose is considered significantly low enough that it would only minorly impair the effectiveness of the biologic therapy. Due to the half-life of most NSAIDs, it is recommended that patients refrain from NSAIDs for approximately 2 weeks prior to the initial injection and for 3–4 weeks after the final treatment. Corticosteroids have even longer half-lives and are therefore recommended to be avoided for at least 1 month prior and for at least 1 month orally after, and in the case of topical steroids, application on or near the site of treatment should be avoided for as long as possible.

Potential Complications and Pitfalls

Intra-articular injections, in general, are typically well-tolerated procedures and have a low likelihood for complication. Aside from inflammation, joint swelling, and localized irritation at the procedure site, post-procedural discomfort should be transient and self-limiting. Infection from an intra-articular injection is unlikely; however, treatments like PRP and BMAC are theoretically more prone introducing a transmissible agent and creating potential infections due to the number of times the injectate is transferred from one receptacle to another. Further, these mediums are rich with growth factors making them a potential medium for culturing a pathogen. Consequently, infection should be considered if the inflammation is progressive or fails to resolve after 5–7 days. Following are the other causes for concern:

- Excessive irritation of site or post-injection soreness
- Soft-tissue or intra-articular infection
- Excessive bleeding, hematoma, or hemarthrosis
- Inadvertent soft tissue or nerve injury
- Allergic reaction to injectate
- Theoretical risk of implantation or seeding of undiagnosed malignancy from harvest site

Clinical Pearls

- Inflammation and joint swelling are to be expected post-procedurally.
- Avoid NSAIDs and corticosteroids before after treatment with regenerative therapies.
- A single treatment of PRP into the knee has been shown to be equivocal or better than entire series of viscosupplementation.
- Off-the-shelf allografts are in clinical trials and not yet commercially available in the United States.
- Intra-articular biologics should only be administered using radiographic guidance.

References

1. Glyn-Jones S, Palmer AJR, Agricola R, Price AJ, Vincent TL, Weinans H, Carr AJ. Osteoarthritis. Lancet. 2015;386(9991):376–87. https://doi.org/10.1016/S0140-6736(14)60802-3.
2. Murphy L, Helmick CG. The impact of osteoarthritis in the United States: a population-health perspective: a population-based review of the fourth most common cause of hospitalization in U.S. adults. Orthop Nurs. 2012;31(2):85–91. Retrieved from https://www.proquest.com/scholarly-journals/impact-osteoarthritis-united-states-population/docview/1010319767/se-2?accountid=10639.
3. Vina ER, Kwoh CK. Epidemiology of osteoarthritis: literature update. Curr Opin Rheumatol. 2018;30(2):160–7. https://doi.org/10.1097/BOR.0000000000000479. PMID: 29227353; PMCID: PMC5832048
4. Hunter DJ, Guermazi A. Imaging techniques in osteoarthritis. PM R. 2012;4(5 Suppl):S68–74. https://doi.org/10.1016/j.pmrj.2012.02.004.
5. Neogi T, Zhang Y. Epidemiology of osteoarthritis. Rheum Dis Clin North Am. 2013;39(1):1–19. https://doi.org/10.1016/j.rdc.2012.10.004. Epub 2012 Nov 10. PMID: 23312408; PMCID: PMC3545412.
6. O'Neill TW, McCabe PS, McBeth J. Update on the epidemiology, risk factors and disease outcomes of osteoarthritis. Best Pract Res Clin Rheumatol. 2018;32(2):312–26. https://doi.org/10.1016/j.berh.2018.10.007. Epub 2018 Nov 22
7. Rossi LA, Piuzzi NS, Shapiro SA. Glenohumeral osteoarthritis: the role for orthobiologic therapies: platelet-rich plasma and cell therapies. JBJS Rev. 2020;8(2):e0075. https://doi.org/10.2106/JBJS.RVW.19.00075. PMID: 32015271; PMCID: PMC7055935
8. Michael JW, Schlüter-Brust KU, Eysel P. The epidemiology, etiology, diagnosis, and treatment of osteoarthritis of the knee. Dtsch Arztebl Int. 2010;107(9):152–62. https://doi.org/10.3238/arztebl.2010.0152. Epub 2010 Mar 5. Erratum in: Dtsch Arztebl Int. 2010 Apr;107(16):294. PMID: 20305774; PMCID: PMC2841860.
9. Hussain SM, Neilly DW, Baliga S, Patil S, Meek R. Knee osteoarthritis: a review of management options. Scott Med J. 2016;61(1):7–16. https://doi.org/10.1177/0036933015619588. Epub 2016 Jun 21
10. Richards MM, Maxwell JS, Weng L, Angelos MG, Golzarian J. Intra-articular treatment of knee osteoarthritis: from anti-inflammatories to products of regenerative medicine. Phys Sportsmed. 2016;44(2):101–8. https://doi.org/10.1080/0091384

7.2016.1168272. Epub 2016 Apr 4. PMID: 26985986; PMCID: PMC4932822.
11. Robinson DM, Eng C, Makovitch S, Rothenberg JB, DeLuca S, Douglas S, Civitarese D, Borg-Stein J. Non-operative orthobiologic use for rotator cuff disorders and glenohumeral osteoarthritis: a systematic review. J Back Musculoskelet Rehabil. 2021;34(1):17–32. https://doi.org/10.3233/BMR-201844.
12. Rabago D, Nourani B. Prolotherapy for osteoarthritis and tendinopathy: a descriptive review. Curr Rheumatol Rep. 2017;19(6):34. https://doi.org/10.1007/s11926-017-0659-3.
13. Kingery MT, Manjunath AK, Anil U, Strauss EJ. Bone marrow mesenchymal stem cell therapy and related bone marrow-derived orthobiologic therapeutics. Curr Rev Musculoskelet Med. 2019;12(4):451–9. https://doi.org/10.1007/s12178-019-09583-1. PMID: 31749105; PMCID: PMC6942063.
14. Van der Kraan PM. The changing role of TGFb in healthy, ageing and osteoarthritic joints. Nat Rev Rheumatol. 2017;13(3):155–63. Epub 2017 Feb 2
15. Cotter EJ, Wang KC, Yanke AB, Chubinskaya S. Bone marrow aspirate concentrate for cartilage defects of the knee: from bench to bedside evidence. Cartilage. 2018;9(2):161–70. https://doi.org/10.1177/1947603517741169. Epub 2017 Nov 10. PMID: 29126349; PMCID: PMC5871125.
16. Sampson S, Botto-van Bemden A, Aufiero D. Autologous bone marrow concentrate: review and application of a novel intra-articular orthobiologic for cartilage disease. Phys Sportsmed. 2013;41(3):7–18. https://doi.org/10.3810/psm.2013.09.2022.
17. Centeno CJ, Al-Sayegh H, Bashir J, Goodyear S, Freeman MD. A prospective multi-site registry study of a specific protocol of autologous bone marrow concentrate for the treatment of shoulder rotator cuff tears and osteoarthritis. J Pain Res. 2015;5(8):269–76. https://doi.org/10.2147/JPR.S80872. PMID: 26089699; PMCID: PMC4463777
18. Hauser RA, Lackner JB, Steilen-Matias D, Harris DK. A systematic review of dextrose prolotherapy for chronic musculoskeletal pain. Clin Med Insights Arthritis Musculoskelet Disord. 2016;7(9):139–59. https://doi.org/10.4137/CMAMD.S39160. PMID: 27429562; PMCID: PMC4938120
19. Rahimzadeh P, Imani F, Faiz SHR, Entezary SR, Zamanabadi MN, Alebouyeh MR. The effects of injecting intra-articular platelet-rich plasma or prolotherapy on pain score and function in knee osteoarthritis. Clin Interv Aging. 2018;4(13):73–9. https://doi.org/10.2147/CIA.S147757. PMID: 29379278; PMCID: PMC5757490
20. Tan J, Chen H, Zhao L, Huang W. Platelet-rich plasma versus hyaluronic acid in the treatment of knee osteoarthritis: a meta-analysis of 26 randomized controlled trials. Arthroscopy. 2021;37(1):309–25. https://doi.org/10.1016/j.arthro.2020.07.011. Epub 2020 Jul 15
21. Kavadar G, Demircioglu DT, Celik MY, Emre TY. Effectiveness of platelet-rich plasma in the treatment of moderate knee osteoarthritis: a randomized prospective study. J Phys Ther Sci. 2015;27(12):3863–7. https://doi.org/10.1589/jpts.27.3863. Epub 2015 Dec 28. PMID: 26834369; PMCID: PMC4713808.
22. Abbas A, Du JT, Dhotar HS. The effect of leukocyte concentration on platelet-rich plasma injections for knee osteoarthritis: a network meta-analysis. J Bone Joint Surg Am. 2021; https://doi.org/10.2106/JBJS.20.02258. Epub ahead of print
23. Manchikanti L, Centeno CJ, Atluri S, Albers SL, Shapiro S, Malanga GA, Abd-Elsayed A, Jerome M, Hirsch JA, Kaye AD, Aydin SM, Beall D, Buford D, Borg-Stein J, Buenaventura RM, Cabaret JA, Calodney AK, Candido KD, Cartier C, Latchaw R, Diwan S, Dodson E, Fausel Z, Fredericson M, Gharibo CG, Gupta M, Kaye AM, Knezevic NN, Kosanovic R, Lucas M, Manchikanti MV, Mason RA, Mautner K, Murala S, Navani A, Pampati V, Pastoriza S, Pasupuleti R, Philip C, Sanapati MR, Sand

T, Shah RV, Soin A, Stemper I, Wargo BW, Hernigou P. Bone marrow concentrate (BMC) therapy in musculoskeletal disorders: evidence-based policy position statement of American Society of Interventional Pain Physicians (ASIPP). Pain Phys. 2020;23(2): E85–E131.

24. Chahla J, Dean CS, Moatshe G, Pascual-Garrido C, Serra Cruz R, LaPrade RF. Concentrated bone marrow aspirate for the treatment of chondral injuries and osteoarthritis of the knee: a systematic review of outcomes. Orthop J Sports Med. 2016;4(1):2325967115625481. https://doi.org/10.1177/2325967115625481. PMID: 26798765; PMCID: PMC4714134.

25. Cavinatto L, Hinckel BB, Tomlinson RE, Gupta S, Farr J, Bartolozzi AR. The role of bone marrow aspirate concentrate for the treatment of focal chondral lesions of the knee: a systematic review and critical analysis of animal and clinical studies. Arthroscopy. 2019;35(6):1860–77. https://doi.org/10.1016/j.arthro.2018.11.073. Epub 2019 Mar 11

26. Keeling LE, Belk JW, Kraeutler MJ, Kallner AC, Lindsay A, McCarty EC, Postma WF. Bone marrow aspirate concentrate for the treatment of knee osteoarthritis: a systematic review. Am J Sports Med. 2021;8:3635465211018837. https://doi.org/10.1177/03635465211018837. Epub ahead of print

27. Cavallo C, Boffa A, Andriolo L, Silva S, Grigolo B, Zaffagnini S, Filardo G. Bone marrow concentrate injections for the treatment of osteoarthritis: evidence from preclinical findings to the clinical application. Int Orthop. 2021;45(2):525–38. https://doi.org/10.1007/s00264-020-04703-w. Epub 2020 Jul 13. PMID: 32661635; PMCID: PMC7843474.

28. Berkoff DJ, Miller LE, Block JE. Clinical utility of ultrasound guidance for intra-articular knee injections: a review. Clin Interv Aging. 2012;7:89–95. https://doi.org/10.2147/CIA.S29265.

29. Daley EL, Bajaj S, Bisson LJ, Cole BJ. Improving injection accuracy of the elbow, knee, and shoulder: does injection site and imaging make a difference? A systematic review. Am J Sports Med. 2011;39(3):656–62. https://doi.org/10.1177/0363546510390610. Epub 2011 Jan 21

30. Fang WH, Chen XT, Vangsness CT Jr. Ultrasound-guided knee injections are more accurate than blind injections: a systematic review of randomized controlled trials. Arthrosc Sports Med Rehabil. 2021;3(4):e1177–87. https://doi.org/10.1016/j.asmr.2021.01.028.

31. Amber KT, Landy DC, Amber I, Knopf D, Guerra J. Comparing the accuracy of ultrasound versus fluoroscopy in glenohumeral injections: a systematic review and meta-analysis. J Clin Ultrasound. 2014;42(7):411–6. https://doi.org/10.1002/jcu.22154. Epub 2014 Mar 25

32. Cianca JC. Intraarticular injections of the knee. In: Malanga GA, Mautner KR, editors. Atlas of ultrasound-guided musculoskel-etal injections. Chicago: McGraw-Hill Education Medical; 2014. p. 252–7.

33. Mattie R, McCormick ZL, Fogg B, Cushman DM. The effect of body mass index on fluoroscopy time and radiation dose in intra-articular glenohumeral joint injections. Clin Imaging. 2017;42: 19–24. https://doi.org/10.1016/j.clinimag.2016.11.008. Epub 2016 Nov 16

34. Batson, J. P., Johnson, S. C., Furman, M. B., Intraarticular shoulder injections-anterior approach: fluoroscopic guidance. (2018). Furman, M. B., Berkwits, L., Cohen, I., Goodman, B. S., Kirschner, J. S., Lee, T. S., & Lin, P. S. Atlas of image-guided spinal procedures. 2nd Ed. Elsevier, Inc, Philadelphia. pgs 535–542.

35. Henning T. Glenohumeral joint injection. In: Malanga GA, Mautner KR, editors. Atlas of ultrasound-guided musculoskeletal injections. Chicago: McGraw-Hill Education Medical; 2014. p. 24–7.

36. Wu T, Nie H, Dietz AB, Salek DR, Smith J, van Wijnen AJ, Qu W. Cytotoxic effects of nonionic iodinated contrast agent on human adipose-derived mesenchymal stem cells. PM & R. 2018;S1934–1482(18)30294–6. Advance online publication. https://doi.org/10.1016/j.pmrj.2018.05.022

37. McKee C, Beeravolu N, Bakshi S, Thibodeau B, Wilson G, Perez-Cruet M, Rasul CG. Cytotoxicity of radiocontrast dyes in human umbilical cord mesenchymal stem cells. Toxicol Appl Pharmacol. 2018;15(349):72–82. https://doi.org/10.1016/j.taap.2018.04.032. Epub 2018 Apr 26

38. Dallaudiere B, Crombé A, Gadeau AP, Pesquer L, Peuchant A, James C, Silvestre A. Iodine contrast agents do not influence platelet-rich plasma function at an early time point in vitro. J Exp Orthop. 2018;5(1):47. https://doi.org/10.1186/s40634-018-0162-4. PMID: 30374787; PMCID: PMC6206314

39. Ha CW, Cho JJ, Elmallah RK, Cherian JJ, Kim TW, Lee MC, Mont MA. A multicenter, single-blind, phase IIa clinical trial to evaluate the efficacy and safety of a cell-mediated gene therapy in degenerative knee arthritis patients. Hum Gene Ther Clin Dev. 2015;26(2):125–30. https://doi.org/10.1089/humc.2014.145. Epub 2015 Apr 17

40. Cherian JJ, Parvizi J, Bramlet D, Lee KH, Romness DW, Mont MA. Preliminary results of a phase II randomized study to determine the efficacy and safety of genetically engineered allogeneic human chondrocytes expressing TGF-β1 in patients with grade 3 chronic degenerative joint disease of the knee. Osteoarthritis Cartil. 2015;23(12):2109–18. https://doi.org/10.1016/j.joca.2015.06.019. Epub 2015 Jul 16

41. Guermazi A, Kalsi G, Niu J, Crema MD, Copeland RO, Orlando A, Noh MJ, Roemer FW. Structural effects of intra-articular TGF-β1 in moderate to advanced knee osteoarthritis: MRI-based assessment in a randomized controlled trial. BMC Musculoskelet Disord. 2017;18(1):461. https://doi.org/10.1186/s12891-017-1830-8. PMID: 29145839; PMCID: PMC5689208

Axial Spine and Sacroiliac Joint

Annu Navani, and Joshua Chrystal

Introduction

The field of regenerative medicine has attracted increasing researchers' and practitioners' interest in the treatment of various spinal and orthopedic conditions. A natural trend toward minimally invasive and nonsurgical treatment options has grown in popularity, not only among practitioners but also among patients, who are becoming more aware of these options and are increasingly requesting such treatments in clinical settings. The utilization of both autologous and allograft biological material to help facilitate healing has been described for decades; however, it has undergone increasing research and scrutiny in more recent years.

This has driven a focused analysis of the mechanisms of action and potential applications of these various treatments, which has produced a plethora of information on best practices regarding the use of various biological treatments in spinal and orthopedic conditions, alike.

In this chapter, we will focus on two main types of biological treatments called *platelet-rich plasma (PRP)* and *mesenchymal stem cells (MSCs)*. The essence of platelet-rich plasma (PRP) is to boost the repair processes in damaged tissue by delivering a concentrated dose of autologous growth factors, thereby activating local mesenchymal stem cells (MSCs) at the site of injury. Some of the specific growth factors released, such as platelet-derived growth factor, transforming growth factor-beta 1, insulin-like growth factor-1, vascular endothelial growth factor, fibroblastic growth factor, and epidermal growth factors, have been shown to control the mechanism of tissue repair and restoration [1]. Through carefully orchestrated chemotaxis, angio-genesis, cellular migration, proliferation, and differentiation and extracellular matrix production, regeneration is facilitated [1].

As with PRP therapy, there is evidence supporting the use of MSCs in musculoskeletal, orthopedic, and spinal conditions. MSCs are self-renewing and relatively undifferentiated. However, upon induction by certain growth factors, these cells can differentiate into osteoblasts, chondroblasts, and adipocytes [1]. They have demonstrated an ability to secrete growth factors and induce cell proliferation, angiogenesis, anti-inflammatory effects, antiapoptotic effects, and immunomodulation. The utilization and exploitation of these properties have shown promise in regenerating the tissue of degenerated *intravertebral disks (IVDs)* by increasing proteoglycan synthesis and type II collagen production and providing pain relief [1].

Indications

Low back *pain* affects large portions of the population and is associated with major social and economic costs. For example, the related healthcare utilization costs related to chronic low back pain are estimated to be $96 million per year in the United States [2]. Discogenic low back pain is the most common cause of chronic low back pain, accounting for 39% of cases. Other common causes of low back pain, such as zygapophysial joint pain, myofascial pain, sacroiliac (SI) joint pain, and others, have lower prevalence rates [3]. In this chapter, we will discuss the use of biologics to treat some of the most common pathologies and sources of pain in the spine, including the following:

- Intervertebral disk (IVD)
- Facet joints
- Sacroiliac (SI) joints

There are many sources of pain generation throughout the axial spine. This chapter will focus on three of the most *com-*

A. Navani (✉)
Comprehensive Spine and Sports Center, Department of Regenerative Medicine, Campbell, CA, USA
e-mail: anavani@cssctr.com

J. Chrystal
Comprehensive Spine and Sports Center, Department of Regenerative Medicine and Department of Physical Medicine, Campbell, CA, USA

mon areas of pain generation (IVDs, facet joints, and SI joints) and the ability to treat these areas utilizing biologics.

Biomechanical changes can initiate pathological processes, leading to biochemical changes and maladaptation in the healing process. Many factors contribute to pain and long-standing injuries, including repetitive trauma, aberrant biomechanical adaptations and changes, health status, and psychosocial factors, to name a few. In modern societies, dietary, nutritional, and *environmental* factors may interfere with the body's ability to fully heal and contribute to an inability to reduce chronic inflammation, leading to various disease processes and chronic pain. Regenerative therapies help focus healing by introducing new biochemical signaling initiators to restart or capacitate the body's ability to complete healing utilizing its own internal mechanisms.

Intervertebral Disks

It is estimated that discogenic pathology, with or without internal derangement, contributes to approximately 16.9–39% of chronic low back pain without radiculopathy [1]. Damage to the IVD ranges from disk degeneration to disk extrusion and sequestration. Current interventional treatments focus on targeting the symptoms without addressing the underlying cause of the pathology, and their palliative effects are transient. In contrast, newer therapies focus on alleviating pain through the restoration of the IVD structure and function.

The IVDs are essential for the health and well-being of the spine. They provide structural support, shock absorption, and a biomechanical pivot point to allow proper range of motion of the vertebral motor unit. The nucleus pulposus at the center of the disk is composed mainly of water and is, therefore, *non-compressible*. Various mechanisms of injury may cause the nucleus pulposus to push off-center, resulting in a deformed disk. Should the disk be deformed in a way that effaces or compresses either the central spinal canal or the nerve roots exiting the neuroforamen, myelopathy- or radiculopathy-type symptoms may occur.

The outer third of the IVD is a ring that is rich in nerve fibers, called the annulus. It surrounds the disk and holds the nucleus pulposus in place. Any tear off in the annulus is a potential pain generator and may be treated utilizing biological therapies such as PRP and MSCs. Recent guidelines that *focused* on the use of biologics to treat LBP support the use of PRP and MSC in the treatment of IVD disease, based on level 3 evidence [1].

Evidence

PRP and MSC injections *have* been utilized with increasing frequency in disk-related disorders of the spine, particularly degenerative disk disease. However, MSCs are used more commonly for IVD degeneration and disk repair.

Intradiscal Platelet Rich Plasma (PRP)

A summary of six studies focusing on the use of PRP to treat IVD degeneration is shown in Table 24.1 [1, 4–8]. The number of patients included in these studies ranged from 8 to 86. Length of follow-up ranged from 8 weeks to 2 years. The largest of the studies included 47 and 86 patients, and the rest were significantly smaller. All the studies reported that intradiscal PRP injection was safe and effective in relieving pain and increasing function. However, because most of the studies were *small* observational reports or case series, more evidence is needed regarding the safety and efficacy of PRP injection therapy in managing discogenic pain.

Navani et al. [9] studied 20 patients, of which 15 were available for follow-up for 18 months post-PRP injection therapy. At the end of the 18-month period, no patients experienced adverse events, except for the initial post-injection flare-up pain, which lasted for 2–4 weeks. At the 6-month follow-up, 94% of patients reported >50% pain relief on the visual analog scale (VAS) and 100% had improvements in SF-36 scores. At 18 months, 93% of patients showed continuous improvement in their SF-36 scores.

In a prospective trial of eight humans, Levi et al. [7] demonstrated that IVD degeneration of up to 5 levels could be clinically and radiographically improved with a single PRP injection at each level. Monfett et al. [5] demonstrated safety and statistically significant improvement in pain and function through the 24-month follow-up period after intradiscal PRP in 29 patients. Tuakli-Wosornu et al. [4] *also* reported significant improvement in 47 patients with discogenic pain who received a single intradiscal injection of PRP and were followed for 1 year.

Kirchner and Anitua [8], in an observational retrospective pilot study of 86 patients with a history of chronic low back pain and degenerative disk disease of the lumbar spine, showed a statistically significant improvement in VAS scores at 1, 3, and 6 months after PRP treatment. VAS score analysis over time showed that, at the end point of the study (6 months), 91% of patients had an excellent score, 8.1% experienced moderate improvement, and 1.2% were in the inefficient *score*. Fluoroscopy-guided injections of plasma-rich growth factors (PRGF) into the IVDs and facet joints of patients with chronic low back pain resulted in significant pain reduction, as assessed using the VAS.

Akeda et al. [6], in a preliminary clinical trial that included 14 patients, demonstrated that average pretreatment VAS scores significantly decreased at 1 month following PRP treatment, which was *sustained* throughout the full 6-month observation. This study showed that intradiscal injection of

Table 24.1 Studies of PRP therapy for IVD degeneration

Study	Conclusions
Tuakli-Wosornu et al. [4], 2016 Chronic lumbar discogenic pain Prospective, double-blind, randomized controlled study, n = 47	Intradiscal injections of PRP × 1 showed significant improvement at 8-week follow-up, with maintained improvement compared to controls at 1-year follow-up
Monfett et al. [5], 2016 Chronic lumbar discogenic pain, lumbar disk degeneration Prospective trial, n = 29	Intradiscal PRP injections showed continuous safety and improvements in pain and function at 2 years postprocedure
Akeda et al. [6], 2017 Chronic lumbar discogenic pain Preliminary clinical trial, n = 14	Intradiscal injection of autologous PRP releasate in patients with low back pain was determined to be safe, with no adverse events observed during follow-up The results showed reduction in mean pain scores at 1 month, which was sustained throughout the 6-month and 12-month follow-up periods
Levi et al. [7], 2016 Chronic lumbar discogenic pain Prospective trial, n = 8	Single or multiple levels (up to 5) of discogenic pain injected with PRP showed encouraging improvement, with more patients developing improvement over time. Cohort up to 6 months
Kirchner and Anitua [8], 2016 Chronic lumbar disk degeneration Observational retrospective pilot study, n = 86	Fluoroscopy-guided infiltrations of IVDs and facet joints with PRGF-Endoret in patients with chronic low back pain were associated with significant pain reduction, as assessed by VAS. The study reported a reduction of VAS scores over time. After 6 months of follow-up, 91% of patients had an excellent score, 8.1% experienced moderate improvement, and 1.2% did not respond to the treatment
Navani et al. [9], 2018 Chronic lumbar discogenic pain Prospective case series n = 20	At 18 months, 15 patients were available for follow-up, compared to 18 patients at 6 months: >50% relief in VAS in 93% of patients (n = 14/15) at 18 months and in 94% of patients (n = 17/18) at 6 months. Improvement in SF-36 scores in 93% of patients (n = 14/15) at 18 months, compared to 100% (n = 18/18) at 6 months

PRP platelet-rich plasma, *PRGF* plasma rich in growth factors, *VAS* visual analog scale, *SF-36* 36-item short form survey

autologous PRP releasate in patients with low back pain was safe and there were no adverse events observed during follow-up.

Intradiscal Stem Cells and Allogeneic Grafts

There are more preclinical and clinical studies of MSC therapy than there are of PRP therapy for disk-related disorders

of the spine [10–16]. There are also guidelines focusing on the use of MSCs for intervertebral disk disease [1].

Table 24.2 summarizes the studies published to date that focus on the use of MSC therapy to treat IVD disease [1]. Wu et al. [19] conducted a systematic review and a single-arm meta-analysis that included six studies meeting their selection criteria.

The authors reported that the pooled mean difference in pain score from baseline to follow-up points of 44.2 points decreased and the pooled mean difference in *Oswestry Disability Index (ODI)* from baseline to follow-up points was 32.2 points. There were no adverse effects, and they concluded that cell-based therapy is *appropriate* for patients with discogenic low back pain and is associated with improved pain relief and disability scores. However, they recommended that more stringently designed, randomized, double-blind clinical trials with appropriately determined sample sizes are needed to confirm their findings.

Wu et al. [19] reported that three of the studies in their review used stem cells and another three studies used chondrocytes. They also reported that five studies used expanded cells and one used unexpanded cells. In these studies, lumbar disks were infiltrated with cells ranging in dosage from 1 to 23 ± five million. Patients had a mean follow-up of 22 months among the six trials.

The pooled mean difference in pain score from baseline to follow-up points was 44.2 points decreased and ODI 32.2 points decreased. No related adverse effects were reported by the included studies. Subgroup analysis was used to explore whether cell-preprocessing conditions (i.e., expanded vs. nonexpanded) were associated with a difference in pain scores [19]. Subsequent meta-regression analysis to determine factors related to improvement in pain scores after stem-cell therapy demonstrated that stem cells were more effective than chondrocytes in reducing pain.

Mochida et al. [15] used the Japanese Orthopedic Association scoring system to demonstrate that patients experienced improvement from 14.2 ± 4.8 points preoperatively to 27.2 ± 1.6 points 3 years *following* transplantation of activated nucleus pulposus cells [15]. Furthermore, patients who were working prior to the treatment were able to return to their original job after an average of 5.8 weeks following treatment.

In the study by Coric et al. [12], the mean SF-36 physical component summary scores improved significantly from baseline after treatment with MSCs. The study also demonstrated a positive effect on *high-intensity* zones that were consistent with posterior annular tears. High-intensity zones present at baseline were either absent or improved in 89% of the patients within 6 months following treatment with MSCs [12].

Table 24.2 Studies of stem-cell therapy for IVD degeneration

Study details	Population	Cell or solution dose and delivery pathway	Results	Conclusion
Noriega et al., 2017 [10] Sample size = 24 Follow-up = 12 months RCT	24 patients with chronic LBP with lumbar disk degeneration and unresponsive to conservative treatments were randomized into two groups Patient mean age (yrs) ± SE = 38 ± 2 s	The intervention group received allogeneic bone marrow MSCs by intradiscal injection of 25×10, 6 cells per segment under local anesthesia	Compared to controls, MSC-treated patients displayed a rapid and significant improvement in all algofunctional indices Both lumbar pain and disability were significantly reduced at 3 months, and improvement was maintained at 6 and 12 months. Overall, there was an average 28% improvement in pain and disability one-year after the intervention Only 5 of the 12 outcomes (40%) in patients receiving MSCs were described as perfect treatment with 100% improvement	28% improvement observed in all patients 40% of patients had perfect result Positive result
Pettine et al., 2015, 2016, 2017 [11] Sample size = 26 Follow-up = 3 years Prospective, open-label, nonrandomized, 2-arm study	26 patients presented with symptomatic moderate to severe discogenic LBP. Patient age = 18–61 years (median 40)	2–3 mL of bone marrow concentrate was injected in lumbar disk (1.66_106/mL)	The average ODI and VAS scores were reduced to 22.8 and 24.4, respectively, at 3 months. After 36 months, 6 patients proceeded to surgery After 36 months, 20 of the 26 patients reported average ODI and VAS improvement to 17.5 ± 32 and 21.9 ± 4.4, respectively One-year MRI indicated 40% of patients improved one modified Pfirrmann Grade and no patient worsened radiographically	At 36-month follow-up, 6 of 26 patients progressed to surgery. The remaining 20 patients (77%) reported significant ODI and VAS improvements Authors concluded that there were no adverse effects, and the study provided evidence of safety and feasibility of intradiscal bone marrow concentrate therapy
Coric et al., 2013 [12] Sample size = 15 Follow-up = 1 year Prospective cohort	15 patients with single-level, symptomatic lumbar degenerative disk disease from L-3 to S-1 and medically refractory LBP Patient age (yrs) = 19–47 years (median 40)	Mean 1.3 mL (1–2 mL, 107/mL) cells solution was injected in the center of the disk space	The mean ODI, NRS, and Short-form-36 physical component summary scores all improved significantly from baseline Ten of the 13 patients (77%) exhibited improvements on MRI. Of these, the HIZ was either absent or improved in 8 patients (89%) by 6 months Of the ten patients who exhibited radiological improvement at 6 months, 8 experienced continued or sustained improvement at the 12-month follow-up Only 3 patients (20%) underwent total disk replacement by the 12-month follow-up due to persistent, but not worse than baseline, LBP	The results of this prospective cohort are promising, with 77% of patients improving Positive result

Table 24.2 (continued)

Study details	Population	Cell or solution dose and delivery pathway	Results	Conclusion
Orozco et al., 2011 [13] Sample size = 10 Follow-up = 1 year Pilot phase 1 trial	Ten patients with degenerative disk disease and persistent LBP (>6 months; decrease of disk height > 50%) Patient age (yrs) = 35_7 (mean_SD)	$23 \pm 5 \times 10^6$ autologous expanded BMSCs were injected into the NP area	Patients exhibited rapid improvement of pain and disability (85% of maximum in 3 months) that approached 71% of optimal efficacy This study confirmed feasibility and safety with identification of strong indications of clinical efficacy	Authors concluded that MSC therapy may be a valid alternative treatment for chronic back pain caused by degenerative disk disease They also concluded that advantages over current gold standards include simpler and more conservative intervention without surgery, preservation of normal biomechanics, and the same or better pain relief

Kumar et al., 2017 [14]
Sample size = 10
Follow-up = 1 year
Phase 1 study
	Ten patients with chronic LBP lasting for more than 3 months with a minimum intensity of 4/10 on a VAS and disability level ≥ 30% on the ODI
	Patient age (yrs) = between 19 and 70
		A single intradiscal injection at a dose of 2×10^7 cells/disk ($N = 5$) or 4×10^7 cells/disk ($N = 5$)	VAS, ODI, and SF-36 scores significantly improved in groups receiving either low or high cell doses and did not differ significantly between the two groups
			  At 12-month follow-up, 7 patients reported 50% or greater improvement in VAS
			  Six patients achieved treatment success, with pain reduction of 50% or greater and improvement on ODI
			Among six patients who achieved significant improvement in VAS, ODI, and SF-36, three patients were determined to have increased water content based on an increased apparent diffusion coefficient on diffusion MRI
				60% significant improvement, with no adverse effect
				Authors concluded that combined implantation of AT-MSCs and hyaluronic acid derivative in chronic discogenic LBP is safe and tolerable
				Positive result

Mochida et al. [15] Sample size = 9 Follow-up = 3 years Prospective clinical study	Nine patients with Pfirrmann grade III disk degeneration and posterior lumbar intervertebral fusion. Patient age = 20–29 years	One million activated autologous NP cells were injected into the degenerated disk 7 d after fusion surgery	Clinical outcomes based on JOA scoring system for LBP showed significant improvement from 14.2 ± 4.8 points preoperatively to 27.2 ± 1.6 points at 3 years after transplantation of the activated NP cells (maximum possible score of 29 points) The JOA scoring system also showed improvement in LBP subscale from 1.2 ± 0.5 points preoperatively to 2.7 ± 0.2 points at 3 years after the transplantation, with maximum possible score of 3 points for no pain No adverse effects were observed during the 3-year follow-up period	Significant improvement in function and pain scores was reported This study confirmed the safety of activated NP cell transplantation, and the findings suggest the minimal efficacy of this treatment to slow the further degeneration of human IVDs

(continued)

Table 24.2 (continued)

Study details	Population	Cell or solution dose and delivery pathway	Results	Conclusion
Meisel et al., 2006 [16] Sample size = 12 Follow-up = 2 years	Patients with discogenic pain after repeat discograms. Patients were treated with cell therapy at least 3 months after the endoscopy Patient age = 18–75 years	Cells were injected into disk approximately 12 weeks following discectomy. The cell dose was not mentioned	The median total ODI score was 2 in the autologous disk chondrocyte transplantation (ADCT) group compared with 6 in the control group. Decreases in the ODI in ADCT-treated patients correlated with the reduction of LBP Decreases in disk height over time were found only in the control group, and of potential significance, IVDs in adjacent segments appeared to retain hydration when compared to those adjacent to levels that had undergone discectomy without cell intervention	Significant improvement Positive result
Beall et al., 2021 [17], sample size = 218, follow-up = 12 months	Patients with ≥6 months of low back pain associated with DDD, Pfirrmann levels 3 to 6, and type 1 or 2 Modic changes at 1–2 levels from L1-S1; Patient age = 18 to 60 yrs	1 ml of cells (~6×10^6) mixed with the reconstituted micronized nucleus allograft, control groups = saline or conservative medical management	Clinically meaningful improvements in mean VASP & ODI scores were achieved and reported in the allograft and saline groups at 12 months. A responder analysis demonstrated a clinically meaningful reduction in ODI of > = 15 points at 12 months that was statistically significant; 76.5% of patients randomized to allograft were responders ($P = 0.03$) compared to 56.7% in the saline group. Additionally, a responder group characterized by a 20 point reduction in pain at 12 months achieved a statistically significant reduction in pain compared to the saline group ($P = 0.022$)	Intradiscal injections of MPCs mixed with reconstituted micronized disc nucleus is a safe and effective treatment for DDD in patients with 1 to 2 affected levels
Amirdelfan et al., 2021 [18], sample size = 100, follow-up = 3 years	Patients with ≥6 months of low back pain associated with DDD and Pfirrmann levels 3 to 6 at 1 level from L1-S1; patient age = ≥18 yrs	2 study groups (18×10^6 MPCs in HA and 6×10^6 MPCs in HA) and 2 control groups (HA alone and saline alone)	Significant differences between the control and MPC groups for improvement in VAS and ODI; correcting for post-treatment interventions, the proportion of subjects with VAS ≥30% and ≥ 50% improvement from baseline, absolute VAS score ≤ 20, and ODI reduction ≥10 and ≥ 15 points from baseline showed MPC therapy superior to controls at various time points through 36 months	Intradiscal injection of MPCs mixed with HA is safe, effective and durable, with improvements shown out to 36 months in patients with single level DDD

RCT randomized controlled trial, *BMSCs* bone marrow-derived stem cells, *MSCs* medicinal signaling cells or mesenchymal stem cells, *JOA* Japanese Orthopedic Association, *MRI* magnetic resonance imaging, *NP* nucleus pulposus, *DDD* degenerative disk disease, *BMC* bone marrow concentrate, *LBP* low back pain, *NRS* numerical rating scale, *ODI* Oswestry Disability Index, *VAS* visual analog scale, *SD* standard deviation, *SF-12* 12-item short-form survey, *HIZ* high intensity zone, *AT-MSCs* adipose tissue-derived mesenchymal stem cells

In 2020, Beall et al. published level-I data on the use of allogeneic grafts for the treatment of degenerative disc disease (DDD) using a combination of MSCs mixed with micronized nucleus pulposus as the vehicle ("viable structural allograft," or VIA Disc Matrix™) [20]. The allograft consisted of 1 ml of cells (~6×10^6) mixed with the nucleus allograft which is reconstituted with normal saline (1:0.75:1) resulting in 1.5 to 1.9 ml of injectate. This study was an extension of the VAST trial (NCT03709901) which meant to establish preliminary safety and efficacy of this particular allogeneic graft. The authors reported on 24 subjects who demonstrated DDD at 1 or 2 vertebral levels from L1 to S1 with a modified Pfirrmann levels 3 to 6 and type 1 or type 2 Modic changes on MRI. Patients were included who had low back pain for ≥6 months, moderate to severe disability (ODI 40%) and VAS of at least 40 mm) that was chronic during the screening Phase and demonstrated. Patients were randomized 3.5:5:1:1 to receive allograft at up to 2 levels, saline as placebo or conventional/conservative medical management. At 6 and 12 months, the subjects showed improvements in VAS (54.8 at baseline, 55.25 at 6 months, and 62.26 at 12 months) and ODI (53.7, 18.5, and 28.7) with only transient adverse events that resolved in all cohorts.

The results of the full VAST study were published in 2021 – a prospective, multicenter, blinded, randomized controlled trial of 218 subjects [17]. The 3 cohorts were as follows: allograft, saline placebo, and conservative medical management; the latter was allowed to crossover to the allograft group at 3 months. There were clinically meaningful improvements in VAS and ODI in both the allograft and saline groups at 12 months. A responder analysis showed a clinically meaningful reduction in ODI of ≥15 points at 12 months, which was statistically significant; 76.5% of the allograft subjects were responders ($p = 0.03$) compared to 56.7% in the saline group.

In the same year, Amirdelfan et al. published on the use of adult allogeneic mesenchymal precursor cells (MPCs), this time mixed with hyaluronic acid (HA) as the vehicle [18]. Similar to the initial publication on the VAST study, this was a preliminary, safety and efficacy study. This study was a multicenter, randomized controlled trial with a total of 100 subjects enrolled and followed out to 36 months. Patients also had low back pain associated with DDD for ≥6 months with modified Pfirrmann levels 3 to 6 from L1-S1 – the major difference here was the subjects could only have 1 affected level, as opposed to 2 in the VAST trial. Another major difference was there were 4 cohorts in this study: 2 study groups (18×10^6 MPCs in HA and 6×10^6 MPCs in HA) and 2 control groups (HA alone and saline alone); cohorts were randomized 3:3:2:2. There were significant difference between the study and control groups in VAS and ODI; additionally, there were no reports of allograft rejection.

Fibrin

Refer to Chap. 6 for a more in-depth discussion of intradiscal fibrin and the supporting evidence

Facet Joints

The facet or zygapophyseal *joints* are synovial-type articular structures that, along with posterior ligamentous structures, provide structure and allow dynamic movement of the spine. In the case of low back pain, controlled studies have shown that the facet joints are the sole sources of pain generation in 16–41% of patients with chronic, non-discogenic, and non-radicular LBP [1, 21].

As discussed earlier, these structures can undergo changes associated with various disease states, trauma, and biomechanical factors. These joints are under an immense amount of stress, especially in the presence of adjacent discogenic pathology due to the changes in biomechanical stress causing those forces to move posteriorly onto these joints and posterior ligaments of the spine. Any trauma to the spine, especially of a shearing nature, may cause fluid to develop within the *joint*, itself. This will often be a source of pain and can be best seen on a T2-weighted MRI image in the axial plane. Traditionally, corticosteroids have been used to decrease the swelling or effusion within the facet joint. Intra-articular injections of PRP and MSCs have also been demonstrated to be effective, providing an anabolic, as opposed to catabolic, means of treating and healing the source of the swelling or effusion.

The facet joints are innervated by the recurrent meningeal nerve. Sensation to these joints is provided by the medial branch. Advances in various technologies, such as radiofrequency ablation, have resulted in a method of treatment that will block pain signals from being transmitted into the central nervous system, often providing excellent relief of pain for those patients diagnosed *with* facetogenic pain generators. This method, however, simply blocks the pain signals and does not address or heal the source of the problem. However, such techniques are very effective and can allow patients to be more active and undergo more aggressive conservative treatments, such as chiropractic or physical therapy, to address the structural support and allow internal mechanisms of healing more time to work. However, pathological processes will ensue, should these aberrant stresses continue. As discussed earlier, the stress put on the bone will cause bone to grow (Wolff's Law) [22].

Facet *hypertrophy* is very common in the aging or traumatic/pathological spine. This hypertrophy can become pathological or pain generating when one or both of the following occur: (1) the facet joint stretches the surrounding capsule and triggers a ligamentous or sclerotogenous type of referred pain and (2) the facet joint grows, pushing the superior adjacent vertebral segment posteriorly, creating a retrolisthesis. Should the facet joint grow and form osteophytes, these areas, themselves, can cause pain. Even worse, they could grow into the neuroforamen, encroaching the exiting nerve root, and contribute to the development of radiculopathy.

Evidence

Targeted use of biologics in the treatment of facet pathology is very promising. Both PRP and MSCs have been studied for the treatment of facet-mediated pain, although PRP are somewhat preferred due to the idea that stem cells could potentially lead to increased regional osteophyte formation and/or enhance growth of the facet joint theoretically causing it to hypertrophy.

Intra-Articular PRP

Recent guidelines that focused on the use of biologics to treat LBP support the use of PRP in the treatment of facet joint pathology, based on level 4 evidence [1]. The rationale of using PRP within the facet joints is the same as that of the knee or the hip – repair relining of the joint and restore biochemical equilibrium of the synovial fluid. The challenge of using PRP within the facet as opposed to a more conventional joint like the knee or hip is the fact that the facet is considerably smaller and therefore it is much more difficult to evaluate what the proper volume of injectate should be such that enough is present to actually make a difference and not too much that could make the capsule to rupture. PRP is not known to promote osteophyte formation or pathological bone growth. This is an important distinction over the use of stem cells within the facet joints as the latter has been known to lead to osteophyte growth when cells leak out of the target and track back along the needle path.

To date, three studies have performed facet injections and treatments with PRP and are summarized in Table 24.3 [1]. These *studies* had follow-up periods ranging from 1 to 6 months, with sample sizes of 46, 19, and 86. Wu et al. compared traditional corticosteroid treatments with PRP treatments. While both had benefits, when compared to corticosteroids, PRP was associated with longer pain relief and reduction of symptoms up to the 6-month follow-up visit. Furthermore, 81% of subjects in the PRP group reported >50% *improvement* in their symptoms at 3 and 6 months, while peak relief was reported by 85% of subjects. Relief was limited to 1 month with the corticosteroid/local anesthetic group [23].

In a smaller study, Wu et al. injected PRP into the facets of 19 subjects. At the 3-month follow-up visit, 79% of sub-

Table 24.3 Studies of PRP therapy for lumbar facet syndrome

Study details	Methods	Results	Conclusion
Wu et al., 2017 [23] Sample size = 46 Follow-up = 6 months Prospective randomized trial Chronic facet joint pain	46 patients with lumbar facet syndrome were randomized to receive either intra-articular injections of PRP or local anesthetic/corticosteroid Outcomes were assessed with VAS, ODI, and RMDQ	Back pain improved in both groups immediately and at one-month follow-up At 3 months, back pain relief was superior in PRP injection group compared to steroid group Functional status improvement was observed in both groups; however, at 3 months, there was significant improvement in PRP group compared to steroid group Highest objective success rate with over 50% pain relief in 81% was found at 3 and 6 months after treatment, whereas highest success rate in 85% of the patients in the steroid group dissipated after 1 month	There was significant improvement in both groups in short-term. However, improvement was long-lasting for 6 months in PRP group Positive study Limited by a small number of patients
Wu et al., 2016 [24] Sample size = 19 Follow-up = 3 months Prospective clinical evaluation Chronic facet joint pain	19 patients with lumbar facet syndrome given intra-articular injections of PRP Outcomes were assessed with VAS, ODI, and RMDQ	79% of the patients reported satisfactory improvement with good or excellent at 3-month follow-up after injection of PRP ODI and RMDQ were also significantly improved. There were no adverse events. A positive small study of intra-articular injection of autologous PRP	Positive results in a study with a small number of patients and relatively short follow-up of 3 months
Kirchner and Anitua, 2016 [25] Sample size = 86 Follow-up = 6 months Observational retrospective pilot study, *n* = 86 humans Facet joint syndrome	One intradiscal, one intra-articular facet, and one transforaminal epidural injection of PRGF under fluoroscopic guidance-control was carried out in 86 patients with chronic LBP	VAS showed a statistically significant decrease at 1, 3, and 6 months after the treatment ($P < 0.0001$), except for the pain reduction between the third and sixth month whose signification was lower ($P < 0.05$)	Positive study with multiple drawbacks with multiple injections in each setting with injection into disk, facet joint, and epidural space Extremely positive results in a low-quality observational study

VAS visual analog scale, *ODI* Oswestry Disability Index, *RMDQ* Roland Morris Disability Questionnaire, *PRP* platelet-rich plasma, *PRGF* platelet-rich growth factor, *LBP* low back pain

jects reported "good" or "excellent" results, with significant improvements in both ODI and RMDQ scores [24].

Finally, Kirchner and Anitua [25] injected platelet-rich growth factors into a single facet and the adjacent intradiscal and epidural space in 86 subjects. While the results were extremely promising *and* demonstrated a significant reduction in VAS scores through the 6-month follow-up visit, the study had several limitations. For example, the injections targeted a number of different sites, thereby reducing the specificity of the treatment outcomes. There is, therefore, a need for additional high-quality, larger-scale studies of the effectiveness of PRP, especially with respect to stem cells.

Intra-Articular Stem Cells

There were no well-controlled studies of the effectiveness of MSCs in the treatment of facet joint pathology; however, given the parameters and what we know about the mechanisms of healing *utilizing* MSCs, positive conclusions can be inferred from their use, and further studies demonstrating the effectiveness of MSCs in facetogenic pathology are needed [1].

Sacroiliac (SI) Joints

The SI joint has been relatively ignored as a major source of low back pain until recent decades, when there was a dramatic uptick in SI joint-mediated pain treatments and therapies. The limits of this joint normally do not exceed approximately 4% of total motion. According to previous studies, SI joints account for approximately 15–30% of the non-radicular type of LBP [26].

Similarly, studies focused on diagnostic blocks of the SI joint have implicated this joint to be directly involved in LBP in approximately 10–25% of cases without disk herniation, discogenic pain, or radiculopathy [1, 26].

The SI joints act as *major* pelvic shock absorbers and connect the posterior pelvic ring. The joints are supported by a dense fibrous network of ligaments that provide stability to the region and account for the relatively low rates of SI dislocations [27]. These ligaments, like all others in the body, can become sources of pain generation. As a ligament (or tendon) inserts into the bone (the enthesis), it becomes confluent with the periosteum of the bone, which is rich in nerve fibers. An avulsion fracture is characterized by an injury to a bone at a location where a ligament or tendon attaches to the bone and pulls off a piece of the bone.

Factors such as traumatic injury, repetitive stress, aging, and arthritic conditions can cause so-called *"microavulsions."* These are microscopic tears in the enthesis of the ligament or tendon and also at the insertion point of the enthesis into the periosteum of the bone, thereby causing aberrant nerve transmission and chronic pain in the area.

Additionally, SI joint pathology, especially when accompanied by increased motion from ligamentous loosening (e.g., due to trauma, childbirth, and/or added stress caused by degenerative conditions), has been linked to increased firing and spasms of the lumbar trunk muscles [27]. This study demonstrated significantly increased SI joint motion in patients with degenerative lumbar spine disorders. The increase in motion was most dramatic for those patients who had spinal deformities. It is well established that, after intervertebral body fusion surgeries, the vertebral *levels* above and below the fusion take on added stress due to the fixations, which cause reduced movement and reduced shock absorption in that area. Given that most lumbar pathology and, therefore, fusion surgeries occur in the L4-L5 and L5-S1 regions, the added stress and subsequent compensatory motion often occurs within the SI joints. As this increased aberrant motion persists, pain can ensue. There has been a recent increase in SI joint fixation/fusion surgeries and debate about the subsequent effect on the biomechanics of the spine. However, there have been positive reports from both surgeons and patients about significant pain relief following SI fusion, independent of any improvement following prior lumbar intervertebral interbody fusion surgeries.

Evidence

Recent guidelines that focused on the use of biologics to treat LBP support the use of PRP in the treatment of SI joint pathology, based on level 4 evidence, as is noted in Table 24.1 [1]. There were no well-controlled studies that demonstrated the effectiveness of MSCs for this treatment; however, given the parameters and what we know about the mechanisms of healing utilizing MSCs, *positive* conclusions can be inferred from their use, and further studies demonstrating MRC effectiveness in the treatment of SI joint pathology are needed.

Intra-articular PRP

PRP has been extensively studied in spine and orthopedics in the context of IVD degeneration, spinal fusion, and osteoarthritis and cartilage repair of major joints. Despite its extensive use in major joints, there have been few reports of its use in SI joints. There is extensive literature on prolotherapy over the SI ligaments, including a recent case study by Ko et al. focusing on four patients with a series of 2 PRP injections via prolotherapy technique at the Hackett points A-C under ultrasound guidance [28]. Ko et al. followed their subjects for 4 years after SI joint PRP infiltration for stability, pain reduction, and quality of life using the SF-36, ODI, McGill pain *questionnaire*, and NRS and demonstrated positive outcomes [28].

The study reported statistically significant pain reduction and improvement in quality of life at both 12-month and 4-year follow-up visits. The PRP, however, was injected at the ligament bone junction at the Hackett points A-C and not directly into the joint, further supporting the notion of ligamentous pain generators surrounding the SI joint. Intra-articular treatments for SI joint-mediated pain have been limited to diagnostic infiltrations of local anesthetic or use of corticosteroids to reduce inflammatory SI joint-mediated pain. Intra-articular PRP and MSC *therapies* have been performed in clinical settings; however, they have not undergone rigorous study. To our knowledge, there are no published reports about the use of *bone marrow aspirate concentrate (BMAC)* for SI joint-mediated pain. Preliminary reports of the safety and efficacy of BMAC intra-articular SI joint injections appear promising.

The PRP study results appear to be an assuring option for SIJ-mediated pain.

There have been few studies published, and a summary of the results is presented in Table 24.4. The studies had sample sizes ranging from 4 to 40, with follow-up periods ranging from 1 to 12 months. Singla et al. [29] compared traditional corticosteroid and local anesthetic SI joint intra-articular infiltration with leukocyte-poor PRP activated by calcium chloride SI joint intra-articular *infiltration*. At the 3-month follow-up visit, 90% of patients in the PRP group reported significant pain relief, whereas only 25% of patients in the corticosteroid/local anesthetic group reported satisfactory results.

Navani and Gupta performed SI joint PRP infiltrations for ten subjects. All subjects experienced pain relief and low VAS scores through the 6-month follow-up visit, during which time no additional interventions were required [30].

The use of biologics to treat chronic SI joint pain at various points in the disease process has improved patients' *pain* and function for as long as 8 years. We recognize that these studies have shortcomings, including small sample sizes. Larger, well-designed, randomized clinical trials are needed to understand the full effect of PRP and stem-cell therapies for SI joint-mediated pain.

Microanatomy and Biochemistry

The source and quality of the biological material used in regenerative medicine treatments can be just as important as the targeted applications of such materials. A tissue donation sample from a source, which is then readministered to the same source, is referred to as an autologous sample.

Table 24.4 Studies of PRP therapy for SI joint-mediated pain

Study details	Methods	Results	Conclusion
Ko et al., 2017 [28] Sample size = 4 Follow-up = 2 yrs. Case series	SI joint injection with PRP under ultrasound guidance Outcomes were assessed with SF-36, McGill Pain Questionnaire, NRS, and ODI	At 12-month follow-up, there was marked improvement in joint stability, a statistically significant reduction in pain, and improvement in quality of life The clinical benefits of PRP were still significant at the 2-year follow-up visit	PRP therapy showed long-lasting, positive results in this short case series of four patients
Singla et al. 2017 [29] Sample size = 40 Follow-up = 3 months Prospective, randomized double-blinded end point study Chronic low back pain with SI joint pathology	Patients were randomized into two groups, with one group receiving 1.5 mL of methylprednisolone 40 mg/mL and 1.5 mL of 2% lidocaine with 0.5 mL of saline and the other group receiving 3 mL of leukocyte-free PRP with 0.5 mL of calcium chloride with ultrasound-guided SI joint injection Outcomes were assessed with VAS scores, ODI, and SF-12	At 3-month follow-up, 90% of the patients reported satisfactory relief with PRP, whereas 25% of the patients reported satisfactory relief with steroids PRP therapy was strongly associated with a VAS reduction of greater than 50% from baseline	Positive prospective, randomized study Small number of patients
Navani and Gupta, 2015 [30] Sample size = 10 (4 males and 6 females) with SI joint pain of greater than 6 months duration Age Distribution= 5 patients below 40 and 5 patients over 40 SI joint pain	PRP injection into the SI joint under fluoroscopic guidance	All patients showed improvement at the 3-month follow-up visit and maintained low pain levels that did not require any additional treatment up to the 6-month follow-up visit SF-36 demonstrated improvement in both physical component summary scores and mental component summary scores in all patients No adverse events	A positive case series of ten patients

PRP platelet-rich plasma, *SF-36* 36-item short form health survey, *VAS* visual analog scale, *ODI* oswestry disability index, *SF-12* 12-item short form health survey, *NRS* numeric rating scale

Readministration may occur after manipulating the cells to enhance concentrations *and* promote the maturation of various cell lineages and/or healing properties; however, local laws and regulations govern this type of cellular manipulation. In the United States, cells may be only minimally manipulated. This is usually limited to centrifugation and point-of-care utilization. For our purposes, a good example of an autologous process includes collecting blood, bone marrow, or adipose tissue from a patient, after concentrating the collected material via minimal manipulation (e.g., centrifugation) and injecting that solution back into the same donor. Autologous *sourcing* is most popular, due to the ease of point-of-care administration. It also circumvents the risk of *graft-versus-host disease (GvHD)* or graft rejection due to immunological incompatibility.

Allogeneic or allograft samples, on the other hand, refer to tissue derived from one source and used to treat another member of the same species. These tissues usually need to be processed to clean and remove the surface proteins from the donation to reduce the chance of GvHD, as the major cause for graft rejection is *human leukocyte antigen (HLA)* mismatching. The final type of *sample* source is known as xenogeneic. This is when tissue is taken from one species and, after processing, is transplanted into a different species. All tissue donated by non-autologous sources will need to undergo some form of manipulation and processing before it is safe (or at least safer) for use in humans.

PRP has become one of the most commonly used autologous types of treatments in regenerative medicine. It represents the easiest and most accessible point-of-care type of biological material used in clinical settings. The sample is collected via a simple blood draw. Following centrifugation *and* sometimes additional minimal manipulation, all of the platelets and various growth factors within the blood plasma are concentrated and reinjected into the areas of pathology in an attempt to trigger, refocus, and thereby promote internal healing mechanisms.

In contrast, MSCs can be harvested from an allographic/allogenic source, since they possess unique characteristics that prevent immunological reactions. For example, because such cells are young and *relatively* undifferentiated, they lack the surface antigens that may trigger HLA mismatching. They also lack major histocompatibility complex-2 expression, as well as the immunological suppression activity mediated by prostaglandin E2 [31]. Therefore, theoretically, these stem cells can be harvested from a number of different sources and allow compatibility across hosts.

Basic Concerns and Contraindications

Although PRP and MSC therapies show promise in preclinical and clinical studies, they are considered off label and investigational. While fibrin, itself, is an FDA-approved product, it is not approved for use intradiscally and is therefore considered off label when used in this fashion. As such, the authors recommend conventional therapies that are FDA-approved for low back pain of disc, facet and SI joint origin be recommended and utilized prior to offering regenerative therapies.

Axial Biomechanics

Before discussing the application of biologics, it is important to understand the underlying mechanisms of various spinal pathologies associated with pain generation. The biomechanics of the axial spine can portend degenerative and other pathological processes. The development of lordotic curves in both the cervical and lumbar spines has allowed humans to become *bipedal* and stand upright, which facilitated hunting, gathering, and climbing and has contributed to our ability to survive and thrive. However, due to an aging population, modern lifestyles, and, of course, gravity, we develop aberrant, asymmetrical postures that are further complicated and exacerbated by muscular imbalances, congenital anomalies, previous trauma, repetitive trauma, and various disease processes. The activities such as sitting in a chair, using a computer or tablet, looking down while reading, etc., all contribute to imbalances that lead to aberrant postural adaptations. The *average* person has rounded shoulders, an anterior head carriage, and straightening of the lordotic curves or over-accentuation of these curves. Each of these factors displaces the biomechanical stresses that are translated through the spine. The IVDs act as "shock absorbers" and attempt to dampen and combat the effects of gravity.

Surrounding the spine is a great network of ligaments that help to support the structure and provide stability. When we are young and have healthy and full IVD's, those ligaments are taught and very stable. As degenerative processes ensue, e.g., due to biomechanical changes, trauma, and aging, there may be a loss of disk height. This creates a scenario where those (noncontractile) ligaments become less taught, allowing for added intersegmental movement and a translation of the biomechanical stresses to move more *posteriorly*. Typically, this is met with challenge by the body. Muscle spasms occur in the posterior elements and paraspinal muscles in an, albeit futile, attempt to hold and stabilize the affected areas. Ultimately, Wolff's law dictates that, when added stress is placed on bone, the bone will undergo adaptive changes in an effort to increase its density [22]. This is the reason why it is so important for aging patients to perform weight-bearing exercises to help prevent osteoporosis.

Unfortunately, this process can also become pathological. As the biomechanical stresses are translated posteriorly, other areas of the spine begin to assume added stress. For instance, both the facet joints and *ligamentum flavum* can become hypertrophic. Furthermore, as the body attempts stabilization, bone will start to grow osteophytes, which are also known as bone spurs. These osteophytes can cause pain and expand into tight places, such as the neuroforamen. Hypertrophic growth of *the* ligamentum flavum can cause central canal stenosis and bony overgrowth in the areas surrounding the facet joints and can contribute to central and neuroforaminal stenosis. Additionally, hypertrophic growth of the facet joints can increase friction and stretching-type stresses on surrounding tissues and nerves, such as the medial *branches*. This will cause irritation, which may lead to swelling and/or continuous/spontaneous firing of these sensory nerves, causing pain.

Preoperative Considerations

The success of the injection therapy depends on these 5 key variables:

- Patient selection
- Indication
- Provider skill level
- Technique of the procedure
- Quality of product

It is imperative to include the patient as a partner in treatment planning process making sure the patient understands the risks and benefits of the treatments and has realistic expectations from the procedure.

Diagnostic Injections

Establishing a diagnosis of facet-mediated pain and sacroiliitis is relatively straightforward and well-described in multiple textbooks. Physical exam and diagnostic injections are the preferred means of ascertaining a facet joint or SI joint is a pain generator. If diagnostic injections with sodium-channel blockers are utilized, one should wait a minimum of 10–14 days before utilizing a regenerative product as these medications can be cytotoxic and cause damage to the injectate (particularly platelets). If corticosteroids are used in the diagnostic injection and PRP is the chosen treatment, one should wait a minimum of 14 days. Similarly, a discogram can be utilized to determine whether discogenic pain is present and which disc is the culprit.

Practitioners may be concerned about whether or not to perform discography prior to these procedures. This is, of course, a clinical decision and may provide diagnostic benefits; however, there are drawbacks to be considered. First, the procedure, itself, is unpleasant for the patient. If given enough clinical evidence from examination and diagnostic imaging (typically, a T2-weighted MRI image in the axial plane is easiest to visualize), demonstrating a contained annular tear can be *adequate* information to proceed with the treatment. Next, the disk is able to receive only a small amount of volume. By adding fluid or gas from a discogram, we are further limiting the volume of biological substrate that can be utilized point of care, thereby perhaps limiting the full effectiveness of the procedure. Therefore, significant time must be permitted to pass if a discogram is utilized to allow the volume from this procedure to dissipate and permit the disc the ability to accept the intended volume of the treatment, itself. This would necessitate two separate procedures and, therefore, cause the patient extra pain, as well as increase the likelihood of any adverse events. Additionally, the insertion of a large-bore needle into the disk can cause perforation and the release of intradiscal contents. The containment of the disk material with some surviving structural annular fibers is important to keep all amount of biological substrate contained within the disk to promote healing. If there is a full-thickness tear in the annulus, whether from the initial trauma/pathology or from the introduction of a large-bore needle into the disk, it may allow any medication, biological or otherwise, to simply spill out the disk and may not provide adequate time for directed healing within the targeted area.

General Considerations

The following risk factors in patients need to be specifically evaluated and addressed prior to the procedure for meaningful response:

- Dependence, addiction
- Medical or psychiatric issues
- Non-compliance
- Comorbidities
- Multiple nonspecific generators of pain
- Immunocompromised patients

The following are generally accepted precautions prior to the spine injection therapies with PRP or MSCs:

- NPO for 6–8 hours preprocedure depending on the risk factors.
- No use of non-steroidal anti-inflammatory drugs (NSAIDs) minimum of 1 week before the procedure.
- No corticosteroids for 10 days prior to the procedure.
- No anticoagulation 7–10 days prior to the procedure depending on the half-life of the anti-coagulant.
- Have a ride back home after the spine procedure.

The following are generally accepted contraindications for the biologic therapy:

- Patient refusal
- Hematologic blood dyscrasias
- Platelet dysfunction
- Generalized or local infection
- Septicemia or fever
- Malignancy, particularly with hematologic or bony involvement
- Allergy to bovine products if bovine thrombus is to be used

Radiographic Guidance

These injections should be performed under direct visualization with the aid of fluoroscopy or CT scan. Ultrasound can be utilized for the SI joint; however, this technique is not widely utilized.

Equipment

The required equipment depends on both the biologic being used and the target being treated. The target-related equipment is essentially the same as what would be used for a typical injection – in other words, an SI injection or intra-articular injection with PRP or MSCs would not be any different than one with corticosteroid. The only differences would be that one must maintain sterile technique throughout and the minimum gauge of the needles should be 22 g. Primer anticoagulant solution EDTA is typically utilized within the commercially available kits for acquiring both PRP and MSCs from patient tissue; however, some recommend predraw priming of equipment with heparin in case of bone marrow–related draw. Traditional or premanufactured kits can both be used successfully as long as sterility and concentration are kept forefront in preparation of the final biologic injectate.

Once the practitioner has decided upon which biologic they will employ for treatment, the next consideration becomes acquisition. While fibrin is available "off-the-shelf," PRP is not and must be acquired from the patient. MSCs and allogeneic grafts are somewhat of a hybrid as they can be acquired from the patient in the same fashion as PRP and will eventually be available for purchase on a patient-by-patient basis.

PRP

While PRP, itself, cannot be purchased, the kits for acquiring it can be. Refer to Chap. 3 for a more in-depth discussion on how it is created from whole blood, how it works, and how it should be handled.

MSCs and Allogeneic Grafts

There are a number of different sources available for MSCs – the most common sources are bone marrow aspirate, aka BMAC, and adipose, aka stromal vascular fraction (refer to Chaps. 4 and 5 for a more detailed discussion of BMAC and adipose-acquired MSCs, respectively). The choice of the particular source type becomes a very important consideration, since every step toward differentiation leads the cell down a path to ultimately become part of a particular tissue type. For example, embryonic stem cells will further differentiate to the ectoderm, mesoderm, and endoderm lineages, which then *further* differentiate into specific cell types, such as neuronal, connective, or organ tissue. Because stem cells are immature by definition, early sourcing focused on using embryological samples. Embryonic stem cells represent true pluripotent cells that are capable of differentiating across all cell lines. While this represents a significant source of these undifferentiated cells, political, ethical, and religious concerns present challenges for research and clinical use of these types of cells. On the other hand, umbilical cord stem cells represent an excellent source of multipotent stem cells that have the potential to differentiate into more than one type of cell, but not *all* cell lines. Stem cells derived from cord blood are rich in *hematopoietic stem cells (HSCs)* and can differentiate and give rise to all blood cell types, whereas cord tissue is rich in MSCs, which can differentiate and give rise to a number of cell lines, such as connective tissue, bone, cartilage, muscle, and ligament [32]. Placental donations represent another viable source of stem cells. Placental tissue is rich in both hematopoietic stem cells (HSCs) and *mesenchymal stem cells (MSCs)* . The latter can be isolated from placental villous tissue, amniotic fluid, and fetal membranes [33].

Although the current use of biologics in clinical settings may be restricted by the human limitations of cellular differentiation, laboratory scientists have demonstrated the ability to take somatic cell types and "dedifferentiate" them into a type of pluripotent cell known as *induced pluripotent stem cells (iPSCs)* via a process called transfection. This is a very exciting development that can potentially yield an unlimited supply of these building blocks, which can then be used to help repair unlimited tissue types. However, this requires a great deal of "manipulation" and, as stated above, there are various local laws that govern the extent to which cells may be manipulated prior to their use in clinical settings. More *importantly*, unlike the other viable sources of stem cells, transfection has been associated with oncogenic properties (e.g., the formation of teratomas and other tumor types in vivo) [34]. Research is underway to address this limitation.

Clinically, however, all of the abovementioned sources of MSCs represent challenges to the practicing physician in terms of both accessibility and point-of-care utilization. In current practice, most physicians are taking advantage of easily accessible methods to harvest and concentrate these cells, such as lipo-aspiration and bone marrow aspiration techniques. While such sources contain a relatively *lower* ratio of stem cells to fully differentiated cells, these techniques represent a true point-of-care method for collecting and concentrating viable autologous stem cells. Due to MSC's specific ability to differentiate into these cell types and because of practical considerations, such as ease of extraction and point-of-care treatment, these are the autologous types of stem cell therapies that are most often utilized in clinical settings.

* At the time of this publication, there are no FDA-approved sources for MSCs or allogeneic grafts that are readily available for "off-the-shelf" purchase for intradiscal therapy; however, there are a number of products in clinical trials (i.e., Discgenics™, Mesoblast™, Vivex™) that are aimed at that very utility.

Fibrin

Refer to Chap. 6.

Technique

The overall technique for injecting a biologic into these targets is virtually the same as any other conventional injection into them for any other purpose. There are some subtle differences and important points to keep in mind:

Facets

- Strict sterile technique throughout – the technique is otherwise the same as any other intra-articular facet injection.
- Volume: 0.3–0.5 cc per joint.
- Do not allow any sodium-channel blocker (i.e., lidocaine or bupivacaine) to enter the joint. If any is used within the spinal needle, make sure to flush it with preservative-free normal saline before the tip makes entry within the joint capsule.
- Try to avoid using any contrast to perform a confirmatory arthrogram as this will limit the amount of volume the joint can accommodate.
- Once the injection is complete, wait 45–60 seconds – with the needle still in place, unscrew the syringe from the luer lock, place the stylet back within the needle, and then slowly withdraw it.

- Replace the stylet within the needle prior to removing it from the facet joint. This will limit the amount of injectate that escapes out once the needle is removed

Sacroiliac Joint

- Strict sterile technique throughout.
- Volume: 3–5 cc per joint.
- The injection should be performed at 2 points – the inferior one-third as well as the middle-most aspect of the joint. The middle aspect of the SIJ may not allow as much of the biologic to be injected here as the inferior portion, however attempt to deliver at least 25% of the total volume here.
- Do not allow any sodium-channel blocker (i.e., lidocaine or bupivacaine) to enter the joint. If any is used within the spinal needle, make sure to flush it with preservative-free normal saline before the tip enters the joint.
- As with the facet injections, try to limit the use of contrast. However, as the SIJ is much larger, a small amount of up to 0.3 cc of contrast may be used.

Intervertebral Discs

- Do not mix antibiotic with the injectate as one would do in a discogram – the technique is otherwise the same as any other with respect to needle trajectory, target, and fluoroscopic view.
- Volume: maximum of 1.8 cc per lumbar disk.
- Do not allow any sodium-channel blocker (i.e., lidocaine or bupivacaine) to enter the joint. If any is used within the spinal needle, make sure to flush it with preservative-free normal saline before the tip makes entry within disk – some lidocaine may be used just outside the annulus to limit the discomfort that is typically encountered when a needle punctures the disk.
- Do not use any contrast – so long as the needle tip is within the middle most aspect of the disc in anterior-posterior and lateral views, there should be little doubt as to whether or not it is within the nucleus pulposus.
- A manometer should be used to ensure that injection pressure is steady and does not exceed 100 psi at any point. Should the injection pressure reach or exceed 80 psi, reduce pressure on the syringe or pause injection altogether until the psi drops below 80 psi.
- Injection should take place over at least 10 seconds.
- Once the injection is complete, wait 60 seconds – with the needle still in place, unscrew the syringe from the luer lock, place the stylet back within the needle, and then slowly withdraw it.

- Replace the stylet within the needle prior to removing it from the facet joint. This will limit the amount of injectate that escapes out once the needle is removed.
- When using intradiscal PRP, sterility should be the absolute priority. PRP is handled extensively throughout the process of its creation from the venous blood draw, centrifuging, and then ultimately drawing it up from the receptacle provided within the kit after it is spun down. During this process, there are number of times that a pathogen may be introduced. As stated above, intradiscal antibiotics are to be avoided; therefore, one must ensure that sterility is maintained as much as possible each time the product changes hands.

Post-procedure Considerations

As mentioned previously, non-steroidal anti-inflammatory drugs (NSAIDs) are contraindicated prior to and following these injections, as the effects of these NSAIDs are thought to reduce the effectiveness of the biologics, by limiting the initial inflammatory phase after the injection, reducing localized debridement, and decreasing local collagen synthesis and subsequent repair of the affected tissues. Baby aspirin used for cardiac prevention is generally not considered contraindicated, as removing the therapy may have unrelated consequences and the dose is considered significantly low enough to marginally impair or at least limit any detriments to the effectiveness of the biologic therapy. Due to the half-life of these medications, it is recommended that the patient refrain from NSAIDs for 2 weeks prior to the treatment and for 3–4 weeks after the treatment for maximum benefits. Corticosteroids, having even longer half-lives, are recommended to be avoided for at least 1 month prior to biologic injections and avoided for at least 1 month systemically and as long as possible for any local injections near the area of treatment.

Consideration to rehabilitate the affected area prior to (prehab) and after the initial soreness from the injection wears off should be taken. Similarly, in orthopedic surgical cases, prehab and rehabilitation of the affected area will result in reduced post-op symptoms, faster recovery, and eventual realignment of muscular fibers that have undergone tendinopathic morphic changes. These additional steps aim for eventual stabilization of the affected area, reduced pain, to help prevent future injury and to help recover faster from future injury. Stabilization of the affected areas after the procedures can be considered and is especially important when dealing with unstable joints and lax ligaments. This laxity can be addressed by stabilization of the area for 3–6 weeks post-op. Typically in the lumbar spine, this is accomplished by using lumbar, thoraco-lumbar, and/or pelvic stabilizing durable medical equipment, depending on the patient's condition and treatment area. As with other durable medical equipment, the patient is advised to wear the bracing when out and about, performing weight-bearing activity and typically avoiding bracing to sleep (can be different for orthopedic joint treatments) or while stationary around the house. Depending on the amount of laxity present, physical activity, and fitness levels of the individual patient, bracing may be unnecessary all together. Use of this type of bracing is typically best serving for 3–6 weeks post-op and combined with rehabilitation to realign muscle fibers and add additional stabilization to the area. Addressing shortened muscles such as glutes, psoas, hamstrings, paraspinals, etc., and strengthening weak areas such as core muscles can help balance forces through the affected joints, minimize future flares, as well as hasten repair of tissue and resolution of symptoms.

Finally, consideration toward the patient's nutritional status may be addressed. As one of the major goals of the biological injections is collagen synthesis and tissue repair, enhancing the body's ability to synthesize and create the collagen molecule may be beneficial. Vitamin C is a rate-limiting co-enzyme for collagen synthesis. An oral dose of approximately 3 grams daily, in divided doses, is typically tolerated well. If the patient is sensitive, or has gastrointestinal issues, a buffered vitamin C can be used to minimize any GI discomfort. Amino acids such as proline, protein supplements containing various free amino acids, and/or collagen products may also be used to theoretically enhance the availability of chemical constituents necessary for tissue repair.

Potential Complications and Pitfalls

The complications from lumbar biologic injections in hands of skilled and diligent interventionist are rare. The most common complications can be categorized in a twofold fashion: complications related to the placement of the needle and complications related to the administration of various drugs including biologics.

Most problems, such as local swelling, pain at the site of the needle insertion, and pain in the low back, are short-lived and self-limited. More serious complications may include thromboembolism, dural puncture, spinal cord trauma, subdural injection, neural trauma, injection into the intervertebral foramen, and hematoma formation; infectious complications include discitis, epidural abscess, and bacterial meningitis; and side effects related are to the administration of steroids, local anesthetics, biologics and other drugs.

Clinical Pearls

Results are often delayed in nature as it will take time for the biologics to reduce swelling and induce repair. PRP often takes 6–8 weeks to realize the benefits of the injections, with more healing through approximately the next 6 months. While discomfort is the most common complaint from these injections, symptoms are managed by minimally invasive methods such as massage, heat/cold applications, and stretches/exercises.

Conclusion

The use of biologics for spine and orthopedic conditions has been one of the most intriguing innovations in recent times. Regenerative injection therapies are of interest to a wide range of stakeholders, including patients, physicians, researchers, the pharmaceutical industry, and medical device companies. These therapies are being utilized at an exponential rate, which is expected to continue with increased efforts toward precision, personalization and uniform utilization, and conformity.

The absence of technical and ethical oversight for the use of biologics has led to nonuniformity in clinical practice and, therefore, outcomes. Additional research is needed to better standardize the utilization of these biological substances. The newest guidelines put forth by the American Society of Interventional Pain Physicians (ASIPP) are a significant step toward this lofty goal and increase the likelihood that biological regenerative medicine injection therapy will become the standard of care.

References

1. Navani A, Manchikanti L, Albers S, et al. Responsible, safe, and effective use of biologics in the management of low back pain: American Society of Interventional Pain Physicians (ASIPP) guidelines. Pain Physician. 2019;22:S1–S74.
2. Dieleman JL, Squires E, Bui AL, Campbell M, Chapin A, Hamavid H, Horst C, Li Z, Matyasz T, Reynolds A, Sadat N, Schneider MT, Murray CJL. Factors associated with increase in US health care spending, 1996–2013. JAMA. 2017;318:1668–78.
3. Conway PH. Editorial: factors associated with increased US health care spending. Implications for controlling health care costs. JAMA. 2017;318:1657–8.
4. Tuakli-Wosornu YA, Terry A, Boachie-Adjei K, et al. Lumbar intradiscal platelet-rich plasma (PRP) injections: a prospective, double blind, randomized controlled. PM R. 2016;8(1):1–10.
5. Monfett M, Harrison J, Boachie-Adjei K, Lutz G. Intradiscal platelet-rich plasma (PRP) injections for discogenic low back pain: an update. Int Orthop. 2016;40(6):1321–8.
6. Akeda K, Ohishi K, Masuda K, et al. Intradiscal injection of autologous platelet-rich plasma releasate to treat discogenic low back pain: a preliminary clinical trial. Asian Spine J. 2017;11(3):380–9.
7. Levi D, Horn S, Tyszko S, et al. Intradiscal platelet-rich plasma injection for chronic discogenic low back pain: preliminary results from a prospective trial. Pain Med. 2016;17(6):1010–22.
8. Kirchner F, Anitua E. Intradiscal and intra-articular facet infiltrations with plasma rich in growth factors in patients with chronic low back pain. J Craniovertebr Junction Spine. 2016;7(4):250–6.
9. Navani A, Ambach M, Navani R, Wei J. Biologics for lumbar discogenic pain: 18-month follow up for safety and efficacy. Interv Pain Manag Rep. 2018;2(3):111–8.
10. Noriega DC, Adura F, Hernández-Ramajo R, Martín-Ferrero MA, Sánchez-Lite I, Toribio B, Alberca M, García V, Moraleda JM, Sánchez A, Garcia-Sancho J. Intervertebral disc repair by allogeneic mesenchymal bone marrow cells: a randomized controlled trial. Transplantation. 2017;10:1945–51.
11. Pettine K, Suzuki R, Sand T, Murphy M. Treatment of discogenic back pain with autologous bone marrow concentrate injection with minimum two-year follow-up. Int Orthop. 2016;40:135–40.
12. Coric D, Pettine K, Sumich A, Boltes MO. Prospective study of disc repair with allogeneic chondrocytes presented at the 2012 joint spine section meeting. J Neurosurg Spine. 2013;18:85–95.
13. Orozco L, Soler R, Morera C, Alberca M, Sánchez A, García-Sancho J. Intervertebral disc repair by autologous mesenchymal bone marrow cells: a pilot study. Transplantation. 2011;92:822–8.
14. Kumar H, Ha DH, Lee EJ, Park JH, Shim JH, Ahn TK, Kim KT, Ropper AE, Sohn S, Kim CH, Thakor DK, Lee SH, Han IB. Safety and tolerability of intradiscal implantation of combined autologous adipose-derived mesenchymal stem cells and hyaluronic acid in patients with chronic discogenic low back pain: 1-year follow-up of a phase I study. Stem Cell Res Ther. 2017;8:262.
15. Mochida J, Sakai D, Nakamura Y, Watanabe T, Yamamoto Y, Kato S. Intervertebral disc repair with activated nucleus pulposus cell transplantation: a three-year, prospective clinical study of its safety. Eur Cell Mater. 2015;29:202–12.
16. Meisel HJ, Ganey T, Hutton WC, Libera J, Minkus Y, Alasevic O. Clinical experience in cell based therapeutics: intervention and outcome. Eur Spine J. 2006;15:S397annular-S405.
17. Beall DP, Davis T, DePalma MJ, Amirdelfan K, Yoon ES, Wilson GL, Bishop R, Tally WC, Gershon SL, Lorio MP, Meisel HJ, Langhorst M, Ganey T, Hunter CW. Viable disc tissue allograft supplementation; one- and two-level treatment of degenerated intervertebral discs in patients with chronic discogenic low back pain: one year results of the VAST randomized controlled trial. Pain Physician. 2021;24(6):465–77. PMID: 34554689.
18. Amirdelfan K, Bae H, McJunkin T, DePalma M, Kim K, Beckworth WJ, Ghiselli G, Bainbridge JS, Dryer R, Deer TR, Brown RD. Allogeneic mesenchymal precursor cells treatment for chronic low back pain associated with degenerative disc disease: a prospective randomized, placebo-controlled 36-month study of safety and efficacy. Spine J. 2021;21(2):212–30. https://doi.org/10.1016/j.spinee.2020.10.004. Epub 2020 Oct 9. PMID: 33045417.
19. Wu T, Song HX, Dong Y, Li J. Cell-based therapies for lumbar discogenic low back pain: systematic review and single-arm meta-analysis. Spine (Phila Pa 1976). 2018;43:49–57.
20. Beall D, Wilson G, Bishop R, Tally W. VAST clinical trial: safely supplementing tissue lost to degenerative disc disease. Int J Spine Surg. 2020;14(2):239–53. https://doi.org/10.14444/7033.
21. Manchikanti L, Hirsch JA, Falco FJ, Boswell MV. Management of lumbar zygapophyseal (facet) joint pain. World J Orthop. 2016;7:315–37.
22. Teichtahl AJ, Wluka AE, Wijethilake P, et al. Wolff's law in action: a mechanism for early knee osteoarthritis. Arthritis Res Ther. 2015;1:17207.
23. Wu J, Zhou J, Liu C, Zhang J, Xiong W, Lv Y, Liu R, Wang R, Du Z, Zhang G, Liu Q. A prospective study comparing platelet-rich

plasma and local anesthetic (LA)/corticosteroid in intra-articular injection for the treatment of lumbar facet joint syndrome. Pain Pract. 2017;17:914–24.

24. Wu J, Du Z, Lv Y, Zhang J, Xiong W, Wang R, Liu R, Zhang G, Liu Q. A new technique for the treatment of lumbar facet joint syndrome using intra-articular injection with autologous platelet rich plasma. Pain Physician. 2016;19:617–25.

25. Kirchner F, Anitua E. Intradiscal and intra-articular facet infiltrations with plasma rich in growth factors reduce pain in patients with chronic low back pain. J Craniovertebr Junction Spine. 2016;7:250–6.

26. Simopoulos TT, Manchikanti L, Gupta S, et al. Systematic review of the diagnostic accuracy and therapeutic effectiveness of sacroiliac joint interventions. Pain Physician. 2015;18:E713–56.

27. Nagamoto Y, Iwasaki M, Sakaura H, et al. Sacroiliac joint motion in patients with degenerative lumbar spine disorders. J Neurosurg Spine. 2015;23(2):209–16.

28. Ko GD, Mindra S, Lawson GE, et al. Case series of ultrasound-guided platelet- rich plasma injection for sacroiliac joint dysfunction. J Back Musculoskelet Rehabil. 2017;30(2):363–70.

29. Singla V, Batra YK, Bharti N, Goni VG, Marwaha N. Steroid vs. platelet-rich plasma in ultrasound-guided sacroiliac joint injection for chronic low back pain. Pain Pract. 2017;17:782–91.

30. Navani A, Gupta D. Role of intra-articular platelet-rich plasma in sacroiliac joint pain. Reg Anesth Pain Med. 2015;19:54–9.

31. Lin C-S, Lin G, Lue TF. Allogenic and xenogenetic transplantation of adipose- derived stem cells in immunocompetent recipients without immunosuppressants. Stem Cells Dev. 2012;21(15):2770–8.

32. O'Donoghue K, Fisk NM. Fetal stem cells. Best Pract Res Clin Obstet Gynaecol. 2004;18(6):835–75.

33. Wang Y, Zhao S. Vascular biology of the placenta. San Rafael: Morgan and Claypool Life Sciences; 2010.

34. Rodolfa K, di Giorgio FP, Sullivan S. Defined reprogramming: a vehicle for changing the differentiated state. Differentiation. 2007;75(7):577–9.

Adipose-Derived Regenerative Cellular Therapy of Chronic Wounds

Joel A. Aronowitz and Bridget Winterhalter

Introduction

Wounds are among the most timeless of medical conditions. A wound by definition is a defect in the integrity of the integument and represents a fundamental threat to health. It is therefore unsurprising that healing is a basic process that is frustrated only by significant negative local or systemic factors. As a consequence of one or more of these factors, either singularly or in combination, some wounds fail to heal in a timely fashion. The term chronic wound refers to an integumentary defect that does not heal in the expected way or within 3 months using conventional wound treatments. A chronic wound is characterized by the persistent loss of skin integrity with exposure and often loss of underlying soft tissue over an extended period of time. A chronic wound is typically defined as failure of the healing process to restore skin integrity after 3 months with conventional wound care. Chronic wounds negatively affect the quality of life and complicate the health of approximately six million people in the United States [1]. These chronic wounds are most often the result of the interplay of systemic factors such as diabetes and obesity with local factors such as arterial insufficiency, chronic edema, venous congestion, and trophic changes [1, 2].

Diabetes is a chronic disease that affects roughly one-third of the adult population in the United States [3]. Patients with diabetes are especially vulnerable to the development of lower extremity chronic wounds. The current treatments available for diabetic chronic wounds usually involve systemic glucose control, ensuring adequate extremity perfusion, debridement of nonviable tissue, off-loading, control of infection, local wound care, and patient education administered by a multidisciplinary team [4]. It has been reported that about fifteen percent of diabetic patients will develop a foot ulcer in their lifetime, and approximately fifteen to twenty percent of these ulcers will result in lower extremity amputation [5]. These amputations result from complications of diabetes that make it difficult for wounds to heal, such as neuropathy, impaired immunity, and vascular deficits.

Similarly, venous disease accounts for a large percentage of lower extremity chronic wounds. Venous hypertension secondary to various causes results in distension and incontinence of low compliance vessel walls and ultimately leads to trophic changes and ultimately skin breakdown. Standard treatment for venous ulcers typically includes the use of mechanical compression and limb elevation to reverse tissue edema and improve venous blood flow. Arterial ulcers, which are less common, are a result of impaired circulation which affects healing and leads to ulceration. Care for chronic wounds caused by arterial insufficiency is centered on reestablishing blood flow and minimizing further loss of tissue perfusion. Failure to treat venous or arterial ulcers can also lead to lower extremity amputation.

Wound healing is a complex multifactorial process involving the interaction of inflammation, wound contraction, granulation tissue formation, re-epithelialization, and angiogenesis [6]. These stages of wound healing are regulated by platelets and immune cells, such as monocytes and neutrophils, which secrete chemokines and cytokines that attract and instruct other locally present cell types in the skin [7]. Unfortunately, most of the available treatments mentioned are limited in effectiveness and are often not sufficient to guarantee adequate healing [6]. The standard rate of healing is low for chronic wounds, only 24% or 30% of diabetic foot ulcers will heal at weeks 12 or 20 respectively [8]. Untreated chronic wounds run the risk of infection, further tissue and bone damage, and pain that collectively may result in the need for amputation as mentioned earlier. As a result, there is a growing need to develop improved techniques in chronic wound management, specifically techniques based on the concept of converting a chronic wound into an acute wound.

J. A. Aronowitz (✉) · B. Winterhalter
University Stem Cell Centers, Cedars-Sinai Medical Center,
Department of Plastic Surgery, Los Angeles, CA, USA
e-mail: dra@aronowitzmd.com

© Springer Nature Switzerland AG 2023
C. W Hunter et al. (eds.), *Regenerative Medicine*, https://doi.org/10.1007/978-3-030-75517-1_25

The skin consists of multiple layers starting from the outside with the epidermis, dermis, and a subcutaneous layer (a layer of adipose tissue containing adipocytes embedded in the stromal vascular fraction (SVF)) [9]. The epidermis is organized into hair follicles containing the interfollicular epidermis as well as sebaceous glands. The interfollicular epidermis is maintained by hair follicle stem cells and associated progenitor cells, which are also responsible for hair growth [10]. After physical damage to the epidermis, hair follicle stem cells migrate to the wound area to regenerate this damaged tissue [11]. Besides hair follicle stem cells and associated progenitor cells, the epidermis consists mainly of keratinocyte in various stages of differentiation. The dermis comprises vasculature and nerves, as well as specialized extracellular matrix. Elastin and collagen are pivotal, which provide elasticity and tear-resistance to the dermis [12]. Lastly, the subcutaneous layer comprises adipocytes and SVF. The SVF consists of all non-parenchymal (adipocyte) cell types, such as fibroblasts, immune cells, endothelial cells, pericytes, and adipose tissue-derived stromal cells (ASCs) [13].

Amniotic tissue has been used in wound care for over 100 years and experienced a resurgence in recent years due to the ready accessibility for "off-the-shelf" purchase. Amniotic tissue has a number of qualities that make it ideal for wound healing:

- Contains an impressive number of cytokines and essential growth factors in concentrated quantities – all of which increase the rate wound healing and enchances it
- Has analgesic properties when applied to the wound itself
- Acts as a biological barrier which can prevent work to prevent infection
- Can reduce inflammation and scar tissue
- Works as a scaffold/matrix for cell growth and proliferation

Currently, "amnio" can be purchased in ready-to-use, thin, dehydrated, sterile sheets that can be applied directly over a wound like a dressing which makes the application relatively straightforward.

Adipose-derived stem cells are of emerging interest in the application of wound healing due to their prolonged self-renewal capacity and their ability to proliferate and induce differentiation into various cell types [6, 14]. Recent studies have suggested that human adipose-derived stem cells (ASCs) may play a supportive role in wound healing by converting chronic wounds into acute wounds through the formation of vascular structures via direct and indirect mechanisms [14]. ASCs promote wound healing by increasing vessel density, granulation tissue thickness, and collagen deposition while simultaneously improving the cosmetic appearance of the resulting scar [15]. Additionally, ASCs secrete nearly all of the growth factors that are involved in normal wound healing, including vascular endothelial growth factor (VEGF), hepatocyte growth factor (HGF), PDGF, and TGFb1, which promote angiogenesis and accelerate wound healing [15]. ASCs are an advantageous resource because they can be found in abundant quantities, they can be harvested with a minimally invasive procedure, they can be differentiated along multiple cell lineage pathways, they are immunocompatible, and they can be safely and effectively transplanted into an autologous or allogeneic host [16].

Commonly Used Treatments

Currently, the treatment of chronic wounds is directed at correcting precipitating and perpetuating factors and optimizing the wound bed [17]. Many wounds heal when basic principles of wound healing are implemented, but unfortunately some wounds do not heal; therefore, innovative approaches are needed. Popular general therapeutic modalities can be seen in debridement, compression, pressure-relieving devices and surfaces, negative pressure, electrical stimulation, hyperbaric oxygen therapy, antimicrobials, and dressings [18].

Debridement, one of the most commonly used techniques, eliminates foreign material and non-viable tissues from a wound bed to allow healthy tissues to be exposed and stimulate healing [18]. Compression therapy, often achieved with compression bandages, is the mainstay of treatment for stasis leg ulcer. However, there is no data to recommend one type of compression over another and require patient adherence [18]. Hyperbaric oxygen therapy, by exposure to 100% oxygen at a pressure > 1 atmosphere, has been used in the treatment of chronic wounds. Its efficacy may involve improved neovascularization, reduced proinflammatory cytokines, and increased production of growth factors and collagen [18]. There has been increasing acceptance of wound dressings that promote moist wound healing. Dressings include the following: transparent films, hydrocolloids, foams, alginates, gels, and collagen-based products. Presently, there is not enough evidence favoring a specific wound dressing. Skin substitutes are often a combination of sophisticated matrix products and cell therapy, ranging from purely synthetic compounds to both cellular and acellular constructs derived from human and animal sources [18]. The use of skin substitutes has improved prognosis and reduced morbidity in the treatment of chronic wounds.

Allografts from human placenta (aka amniotic tissue) have several angiogenic growth factors capable of stimulating angiogenesis and amplifying the angiogenic response from human endothelial cells in vitro. Randomized clinical trials have shown some substitutes to be effective for healing chronic wounds [19]. Despite huge advances in medical care and nutrition, which have resulted in commendable change in the outcome of wound management, new therapies in this area are required to optimize outcomes for patients. Adipose stromal cells, with their unique properties to self-renew and undergo differentiation, are emerging as promising candidate for cell-based therapy for treatment of chronic wounds.

Indications

The most common wound type for which regenerative therapy is used (particularly amniotic tissue) is burns. Other wound types include the following:

- Chronic wounds
- Pressure ulcers
- Diabetic foot ulcers
- Venous leg ulcers

Evidence

Dongen et al. discuss the most recent research, including two clinical trials directed at safety of revascularization after critical limb ischemia. In both studies, cultured ASCs were injected intramuscularly to treat patients with non-healing ischemic ulcers. In Lee et al.'s study, ASCs were used in fifteen patients with critical limb ischemia [28]. The ASCs were derived from the abdominal subcutaneous adipose tissue and injected at a total of 3×10^8 cells/ml at 60 points into the lower extremity muscles. Clinical improvement occurred in 66.7% of the patients. Clinical outcomes included lower pain scores, improved claudication walking distances, and no changes in the ankle brachial index (ABI) due to formation of numerous collaterals [28].

In the second trial by Bura et al., only seven patients were treated, of whom four patients underwent amputation within 5 months after injection of ASCs [29]. They isolated ASCs from the abdominal fat and cultured them for 2 weeks, which yielded more than 200 million cells with almost total homogeneity and no karyotype abnormalities [30]. The investigators later injected ASCs intramuscularly into the ischemic leg of patients with no complications. Three non-amputated patients reported a decrease of pain 6 months postoperative. Although the results from these non-controlled, small-scaled studies are promising for part of the treated patients, the therapeutic effect needs to be corroborated in randomized, placebo-controlled large trials. This is critical, if alone to distinguish responders from non-responders and to determine

optimal dosing, frequency of dosing, and to identify parameters that dictate efficient and effective wound healing in adipose tissue [9].

In both studies, a large proportion of patients did not respond after administration of cultured ASCs [9]. This lack of effect of ASCs might relate to disturbed migration and/or disposal of ASCs via the circulation and lymph system within the first 24 h after injection [31]. Other factors that influence the therapeutic impact of ASCs are donor characteristics (e.g., age or comorbidity such as diabetes mellitus), as well as the induction of senescence of ASCs after enzymatic isolation and culture [32, 33]. The two studies on the therapeutic benefit of ASCs on critical limb ischemia, however, showed that age or comorbidity did not influence the therapeutic impact of ASCs [28, 29].

A third study by Marino et al. examined 20 patients with peripheral arterial disease, treating 10 patient's leg ulcers with ASC transplantation, while the other 10 patients remained as controls. The authors isolated ASCs from the abdominal wall fat and used an advanced Celution system for cell expansion. Then, a 10-mm syringe was used to inject the ASCs at the edges of the ulcer. Compared to the untreated controls, wounds that were treated with stem cells showed marked reductions in the ulcers' diameter, depth, and associated pain, with no recorded adverse events. Complete ulcer healing occurred in six of the ten ASCs-treated patients [34].

Microanatomy and Biochemistry

Adipose-Derived Tissue

Cultured ASCs secrete a plethora of angiogenic, anti-fibrotic, and anti-apoptotic growth factors [9]. There are several potential mechanisms by which stem cells could significantly contribute to wound healing, but the exact mechanism is still under investigation [18]. ASCs are mesenchymal stromal cells that are present in SVF of adipose tissue, attached around vessels as precursor cells (i.e., pericytes and periadventitial cells) [20, 21]. In vitro, ASCs have the ability to differentiate into multiple lineages under ectodermal, endodermal, and mesenchymal lineages [9]. Different studies propose that mesenchymal stem cells' mechanism of action include immune modulation, differentiation into epidermal and dermal cells to replace the damaged skin, and paracrine signaling pathways [18].

Adipose-derived stromal cells have recently been demonstrated to hold several immunomodulatory effects on host immune cells in both wound and transplant biology contexts, secreting a plethora of angiogenic, anti-fibrotic, and anti-apoptotic growth factors [9]. To release immunosuppressive factors, ASCs can downregulate the immune response, making them immune privileged and more ideal for allogeneic transplantation [18]. Another consideration in wound repair is scarring caused by deposition of excessive extracellular matrix. The anti-inflammatory effects of ASCs may decrease fibrosis and therefore reduce scar formation [22]. It is possible that ASCs can sense that degree of inflammation in the microenvironment and respond by releasing growth factors, cytokines, and other mediators to reduce inflammation using real-time biochemical cues. If effective, reducing inflammation to a better level to allow healing to proceed should also result in improved tensile strength and scar quality, thereby reducing wound recurrence [23]. ASCs injected directly into excisional wounds were shown to differentiate into a number of skin cell types and therefore repopulate the wound bed [23]. In tissues with a high rate of cell turnover, such as skin, differentiation or transdifferentiation rather than cell fusion is the principal mechanism [24]. However, the transdifferentiation ability of adipose-derived mesenchymal cells has been questioned. They are also rapidly mobilized in response to hypoxia, which is commonly found in acute and chronic wounds, with poor vascularization. MSC differentiation into epidermal cells and their migratory capacity contribute to the cell repopulation of the wound bed [18].

In addition, paracrine effects of ASCs in several skin cells have been studied for initiating the tissue regeneration process [25, 26]. In one animal study, the use of ASCs seeded on an acellular dermal matrix found that it enhanced wound healing, promoted angiogenesis, and contributed to newly formed vasculature in murine mouse models [27]. Finally, using these cells in treatment of wound reduces inflammation, accelerates wound healing, promotes granulation tissue, and increases angiogenesis [16].

Amniotic Tissue

The amniotic membrane consists of 2 distinct layers:

- Amnion – innermost layer of the placenta
- Chorion – maternal side

The amnion consists of a thick basement layer and an avascular stromal matrix. It surrounds the fetus, separating it from the mother, and protects it during development. The amnion is separated from the chorion by a jelly-like layer. It has a layer of epithelial cells which can be easily removed by basic cell scraping. The cells of amnion are pluripotent which makes it particularly useful for tissue transplantation and surface reconstructive surgery (i.e., ocular and genitourinary surgeries). Amniotic tissue has the following properties:

- Anti-bacterial and anti-viral – expression of β-defensins, elafin, and cystatin E
- Anti-inflammatory – suppressing inflammatory cytokines (i.e., IL-1α and β) and expression of migration inhibitory factor (MIF), IL-1 receptor antagonist, PGE2, TGF-β, HGF, and TNF-α
- Promotes epithelialization
- Supports cellular adhesion
- Promotes angiogenesis – secretion of VEGF, IL-8, angiogenin, interferon-γ, IL-6, bFGF, EGF, and PDGF from amniotic mesenchymal cells
- Secretes growth factors – EGF, KGF, and HGF
- Pro-apoptotic properties toward inflammatory cells – secretion of IL-1, IL-2 receptor antagonist, IL-10, and endostatin (all of which inhibit tumor growth as well)

Basic Concerns and Contraindications

As with most regenerative therapies, conventional therapies for wound repair should be attempted first.

Preoperative Considerations

Adipose-Derived Tissue – Enzymatic Versus Mechanical Isolation

In general, the most efficient methods can isolate about 500,000-1,000,000 cells per gram of lipoaspirate tissue with a > 80% viability [35]. The number of viable cells required for treatment of a particular condition is unknown because there is insufficient data to establish a reliable dose vs effective relationship. Because no additional adverse effects are reported with the use of autologous ASCs in fat grafting, the largest number of cells isolated at the point of care without expansion in culture is typically used.

In clinical practice, adipose-derived stem cells are often not administered as a pure isolate, but rather as one constituent of stromal vascular fraction (SVF), a heterogenous mixture of cells resulting from the mechanical or enzymatic processing of aspirated adipose tissue. SVF contains a variety of cells, including macrophages, various blood cells, pericytes, fibroblasts, smooth muscle cells, vascular endothelial progenitors, and adipose-derived stem cells [36]. SVF is one component of the heterogenous mixture of adipose tissue fragments, stromal tissue, blood, and tumescent fluid which constitutes lipoaspirate. The ASC content of SVF varies substantially depending on the method employed.

Enzymatic methods of isolating SVF cells from adipose tissue at the point of care are based on a commonly used laboratory method of obtaining cells. The methods used to manually process adipose tissue using collagenase follow the same basic steps, but vary slightly in technique and reagents used. Lipoaspirate is washed 2–3 times using an aqueous salt solution such as lactated Ringer's solution, phosphate-buffered saline, Hank's balanced salt solution. It is important to select a buffer which is both suitable for optimal enzymatic activity and preserving the viability of the cell population. Proteolytic enzymes such as collagenase, dipase, *Clostridium histiolyticum* neutral protease (CHNP), thermolysin, clostripain, and trypsin all include calcium ions as an activator of catalytic activity [37]. Therefore, it is important to have an adequate concentration of calcium in the buffer used.

The washed lipoaspirate is then incubated with the enzymatic solution of variable concentration and composition, depending on the method and tissue dissociation enzyme product used. Enzymatic digestion is typically carried out in a heated shaker to provide constant agitation at 37 degrees C for 30 minutes to 2 hours. The digested adipose tissue is then centrifuged, separating the processed lipoaspirate into three main layers: the oil/adipose tissue layer, the aqueous layer, and the pellet. The SVF is contained within the pellet, allowing the other layers to be discarded. The pellet is washed to remove any residual enzyme and filtered to remove tissue fragments and detritus. Collagenase-based enzymatic methods can be up to 1000 times more effective in recovering SVF cells than mechanical methods [35]. Enzymatic methods are more efficient in isolating SVF cells because of the disruption of the collagen-based extracellular matrix (ECM) which binds together adipocytes and other cells of adipose tissue. Mixtures of enzymes have been shown to yield more nucleated cells than using only one enzyme, a quality attributed to the synergistic effect of the proteolytic enzymes in the breakdown of ECM [38]; however, collagenase is still frequently used as the sole proteolytic enzyme in methods using products such as Collagenase NB6.

One topic which is debated across the literature is the effect that enzymatic digestion has on the resulting phenotype of the cellular population as well as on the other cellular properties such as proliferation rate and differentiation capacity. In 2014, Busser et al. compared the phenotypic characterization and cellular functions of resulting SVF cells isolated by two different methods: enzymatic digestion with collagenase and explant culture [39]. They noted similar phenotypes and functions of the ASCs obtained with both methods in terms of surface marker characterization, trilineage differentiation, hematopoiesis supporting activities, population doubling-time, and CFU-F formation. Similar findings were observed by Gittel et al. in 2013 when they compared the characteristics of cells from equine adipose tissue isolated enzymatically versus by explant culture [40]. Overall, it appears that the use of collagenase as the primary proteolytic enzyme with an incubation time of 2 hours or less does not alter the pheno-

typic or functional characteristics of adipose derived stem cells (ADSC) populations of SVF when compared to cells isolated using non-enzymatic methods [37].

According to Aronowitz et al., published yields of viable, nucleated SVF cells achieved using manual, collagenase-based digestions range from 100,000 nucleated cells/cc to 1,300,000 nucleated cells/cc of lipoaspirate processed. Equipment like the PNC Multi-Station (PNC International, Gyeonggi-do, Republic of Korea) is commercially available for use in the manual preparation of SVF. The PNC Multi-Station contains a centrifuge and heated shaker inside of a sterile biohood, which allows the processing to be conducted in sterile conditions.

Mechanical methods for SVF isolation report significantly lower yields of nucleated cells/cc of lipoaspirate processed. Cell yields are reported from 10,000 nucleated cells/cc of lipoaspirate to 240,000 nucleated cells/cc of lipoaspirate [41]. Mechanical methods seek alternative non-enzymatic means of removing SVF cells from the adipose tissue and tend to be focused around washing and vibrating lipoaspirate, followed by centrifugation in order to concentrate the SVF cells. The composition of the cell populations recovered through simple centrifugation and other non-enzymatic methods has been shown to contain a greater frequency of peripheral blood mononuclear cells and a substantially lower number of progenitor cells [42].

Although enzymatic methods consistently yield higher cell counts with a higher frequency of progenitor cells, mechanical methods offer some distinct advantages. Mechanical methods tend to offer a faster processing time, some less than 15 minutes, because they do not require additional time for enzymatic digestion to occur, which sometimes can take up to 120 minute. In addition to time, mechanical methods can be fairly expensive, with costs of $2–$5 per gram of tissue processed using GMP grade enzymes [43].

Equipment

Amniotic Tissue

The only equipment needed when using amniotic tissue for wound care are sterile forceps to handle and place the graft on the wound, materials for dressing to place over the graft, and Steri-Strips to potentially secure the graft in place if movement is a concern.

Adipose-Derived Tissue

Refer to Chap. 5 for an in-depth discussion on obtaining and processing SVF from adipose.

Technique

Amniotic Tissue

The technique for applying amnio depends on the disease process, wound type, and manufacturer recommendations. The initial steps that one would follow prior to initiating treatment are the same as those that would be implemented for any wound treatment and not specific to amniotic tissue (i.e., documentation of wound size, depth, presence of necrotic tissue, assessment of circulatory status, barriers to healing, nutritional status, etc.). Next the wound should be clear of any necrotic tissue (i.e., debridement) and have no signs of infection. Most amniotic tissue is available in small ready-to-use sheets that can be trimmed to fit the wound. It can be placed on the wound wet or dry and the stromal collagen layer MUST be facing the wound surface. Thereafter, steps should be taken to ensure that the AmnioGraft® is held in place – one can use Steri-Strips and/or a dressing. The graft should then be left in place for at least 7–10 days before it is disturbed. A second graft can be placed after 14 days if needed.

Adipose-Derived Tissue

The mode of delivery for adipose-derived MSCs will differ depending on the disease process and wound. Progress has been made with ASCs alone, ASCs with scaffolds, as well as skin graft models. The routes of delivery of ADSCs alone into the wound vary between topical administration over the wound bed, intradermal injection around the wound, and intrafascial and intramuscular injection [30]. Amos and colleagues reported that the formulation of ASCs delivery can affect their therapeutic potential. For example, ASCs delivery as multicellular aggregates caused a significant increase in wound closure rate, compared to wounds treated with an equal number of ASCs, delivered by suspension [44].

Scaffolds are engineered materials designed to cause desirable cellular interactions and contribute to forming new functional tissues [45]. Biosynthetic scaffolds have been used in wound healing and skin regeneration either alone or in combination with cells [46]. Adding a scaffold to ASCs provides the wound tissue with an extracellular matrix-like architecture, which structurally supports these cells, guiding their growth in three-dimensional manner in the wound space and recreating the exact size of the wound defect [45, 46]. This method also facilitates revascularization and integration of the implanted cells in the area [30]. Several biosynthetic materials have been used in experimental studies with a special focus over those derived from the ECM, owing to easier handling and enhanced resistance to infection [47].

The various supporting scaffolds used in ASC wound healing and those that have shown promising results are decellularized silk fibroin scaffold, PLGA scaffold, and atelocollagen matrix silicon membrane [30].

Skin substitutes are currently being used as scaffolds to ensure optimal delivery of stem cells into the wound defect [48]. In comparison to control mice, hyperglycemic mice, treated with ASCs, showed a significant improvement in graft intake [49]. Trottier et al. reported their ability to construct a trilayer human skin substitute from keratinocytes, dermal fibroblasts, and ASCs without using a scaffold. The resulting substitute was grafted into athymic mice and was able to create a well-differentiated epidermis in 21 days [50].

Potential Complications and Pitfalls

Despite the rapid progress in evaluating the efficacy of mesenchymal cells from adipose tissue in wound healing and the growing research reporting the promising outcomes in chronic wound care, there remain issues that must be addressed, most importantly, concerns with the risk of malignant transformation and cancer induction. In a study by Yu et al., subcutaneous injection of ASCs with tumor cells in nude mice stimulated tumor outgrowth [51]. In another study by Rubio et al., spontaneous transformation of stem cells into malignant phenotypes was reported [52]. However, the authors later retracted their article due to suspicion of contamination and inability to replicate the transformation events [30]. Contrary, all conducted clinical trials to date confirmed the safety of ASCs treatment in diabetic ulcer patients; however, they were limited by small sample size and short follow-up period [28, 29, 34]. Studies exploring the long-term safety of ASCs-derived therapeutics are necessary.

Most SVF cell isolation protocols follow the same basic steps which involve washing the lipoaspirate to remove excess blood and tumescent solution, enzymatic digestion to dissociate the tissue, followed by centrifugation and additional washing to collect the SVF cells, but there is significant variation in how these methods are executed across different isolation methods. In obtaining a protocol that is standardized, it will allow for more research with larger populations. These processes of obtaining cells can be complex and must be done correctly to ensure a safe application of ASCs.

There still remains questions in the mechanism of action for ASCs. There appears to be more investigations to define the interactions between adipose-derived cells and the other cell types present in the wound. Further identification of the ASC-derived factors responsible for tissue responses to injury are crucial for the understanding of how ASCs are affected by the wound environment. Determining whether ASC production of cytokines and growth factors is regulated during wound healing will explore whether ASC supernatant can be used to enhance healing [18]. Without knowing the full mechanism of action, it is difficult to determine the degree of cell survival following implantation, which may adversely affect long-term treatment.

Orteo-Vinas reports other considerations such as the following: (1) cost-effectiveness of acquisition, testing, storage, and subsequent use in humans; (2) easy accomplishment of stem cell delivery without the use of sophisticated laboratory techniques; (3) the treatment must be such that it can be used in the clinic, and not just within the confines of a laboratory-associated clinical research team; (4) having competent laboratory personnel; and (5) the availability of off-the-shelf stem cell product for immediate use in cases involving burns and/or trauma [18].

Conclusion

The administration of ASCs is promising as new therapy for the treatment of non-healing dermal wounds. Currently common treatments include debridement, compression, pressure-relieving devices and surfaces, negative pressure, electrical stimulation, hyperbaric oxygen therapy, antimicrobials, and various dressings. Although the mechanism is not fully understood, it is well known and understood that cultured ASCs secrete a plethora of angiogenic, anti-fibrotic, and anti-apoptotic growth factors. There are currently a few studies that explore the topic of ASCs in wounds, and they appeared promising and safe, but population sizes remain small. Mechanical techniques, such as simple washing or centrifugation of lipoaspirate, are effective in isolating ASCs. Mechanical methods are appealing because they are simple, quick, and generally not associated with expensive equipment or disposables. While more expensive than mechanical options, enzymatic methods for the isolation of stromal vascular fraction cells from adipose tissue yield more nucleated cells with a higher number of progenitor cells.

References

1. King H, Aubert RE, Herman WH. Global burden of diabetes, 1995-2025: prevalence, numerical estimates, and projections. Diabetes Care. 1998;21:1414–31.
2. Williams R, van Gaal L, Lucioni C. Assessing the impact of complications on the costs of type II diabetes. Diabetologia. 2002;45:S13–7.
3. Reiber GE. The epidemiology of diabetic foot problems. Diabet Med. 1996;13(Suppl 1):S6–11.
4. Reiber GE, Vileikyte L, Boyko EJ, et al. Causal pathways for incident lower-extremity ulcers in patients with diabetes from two settings. Diabetes Care. 1999;22:157–62.
5. Hanft JR, Surprenant MS. Healing of chronic foot ulcers in diabetic patients treated with a human fibroblast-derived dermis. J Foot Ankle Surg. 2002;41:291–9.
6. www.emedicine.com/orthoped/topic387.htm. 2011.
7. Roupé KM, Nybo M, Sjöbring U, Alberius P, Schmidtchen A, Sørensen OE. Injury is a major inducer of epidermal innate immune responses during wound healing. J Investig Dermatol. 2010;130:1167–77.
8. Marston WA, Hanft J, Norwood P, et al. The efficacy and safety of Dermagraft in improving the healing of chronic diabetic foot ulcers: results of a prospective randomized trial. Diabetes Care. 2003;26:1701–5.
9. Dongen JV, Harmsen M, Lei BVD, Stevens H. Augmentation of dermal wound healing by adipose tissue-derived stromal cells (ASC). Bioengineering. 2018;5:91.
10. Clayton E, Doupé DP, Klein AM, Winton DJ, Simons BD, Jones PH. A single type of progenitor cell maintains normal epidermis. Nature. 2007;446:185–9.
11. Mascré G, Dekoninck S, Drogat B, Youssef KK, Brohée S, Sotiropoulou PA, et al. Distinct contribution of stem and progenitor cells to epidermal maintenance. Nature. 2012;489:257–62.
12. Makrantonaki E, Zouboulis CC. Molecular mechanisms of skin aging: state of the art. Ann N Y Acad Sci. 2007;1119:40–50.
13. Bourin P, Bunnell BA, Casteilla L, Dominici M, Katz AJ, March KL, et al. Stromal cells from the adipose tissue-derived stromal vascular fraction and culture expanded adipose tissue-derived stromal/stem cells: a joint statement of the International Federation for Adipose Therapeutics and Science (IFATS) and the International Society for Cellular Therapy (ISCT). Cytotherapy. 2013;15:641–8.
14. Wieman TJ, Smiell JM, Su Y. Efficacy and safety of a topical gel formulation of recombinant human platelet-derived growth factor-BB (becaplermin) in patients with chronic neuropathic diabetic ulcers. A phase III randomized placebo-controlled double-blind study. Diabetes Care. 1998;21:822–7.
15. Damour O, Gueugniaud PY, Berthin-Maghit M, et al. A dermal substrate made of collagen--GAG--chitosan for deep burn coverage: first clinical uses. Clin Mater. 1994;15:273–6.
16. Coulomb B, Dubertret L. In vitro reconstruction of the human skin. Rev Laryngol Otol Rhinol (Bord). 1987;108:55–7.
17. Panuncialman J, Falanga V. The science of wound bed preparation. Surg Clin North Am. 2009;89:611–26.
18. Otero-Vinas M, Falanga V. Mesenchymal stem cells in chronic wounds: the spectrum from basic to advanced therapy. Wound Heal Soc. 2015;5:4.
19. Koob TJ, Lim JJ, Massee M, et al. Angiogenic properties of dehydrated human amnion/chorion allografts: therapeutic potential for soft tissue repair and regeneration. Vasc Cell. 2014;6:10.
20. Corselli M, Chen C-W, Sun B, Yap S, Rubin JP, Péault B. The tunica adventitia of human arteries and veins as a source of mesenchymal stem cells. Stem Cells Dev. 2012;21:1299–308.
21. Lin G, Garcia M, Ning H, Banie L, Guo Y-L, Lue TF, et al. Defining stem and progenitor cells within adipose tissue. Stem Cells Dev. 2008;17:1053–63.
22. Nuschke A. Activity of mesenchymal stem cells in therapies for chronic skin wound healing. Organogenesis. 2013;10:29–37.
23. Ennis WJ, Sui A, Bartholomew A. Stem cells and healing: impact on inflammation. Adv Wound Care. 2013;2:369–78.
24. Li H, Fu X. Mechanisms of action of mesenchymal stem cells in cutaneous wound repair and regeneration. Cell Tissue Res. 2012;348:371–7.
25. Kokai LE, Marra K, Rubin JP. Adipose stem cells: biology and clinical applications for tissue repair and regeneration. Transl Res. 2014;163:399–408.
26. Hassan WU, Greiser U, Wang W. Role of adipose-derived stem cells in wound healing. Wound Repair Regen. 2014;22:313–25.

27. Huang S-P, Hsu C-C, Chang S-C, Wang C-H, Deng S-C, Dai N-T, et al. Adipose-derived stem cells seeded on acellular dermal matrix grafts enhance wound healing in a murine model of a full-thickness defect. Ann Plast Surg. 2012;69:656–62.

28. Lee HC, An SG, Lee HW, Park J-S, Cha KS, Hong TJ, et al. Safety and effect of adipose tissue-derived stem cell implantation in patients with critical limb ischemia. Circ J. 2012;76:1750–60.

29. Bura A, Planat-Benard V, Bourin P, Silvestre J-S, Gross F, Grolleau J-L, et al. Phase I trial: the use of autologous cultured adipose-derived stroma/stem cells to treat patients with non-revascularizable critical limb ischemia. Cytotherapy. 2014;16:245–57.

30. Gadelkarim M, Abushouk AI, Ghanem E, Hamaad AM, Saad AM, Abdel-Daim MM. Adipose-derived stem cells: effectiveness and advances in delivery in diabetic wound healing. Biomed Pharmacother. 2018;107:625–33.

31. Parvizi M, Harmsen MC. Therapeutic Prospect of adipose-derived stromal cells for the treatment of abdominal aortic aneurysm. Stem Cells Dev. 2015;24:1493–505.

32. Fossett E, Khan WS, Longo UG, Smitham PJ. Effect of age and gender on cell proliferation and cell surface characterization of synovial fat pad derived mesenchymal stem cells. J Orthop Res. 2012;30:1013–8.

33. Pozzi A, Zent R, Chetyrkin S, Borza C, Bulus N, Chuang P, et al. Modification of collagen IV by glucose or methylglyoxal alters distinct mesangial cell functions. J Am Soc Nephrol. 2009;20:2119–25.

34. Marino G, Moraci M, Armenia E, Orabona C, Sergio R, Sena GD, et al. Therapy with autologous adipose-derived regenerative cells for the care of chronic ulcer of lower limbs in patients with peripheral arterial disease. J Surg Res. 2013;185:36–44.

35. Aronowitz J, Lockhart R, Hakakian C. Mechanical versus enzymatic isolation of stromal vascular fraction cells from adipose tissue. Springerplus. 2015;4:713.

36. Yoshimura K, Shiguera T, Matsumoto D, et al. Characterization of freshly isolated and cultured cells derived from the fatty and fluid portions of liposuction aspirates. J Cell Physiol. 2006;208:64–76.

37. Lockhart RA, Hakakian CS. Tissue dissociation enzymes for adipose stromal vascular fraction cell isolation: a review. J Stem Cell Res Thera. 2015;5:12.

38. McCarthy RC, Breite AG, Dwulet FE. Biochemical analysis of crude collagenase products used in adipose derived stromal cell isolation procedures and development of a purified tissue dissociation enzyme mixture. Available at http://www.vitacyte.com/wp-content/uploads/2009/01/ifats-vitacyte.pdf. Accessed 6 Nov 2018.

39. Busser H, Najar M, Raicevic G, Pieters K, Pombo RV, Philippart P, et al. Isolation and characterization of human mesenchymal stromal cell subpopulations: comparison of bone marrow and adipose tissue. Stem Cells Dev. 2015;24:2142–57.

40. Gittel C, Brehm W, Burk J, Juelke H, Staszyk C, Ribitsch I. Isolation of equine multipotent mesenchymal stromal cells by enzymatic tissue digestion or explant technique: comparison of cellular properties. BMC Vet Res. 2013;9:221.

41. Aronowitz JA, Lockhart RA, Hakakian CS, Hicok KC. Clinical safety of stromal vascular fraction separation at the point of care. Ann Plast Surg. 2015;75:666–71.

42. Conde-Green A, Rodriguez RL, Slezak S, et al. Enzymatic digestion and mechanical processing of aspirated adipose tissue. Plast Reconstr Surg. 2014;54

43. Aronowitz J, Ellenhorn JD. Adipose stromal vascular fraction isolation: a head to head comparison of four commercial cell separation systems. Plast Reconstr Surg. 2013;132(6):932e–9e.

44. Amos PJ, Kapur SK, Stapor PC, Shang H, Bekiranov S, Khurgel M, et al. Human adipose-derived stromal cells accelerate diabetic wound healing: impact of cell formulation and delivery. Tissue Eng Part A. 2010;16:1595–606.

45. Murugan R, Ramakrishna S. Design strategies of tissue engineering scaffolds with controlled fiber orientation. Tissue Eng. 2007;13:1845–66.

46. Ma K, Liao S, He L, Lu J, Ramakrishna S, Chan CK. Effects of nanofiber/stem cell composite on wound healing in acute full-thickness skin wounds. Tissue Eng Part A. 2011;17:1413–24.

47. Cherubino M, Marra KG. Adipose-derived stem cells for soft tissue reconstruction. Regen Med. 2009;4:109–17.

48. Vermette M, Trottier V, Menard V, Saintpierre L, Roy A, Fradette J. Production of a new tissue-engineered adipose substitute from human adipose-derived stromal cells. Biomaterials. 2007;28:2850–60.

49. Hu M, Hong W, Senarath-Yapa K, Zimmermann A, Chung M, Esquivel M, et al. Adipose-derived stem cells promote engraftment of autologous skin grafts in diabetic mouse models. J Surg Res. 2014;186:578–9.

50. Trottier V, Marceau-Fortier G, Germain L, Vincent C, Fradette J. IFATS collection: using human adipose-derived stem/stromal cells for the production of new skin substitutes. Stem Cells. 2008;26:2713–23.

51. Yu JM, Jun ES, Bae YC, Jung JS. Mesenchymal stem cells derived from human adipose tissues favor tumor cell growth in vivo. Stem Cells Dev. 2008;17:463–74.

52. Rubio D, Garcia-Castro J, Martín MC, Fuente RDL, Cigudosa JC, Lloyd AC, et al. Spontaneous human adult stem cell transformation. Cancer Res. 2005;65:3035–9.

Esthetic Surgery Applications for Adipose-Derived Stem Cells

Joel A. Aronowitz, Daniel Oheb, Nathan Cai, Asli Pekcan, Bridget Winterhalter, and Joseph Clayton

Overview

Adipose-derived stem cells (ADSCs) are small stellate-shaped pluripotential cells, which reside in large numbers on the walls of small vasculature. It is estimated that each gram of adipose tissue contains approximately three million of these mesenchymal cells. Although the HLA surface protein expression of these cells is heterogenous, including CD 34, the sine qua non of definitive identification is their tendency to adhere and form colonies in vitro [1, 2]. This tendency is reported in a standard laboratory test known as Colony Forming Units (CFU). Adipose cells expressing this ability are further categorized by a range of tests including mRNA arrays, flow cytometry, and transcriptomic analyses [1, 2]. Interest in these cells has resulted in publications from a wide range of researchers from cell biologists to practicing clinicians. This profusion of scientific papers often produces a confusing terminology as authors frequently suggest updated nomenclature to reflect an evolving understanding of the behavior of these complex cells. In the final analysis, ADSCs may be properly classified as mesenchymal, connective tissue pluripotential messenger cells found most abundantly in peripheral adipose tissue. They possess the potential to repeatedly divide 25–28 times to produce daughter stem cells. The somatic daughter cells may differentiate to virtually any connective tissue cell type, recruit circulating cells such as the macrophage, and direct the differentiation of these and locally resident cells. The term messenger is appended to ADSCs because they are well known to direct the behavior of other cells through a protean range of paracrine effects.

A large body of preclinical research elucidates the cellular characteristics and behavior of these pluripotential cells, which are found concentrated in the stroma of adipose tissue and support the concept of utilizing these cells for cosmetic purposes to replace senescent volume loss and reverse other esthetic changes in the skin and subcutaneous tissue observed with age such as attenuation of dermal thickness, loss of subcutaneous adipose tissue, loss of dermal elasticity, and pigment changes.

Based on the preclinical literature, clinicians applied autologous adipose-derived stem cells in the treatment of a wide variety of esthetic purposes. The ease of accessibility of autologous cells from adipose tissue, their regenerative properties, and a proven clinical safety profile suggest that ADSCs offer the possibility of a unique treatment modality for many age-related and elective cosmetic treatments [3]. These applications include restoration of volume loss and redistribution due to aging, improvement of thin and fragile skin, and reversal of unsightly pigmentary changes associated with age and sun exposure. Most of these reports pertain to treatment of facial senescent changes, but there are many applications related to other areas of the body as well.

A brief summary of age-associated and environmental exposure-induced changes in the skin and subcutaneous layer is necessary to understand stem cell applications in treatment.

Age-related changes to the skin include a loss of skin elasticity, wrinkling, decreased amount of hair follicles, and a reduction in sweat and sebaceous glands. Overall, the epidermis regenerates at a slower rate with age, influenced in part by slowed collagen synthesis [4]. Skin thickness is decreased due to loss of vascularization and cell proliferation, particularly of keratinocytes and collagen [5, 6]. Changes in the structure and adhesion levels of subcutaneous white adipose tissue reduce skin elasticity [7]. As the number of adipose-derived stem cells decreases, the number of

J. A. Aronowitz (✉) · B. Winterhalter
University Stem Cell Centers, Cedars-Sinai Medical Center, Department of Plastic Surgery, Los Angeles, CA, USA
e-mail: dra@aronowitzmd.com

D. Oheb
Tower Outpatient Surgery Center, Los Angeles, CA, USA

N. Cai · J. Clayton
University Stem Cell Center, Los Angeles, CA, USA

A. Pekcan
University Stem Center, Los Angeles, CA, USA

adipocytes in a given area of the face decreases, reducing skin stiffness and increasing wrinkles [7].

Environmental factors such as tobacco smoking, infrared radiation, and ultraviolet exposure can accelerate the skin aging process and further contribute to wrinkles and loss of epidermal thickness [4, 8]. Environmental damage to the skin, including photoaging, tends to occur more often in males than in females [9]. Ultraviolet exposure in particular alters skin pigmentation and texture as well [8].

Beyond aging, the skin can undergo esthetic changes such as vitiligo, a skin condition in which loss of melanocytes leads to skin discoloration [10]. Although it only affects about 1% of the population, it serves as a good example of a disorder that drives patients to seek cosmetic treatments to preserve their appearance [11]. Traditionally, vitiligo is treated with melanocyte transplantation [10].

With respect to the subcutaneous fat layer, the aging face exhibits a change in the distribution of adipose tissue with age. Three-dimensional modeling of young and old facial shapes shows an increase in the mobility of facial tissues with age, specifically toward the lower anterior portion of the face [12]. Overall migration of facial fat to the inferior portion of the face contributes to a more masculine, square face, and more prominent fat pads below the eyelid [13, 14]. The nasolabial fold also becomes more defined, and loss of subcutaneous fat in the orbital area leads to a deeper orbital rim that makes the eyes appear sunken [14]. There has been an increase in efforts to identify more ways to address the facial volume loss a patient might experience as they age [15].

Historically, tissue volume concerns have been addressed with fat grafting. This technique is not only used for facial procedures but can also be useful for breast reconstruction purposes in breast cancer patients following mastectomy [16]. However, traditional autologous fat grafting is limited by unpredictable rates of reabsorption, causing patients to undergo multiple fat grafting procedures in order to maintain the desired result [17–19]. Furthermore, there is an apparent lack of consensus regarding good clinical practices related to autologous fat transfer procedures that further contribute to the large variation in results [17].

Traditional treatments for aging and environment-induced changes to the skin include topical treatments, such as retinoid creams. These creams are meant to increase collagen synthesis and counteract the degradation of skin cells that causes a loss of skin thickness and an increase in skin discoloration [20, 21]. Chemical peeling is another common treatment for dermatological concerns such as photoaging to stimulate keratinocyte regeneration [22]. However, chemical peels run the risk of a wide range of complications, including but not limited to hyperpigmentation, skin irritation, and scarring [23]. Preclinical and clinical research in recent years suggests that the efficacy of these treatments rest, at least in part, from initiation of natural tissue regeneration mecha-

nisms through the controlled application of a chemical, mechanical, or thermal trauma. ADSCs are rapidly attracted to the zone of injury, and adipose derived stem cell's normal regenerative activities are likely key to the desired effects of traditional cosmetic treatments such as dermabrasion, laser therapy, intense pulsed light, and chemical peels with caustics or acids. The idea of activating the regenerative and anti-aging effects of pluripotential cells without the negative effects of an inciting trauma is fundamental to the use of cosmetic cellular therapies.

The use of mesenchymal stem cells (MSCs) as a supplement and/or substitute for existing esthetic treatments is promising. MSCs possess the benefit of high availability from autologous sources, ease of isolation, and ease of in vitro expansion. Adipose tissue and bone marrow are two major sources of MSCs, both of which have similar properties to each other [24]. However, adipose tissue has proven to be a more favorable source of mesenchymal stem cells, as there are a higher percentage of MSCs present in adipose tissue in comparison with bone marrow. MSCs make up approximately 1% of adipose tissue, while they make up only about 0.001% of bone marrow [25]. Adipose-derived stem cells (ADSCs) are isolated from Stromal Vascular Fraction (SVF), a heterogeneous population of cells isolated from adipose tissue using a simple and safe protocol [3, 26]. The umbilical cord is also a viable source of MSCs [27]. These cells, though not autologously sourced like adipose stem cells, do appear to enjoy the same immunologic tolerance of all mesenchymal stem cells, and clinical experience shows they can be administered with a high safety profile. The clinical safety of these cells depends on the integrity of the tissue bank that sources and prepares the cells, and thus, FDA tissue bank accreditation is essential.

Applications of Adipose Stem Cells for Esthetic Therapies

Tissue Volume

Traditional fat grafting techniques aiming to restore volume to the face struggle with the limitation of a lack of volume retention following the procedure [17–19]. This limitation may be attributed to the hypoxic conditions the adipose graft experiences following transplantation. The process of revascularization of an adipose tissue graft can be inefficient, causing the graft to suffer hypoxic conditions beyond its upper limit, which is around 24 hours. The signaling pathway that connects hypoxia to apoptosis of adipose cells is also not clearly defined [28]. The use of mesenchymal stem cells (MSCs) in conjunction with autologous fat grafting might address this issue [29]. Studies have identified that fat grafts enriched with adipose-derived stem cells might mitigate that drawback of fat grafting. In one such study comparing tradi-

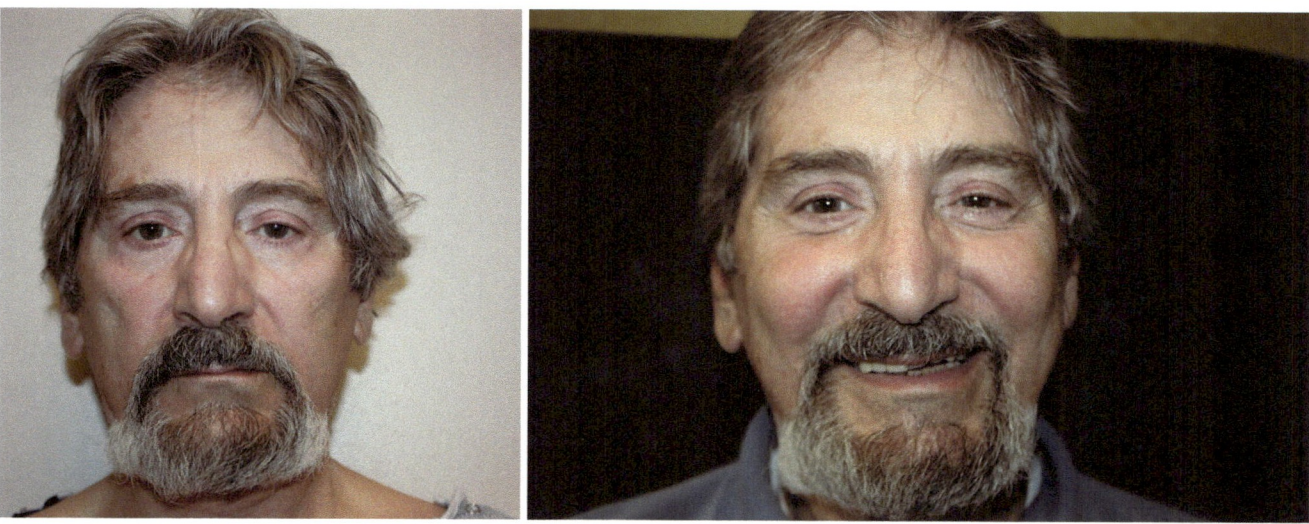

Fig. 26.1 Sixty-four-year-old man before and 1 year post-op CAL bilateral cheeks with 30 cc enhanced fat graft per side

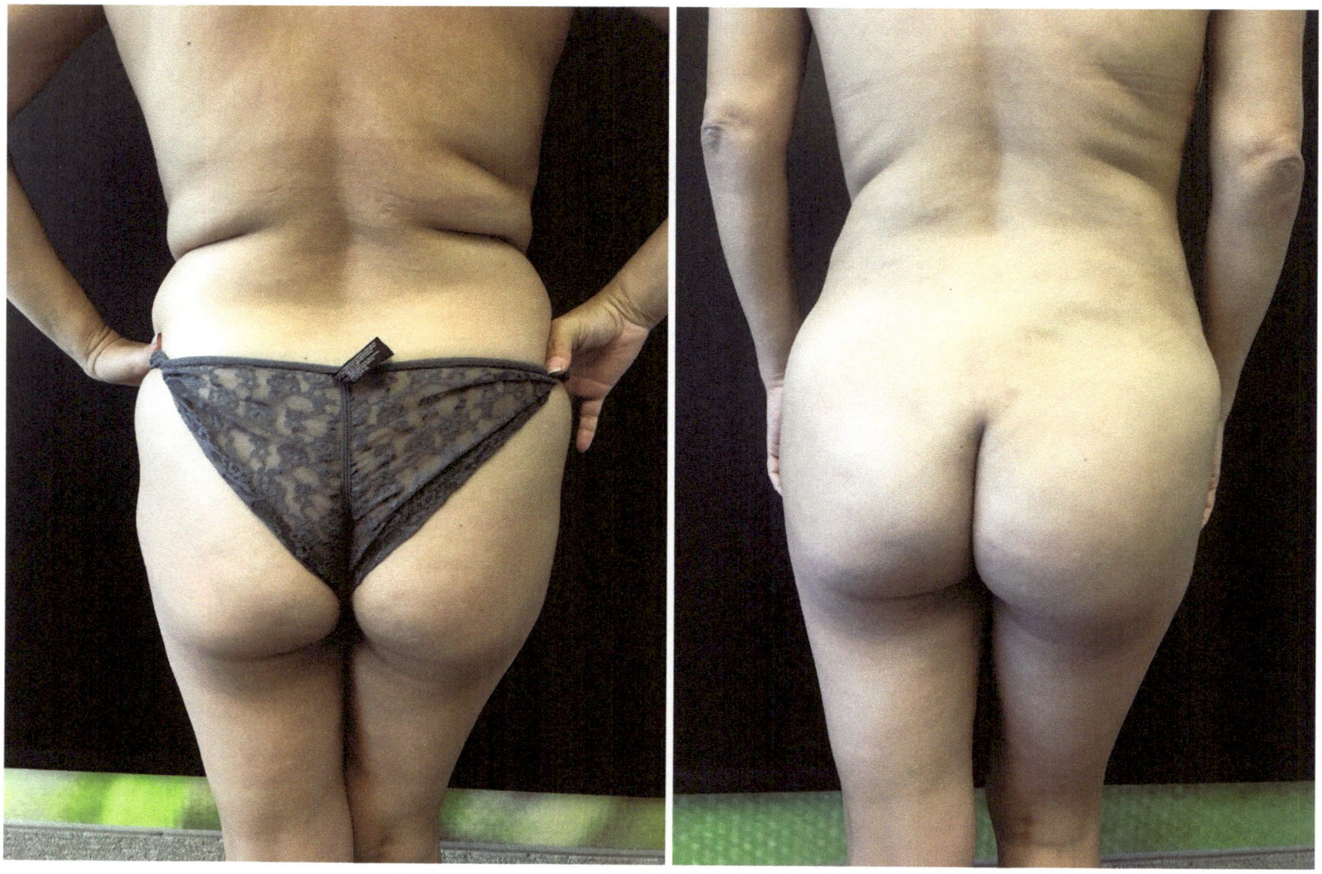

Fig. 26.2 Forty-eight-year-old woman, 4 months post-op CAL fat graft to buttock

tional fat grafts with those supplemented with ADSCs, no patients who received stem cell therapy had to return for another grafting procedure to preserve volume [19].

ADSCs can be used in conjunction with lipoaspirated fat in a method referred to as cell-assisted lipotransfer (CAL) [30]. This can be a safe and effective option for patients interested in an increase in facial volume (Fig. 26.1), as well as patients interested in breast reconstruction or buttock augmentation (Figs. 26.2, 26.3, and 26.4) [28–30].

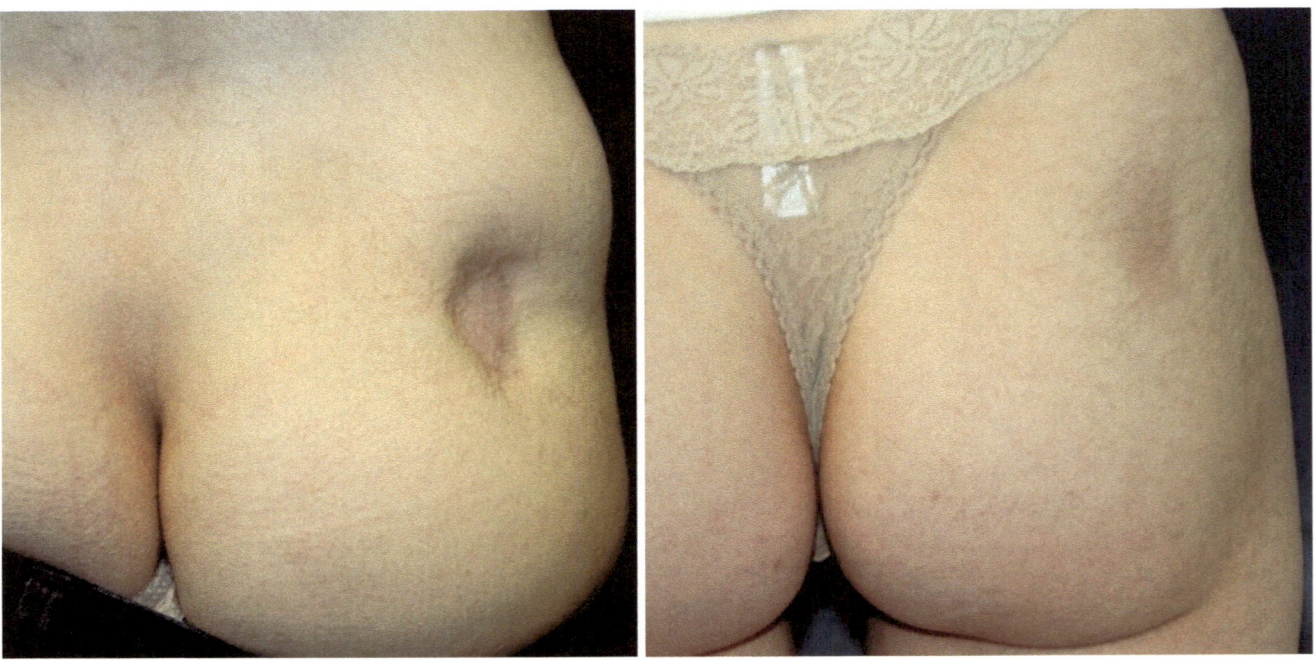

Fig. 26.3 Thirty-year-old woman, 6 months post-op, CAL fat graft to buttock for steroid injection defect

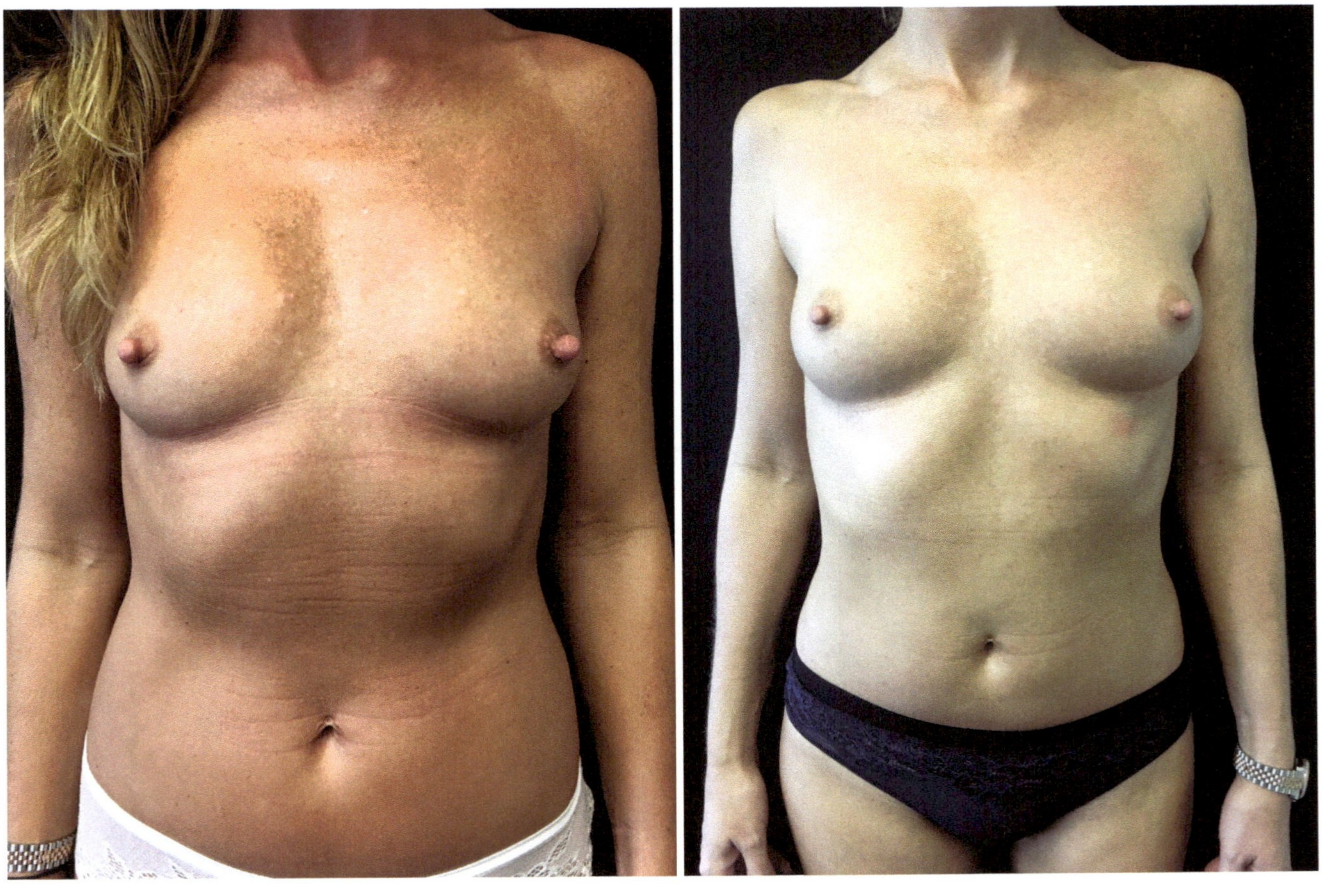

Fig. 26.4 Thirty-six-year-old woman receiving CAL Fat Transfer to breast, 1 year post-op

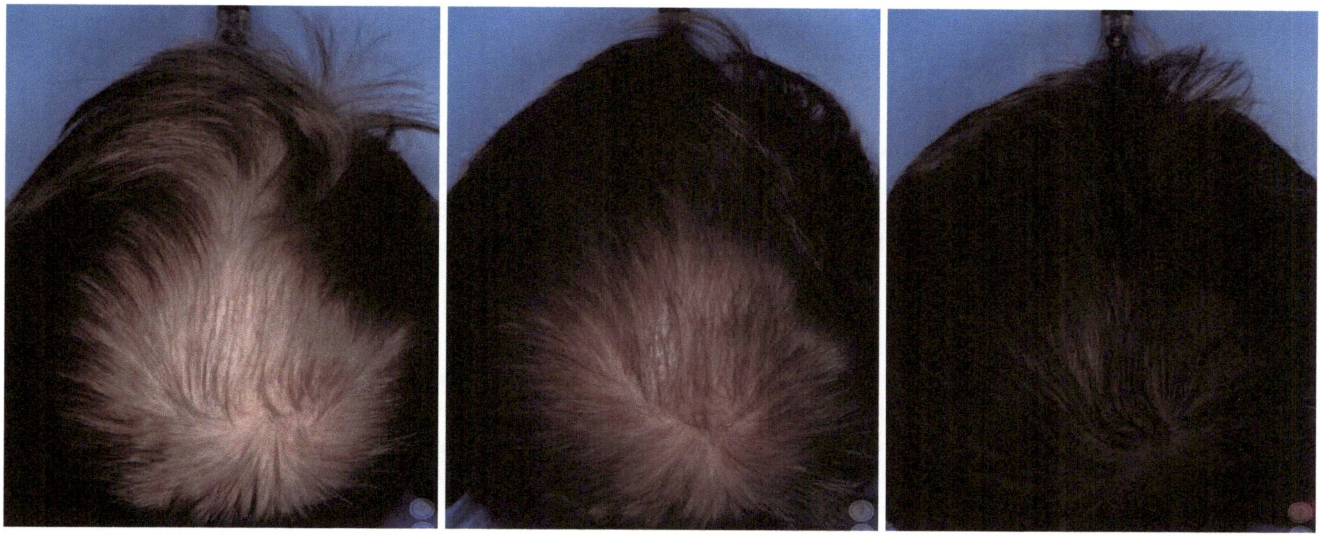

Fig. 26.5 STYLE Trial, adipose stem cells injected into scalp for early-stage androgenic alopecia, pre-op, and 6 months and 12 months post-op

Dermatological Applications (*Pigmentation, Thickness, Elasticity, and Vitiligo)

The uses of ADSCs extend beyond preservation of tissue volume to esthetic therapies related to the skin. ADSCs have been suggested to kickstart the re-epithelialization process, stimulating keratinocyte production and organizing these newly formed cells [31, 32]. ADSCs can also promote collagen synthesis to further counteract the loss of dermal thickness that occurs due to aging [6].

ADSCs reduce cell death related to UVB ray exposure, indicating they could play a role in reducing wrinkles caused by photoaging [33]. The antioxidative effects of ADSCs could combat the oxidative stress that might lead to skin discoloration. Proteomic analysis of cultured ADSCs shows a wide range of antioxidant proteins in the culture, such as SOD2, PEDF, and HGF [8]. Since antioxidants can influence melanin production, ADSCs can be a useful skin whitening agent to counteract skin discoloration [8].

With respect to vitiligo, one study of nude mice found that a skin graft containing both melanocytes and ADSCs was significantly more effective than a graft of melanocytes alone, as shown by a higher increase in overall melanocytes [10]. These findings can be reconciled with the aforementioned skin-whitening potential of ADSCs. While ADSCs reduce the number of Trp-1-positive mature melanocytes, they can increase the number of Trp-2 positive precursor cells that are later differentiated into melanocytes. Preparing melanocyte grafts supplemented with ADSCs requires a particular balance of ADSCs and melanocytes in order to combat the effects of vitiligo [34].

Hair (Kerastem STYLE Trial)

Hair loss and the decrease in subcutaneous scalp tissue occur simultaneously, highlighting the relationship of hair loss to adipose tissue [35]. Conveniently, ADSCs show promise for the treatment of early androgenetic alopecia. Results from an FDA-approved Phase II clinical trial determined that injection of fat supplemented with a low dose of ADSCs into the scalp led to an increase in the amount of hair in subjects exhibiting the early stages of hair loss (Fig. 26.5) [36].

Improvement of Wound Healing and Scarring

ADSCs show promise for a wide array of applications in the field of wound care. Wound care entails regular treatment of acute and chronic wounds, often involving skin grafts. MSCs are particularly useful for this purpose due to their immunoprivileged status as well as their ability to signal cell proliferation [37].

One application of ADSCs in wound care is the treatment of wound patients suffering from Diabetes Mellitus (DM). Wound healing in these patients is challenging due to a lack of re-epithelialization of tissue as a result of deficient cellular proliferation, similar to the lack of proliferation that causes wrinkles in aging patients. Wounds treated with ADSCs have shown an increased level of some of the growth factors necessary for cellular proliferation, such as transforming growth factor-Beta1 (TGF-β1) and transforming growth factor-Beta3 (TGF-β3) [38]. TGF-β1 is implicated in cellular migration to a wound site, while TGF-β3 enhances collagen organization, indicating that ADSCs might accelerate wound healing through these mechanisms [8, 38]. Moreover, ADSCs that are introduced systemically seem to migrate to an injury site to promote growth in that area [25]. Studies have also suggested the effectiveness of ADSCs in treatment for fibrotic scars that might arise after a wound has healed [39].

Benefits of Using Adipose Stem Cells for Esthetic Purposes

The benefits of using adipose-derived stem cells for esthetic therapies are vast. An autologous and abundant sample can be extracted from a patient's abdominal area [38]. ADSCs possess the most promise for isolation in a safe and minimally invasive manner [3, 40]. SVF extraction yields a higher amount of mesenchymal stem cells compared to bone marrow and can also be expanded in vitro in order to increase the amount of cells available for transplantation into patients [26, 38]. Adipose-derived stem cells are also immunoprivileged, lacking the class II major histocompatibility complex (MHC-II) [25]. This makes ADSCs ideal candidates for fat grafting procedures, minimizing the adverse immune response of the patient [25].

Current Limitations and Future Directions for Adipose Stem Cell Treatments

Although ADSCs show a great deal of promise for esthetic therapies, limitations exist that merit consideration. While SVF can be isolated easily from a patient, the resulting sample is heterogeneous, containing fibroblasts, pericytes, smooth muscle cells, preadipocytes, endothelial cells, and immune cells [40]. Isolation of ADSCs requires further effort and runs the risk of an ADSC culture that is not exclusively made up of ADSCs.

Another limitation of these therapies lies in the possible tumorigenic effects of ADSCs. One study suggests that certain signaling molecules such as CXCL1 that can be derived from ADSCs might increase the risk of breast cancer recurrence [41]. Obesity also seems to play a role, altering the properties of ADSCs [42]. One study identified that leptin secreted from obesity-altered ADSCs might stimulate breast cancer growth through estrogen-dependent pathways [42, 43]. Potential cross-contamination of stem cell samples must also be taken into account and underscores the importance of clearly defined and closely followed protocols for cell extraction and maintenance [44]. Further consideration should be given to these risks before implementing ADSC-related esthetic treatments.

At the time of this publication, clinical trials are currently conducted to explore some of the benefits of ADSCs for esthetic purposes. One such study sponsored by the Medical University of Warsaw [45] aims to explore the use of ADSC injections to treat scars. Another study, also sponsored by the Medical University of Warsaw [46], aims to explore the effectiveness of using ADSCs to treat chronic wounds related to diabetic foot syndrome.

Conclusions

Adipose-derived stem cells possess a great deal of promise for applications in esthetic therapies ranging from antiaging treatments to tissue reconstruction and wound care. They have the potential to address issues of fat graft volume retention, skin elasticity, skin pigmentation, and skin thickness. Moreover, ADSCs can supplement wound care procedures to provide more effective treatment. Although further research is necessary to solidify the safety and efficacy of these treatments, current findings suggest that ADSCs may play a pivotal role in esthetic therapies in the near future.

References

1. Yu G, Wu X, Dietrich MA, Polk P, Scott LK, Ptitsyn AA, Gimble JM. Yield and characterization of subcutaneous human adipose-derived stem cells by flow cytometric and adipogenic mRNA analyzes. Cytotherapy. 2010;12(4):538–46.
2. Gronthos S, Franklin DM, Leddy HA, Robey PG, Storms RW, Gimble JM. Surface protein characterization of human adipose tissue-derived stromal cells. J Cell Physiol. 2001;189(1):54–63.
3. Aronowitz JA, Lockhart RA, Hakakian CS, Hicok KC. Clinical safety of stromal vascular fraction separation at the point of care. Ann Plast Surg. 2015;75(6):666–71.
4. Kohl E, Steinbauer J, Landthaler M, Szeimies RM. Skin ageing. J Eur Acad Dermatol Venereol. 2011;25(8):873–84.
5. Zouboulis CC, Makrantonaki E. Clinical aspects and molecular diagnostics of skin aging. Clin Dermatol. 2011;29(1):3–14.
6. Gaur M, Dobke M, Lunyak VV. Mesenchymal stem cells from adipose tissue in clinical applications for dermatological indications and skin aging. Int J Mol Sci. 2017;18(1):208.
7. Wollina U, Wetzker R, Abdel-Naser MB, Kruglikov IL. Role of adipose tissue in facial aging. Clin Interv Aging. 2017;12:2069.
8. Kim WS, Park BS, Sung JH. Protective role of adipose-derived stem cells and their soluble factors in photoaging. Arch Dermatol Res. 2009;301(5):329–36.
9. Durai PC, Thappa DM, Kumari R, Malathi M. Aging in elderly: chronological versus photoaging. Indian J Dermatol. 2012;57(5):343.
10. Lim WS, Kim CH, Kim JY, Do BR, Kim EJ, Lee AY. Adipose-derived stem cells improve efficacy of melanocyte transplantation in animal skin. Biomol Ther. 2014;22(4):328.
11. Ezzedine K, Eleftheriadou V, Whitton M, van Geel N. Vitiligo. Lancet. 2015;386:74–84.
12. Iblher N, Gladilin E, Stark BG. Soft-tissue mobility of the lower face depending on positional changes and age: a three-dimensional morphometric surface analysis. Plast Reconstr Surg. 2013;131(2):372–81.
13. Cotofana S, Fratila AA, Schenck TL, Redka-Swoboda W, Zilinsky I, Pavicic T. The anatomy of the aging face: a review. Facial Plast Surg. 2016;32(03):253–60.

14. Ko AC, Korn BS, Kikkawa DO. The aging face. Surv Ophthalmol. 2017;62(2):190–202.
15. Lambros V. Models of facial aging and implications for treatment. Clin Plast Surg. 2008;35(3):319–27.
16. Kolasinski J. Total breast reconstruction with fat grafting combined with internal tissue expansion. Plast Reconstr Surg Glob Open. 2019;7(4):e2009.
17. Gir P, Brown SA, Oni G, Kashefi N, Mojallal A, Rohrich RJ. Fat grafting: evidence-based review on autologous fat harvesting, processing, reinjection, and storage. Plast Reconstr Surg. 2012;130(1):249–58.
18. Mazzola RF, Mazzola IC. History of fat grafting: from ram fat to stem cells. Clin Plast Surg. 2015;42(2):147–53.
19. Sterodimas A, de Faria J, Nicaretta B, Boriani F. Autologous fat transplantation versus adipose-derived stem cell–enriched lipografts: a study. Aesthet Surg J. 2011;31(6):682–93.
20. Poon F, Kang S, Chien AL. Mechanisms and treatments of photoaging. Photodermatol Photoimmunol Photomed. 2015;31(2):65–74.
21. Hubbard BA, Unger JG, Rohrich RJ. Reversal of skin aging with topical retinoids. Plast Reconstr Surg. 2014;133(4):481e–90e.
22. Truchuelo M, Cerdá P, Fernández LF. Chemical peeling: a useful tool in the office. Actas Dermosifiliogr. 2017;108(4):315–22.
23. Nikalji N, Godse K, Sakhiya J, Patil S, Nadkarni N. Complications of medium depth and deep chemical peels. J Cutan Aesthet Surg. 2012;5(4):254.
24. Izadpanah R, Trygg C, Patel B, Kriedt C, Dufour J, Gimble JM, Bunnell BA. Biologic properties of mesenchymal stem cells derived from bone marrow and adipose tissue. J Cell Biochem. 2006;99(5):1285–97.
25. Ong WK, Sugii S. Adipose-derived stem cells: fatty potentials for therapy. Int J Biochem Cell Biol. 2013;45(6):1083–6.
26. Aronowitz JA, Lockhart RA, Dos-Anjos VS. Use of freshly isolated human adipose stromal cells for clinical applications. Aesthet Surg J. 2017;37(suppl_3):S4–8.
27. Kamolz LP, Kolbus A, Wick N, Mazal PR, Eisenbock B, Burjak S, Meissl G. Cultured human epithelium: human umbilical cord blood stem cells differentiate into keratinocytes under in vitro conditions. Burns. 2006;32(1):16–9.
28. Landau MJ, Birnbaum ZE, Kurtz LG, Aronowitz JA. Proposed methods to improve the survival of adipose tissue in autologous fat grafting. Plast Reconstr Surg Glob Open. 2018;6(8):e1870.
29. Salibian AA, Widgerow AD, Abrouk M, Evans GR. Stem cells in plastic surgery: a review of current clinical and translational applications. Arch Plast Surg. 2013;40(6):666.
30. Yoshimura K, Sato K, Aoi N, Kurita M, Hirohi T, Harii K. Cell-assisted lipotransfer for cosmetic breast augmentation: supportive use of adipose-derived stem/stromal cells. Aesthet Plast Surg. 2008;32(1):48–55.
31. Aronowitz JA, Lockhart AR. Cell-assisted lipotransfer for the correction of facial contour deformities. Stem Cell Res Th. 2016;1(1):23–9.
32. Ito S, Kai Y, Masuda T, Tanaka F, Matsumoto T, Kamohara Y, Hayakawa H, Ueo H, Iwaguro H, Hedrick MH, Mimori K. Long-term outcome of adipose-derived regenerative cell-enriched autologous fat transplantation for reconstruction after breast-conserving surgery for Japanese women with breast cancer. Surg Today. 2017;47(12):1500–11.
33. Kim WS, Park BS, Park SH, Kim HK, Sung JH. Antiwrinkle effect of adipose-derived stem cell: activation of dermal fibroblast by secretory factors. J Dermatol Sci. 2009;53(2):96–102.
34. Kim JY, Park CD, Lee JH, Lee CH, Do BR, Lee AY. Co-culture of melanocytes with adipose-derived stem cells as a potential substitute for co-culture with keratinocytes. Acta Derm Venereol. 2012;92(1):16–23.
35. Festa E, Fretz J, Berry R, Schmidt B, Rodeheffer M, Horowitz M, Horsley V. Adipocyte lineage cells contribute to the skin stem cell niche to drive hair cycling. Cell. 2011;146(5):761–71.
36. ClinicalTrials.gov [Internet]. Bethesda (MD): *National Library of Medicine* (US). 2000 Feb 29. Identifier NCT02503852, STYLE -- A Trial of Cell Enriched Adipose For Androgenetic Alopecia (STYLE); 2015 July 21 [cited 2019 November 13]. Available from: https://clinicaltrials.gov/ct2/show/NCT02503852
37. Huang L, Burd A. An update review of stem cell applications in burns and wound care. Indian J Plast Surg. 2012;45(2):229.
38. Gadelkarim M, Abushouk AI, Ghanem E, Hamaad AM, Saad AM, Abdel-Daim MM. Adipose-derived stem cells: effectiveness and advances in delivery in diabetic wound healing. Biomed Pharmacother. 2018;107:625–33.
39. Spiekman M, van Dongen JA, Willemsen JC, Hoppe DL, van der Lei B, Harmsen MC. The power of fat and its adipose-derived stromal cells: emerging concepts for fibrotic scar treatment. J Tissue Eng Regen Med. 2017;11(11):3220–35.
40. Nowacki M, Kloskowski T, Pietkun K, Zegarski M, Pokrywczyńska M, Habib SL, Drewa T, Zegarska B. The use of stem cells in aesthetic dermatology and plastic surgery procedures. A compact review of experimental and clinical applications. Postepy Dermatol Alergol. 2017;34(6):526.
41. Wang Y, Liu J, Jiang Q, Deng J, Xu F, Chen X, Cheng F, Zhang Y, Yao Y, Xia Z, Xu X. Human adipose-derived mesenchymal stem cell-secreted CXCL1 and CXCL8 facilitate breast tumor growth by promoting angiogenesis. Stem Cells. 2017;35(9):2060–70.
42. Sabol RA, Giacomelli P, Beighley A, Bunnell BA. Adipose stem cells and cancer: concise review. Stem Cells. 2019;37:1261–6. https://doi.org/10.1002/stem.3050.
43. Strong AL, Strong TA, Rhodes LV, Semon JA, Zhang X, Shi Z, Zhang S, Gimble JM, Burow ME, Bunnell BA. Obesity associated alterations in the biology of adipose stem cells mediate enhanced tumorigenesis by estrogen dependent pathways. Breast Cancer Res. 2013;15(5):R102.
44. Torsvik A, Røsland GV, Svendsen A, Molven A, Immervoll H, McCormack E, Lønning PE, Primon M, Sobala E, Tonn JC, Goldbrunner R. Spontaneous malignant transformation of human mesenchymal stem cells reflects cross-contamination: putting the research field on track–letter. Cancer Res. 2010;70(15):6393–6.
45. ClinicalTrials.gov [Internet]. Bethesda (MD): *National Library of Medicine* (US). 2000 Feb 29. Identifier NCT03887208, Therapy of Scars and Cutis Laxa With Autologous Adipose Derived Mesenchymal Stem Cells (2ABC); 2019 March 22 [cited 2019 November 13]. Available from: https://clinicaltrials.gov/ct2/show/NCT03887208
46. ClinicalTrials.gov [Internet]. Bethesda (MD): *National Library of Medicine* (US). 2000 Feb 29. Identifier NCT03865394, Treatment of Chronic Wounds in Diabetic Foot Syndrome With Autologous Adipose Derived Mesenchymal Stem Cells (1ABC); 2019 March 6 [cited 2019 November 13]. Available from: https://clinicaltrials.gov/ct2/show/NCT03865394

Index